INTRODUCTORY
ALGEBRA

FIFTH EDITION

INTRODUCTORY ALGEBRA

FIFTH EDITION

D. FRANKLIN WRIGHT
CERRITOS COLLEGE

HAWKES
PUBLISHING

Editor: Mandy Glover
Developmental Editor: Marcel Prevuznak
Production Editors: Shaun Gibson, Ann Lucius, Mary Janelle Cady, Kimberly Scott
Answer Key Editors: Amanda Matlock, Lindsay Stevens, Emily Stegman
Editorial Assistants: Priyanka Bihani, Ashley Rankin, Barron Whitehead
Layout: QSI (Pvt.) Ltd.
Art: Kim Cope
Cover Art and Design: Johnson Design

HAWKES
PUBLISHING
A division of Quant Systems, Inc.

Library of Congress Control Number: 2002105615

Printed in the United States of America

ISBN:
Student (paperback): 0-918091-62-4
Student (case): 0-918091-67-5
Student Solution Manual: 0-918091-69-1

CONTENTS

CHAPTER 3 Solving Equations and Inequalities

CHAPTER 4 Graphing Linear Equations in Two Variables

CHAPTER 5 Systems of Linear Equations

CHAPTER 6 Exponents and Polynomials

CHAPTER 7 Factoring Polynomials and Solving Quadratic Equations

CHAPTER 8 Rational Expressions

CHAPTER 9 Real Numbers and Radicals

CHAPTER 10 Quadratic Equations

Appendix Graphing Linear Inequalities and Systems of Inequalities

PREFACE

Purpose and Style

Introductory Algebra (fifth edition) provides a smooth transition from arithmetic (or prealgebra) to the more abstract skills and reasoning abilities developed in a beginning algebra course. With feedback from users, insightful comments from reviewers, and skillful editing and design by the editorial staff at Hawkes Publishing, I have confidence that students and instructors alike will find that this text is indeed a superior teaching and learning tool. The text may be used independently or in conjunction with the software package ***Hawkes Learning Systems: Introductory Algebra*** developed by Quant Systems.

I have assumed that students have basic arithmetic knowledge and no previous experience with algebra. Chapter 1 begins with the algebraic concepts of integers, real number lines, absolute value, and operations with integers. Geometric figures and related formulas are introduced to provide practice with algebraic concepts and as reference for word problems throughout the text.

The writing style is informal and non-technical while providing carefully worded, thorough explanations that are easily understood and mathematically accurate. Each topic is developed in a straightforward step-by-step manner. Each section contains many detailed examples to lead students successfully through the exercises and help them develop an understanding of the related algebraic concepts. Whenever possible, information is presented in list form for organized learning and easy reference. Practice problems with answers are provided in almost every section to allow students to "warm up" and to provide instructors with immediate classroom feedback.

The NCTM and AMATYC curriculum standards have been taken into consideration in the development of the topics throughout the text. In particular:
· there is emphasis on reading and writing skills as they relate to mathematics
· techniques for using a graphing calculator are discussed early and the chapters on graphing linear equations have been moved into Chapters 4 and 5
· a special effort has been made to make the exercises motivating and interesting
· geometric concepts are integrated throughout
· statistical concepts, such as interpreting bar graphs and calculating elementary statistics, are included where appropriate

Features

"Did You Know?": A feature at the beginning of every chapter that presents some interesting math history related to the chapter at hand such as the Pythagoreans discovery of irrational numbers presented at the beginning of Chapter 8.

Introduction: Presented before the first section of every chapter, this feature provides an introduction to the subject of the chapter and its purpose.

Objectives: The objectives provide students with a clear and concise list of skills presented in each section.

Definition Boxes: Definitions are presented in highly visible boxes for easy reference.

Notes: Notes highlight common mistakes and give additional clarification to more subtle details.

Examples: Examples are denoted with titled headers indicating the problem solving skill being presented. Each section contains many carefully explained examples with lots of tables, diagrams, and graphs. Examples are presented in an easy to understand step-by-step fashion and annotated with notes for additional clarification.

Calculator Instruction: Step-by-step instructions are presented to introduce students to basic graphing skills with a TI-83 Plus calculator along with actual screen shots of a TI-83 Plus for visual reference.

Practice Problems: Practice Problems are presented at the end of almost every section with answers giving the students an opportunity to practice their newly acquired skills.

Exercises: Each section includes a variety of paired and graded exercises to give the students much needed practice applying and reinforcing the skills learned in the section. More than 4300 carefully selected and graded exercises are provided in the sections. The exercises proceed from relatively easy to more difficult ones.

Calculator Problems: Each problem is designed to highlight the usefulness of a calculator in solving certain complex problems, but maintain the necessity of understanding the concepts behind the problem.

Writing and Thinking About Mathematics: Provides students an opportunity to independently explore and expand on concepts presented in the chapter.

Index of Key Terms and Ideas: Highlights the main concepts and skills presented in the chapter along with full definitions and page numbers for easy reference.

Chapter Test: Provides an opportunity for the students to practice the skills presented in the chapter in a test format.

Cumulative Review: As new concepts build on previous concepts, the cumulative review provides the student with an opportunity to continually reinforce existing skills while practicing newer skills.

Answers: Answers to odd numbered section exercises and answers to all even and odd numbered exercises in the cumulative reviews and chapter tests.

Teachers' Edition:

Answers: Answers to all the exercises are conveniently located in the margins next to the problems.

Teaching Notes: Suggestions for more in-depth classroom discussions and alternate methods and techniques are located in the margins of every section.

Changes included in the new edition:

· New, reader-friendly, layout
· Rearrangement of chapters for better flow, continuity and progression
· Writing and Thinking About Mathematics
· Calculator Instructions (New emphasis on the graphing calculator)
· Calculator Problems

Content

There is sufficient material for a three- or four- semester-hour course. The topics in Chapters 1 – 9 form the core of the course. Chapter 10 (Quadratic Equations) and the Appendix (Graphing Linear Inequalities and Systems of Linear Inequalities) provide additional flexibility in the course depending on students' background and the goals of the course.

Chapter 1, Integers and Real Numbers, develops the algebraic concept of integers and the basic skills of operating with integers. Real numbers, rational and irrational numbers, are discussed in conjunction with the real number line. Variables, absolute value, and exponents are defined and expressions are evaluated by using the rules for order of operations. The chapter closes with a discussion of the properties of addition and multiplication with real numbers.

Chapter 2, Fractions, Decimals, and Algebraic Expressions, provides a review of arithmetic concepts and, through the use of variables and signed numbers, shows how these concepts can be generalized with algebraic expressions. Algebraic expressions are simplified by combining like terms and evaluated by using the rules for order of operations. A new section discussing change in value and average has been added. The last section involves translating English phrases and algebraic expressions as a lead in to interpreting and understanding word problems.

Chapter 3, Solving Equations and Inequalities, develops the techniques for solving linear (or first-degree) equations in a step-by-step manner over two sections. Techniques include combining like terms and use of the distributive property. Applications relate to number problems, consecutive integers, percent, and work with formulas. Formulas related to the geometric concepts of perimeter, area, and volume are included. Linear inequalities and intervals are discussed in the last section.

Chapter 4, Graphing Linear Equations in Two Variables, allows for the early introduction of a graphing calculator and the ideas and notation related to functions. The chapter begins with an introduction to the Cartesian coordinate system and graphing ordered pairs of real numbers. The sections on graphing linear equations include the standard form $[Ax + By = C]$, the slope-intercept form $[y = mx + b]$, and the point-slope form $[y - y_1 = m(x - x_1)]$. Students should understand that the topic of functions is one of the most important and useful in all of mathematics and they will see it again and again.

Chapter 5, Systems of Linear Equations, shows how to solve systems of linear equations three ways: by graphing (including the use of a graphing calculator), by substitution, and by addition. Applications are related to distance-rate-time, number problems, amounts and costs, interest, and mixture.

Chapter 6, Exponents and Polynomials, studies the properties of exponents in depth and shows how to read and write scientific notation. The remainder of the chapter is concerned with definitions and operations related to polynomials. Included are the FOIL method of multiplication with two binomials, special products of binomials, and the division algorithm.

Chapter 7, Factoring Polynomials and Solving Quadratic Equations, discusses methods of factoring polynomials, including finding common monomial factors, factoring by grouping, factoring trinomials by grouping and by trial-and-error, and factoring special products. Then, the important topic of solving quadratic equations is introduced, and quadratic equations are solved by factoring only. Applications with quadratic equations are included in two sections. (Quadratic equations are discussed in more detail in Chapter 10.)

Chapter 8, Rational Expressions, provides still more practice with factoring and shows how to use factoring to operate with rational expressions (algebraic fractions with polynomials in the numerator and denominator). The topics of solving equations and the use of rational expressions to analyze and solve word problems (proportions, distance-rate-time, work, and variation) complete the chapter.

Chapter 9, Real Numbers and Radicals, discusses the real numbers in detail with emphasis on simplifying radicals. Fractional exponents are discussed and applications with square roots involving solving equations, the Pythagorean Theorem, and the distance between two points.

Chapter 10, Quadratic Equations, develops the quadratic formula over three sections by discussing solving quadratic equations first by the square root method and then by completing the square. Applications are related to the Pythagorean Theorem, work, and distance-rate-time. The last section presents quadratic functions and the graphs of parabolas.

Appendix A.1, Graphing Linear Inequalities and **Appendix A.2, Graphing Systems of Linear Inequalities,** discuss the ideas of half-planes and provide an opportunity to show the power of a graphing calculator. Advanced students will find these topics particularly interesting. **Appendix A.3, Pi** contains a brief discussion on the irrational number π.

I recommend that the topics be covered in the order presented since most sections assume knowledge of the material in a previous section. This is particularly true of the cumulative review sections at the end of each chapter. Of course, time and other circumstances may dictate another sequence of topics. For example, in some programs, Chapters 1 and 2 might be considered review.

Acknowledgements

I would like to thank Editor Mandy Glover and Developmental Editor Marcel Prevuznak for their hard work and invaluable assistance in the development and production of this text.

Many thanks go to the following manuscript reviewers who offered their constructive and critical comments: Russ Baker and Consuelo Stewart at Howard Community College; Virginia Puckett at Miami-Dade Community College; Thom Clark and Laura Hoye at Trident Technical College; Bill Schurter at the University of the Incarnate Word.

Finally, special thanks go to James Hawkes for his faith in this fifth edition and his willingness to commit so many resources to guarantee a top-quality product for students and teachers.

D. Franklin Wright

TO THE STUDENT

The goal of this text and of your instructor is for you to succeed in introductory algebra. Certainly, you should make this your goal as well. What follows is a brief discussion about developing good work habits and using the features of this text to your best advantage. For you to achieve the greatest return on your investment of time and energy you should practice the following three rules of learning.

1. Reserve a block of time to study every day.
2. Study what you don't know.
3. Don't be afraid to make mistakes.

How to use this book

The following seven-step guide will not only make using this book a more worthwhile and efficient task, but it will also help you benefit more from classroom lectures or the assistance that you receive in a math lab.

1. Try to look over the assigned section(s) before attending class or lab. In this way, new ideas may not sound so foreign when you hear them mentioned again. This will also help you see where you need to ask questions about material that seems difficult to you.
2. Read examples carefully. They have been chosen and written to show you all of the problem-solving steps that you need to be familiar with. You might even try to solve example problems on your own before studying the solutions that are given.
3. Work the section exercises faithfully as they are assigned. Problem-solving practice is the single most important element in achieving success in any math class, and there is no good substitute for actually doing this work yourself. Demonstrating that you can think independently through each step of each type of problem will also give you confidence in your ability to answer questions on quizzes and exams. Check the Answer Key periodically while working section exercises to be sure that you have the right ideas and are proceeding in the right manner.
4. Use the "Writing and Thinking About Mathematics" questions as an opportunity to explore the way that you think about math. A big part of learning and understanding mathematics is being able to talk about mathematical ideas and communicate the thinking that you do when you approach new concepts and problems. These questions can help you analyze

your own approach to mathematics and, in class or group discussions, learn from ideas expressed by your fellow students.

5. Use the Chapter Index of Key Ideas and Terms as a recap when you begin to prepare for a Chapter Test. It will reference all the major ideas that you should be familiar with from that chapter and indicate where you can turn if review is needed. You can also use the Chapter Index as a final checklist once you feel you have completed your review and are prepared for the Chapter Test.

6. Chapter Tests are provided so that you can practice for the tests that are actually given in class or lab. To simulate a test situation, block out a one-hour, uninterrupted period in a quiet place where your only focus is on accurately completing the Chapter Test. Use the Answer Key at the back of the book as a self-check only after you have completed all of the questions on the test.

7. Cumulative Reviews will help you retain the skills that you acquired in studying earlier chapters. They appear after every chapter beginning with Chapter 2. Approach them in much the same manner as you would the Chapter Tests in order to keep all of your skills sharp throughout the entire course.

How to Prepare for an Exam

Gaining Skill and Confidence

The stress that many students feel while trying to succeed in mathematics is what you have probably heard called "math anxiety." It is a real-life phenomenon, and many students experience such a high level of anxiety during mathematics exams in particular that they simply cannot perform to the best of their abilities. It is possible to overcome this stress simply by building your confidence in your ability to do mathematics and by minimizing your fears of making mistakes.

No matter how much it may seem that in mathematics you must either be right or wrong, with no middle ground, you should realize that you can be learning just as much from the times that you make mistakes as you can from the times that your work is correct. Success will come. Don't think that making mistakes at first means that you'll never be any good at mathematics. Learning mathematics requires lots of practice. Most importantly, it requires a true confidence in yourself and in the fact that with practice and persistence the mistakes will become fewer, the successes will become greater, and you will be able to say, "I can do this."

Showing What You Know

If you have attended class or lab regularly, taken good notes, read your textbook, kept up with homework exercises, and asked for help when it was needed, then you have already made significant progress in preparing for an exam and conquering any anxiety. Here are a few other suggestions to maximize your preparedness and minimize your stress.

1. Give yourself enough time to review. You will generally have several days advance notice before an exam. Set aside a block of time each day with the goal of reviewing a manageable portion of the material that the test will cover. Don't cram!

2. Work lots of problems to refresh your memory and sharpen your skills. Go back to redo selected exercises from all of your homework assignments.

3. Reread your text and your notes, and use the Chapter Index of Key Ideas and Terms and the Chapter Test to recap major ideas and do a self-evaluated test simulation.

4. Be sure that you are well-rested so that you can be alert and focused during the exam.

5. Don't study up to the last minute. Give yourself some time to wind down before the exam. This will help you to organize your thoughts and feel more calm as the test begins.

6. As you take the test, realize that its purpose is not to trick you, but to give you and your instructor an accurate idea of what you have learned. Good study habits, a positive attitude, and confidence in your own ability will be reflected in your performance on any exam.

7. Finally, you should realize that your responsibility does not end with taking the exam. When your instructor returns your corrected exam, you should review your instructor's comments and any mistakes that you might have made. Take the opportunity to learn from this important feedback about what you have accomplished, where you could work harder, and how you can best prepare for future exams.

HAWKES LEARNING SYSTEMS: INTRODUCTORY ALGEBRA

Overview

This multimedia courseware allows students to become better problem-solvers by creating a mastery level of learning in the classroom. The software includes a "demonstrate", "instruct", "practice", "tutor", and "certify" mode in each lesson, allowing students to learn through step-by-step interactions with the software. These automated homework system's tutorial and assessment modes extend instructional influence beyond the classroom. Intelligence is what makes the tutorials so unique. By offering intelligent tutoring and mastery level testing to measure what has been learned, the software extends the instructor's ability to influence students to solve problems. This courseware can be ordered either seperately or bundled together with this text.

Minimum Requirements

In order to run *HLS: Introductory Algebra*, you will need:

Intel® Pentium® 166 MHz or faster processor or equivalent
Windows® 98SE or later, Windows NT® Workstation 4.0 (SP 6) or later
32 MB RAM (64 MB recommended)
150 MB hard drive space
256 color display (800x600, 16-bit color recommended)
Internet Explorer 4.0 or later
CD-ROM drive

Getting Started

Before you can run *HLS: Introductory Algebra*, you will need an access code. This 30 character code is <u>your</u> personal access code. To obtain an access code, go to **http://www.quantsystems.com** and follow the links to the access code request page (unless directed otherwise by your instructor.)

Installation

Insert the ***HLS: Introductory Algebra*** Installation CD-ROM into the CD-ROM drive. Select the Start/Run command, type in the CD-ROM drive letter followed by \setup.exe. (For example, d:\setup.exe where d is the CD-ROM drive letter.)

The complete installation will use over 140 MB of hard drive space and will install the entire product, except the multimedia files, on your hard drive.

After selecting the desired installation option, follow the on-screen instructions to complete your installation of ***HLS: Introductory Algebra.***

Starting the Courseware

After you installed ***HLS: Introductory Algebra*** on your computer, to run the courseware select Start/Programs/Hawkes Learning Systems/Introductory Algebra.

You will be prompted to enter your access code with a message box similar to the following:

Type your authorization code in the box. When you are finished, press OK.

If you typed in your authorization code correctly, you will be prompted to save the code to disk. If you choose to save your code to disk, typing

If you typed in your authorization code correctly, you will be prompted to save the code to disk. If you choose to save your code to disk, typing in the authorization code each time you run *HLS: Introductory Algebra* will not be necessary. Instead, select the <F1> - Load from disk button when prompted to enter your authorization code and choose the path to your saved authorization code.

Now that you have entered your authorization code and saved it to diskette, you are ready to run a lesson. From the table of contents screen, choose the appropriate chapter and then choose the lesson you wish to run.

Features

Each lesson in *HLS: Introductory Algebra* has five modes: Demonstrate, Instruct, Practice, Tutor, and Certify.

Demonstrate: Demonstrate provides you with a brief overview of the lesson. It presents an example of the type of question you will see in Practice and Certify, shows you how to input an answer, lists some of the specific features of the lesson, and tells you how many correct answers are needed to pass Certify.

Instruct: Instruct provides an expository on the material covered in the lesson in a multimedia environment. This same instruct mode can be accessed via the tutor mode.

Practice: Practice allows you to hone your problem-solving skills. It provides an unlimited number of randomly generated problems. Practice also provides access to the Tutor mode by selecting the Tutor button located by the Submit button.

Tutor: Tutor mode is broken up into several parts: Instruct, Explain Error, Step by Step, and Solution.

1. Instruct, which can also be selected directly from Practice mode, contains a multimedia lecture of the material covered in a lesson.

2. Explain Error is active whenever a problem is incorrectly answered. It will attempt to explain the error that caused you to incorrectly answer the problem.

3. Step by Step is an interactive "step through" of the problem. It breaks each problem into several steps, explains to you each step in solving the problem, and asks you a question about the step. After you answer the last step correctly, you have solved the problem.

4. Solution will provide you with a detailed "worked-out" solution to the problem.

Throughout the Tutor, you will see words or phrases colored green with a dashed underline. These are called Hot Words. Clicking on a Hot Word will provide you with more information on these word(s) or phrases.

Certify: Certify is the testing mode. You are given a finite number of problems and a certain number of strikes (problems you can get wrong). If you answer the required number of questions, you will receive a certification code and a certificate. Write down your certification code and/or print out your certificate. The certification code will be used by your instructor to update your records. Note that the Tutor is not available in Certify.

Integration of Courseware and Textbook

Throughout this text, you will see two icons that help to integrate the Introductory Algebra textbook and **HLS: *Introductory Algebra*** courseware.

 This icon occurs throughout the body of the textbook. It signals that the *HLS: Introductory Algebra* courseware has animation in Instruct that will help you better understand the topic you are studying. Note that not all of the animation in the software is displayed in the textbook in this manner.

 This icon indicates which *HLS: Introductory Algebra* lessons you should run in order to test yourself on the subject material and to review the contents of a chapter.

Support

If you have questions about *HLS: Introductory Algebra* or are having technical difficulties, we can be contacted as follows:

Phone: (843) 571-2825
Email: techsupport@quantsystems.com
Web: www.quantsystems.com

Our support hours are 8:30 am to 5:30 pm, Eastern Time, Monday through Friday.

Integers and Real Numbers

Did You Know?

Arithmetic operations defined on the set of positive integers, negative integers, and zero are studied in this chapter. The integer zero will be shown to have interesting properties under the operations of addition, subtraction, multiplication, and division.

Curiously, zero was not recognized as a number by early Greek mathematicians. When Hindu scientists developed the place-value numeration system we currently use, the zero symbol was initially a place holder but not a number. The spread of Islam transmitted the Hindu number system to Europe where it became known as the Hindu-Arabic system and replaced Roman numerals. The word zero comes from the Hindu word meaning "void," which was translated into Arabic as "sifr" and later into Latin as "zephirum," hence the derivation of our English words "zero" and "cipher."

Almost all of the operational properties of zero were known to the Hindus. However, the Hindu mathematician Bhaskara the Learned (1114 – 1185?) asserted that a number divided by zero was zero, or possibly infinite. Bhaskara did not seem to understand the role of zero as a divisor since division by zero is undefined and hence, an impossible operation in mathematics.

Albert Einstein, in his development of a proof that the universe was stable and unchangeable in time, divided both sides of one of his intermediate equations by a complicated expression that under certain circumstances could become zero. When the expression became zero, Einstein's proof did not hold and the possibility of a pulsating, expanding, or contracting universe had to be considered. This error was pointed out to Einstein and he was forced to withdraw his proof that the universe was stable. The moral of this story is that although zero seems like a "harmless" number, its operational properties are different from those of the positive and negative integers.

"How can it be that mathematics, being after all a product of human thought independent of exercise, is so admirably adapted to the objects of reality?"

Albert Einstein (1879 – 1955)

Numbers and number concepts form the foundation for the study of algebra. In this chapter, you will learn about positive and negative numbers and how to operate with these numbers. Believe it or not, the key number is 0. Pay particularly close attention to the idea of the magnitude of a number, called its absolute value, and the terminology used to represent different types of numbers.

1.1   The Real Number Line and Absolute Value

Objectives

After completing this section, you will be able to:

1. Identify types of numbers.

2. Determine if given numbers are greater than, less than, or equal to other given numbers.

3. Determine absolute values.

The study of algebra requires that we know a variety of types of numbers and their names. The set of numbers

$$N = \{\, 1, 2, 3, 4, 5, 6, 7, 8, 9, 10, 11, \dots \,\}$$

is called the **counting numbers** or **natural numbers**. The three dots indicate that the pattern is to continue without end. Putting 0 with the set of natural numbers gives the set of **whole numbers**.

$$W = \{\, 0, 1, 2, 3, 4, 5, 6, \dots \,\}$$

Thus, mathematicians make the distinction that 0 is a whole number but not a natural number.

To help in understanding different types of numbers and their relationships to each other, we begin with a "picture" called a **number line**. For example, choose some point on a horizontal line and label it with the number 0 (Figure 1.1).

0

Figure 1.1

Now choose another point on the line to the right of 0 and label it with the number 1 (Figure 1.2).

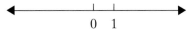

0   1

Figure 1.2

We now have a number line. Points corresponding to all the whole numbers are determined. The point corresponding to 2 is the same distance from 1 as 1 is from 0, 3 from 2, 4 from 3, and so on (Figure 1.3).

Figure 1.3

The **graph** of a number is the point that corresponds to the number and the number is called the **coordinate** of the point. We will follow the convention of using the terms "number" and "point" interchangeably. For example, a point can be called "seven" or "two". The graph of 7 is indicated by marking the point corresponding to 7 with a large dot (Figure 1.4).

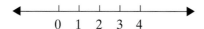

Figure 1.4

The graph of the set $A = \{ 2, 4, 6 \}$ is shown in Figure 1.5.

Figure 1.5

On a horizontal number line, the point one unit to the left of 0 is the **opposite** of 1. It is called **negative** 1 and is symbolized −1. Similarly, the point two units to the left of 0 is the opposite of 2, called negative 2, and symbolized −2, and so on (Figure 1.6).

The opposite of 1 is −1;   The opposite of −1 is −(−1) = +1;
The opposite of 2 is −2;   The opposite of −2 is −(−2) = +2;
The opposite of 3 is −3;   The opposite of −3 is −(−3) = +3;
and so on.   and so on.

NOTES The − sign indicates the opposite of a number as well as a negative number. It is also used, as we will see in Section 1.3, to indicate subtraction. To avoid confusion, you must learn (by practice) just how the − sign is used in each particular situation.

Numbers and their Opposites

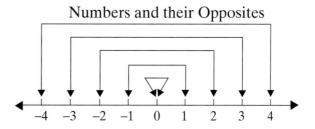

Figure 1.6

Integers

*The set of numbers consisting of the whole numbers and their opposites is called the set of **integers**.*

The natural numbers are also called **positive integers**. Their opposites are called **negative integers**. Zero is its own opposite and is neither positive nor negative (Figure 1.7). Note that the opposite of a positive integer is a negative integer, and the opposite of a negative integer is a positive integer.

Integers:	{ ... , –3, –2, –1, 0, 1, 2, 3, ... }
Positive integers:	{ 1, 2, 3, 4, 5, ... }
Negative integers:	{ ... , –4, –3, –2, –1 }

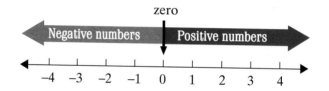

Figure 1.7

Example 1: Opposites

a. Find the opposite of 7.

  Solution: –7

b. Find the opposite of –3.

  Solution: –(–3) or +3

  In words, the opposite of –3 is +3.

Example 2: Number Line

a. Graph the set of integers { –3, –1, 1, 3 }.

  Solution:

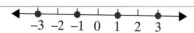

b. Graph the set of integers { ... , –5, –4, –3 }.

  Solution:

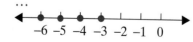

The three dots above the number line indicate that the pattern in the graph continues without end.

The integers are not the only numbers that can be represented on a number line. Fractions and decimal numbers such as $\frac{1}{2}$, $-\frac{4}{3}$, $\frac{3}{4}$, and -2.3, as well as numbers such as π and $\sqrt{2}$, can also be represented (Figure 1.8).

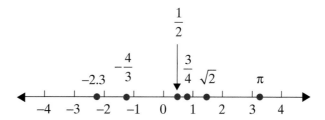

Figure 1.8

Numbers that can be written as fractions and whose numerators and denominators are integers have the technical name **rational numbers**. Positive and negative decimal numbers and the integers themselves can also be classified as rational numbers. For example, the following numbers are all rational numbers:

$$1.3 = \frac{13}{10}, \quad 5 = \frac{5}{1}, \quad -4 = \frac{-4}{1}, \quad \frac{3}{8}, \quad \text{and} \quad \frac{17}{6}.$$

To state general rules, to work with equations, and to form definitions we use variables. A **variable** is a letter (or other symbol) that can represent more than one number.

Rational Numbers

A **rational number** is a number that can be written in the form of $\frac{a}{b}$ where a and b are integers and $b \neq 0$.

OR

A **rational number** is a number that can be written in decimal form as a terminating decimal or as an infinite repeating decimal.

Other numbers on a number line, such as $\sqrt{2}, \sqrt{3}, \pi$, and $\sqrt[3]{5}$, are called **irrational numbers**. These numbers can be written as infinite nonrepeating decimal numbers. All rational numbers and irrational numbers are classified as **real numbers** and can be written in some decimal form. The number line is called the **real number line**.

We will discuss rational numbers (in both decimal form and fractional form) in detail in Chapter 2 and irrational numbers in Chapter 9. For now, we are only interested in recognizing various types of numbers and locating their positions on the real number line.

With a calculator, you can find the following decimal values and approximations:

Examples of Rational Numbers:

$\dfrac{3}{4} = 0.75$ 　　　　　　This decimal number is terminating.

$\dfrac{1}{3} = 0.33333333...$ 　　There is an infinite number of 3's in this repeating pattern.

$\dfrac{3}{11} = 0.27272727...$ 　　This repeating pattern shows that there may be more than one digit in the pattern.

Examples of Irrational Numbers:

$\sqrt{2} = 1.414213562...$ 　　There is an infinite number of digits with no repeating pattern.

$\pi = 3.141592653...$ 　　There is an infinite number of digits with no repeating pattern.
(The TI-83 Plus calculator will show a 4 in place of the ninth place digit 3 because it rounds off decimal numbers.)

$1.41441444144441...$ 　　There is an infinite number of digits and a pattern of sorts. However, the pattern is nonrepeating.

The following diagram (Figure 1.9) illustrates the relationships among the various categories of real numbers.

Animation:
Name That Real
Number

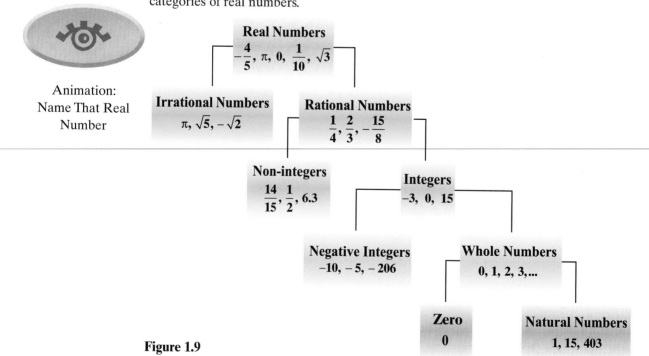

Figure 1.9

Inequality Symbols

On a horizontal number line, **smaller numbers are always to the left of larger numbers**. Each number is smaller than any number to its right and larger than any number to its left. Two symbols used to indicate order are

$$<, \qquad \text{read "is less than"}$$
$$\text{and} \qquad >, \qquad \text{read "is greater than."}$$

Using the real number line in Figure 1.10, you can see the following relationships:

Using $<$		or	Using $>$	
$0 < 3$	0 is less than 3		$3 > 0$	3 is greater than 0
$-2 < 1$	-2 is less than 1		$1 > -2$	1 is greater than -2
$-7 < -4$	-7 is less than -4		$-4 > -7$	-4 is greater than -7
$\dfrac{1}{4} < \dfrac{9}{8}$	$\dfrac{1}{4}$ is less than $\dfrac{9}{8}$		$\dfrac{9}{8} > \dfrac{1}{4}$	$\dfrac{9}{8}$ is greater than $\dfrac{1}{4}$

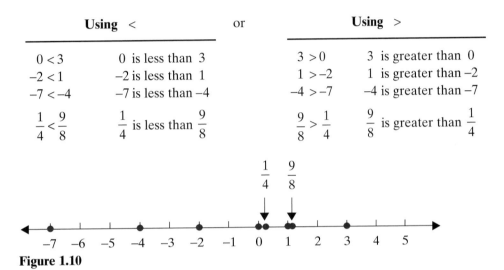

Figure 1.10

Two other symbols commonly used are

$$\leq, \qquad \text{read "is less than or equal to"}$$
$$\text{and} \qquad \geq, \qquad \text{read "is greater than or equal to."}$$

For example, $5 \geq -10$ is true since 5 is greater than -10. Also, $5 \geq 5$ is true since 5 does equal 5.

Table of Symbols

$=$	*is equal to*		$\neq$	*is not equal to*
$<$	*is less than*		$>$	*is greater than*
$\leq$	*is less than or equal to*		$\geq$	*is greater than or equal to*

> **NOTES**
>
> **Special Note About the Inequality Symbols.**
> Each symbol can be read from left to right as was just indicated in the Table of Symbols. However, each symbol can also be read from right to left. Thus, any inequality can be read in two ways. For example, 6 < 10 can be read from left to right as "6 is less than 10", but also from right to left as "10 is greater than 6." We will see that this flexibility is particularly useful when reading expressions with variables in Section 3.7.

Example 3: Inequalities

a.   Determine whether each of the following statements is true or false.

7 < 15	True, since 7 is less than 15.
3 > −1	True, since 3 is greater than −1.
4 ≥ −4	True, since 4 is greater than −4.
2.7 ≥ 2.7	True, since 2.7 is equal to 2.7.
−5 < −6	False, since −5 is greater than −6.

(**Note**: 7 < 15 can be read as "7 is less than 15" or as "15 is greater than 7.")
(**Note**: 3 > −1 can be read as "3 is greater than −1" or as "−1 is less than 3.")

b.   Graph the set of **real numbers** { $-\dfrac{3}{4}$, 0, 1, 1.5, 3 }
Solution:

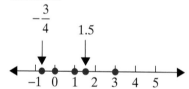

c.   Graph all **natural numbers** less than or equal to 3.
Solution:

Remember that the natural numbers are 1, 2, 3, 4, ...

d.   Graph all **integers** less than 0.
Solution:

Absolute Value

In working with the real number line, you may have noticed that any integer and its opposite lie the same number of units from 0 on the number line. For example, both +7 and −7 are seven units from 0 (Figure 1.11). The + and − signs indicate direction and the 7 indicates distance.

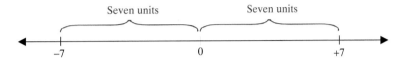

Figure 1.11

The distance a number is from 0 on a number line is called its **absolute value** and is symbolized by two vertical bars, $|\quad|$. Thus, $|+7| = 7$ and $|-7| = 7$. Similarly,

Animation:
Introduction to
Absolute Values

$|3| = 3$

$|-4| = 4$

$|-2| = 2$

$|0| = 0$

$\left|-\dfrac{4}{3}\right| = \dfrac{4}{3}$

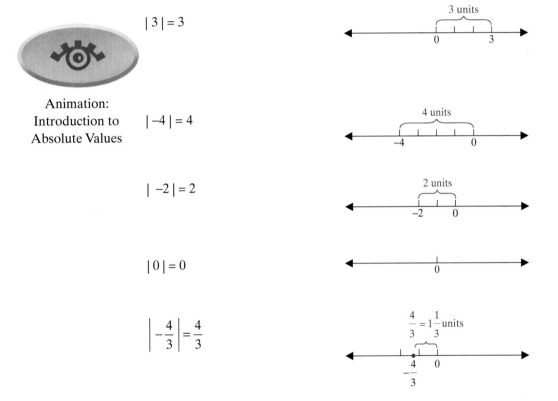

Since distance (similar to length) is never negative, the absolute value of a number is never negative. Or, the absolute value of a nonzero number is always positive.

Absolute Value

The **absolute value** of a real number is its distance from 0. Note that the absolute value of a real number is never negative.

$$|a| = a \text{ if } a \text{ is a positive number or } 0.$$
$$|a| = -a \text{ if } a \text{ is a negative number.}$$

NOTES

The symbol $-a$ should be thought of as the "opposite of a." Since a is a variable, a might represent a positive number, a negative number, or 0. This use of symbols can make the definition of absolute value difficult to understand at first. As an aid to understanding the use of the negative sign, consider the following examples.

If $a = -6$, then $-a = -(-6) = 6.$

Similarly,

If $x = -1$, then $-x = -(-1) = 1.$
If $y = -10$, then $-y = -(-10) = 10.$

Remember that $-a$ (the opposite of a) represents a positive number whenever a represents a negative number.

Example 4: Absolute Value ● ● ● ● ● ● ● ● ● ● ● ● ● ● ● ● ● ●

a. $|6.3| = 6.3$
The number 6.3 is 6.3 units from 0. Also, 6.3 is positive so its absolute value is the same as the number itself.

b. $|-5.1| = -(-5.1) = 5.1$
The number -5.1 is 5.1 units from 0. Also, -5.1 is negative so its absolute value is its opposite.

c. If $|x| = 7$, what are the possible values for x?
Solution: $x = 7$ or $x = -7$ since $|7| = 7$ and $|-7| = 7.$

d. If $|x| = 1.35$, what are the possible values for x?
Solution: $x = 1.35$ or $x = -1.35$ since $|1.35| = 1.35$ and $|-1.35| = 1.35.$

e. True or False: $|-4| \geq 4$

Solution: True, since $|-4| = 4$ and $4 \geq 4$.

f. True or False: $\left|-5\frac{1}{2}\right| < 5\frac{1}{2}$

Solution: False, since $\left|-5\frac{1}{2}\right| = 5\frac{1}{2}$ and $5\frac{1}{2} \not< 5\frac{1}{2}$.

($\not<$ is read "is not less than")

g. If $|x| = -3$, what are the possible values for x?

Solution: There are no values of x for which $|x| = -3$. The absolute value can never be negative. There is no solution.

h. If $|x| < 3$, what are the possible integer values for x? Graph these numbers on a number line.

Solution: The integers are within 3 units of 0: $-2, -1, 0, 1, 2$.

i. If $|x| \geq 4$, what are the possible integer values for x? Graph these numbers on a number line.

Solution: The integers must be 4 or more units from 0: $..., -7, -6, -5, -4, 4, 5, 6, 7, ...$

Fill in the blank with the appropriate symbol: <, >, or =.

1. -2 ____ 1

2. $1\dfrac{6}{10}$ ____ 1.6

3. $-(-4.1\ )$ ____ -7.2

4. *Graph the set of all negative integers on a number line.*

5. *True or False:* $3.6 \leq |-3.6|$

6. *List the numbers that satisfy the equation* $|x| = 8$.

7. *List the numbers that satisfy the equation* $|x| = -6$.

1.1 Exercises

In Exercises 1 – 10 graph each set of real numbers on a real number line. For decimal representations of fractions, use a calculator.

1. $\{1, 2, 5, 6\}$ **2.** $\{-3, -2, 0, 1\}$ **3.** $\{2, -3, 0, -1\}$

4. $\{-2, -1, 4, -3\}$ **5.** $\left\{0,\ -1,\ \dfrac{5}{4},\ 3,\ 1\right\}$ **6.** $\left\{-2,\ -1,\ -\dfrac{1}{3},\ 2\right\}$

7. $\left\{-\dfrac{3}{4},\ 0,\ 2,\ 3.6\right\}$ **8.** $\left\{-3.4,\ -2,\ 0.5,\ 1,\ \dfrac{5}{2}\right\}$ **9.** $\left\{-\dfrac{7}{2},\ -1.5,\ 1,\ \dfrac{4}{3},\ 2\right\}$

10. $\left\{-4,\ -\dfrac{7}{3},\ -1,\ 0.2,\ \dfrac{5}{2}\right\}$

In Exercises 11 – 24, graph each set of integers on a real number line.

11. All positive integers less than 4   **12.** All whole numbers less than 7
13. All positive integers less than or   **14.** All negative integers greater than or
       equal to 3   equal to –3
15. All integers less than –6   **16.** All integers greater than or equal to –1
17. All integers less than 8   **18.** All whole numbers less than or equal to 4
19. All negative integers greater than   **20.** All integers greater than or equal to –6
       or equal to 2
21. All integers more than 6 units from 0 **22.** All integers more than 3 units from 0
23. All integers less than 3 units from 0 **24.** All integers less than 8 units from 0

Answers to Practice Problems: 1. $<$ **2.** $=$ **3.** $>$ **4.**   **5.** True **6.** 8, –8 **7.** None

Fill in the blank in Exercises 25 – 39 with the appropriate symbol: $<$, $>$, or $=$.

25. 4 ___ 6

26. –3 ___ 1

27. –2 ___ –4

28. 5 ___ –(–5)

29. $\dfrac{1}{3}$ ___ $\dfrac{1}{2}$

30. $-\dfrac{2}{3}$ ___ $\dfrac{1}{8}$

31. $-\dfrac{2}{8}$ ___ $-\dfrac{1}{4}$

32. –8 ___ 0

33. 2.3 ___ 1.6

34. $-\dfrac{3}{4}$ ___ –1

35. $\dfrac{9}{16}$ ___ $\dfrac{3}{4}$

36. $-\dfrac{1}{2}$ ___ $-\dfrac{1}{3}$

37. –2.3 ___ $-2\dfrac{3}{10}$

38. 5.6 ___ –(–8.7)

39. $-\dfrac{4}{3}$ ___ $-\left(-\dfrac{1}{3}\right)$

Determine whether each statement in Exercises 40 – 64 is true or false. If a statement is false, rewrite it in a form that is a true statement. (There may be more than one way to correct a statement.)

40. $0 = -0$

41. $-22 < -16$

42. $-9 > -8.5$

43. $11 = -(-11)$

44. $-17 \le 17$

45. $-6 < -8$

46. $4.7 \ge 3.5$

47. $-\dfrac{1}{3} \le 0$

48. $\dfrac{3}{5} > \dfrac{1}{4}$

49. $-2.3 < 1$

50. $|-5| = 5$

51. $|-8| \ge 4$

52. $|-6| \ge 6$

53. $|-7| < |7|$

54. $|-1.9| < 2$

55. $|-1.6| < |-2.1|$

56. $\left|-\dfrac{5}{2}\right| < 2$

57. $\dfrac{2}{3} < |-1|$

58. $3 > \left|-\dfrac{4}{3}\right|$

59. $|-3.4| < 0$

60. $-|-3| < -|4|$

61. $-|5| > -|3.1|$

62. $|-6.2| = 6.2$

63. $-|73| < |-73|$

64. $|2.5| = \left|-\dfrac{5}{2}\right|$

List the numbers, then graph the numbers on a real number line that satisfy the equations in Exercises 65 – 74.

65. $|x| = 4$

66. $|y| = 6$

67. $|x| = 9$

68. $13 = |x|$

69. $0 = |y|$

70. $-2 = |x|$

71. $|x| = -3$

72. $|x| = 3.5$

73. $|x| = 4.7$

74. $|x| = \left|-\dfrac{5}{4}\right|$

On a number line, graph the integers that satisfy the conditions stated in Exercises 75 – 84.

75. $|x| \le 4$ **76.** $|x| < 6$ **77.** $|x| > 5$ **78.** $|x| > 2$ **79.** $|x| < 7$

80. $|x| \le 2$ **81.** $|x| \le x$ **82.** $|x| = -x$ **83.** $|x| = x$ **84.** $|x| > x$

Choose the response that correctly completes each sentence in Exercises 85 – 89. Give one or more examples to illustrate your conclusion for each exercise.

85. $|x|$ is (never, sometimes, always) equal to x.

86. $|x|$ is (never, sometimes, always) equal to $-x$.

87. $|x|$ is (never, sometimes, always) a negative number.

88. $|x|$ is (never, sometimes, always) equal to 0.

89. $|x|$ is (never, sometimes, always) equal to a positive number.

Calculator Problems

A TI-83 Plus calculator has the absolute value command built in. To access the absolute command press **MATH**, go to **NUM** at the top of the screen and press **1** or **ENTER**. The command **abs(** will appear on the display screen. Then enter any arithmetic expression you wish and the calculator will print the absolute value of that expression. (**Note:** The negative sign is on the key marked **(−)** next to the **ENTER** key.)

Follow the directions above and use your calculator to find the value of each of the following expressions.

90. $|34 - 80|$ **91.** $|17.5 + 16.3 - 95.2|$ **92.** $-|10 - 16|$

93. $\dfrac{|-6| - |-3|}{|-6 - 3|}$ **94.** $\dfrac{|4| + |-4|}{|-8|}$

Writing and Thinking About Mathematics

95. Explain, in your own words, how a variable expression such as $-y$ might represent a positive number.

96. Explain, in your own words, the meaning of absolute value.

Hawkes Learning Systems: Introductory Algebra

Name That Real Number
Introduction to Absolute Values

1.2   Addition with Integers

After completing this section, you will be able to:

1. Add integers.

2. Determine if given integers are solutions for specified equations.

3. Complete statements about integers.

Picture a straight line in an open field and numbers marked on a number line. An archer stands at 0 and shoots an arrow to +3, then stands at 3 and shoots the arrow 5 more units in the positive direction (to the right). Where will the arrow land? (Figure 1.12.)

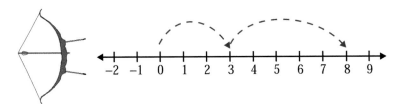

Figure 1.12

Naturally, you have figured out that the answer is +8. What you have done is add the two positive integers, +3 and +5.

$$(+3) + (+5) = +8 \quad \text{or} \quad 3 + 5 = 8$$

Suppose another archer shoots an arrow in the same manner as the first but in the opposite direction. Where would his arrow land? The arrow lands at −8. You have just added −3 and −5 (Figure 1.13).

Animation:
Addition with
Integers

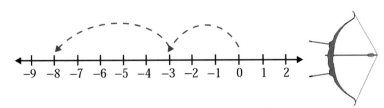

Figure 1.13

If the archer stands at 0 and shoots an arrow to +3 and then the archer goes to +3 and turns around and shoots an arrow 5 units in the opposite direction, where will the arrow stick? Would you believe at −2? (Figure 1.14.)

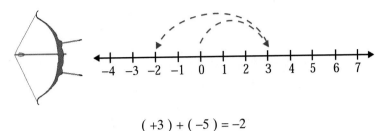

$$(+3) + (-5) = -2$$

Figure 1.14

For our final archer, the first shot is to −3. Then, after going to −3, he turns around and shoots 5 units in the opposite direction. Where is the arrow? It is at +2 (Figure 1.15).

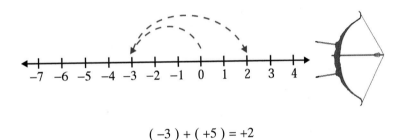

$$(-3) + (+5) = +2$$

Figure 1.15

In summary:

1. The sum of two positive integers is positive.

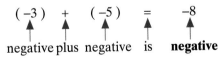

$$(+3) \quad + \quad (+5) \quad = \quad +8$$
positive plus positive is **positive**

2. The sum of two negative integers is negative.

$$(-3) \quad + \quad (-5) \quad = \quad -8$$
negative plus negative is **negative**

3. The sum of a positive integer and a negative integer may be negative or positive (or zero) depending on which number is further from 0.

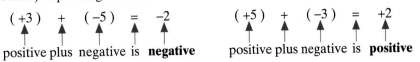

$$(+3) \quad + \quad (-5) \quad = \quad -2 \qquad (+5) \quad + \quad (-3) \quad = \quad +2$$
positive plus negative is **negative**   positive plus negative is **positive**

Practice Problems

Find each sum. Add from left to right if there are more than two numbers.

1. $(-14) + (-6) =$ **2.** $(+16) + (-10) =$

3. $(-12) + (8) =$ **4.** $11 + 7 =$

5. $(-13) + (+13) =$ **6.** $(-11) + (-8) =$

7. $(+6) + (-7) + (-1) =$ **8.** $(+100) + (-100) + (+10) =$

You probably did quite well and understand how to add integers. The rules can be written out in the following rather formal manner.

Rules for Addition with Integers

1. To add two integers with like signs, add their absolute values and use the common sign:

$$(+7) + (+3) = +(|+7| + |+3|) = +(7 + 3) = +10$$
$$(-7) + (-3) = -(|-7| + |-3|) = -(7 + 3) = -10$$

2. To add two integers with unlike signs, subtract their absolute values (the smaller from the larger) and use the sign of the number with the larger absolute value:

$$(-12) + (+10) = -(|-12| - |+10|) = -(12 - 10) = -2$$
$$(+12) + (-10) = +(|+12| - |-10|) = +(12 - 10) = +2$$
$$(-15) + (+15) = (|-15| - |+15|) = (15 - 15) = 0$$

An **equation** is a statement that two expressions are equal. Since equations in algebra are almost always written horizontally, you should become used to working with sums written horizontally. However, there are situations (as in long division) where sums (and differences) are written vertically with one number directly under another. We illustrate this technique in Example 1.

Answers to Practice Problems: **1.** -20 **2.** $+6$ **3.** -4 **4.** 18 **5.** 0 **6.** -19 **7.** -2 **8.** 10

Example 1: Vertical •

Find each sum.

a. $\begin{array}{r} -10 \\ 7 \\ \hline -3 \end{array}$
b. $\begin{array}{r} -4 \\ 6 \\ -15 \\ \hline -13 \end{array}$
c. $\begin{array}{r} -5 \\ -8 \\ -9 \\ \hline -22 \end{array}$
d. $\begin{array}{r} -10 \\ 3 \\ 7 \\ \hline 0 \end{array}$

• •

Now that we know how to add positive and negative integers, we can determine whether or not a particular integer satisfies an equation that contains a variable. Recall, a variable is a letter (or symbol) that can represent one or more numbers. A number is said **to be a solution or to satisfy an equation** if it gives a true statement when substituted for the variable.

Example 2: Solutions •

Determine whether or not the given integer is a solution to the given equation by substituting for the variable and adding.

a. $x + 5 = -2;\ x = -7$
   Solution: $(-7) + 5 = -2$ is true, so -7 is a solution.

b. $y + (-4) = -6;\ y = -2$
   Solution: $(-2) + (-4) = -6$ is true, so -2 is a solution.

c. $14 + z = -3;\ z = -11$
   Solution: $14 + (-11) = -3$ is false since $14 + (-11) = +3$.
   So, -11 is **not** a solution.

• •

1.2 Exercises

Find the sum in Exercises 1 – 46.

1. $4 + 9$
2. $8 + (-3)$
3. $(-9) + 5$
4. $(-7) + (-3)$
5. $(-9) + 9$
6. $2 + (-8)$
7. $11 + (-6)$
8. $(-12) + 3$
9. $-18 + 5$
10. $26 + (-26)$
11. $-5 + (-3)$
12. $11 + (-2)$
13. $(-2) + (-8)$
14. $10 + (-3)$
15. $17 + (-17)$
16. $(-7) + 20$
17. $21 + (-4)$
18. $(-15) + (-3)$
19. $(-12) + (-17)$
20. $24 + (-16)$
21. $-4 + (-5)$
22. $(-6) + (-8)$
23. $9 + (-12)$
24. $-12 + 9$
25. $38 + (-16)$
26. $(-20) + (-11)$
27. $(-33) + (-21)$
28. $(-21) + 18$
29. $-3 + 4 + (-8)$
30. $(-9) + (-6) + 5$
31. $(-9) + (-2) + (-5)$
32. $(-21) + 6 + 15$
33. $-13 + (-1) + (-12)$
34. $-19 + (-2) + (-4)$
35. $27 + (-14) + (-13)$
36. $-33 + 29 + 2$

37. $-43 + (-16) + 27$ **38.** $-68 + (-3) + 42$ **39.** $-38 + 49 + (-6)$
40. $102 + (-93) + (-6)$

41. -21 **42.** -12 **43.** -15
$\underline{-62}$ $\underline{17}$ 8
$\underline{19}$

44. -7 **45.** -163 **46.** -93
23 204 -87
$\underline{-9}$ $\underline{-73}$ $\underline{147}$

In Exercises 47 – 60, determine whether or not the given number is a solution to the given equation by substituting and then evaluating.

47. $x + 4 = 2;\ x = -2$
48. $x + (-7) = 10;\ x = -3$
49. $-10 + x = -14;\ x = -4$
50. $y + 9 = 7;\ y = -2$
51. $17 + y = 11;\ y = -6$
52. $z + (-1) = 9;\ z = 8$
53. $z + (-12) = 6;\ z = 18$
54. $x + 3 = -10;\ x = -7$
55. $x + (-5) = -15;\ x = -10$
56. $|x| + 15 = 17;\ x = -2$
57. $|y| + (-10) = -12;\ y = -2$
58. $-26 + |x| = -8;\ x = -18$
59. $42 + |z| = -30;\ z = -72$
60. $|x| + (-5) = -1;\ x = -4$

61. If x is a positive number and y is a negative number, then $x + y$ is (never, sometimes, always) a negative number.

62. If x and y are integers, then $x + y$ is (never, sometimes, always) equal to 0.

63. If x and y are positive numbers, then $x + y$ is (never, sometimes, always) equal to 0.

64. If y is an integer, then $y + (-y)$ is (never, sometimes, always) equal to 0.

65. If x and y are negative numbers, then $x + y$ is (never, sometimes, always) equal to 0.

66. If x and y are integers, then $x + y$ is (never, sometimes, always) negative.

67. If x and y are integers, then $x + y$ is (never, sometimes, always) positive.

68. If x represents a negative number, then $-x$ is (never, sometimes, always) negative.

69. If x represents a negative number, then $-x$ is (never, sometimes, always) positive.

70. If x represents a positive number, then $-x$ is (never, sometimes, always) negative.

Calculator Problems

Use your TI-83 Plus calculator to find the value of each of the following expressions. (Remember that the key marked $(-)$ next to the ENTER key is used to indicate negative numbers.)

71. $47 + (-29) + 66$

72. $56 + (-41) + (-28)$

73. $(-16,945) + (-27,302) + (-53,467)$

74. $2,932 + 4,751 + (-3,876)$

75. $(-8,154) + 2,147 + (-136)$

Writing and Thinking About Mathematics

76. Describe, in your own words, how the sum of the absolute values of two numbers might be 0. (Is this even possible?)

Hawkes Learning Systems: Introductory Algebra

Addition with Integers

1.3   Subtraction with Integers

Objectives

After completing this section, you will be able to:

1. Find the additive inverse of an integer.

2. Subtract integers.

3. Determine if given integers are solutions for specified equations.

In basic arithmetic, subtraction is defined in terms of addition. For example, we know that the difference 32 − 25 is equal to 7 because 25 + 7 = 32. A beginning student in arithmetic does not know how to find a difference such as 15 − 20, where a larger number is subtracted from a smaller number, because negative numbers are not yet defined and there is no way to add a positive number to 20 and get 15. Now, with our knowledge of negative numbers, we will define subtraction in such a way that larger numbers may be subtracted from smaller numbers. We will still define subtraction in terms of addition, but we will apply our new rules of addition with integers.

Before we proceed to develop the techniques for subtraction with integers, we will state and illustrate an important relationship between any integer and its opposite.

Additive Inverse

*The **opposite** of an integer is called its **additive inverse**. The sum of a number and its additive inverse is zero. Symbolically, for any integer a,*

$$a + (-a) = 0$$

Example 1: Additive Inverse ● ● ● ● ● ● ● ● ● ● ● ● ● ● ●

a. Find the additive inverse (opposite) of 3.
   Solution: The additive inverse of 3 is −3, 3 + (−3) = 0.

b. Find the additive inverse (opposite) of −7.
   Solution: The additive inverse of −7 is −(−7) = +7, (−7) + (+7) = 0.

c. Find the additive inverse (opposite) of 0.
   Solution: The additive inverse of 0 is −0 = 0.
       That is, 0 is its own opposite, (0) + (−0) = 0 + 0 = 0.

● ●

In Section 1.2, we added integers such as

$$5 + (-2) = 3 \quad \text{and} \quad 26 + (-9) = 17.$$

Note that in each case, subtraction will give the same results; that is,

$$5 - 2 = 3 \quad \text{and} \quad 26 - 9 = 17.$$

Thus, it seems that subtraction and addition are closely related and, in fact, they are. From arithmetic, $5 - 2$ is asking, "What number added to 2 gives 5?" In other words,

$$5 - 2 = 3 \quad \text{because} \quad 5 = 2 + 3$$
$$26 - 9 = 17 \quad \text{because} \quad 26 = 9 + 17$$

What do we mean by $4 - (-1)$? Do we mean, "What number added to -1 gives 4?" Precisely. Thus,

$$4 - (-1) = 5 \quad \text{because} \quad 4 = (-1) + 5$$
$$4 - (-2) = 6 \quad \text{because} \quad 4 = (-2) + 6$$
$$4 - (-3) = 7 \quad \text{because} \quad 4 = (-3) + 7$$

But note the following results:

$$4 + (+1) = 5$$
$$4 + (+2) = 6$$
$$4 + (+3) = 7$$

The following relationship between subtraction and addition becomes the basis for subtraction with all numbers:

$$4 - (-1) = 4 + (+1) = 5 \qquad (-4) - (-1) = (-4) + (+1) = -3$$
$$4 - (-2) = 4 + (+2) = 6 \qquad (-4) - (-2) = (-4) + (+2) = -2$$
$$4 - (-3) = 4 + (+3) = 7 \qquad (-4) - (-3) = (-4) + (+3) = -1$$

Example 2: Subtraction ● ● ● ● ● ● ● ● ● ● ● ● ● ● ● ● ● ●

a. $(-1) - (-4) = (-1) + (+4) = +3$
b. $(-1) - (-5) = (-1) + (+5) = +4$
c. $(-1) - (-8) = (-1) + (+8) = 7$
d. $(10) - (-2) = (10) + (+2) = 12$
e. $(-10) - (-5) = (-10) + (+5) = -5$

● ●

Practice Problems

1. $(-4) - (-4) = (-4) + (+4) =$
2. $(-3) - (-8) = (-3) + (+8) =$
3. $(-3) - (+8) = (-3) + (-8) =$
4. $14 - (-5) = 14 + (5) =$
5. $14 - (5) = 14 + (-5) =$

You may have noticed that in subtraction, the **opposite** of the number being subtracted is **added**. For example, we could write

$$(-1) - (-4) = (-1) + [-(-4)] = (-1) + (+4) = +3$$

or

add opposite

$$(-10) - (-3) = (-10) + [-(-3)] = (-10) + (+3) = -7$$

Subtraction

For any integers a and b,

$$a - b = a + (-b)$$

This definition translates as, "To subtract b from a, **add** the **opposite** of b to a." In practice, the notation $a - b$ is thought of as addition of signed numbers. That is, since $a - b = a + (-b)$, we think of the plus sign, +, as being present in $a - b$. In fact, an expression such as $4 - 19$ can be thought of as "four plus negative nineteen." We have

$$4 - 19 = 4 + (-19) = -15$$
$$-25 - 30 = -25 + (-30) = -55$$
$$-3 - (-17) = -3 + (+17) = 14$$
$$24 - 11 - 6 = 24 + (-11) + (-6) = 7$$

Generally, the second step is omitted and we go directly to the answer by computing the sum mentally.

$$4 - 19 = -15$$
$$-25 - 30 = -55$$
$$-3 - (-17) = 14$$
$$24 - 11 - 6 = 7$$

Answers to practice problems: 1. 0 **2.** 5 **3.** −11 **4.** 19 **5.** 9

The numbers may also be written vertically, that is, one underneath the other. In this case, the sign of the number being subtracted (the bottom number) is changed and addition is performed.

Example 3: Subtract and Add ● ● ● ● ● ● ● ● ● ● ● ● ● ● ● ● ●

a. **Subtract**   **Add**   b. **Subtract**   **Add**

$$43 \qquad\qquad 43 \qquad\qquad\qquad -38 \qquad\qquad -38$$

$$\underline{-(-25)} \xrightarrow[\text{change}]{\text{sign}} \underline{+25} \qquad\qquad \underline{-(+11)} \xrightarrow[\text{change}]{\text{sign}} \underline{-11}$$

$$\qquad\qquad\qquad 68 \qquad\qquad\qquad\qquad\qquad\qquad -49$$

c. **Subtract**   **Add**   d. **Subtract**   **Add**

$$-73 \qquad\qquad -73 \qquad\qquad\qquad 17 \qquad\qquad 17$$

$$\underline{-(-32)} \xrightarrow[\text{change}]{\text{sign}} \underline{+32} \qquad\qquad \underline{-(+69)} \xrightarrow[\text{change}]{\text{sign}} \underline{-69}$$

$$\qquad\qquad\qquad -41 \qquad\qquad\qquad\qquad\qquad\qquad -52$$

● ●

Now, using subtraction as well as addition, we can determine whether or not a number is a solution to an equation of a slightly more complex nature.

Example 4: Solutions ● ● ● ● ● ● ● ● ● ● ● ● ● ● ● ● ● ● ●

Determine whether or not the given number is a solution to the given equation by substituting and then evaluating.

a. $x - (-5) = 6; x = 1$
Solution: $1 - (-5) = 1 + (+5) = 6$ is true, so 1 is a solution.

b. $5 - y = 7; y = -2$
Solution: $5 - (-2) = 5 + (+2) = 7$ is true, so −2 is a solution.
    Note that parentheses were used around −2. This should be done whenever substituting negative numbers for the variable.

c. $z - 14 = -3; z = 10$
Solution: $10 - 14 = -4$ and $-4 = -3$ is false, so 10 is not a solution.

● ●

Practice Problems

1. *What is the additive inverse of 85?*

2. *Find the difference:* $-6 - (-5)$

3. *Simplify:* $-6 - 4 - (-2)$

4. *True or false:* $-5 + (-3) < -5 - (-3)$

5. *Is* $x = 15$ *a solution to the equation* $x - 1 = -16?$

1.3 Exercises

Find the additive inverse for each integer given in Exercises 1 – 10.

1. 11	**2.** 17	**3.** −6	**4.** −23	**5.** 47
6. −34	**7.** 0	**8.** 100	**9.** −52	**10.** −257

Simplify the expressions in Exercises 11 – 22.

11. $8 - 3$	**12.** $5 - 7$	**13.** $-4 - 6$	**14.** $3 - (-4)$
15. $5 - (-7)$	**16.** $-18 - 17$	**17.** $-8 - (-11)$	**18.** $0 - (-12)$
19. $-14 - 2$	**20.** $-8 - 7$	**21.** $16 - (-8)$	**22.** $15 - 23$

Subtract the bottom number from the top number in Exercises 23 – 30.

23. 27	**24.** 19	**25.** −23	**26.** −41
42	26	−7	−8
27. −21	**28.** −47	**29.** −27	**30.** −19
36	13	27	−26

31. Find the difference between −5 and −6. (**Hint**: Subtract the numbers in the order given.)

32. Subtract −3 from −10.

33. Subtract −2 from 6.

34. Find the difference between 30 and −12. (**Hint**: Subtract the numbers in the order given.)

35. Subtract 13 from −13.

Answers to Practice Problems: 1. −85 **2.** −1 **3.** −8 **4.** True **5.** Not a solution

Perform the indicated operations in Exercises 36 – 45.

36. $-6 + (-4) - 5$
39. $-3 + (-7) + 2$
42. $-2 - 2 + 11$
45. $-113 + 53 - 79$

37. $-7 - (-2) + 6$
40. $-5 - 2 - (-4)$
43. $-3 - (-3) + (-6)$

38. $6 + (-3) + (-4)$
41. $-8 - 5 - (-3)$
44. $97 - 16 - (81)$

Perform the operations on each side of the blank and then fill in the blank in Exercises 46 – 55 with the proper symbol: <, >, or =.

46. $-6 + (-2)$ _____ $3 + (-8)$
48. $5 - 8$ _____ $8 - 5$
50. $11 + (-3)$ _____ $11 - 3$
52. $-8 - (-8)$ _____ $-14 - 13$
54. $-151 - 86$ _____ $-(107 + 141)$

47. $-4 - (-3)$ _____ $-4 + (-3)$
49. $7 - (-3)$ _____ $-3 - 7$
51. $0 - 6$ _____ $0 - (-6)$
53. $-7 - (-3)$ _____ $4 - 9$
55. $25 - 62$ _____ $-11 - 23$

In Exercises 56 – 70, determine whether or not the given number is a solution to the given equation by substituting and then evaluating.

56. $x + 5 = -3; \ x = -8$
58. $15 - y = 17; \ y = -2$
60. $x - 3 = -7; \ x = 4$
62. $-9 - x = -14; \ x = 5$
64. $x - 2 = -3; \ x = -1$
66. $|x| - (-10) = 14; \ x = -4$
68. $|z| - 5 = 0; \ z = -5$
70. $|x| - |-3| = 25; \ x = -28$

57. $x - 6 = -9; \ x = -3$
59. $11 - x = 8; \ x = 3$
61. $y - 2 = 6; \ y = 4$
63. $x + 13 = 3; \ x = -10$
65. $-18 - y = -30; \ y = 12$
67. $|y| - 12 = -3; \ y = -9$
69. $-16 - |x| = 0; \ x = 16$

Calculator Problems

In Exercises 71 – 73, use your TI-83 Plus calculator to find the value indicated on each side of the blank and then fill in the blank with the proper symbol: <, >, or =.

71. $648 - (-396)$ _____ $124 - 163$
72. $-19,824 - 23,417$ _____ $12,793 - (-14,387)$
73. $-43,931 - (-28,677)$ _____ $-(13,665 + 21,425)$

Temperatures above 0 and below 0 as well as gains and losses can be thought of in terms of positive and negative numbers. Use positive and negative numbers to answer Exercises 74 – 77.

74. Harry and his wife went on a diet plan for 5 weeks. During those 5 weeks, Harry lost 5 pounds, gained 3 pounds, lost 2 pounds, lost 4 pounds, and gained 1 pound. What was his total loss (or gain) for the 5 weeks? If he weighed 210 pounds when he started the diet plan, what did he weigh at the end of the 5-week period? During the same time, his wife lost 10 pounds.

75. In a 5-day week the NASDAQ stock market posted a gain of 38 points, a loss of 65 points, a loss of 32 points, a gain of 10 points, and a gain of 15 points. If the NASDAQ started the week at 2350 points, what was the market at the end of the week?

76. In ten running plays in a football game the fullback gained 5 yards, lost 3 yards, gained 15 yards, gained 7 yards, gained 12 yards, lost 4 yards, lost 2 yards, gained 20 yards, lost 5 yards, and gained 6 yards. What was his net yardage for the game?

77. Beginning at a temperature of $10°$ above 0, the temperature in a scientific experiment was measured hourly for four hours. It dropped $5°$, dropped $8°$, dropped $6°$, then rose $3°$. What was the final temperature recorded?

Hawkes Learning Systems: Introductory Algebra

 Subtraction with Integers

1.4 Multiplication and Division with Integers

Objectives

After completing this section, you will be able to:

1. *Multiply integers.*

2. *Divide integers.*

3. *Complete statements about the products and quotients of integers.*

4. *Determine if equations are true or false.*

Multiplication is shorthand for repeated addition. That is,

$$7 + 7 + 7 + 7 + 7 = 5 \cdot 7 = 35$$

and

$$(-6) + (-6) + (-6) = 3(-6) = -18.$$

Similarly,

$$(-2) + (-2) + (-2) + (-2) + (-2) = 5(-2) = -10$$

Repeated addition with a negative integer results in a product of a positive integer and a negative integer. Since the sum of negative integers is negative, we have the following general rule.

The product of a positive integer and a negative integer is negative.

> **NOTES** Multiplication can be represented by a raised dot, as in $5 \cdot 7$, or by a number next to a parenthesis as in $3(-6)$ or $(3)(-6)$.

Example 1: Positive Times Negative

a. $5(-3) = (-3) + (-3) + (-3) + (-3) + (-3) = -15$
b. $7(-10) = -70$
c. $42(-1) = -42$
d. $3(-5) = -15$

The product of two negative integers can be explained in terms of opposites. We also need the fact that for any integer a, we can think of the **opposite of a**, $(-a)$, as the product of -1 and a. That is, we have

$$-a = -1 \cdot a.$$

Thus,

$$-4 = -1 \cdot 4 = -1(4)$$

and in the product $-4(-7)$ we have,

$$-4(-7) = -1(4)(-7) = -1[(4)(-7)] = -1(-28) = -(-28) = 28.$$

Although one example does not prove a rule, this process can be used in general to arrive at the following correct conclusion:

The product of two negative integers is positive.

Example 2: Negative Times Negative ● ● ● ● ● ● ● ● ● ● ● ●

a. $(-4)(-9) = +36$
b. $-7(-5) = +35$
c. $-2(-6) = +12$
d. $(-1)(-5)(-3)(-2) = 5(-3)(-2) = -15(-2) = +30$

● ●

What happens if a number is multiplied by 0? For example, $3(0) = 0 + 0 + 0 = 0$. In fact,

multiplication by 0 always gives a product of 0.

Example 3: Multiplication by 0 ● ● ● ● ● ● ● ● ● ● ● ● ● ●

a. $6 \cdot 0 = 0$
b. $-13 \cdot 0 = 0$

● ●

The rules for multiplication can be summarized as follows.

Rules for Multiplication with Integers

If a and b are positive integers, then

1.   The product of two positive integers is positive: $a \cdot b = ab$.

2.   The product of two negative integers is positive: $(-a)(-b) = ab$.

3.   The product of a positive integer and a negative integer is negative: $a(-b) = -ab$.

4.   The product of 0 and any integer is 0: $a \cdot 0 = 0$ and $(-a) \cdot 0 = 0$.

Practice Problems

Find the following products.

1. $5(-3) =$ 2. $-6(-4) =$

3. $-8(4) =$ 4. $-12(0) =$

5. $-9(-2)(-1) =$ 6. $3(-20)(5) =$

Answers to Practice Problems: 1. -15 **2.** 24 **3.** -32 **4.** 0 **5.** -18 **6.** -300

The rules for multiplication lead directly to the rules for division since division is defined in terms of multiplication. For convenience, division is indicated in fraction form.

Division with Integers

For integers a, b, and x (where b ≠ 0),

$$\frac{a}{b} = x \ \text{means that} \ a = b \cdot x$$

If a is any integer, then $\dfrac{a}{0}$ *is* **undefined**, *but* $\dfrac{0}{b} = 0.$

Now you may ask, "Why is division by 0 undefined?" The following explanation will answer that question.

Division by 0 is Undefined

1. *Suppose that $a \neq 0$ and $\dfrac{a}{0} = x$. Then, since division is related to multiplication, we must have $a = 0 \cdot x$. But this is not possible because $0 \cdot x = 0$ for any value of x and we stated that $a \neq 0$.*

2. *Suppose that $\dfrac{0}{0} = x$. Then, $0 = 0 \cdot x$ which is true for all values of x. But we must have a unique answer for x.*

*Therefore, in any case, we conclude that **division by 0 is undefined**.*

Example 4: Division

a. $\dfrac{36}{9} = 4$ because $36 = 9 \cdot 4.$

b. $\dfrac{-36}{9} = -4$ because $-36 = 9(-4).$

c. $\dfrac{36}{-9} = -4$ because $36 = -9(-4).$

d. $\dfrac{-36}{-9} = +4$ because $-36 = -9(4).$

The rules for division can be stated as follows.

Rules for Division with Integers

If a and b are positive integers,

1. *The quotient of two positive integers is positive:* $\dfrac{a}{b} = +\dfrac{a}{b}$.

2. *The quotient of two negative integers is positive:* $\dfrac{-a}{-b} = +\dfrac{a}{b}$.

3. *The quotient of a positive integer and a negative integer is negative:*

$$\dfrac{-a}{b} = -\dfrac{a}{b} \quad and \quad \dfrac{a}{-b} = -\dfrac{a}{b}.$$

NOTES

The following common rules about multiplication and division with two non-zero integers are helpful in remembering the signs of answers.

1. If the integers have the same sign, both the product and quotient will be positive.
2. If the integers have different signs, both the product and quotient will be negative.

The quotient of two integers may not always be another integer. Just as with whole numbers, the quotient may be a fraction. For example, $\dfrac{-3}{6} = -\dfrac{1}{2}$ and $\dfrac{2}{-8} = -\dfrac{1}{4}$. We will discuss these ideas more thoroughly in Chapter 2. In this section, the problems are set so that multiplication and division with integers will have an integer result. **Remember that division by 0 is not defined**.

Practice Problems

Find the quotients.

1. $\dfrac{-30}{10}$ **2.** $\dfrac{40}{-10}$ **3.** $\dfrac{-20}{-10}$ **4.** $\dfrac{-7}{0}$ **5.** $\dfrac{0}{13}$

Answers to Practice Problems: 1. -3 **2.** -4 **3.** 2 **4.** undefined **5.** 0

Now that we have all the rules for addition, subtraction, multiplication, and division with integers, we can discuss the solutions to equations that involve any or all of these operations.

To indicate multiplication of a variable by a specific number, the times sign (or raised dot) is optional. For example,

$$23 \cdot x = 23x \quad \text{and} \quad -7 \cdot y = -7y.$$

In each case, the number is called the **coefficient** of the variable.

Example 5: Solutions ● ● ● ● ● ● ● ● ● ● ● ● ● ● ● ● ● ● ●

Determine whether or not the given integer is a solution to the given equation by substituting and then evaluating.

a. $7x = -21; x = -3$
  Solution: $7(-3) = -21$ is true, so -3 is a solution.

b. $-8y = 56; y = -7$
  Solution: $-8(-7) = 56$ is true, so -7 is a solution.

c. $\dfrac{y}{-4} = -10; \ y = -40$

  Solution: $\dfrac{-40}{-4} = 10$ and $10 = -10$ is false, so -40 is not a solution.

d. $-5x + 7 = -3; x = 2$
  Solution: $-5(2) + 7 = -10 + 7 = -3$ is true, so 2 is a solution.

● ●

1.4 Exercises

Find the product in Exercises 1 – 20.

1. $4 \cdot (-3)$ **2.** $(-5) \cdot 6$ **3.** $(-8)(-7)$ **4.** $12 \cdot 4$
5. $19 \cdot 3$ **6.** $(-11)(-2)$ **7.** $(-14)(-4)$ **8.** $(-3)(7)$
9. $(5)(-6)$ **10.** $(-11)(-6)$ **11.** $(-13)(-2)$ **12.** $10(-7)$
13. $(-5)(12)$ **14.** $(-8)(-9)$ **15.** $(-2)(-3)(-4)$ **16.** $(-6)(-3)(-9)$
17. $-8 \cdot 4 \cdot 9$ **18.** $-3 \cdot 2 \cdot (-3)$ **19.** $(-7)(-16) \cdot 0$ **20.** $(-9) \cdot 11 \cdot 4$

Find the quotient in Exercises 21 – 35.

21. $\dfrac{-8}{-2}$ **22.** $\dfrac{-20}{-10}$ **23.** $\dfrac{-30}{5}$ **24.** $\dfrac{-26}{-13}$

25. $\dfrac{39}{-13}$ **26.** $\dfrac{-51}{3}$ **27.** $\dfrac{-91}{-7}$ **28.** $\dfrac{0}{6}$

29. $\dfrac{0}{-7}$ **30.** $\dfrac{-3}{0}$ **31.** $\dfrac{16}{0}$ **32.** $\dfrac{-34}{2}$

33. $\dfrac{44}{-4}$ **34.** $\dfrac{-60}{-12}$ **35.** $\dfrac{-36}{9}$

Correctly complete the sentences in Exercises 36 – 45 with positive, negative, 0, or undefined.

36. If x is a positive integer, then $x(-x)$ is a _____ integer.
37. If x is a negative integer, then $x(-x)$ is a _____ integer.
38. If x is a negative integer and y is a negative integer, then xy is a _____ integer.
39. If x is a positive integer and y is a negative integer, then xy is a _____ integer.

40. If x and y are positive integers, then $\dfrac{x}{y}$ is a _____ number.

41. If x is a negative integer and y is a natural number then $\dfrac{x}{y}$ is a _____ number.

42. If x and y are negative numbers, then $\dfrac{x}{y}$ is a _____ number.

43. If x is any real number, then $x \cdot 0$ is _____.

44. If x is a nonzero real number, then $\dfrac{0}{x}$ is _____.

45. If x is any real number, then $\dfrac{x}{0}$ is _____.

Determine whether each statement in Exercises 46 – 55 is true or false. If a statement is false, rewrite it in a form that is true. (There may be more than one correct new form.)

46. $(-4) \cdot (6) < 3 \cdot 8$ **47.** $(-7) \cdot (-9) = 3 \cdot 21$
48. $(-12) \cdot (6) = 9(-8)$ **49.** $(-6)(9) = (18)(-3)$
50. $6(-3) \geq (-14) + (-4)$ **51.** $7 + 8 > (-10) + (-5)$
52. $-7 + 0 \leq (-7) \cdot (0)$ **53.** $17 + (-3) < (-14) + (-4)$
54. $-4(9) = (-24) + (-12)$ **55.** $14 + 6 \leq -2(-10)$

In Exercises 56 – 65, determine whether or not the given number is a solution to the given equation by substituting and then evaluating.

56. $11x = 55$; $x = 5$

57. $-7x = 84$; $x = -12$

58. $\dfrac{x}{7} = -6$; $x = 42$

59. $\dfrac{y}{-6} = 12$; $y = -72$

60. $\dfrac{y}{10} = -9$; $y = -90$

61. $-9x = -72$; $x = -8$

62. $5x + 3 = 18$; $x = 3$

63. $-3x + 7 = -8$; $x = 5$

64. $4x - 3 = -23$; $x = -5$

65. $7x - 6 = -34$; $x = -4$

Calculator Problems

Use a calculator to find the value of each expression in Exercises 66 – 70.

66. $(273)(-24)(-180)$

67. $(-4{,}613)(-45)(-166)$

68. $(54)(-17)(24)$

69. $(-77{,}459) \div 29$

70. $(-62{,}234) \div (-37)$

Writing and Thinking About Mathematics

---

71. Explain the conditions under which the quotient of two numbers is 0.

72. Explain, in your own words, why division by 0 is not a valid arithmetic operation.

Hawkes Learning Systems: Introductory Algebra

Multiplication and Division with Integers

1.5   Exponents, Prime Numbers, and Order of Operations

After completing this section, you will be able to:

1. Evaluate expressions with exponents.

2. Recognize prime numbers less than 50.

3. Determine the prime factorization of integers.

4. Follow the rules for order of operations to evaluate expressions.

Exponents

Repeated addition with the same number is indicated with multiplication. For example,

$$14 + 14 + 14 + 14 + 14 = 5 \cdot 14 = 70 \qquad \text{and} \qquad 6 + 6 + 6 = 3 \cdot 6 = 18$$

factors   product   factors   product

The result of multiplication is called the **product**, and the numbers being multiplied are called **factors** of the product.

In a similar manner, multiplication by the same number can be indicated by using exponents. For example, if 2 is used as a factor 5 times, we can write

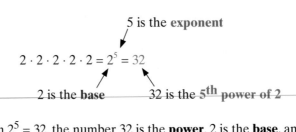

5 is the **exponent**

$$2 \cdot 2 \cdot 2 \cdot 2 \cdot 2 = 2^5 = 32$$

2 is the **base**   32 is the 5th power of 2

In the equation $2^5 = 32$, the number 32 is the **power**, 2 is the **base**, and 5 is the **exponent**. The expression 2^5 is read "two to the fifth power." We can also say that "32 is the fifth power of 2." Because of the wording in discussing exponents and powers, there can be some confusion as to just what number is the power. To help in understanding this rather technical distinction, think that the exponent describes the power.

With Repeated Multiplication	With Exponents
a. $4 \cdot 4 = 16$	$4^2 = 16$
b. $3 \cdot 3 \cdot 3 \cdot 3 = 81$	$3^4 = 81$

If an exponent is 2, then the base is said to be **squared**. If an exponent is 3 then the base is said to be **cubed**. For example, $5^2 = 25$ is read "five squared is equal to twenty-five" and $7^3 = 343$ is read "seven cubed is equal to three hundred forty-three." If no exponent is written on a number or a variable, then the exponent is understood to be 1. Thus,

$$6 = 6^1 \text{ and } a = a^1.$$

The use of 0 as an exponent is a special case and, as we will see later in dealing with properties of exponents, 0 exponents are needed in simplifying algebraic expressions. To help in understanding the meaning of 0 as an exponent, consider the following patterns of powers.

$$2^4 = 16 \qquad 5^4 = 625$$
$$2^3 = 8 \qquad 5^3 = 125$$
$$2^2 = 4 \qquad 5^2 = 25$$
$$2^1 = 2 \qquad 5^1 = 5$$
$$2^0 = ? \qquad 5^0 = ?$$

If you study these patterns of powers, you will see that each power of 2 is found by dividing the previous power by 2, and each power of 5 is found by dividing the previous power by 5. These results lead to the conclusion that $2^0 = 1$ and $5^0 = 1$. We have the following general properties of exponents.

General Properties of Exponents

In general, for any whole number n and any integer a,

1. $\underbrace{a \cdot a \cdot a \cdot a \cdot \ldots \cdot a}_{n \text{ factors}} = a^n$

2. $a^1 = a$

3. $a^0 = 1$ *(for $a \neq 0$)*

(The expression 0^0 is undefined.)

Example 1: Calculator ●

Use a calculator to find the following powers.

a. 16^3 **b.** 52^4 **c.** 12^0 **d.** 85^1

Solutions:

With a Scientific Calculator

Your scientific calculator will have a key marked as $\mathbf{x^y}$ or $\mathbf{y^x}$.

a. To evaluate 16^3:

Enter the base **16**.

Press the key marked $\mathbf{x^y}$.   (Note that nothing will appear on the display. The calculator is waiting for you to enter the exponent.)

Enter the exponent **3**.

Press **=** or **ENTER**.

The display should read **4096**.

b. To evaluate 52^4:

Enter the base **52**.

Press the key marked $\mathbf{x^y}$.   (Note that nothing will appear on the display. The calculator is waiting for you to enter the exponent.)

Enter the exponent **4**.

Press **=** or **ENTER**.

The display should read **7311616**.

c. To evaluate 12^0:

Follow the steps outlined in Examples **a** and **b**.

The display should read **1**.

d. To evaluate 85^1:

Follow the steps outlined in Examples **a** and **b**.

The display should read **85**.

With a TI-83 Plus (or other graphing calculator)

Your TI-83 Plus calculator will have a key marked with a caret (^). This key indicates that an exponent is to follow. Each expression and its value are shown in the display screens.

● ●

Prime Factors

In working with fractions and simplifying algebraic expressions, we sometimes want to see integers factored in their simplest form. That is, we want each number to be factored so that each factor is a **prime number**. Remember that **even integers** are divisible by 2 and **odd integers** are not divisible by 2.

Prime Number

*A **prime number** is a whole number (other than 0 or 1) that has exactly two different factors, itself and 1.*

A whole number (other than 0 or 1) that is not prime is a **composite number**.

You should memorize the prime numbers less than 50. Here is a list of the prime numbers less than 100:

2, 3, 5, 7, 11, 13, 17, 19, 23, 29, 31, 37, 41, 43, 47, 53, 59, 61, 67, 71, 73, 79, 83, 89, 97

NOTES Note the following two facts:
1. 2 is the only even prime number.
2. While all other prime numbers are odd, not all odd numbers are prime. For example, 9 is an odd number that is not prime.

The following quick tests for divisibility can be helpful in finding beginning factors of composite numbers. The concepts of prime factors and divisibility can be applied to negative integers as well as positive integers. We simply treat a negative integer as -1 times a positive integer and factor the positive integer as before.

Tests for Divisibility

An integer is divisible

By 2: *if the units digit is 0, 2, 4, 6, or 8.*

By 3: *if the sum of the digits is divisible by 3.*

By 5: *if the units digit is 0 or 5.*

By 6: *if the number is divisible by both 2 and 3.*

By 9: *if the sum of the digits is divisible by 9.*

By 10: *if the units digit is 0.*

Example 2: Prime Factorization ● ● ● ● ● ● ● ● ● ● ● ● ● ●

Find the prime factorization of each of the following composite numbers.

a. 72   **b.** 60   **c.** 165

Solution: By the tests for divisibility:

a. 72 is divisible by 2, 3, 6, and 9. $\;72 = 8 \cdot 9 = 2 \cdot 4 \cdot 3 \cdot 3 = 2^3 \cdot 3^2$.

b. 60 is divisible by 2, 3, 5, 6, and 10. $\;60 = 2 \cdot 3 \cdot 2 \cdot 5 = 2^2 \cdot 3 \cdot 5$.

c. 165 is divisible by 3 and 5. $\;165 = 5 \cdot 33 = 3 \cdot 5 \cdot 11$.

Therefore, the prime factorization of each number maybe found by using any of these known factors.

● ●

You may have noticed that, regardless of how you begin, there is only one prime factorization for any composite number. This fact is known as the Fundamental Theorem of Arithmetic.

The Fundamental Theorem of Arithmetic

Every composite number has a unique prime factorization.

Order of Operations

Mathematicians have agreed on a set of rules for the order of performing operations when evaluating numerical expressions that contain grouping symbols, exponents, and the operations of addition, subtraction, multiplication, and division. These rules are used throughout all levels of mathematics and science so that there is only one correct answer for the value of an expression, and so that everyone will get that answer. For example, evaluate the following expression:

$$36 \div 4 + 6 \cdot 2^2$$

Did you get 33 or 60? The correct answer is 33, which can be found by following a set of rules called the Rules for Order of Operations.

Rules for Order of Operations

1. *Simplify within grouping symbols, such as parentheses ( ), brackets [ ], and braces { }. (Start with the innermost grouping symbol.)*

2. *Find any powers indicated by exponents.*

3. *Moving from **left to right**, perform any multiplications or divisions in the order they appear.*

4. *Moving from **left to right**, perform any additions or subtractions in the order they appear.*

NOTES Other grouping symbols are the absolute value bars, the fraction bar, and radicals such as the square root symbol.

A well known mnemonic device for remembering the rules for order of operations is the following:

Please	Excuse	My	Dear	Aunt	Sally
P(arentheses)	**E**(xponents)	**M**(ultiplication)	**D**(ivision)	**A**(ddition)	**S**(ubtraction)

 NOTES Remember that, while the mnemonic device is helpful, multiplication might not come before division and addition might not come before subtraction. The priority is from **left** to **right**.

For example:
$$12 \div 3 \cdot 4 = 4 \cdot 4 = 16$$

but,

$$12 \cdot 3 \div 4 = 36 \div 4 = 9$$

The rules for order of operations can be particularly useful in determining the values of expressions involving negative numbers and exponents. For example, the two expressions

$$-7^2 \text{ and } (-7)^2$$

have two different values. By the order of operations, exponents come before multiplication. Thus,

$$-7^2 = -1 \cdot 7^2 = -1 \cdot 49 = -49 \quad \text{but } (-7)^2 = (-7)(-7) = 49$$

Remember that if the base is a negative number, then the negative number must be placed in parentheses. This distinction is particularly important when the exponent is an even integer. Thus,

$$-3^4 = -1 \cdot 3^4 = -1 \cdot 81 = -81 \text{ and } (-3)^4 = (-3)(-3)(-3)(-3) = 81$$

but

$$-3^3 = -1 \cdot 3^3 = -1 \cdot 27 = -27 \text{ and } (-3)^3 = (-3)(-3)(-3) = -27.$$

Example 3 shows several expressions evaluated in a step by step manner following the rules for order of operations. Study these carefully.

Example 3: Order of Operations ● ● ● ● ● ● ● ● ● ● ● ● ● ● ● ●

Use the rules for order of operations to evaluate each of the following expressions.

a. $36 \div 4 - 6 \cdot 2^2$ **b.** $2(3^2 - 1) - 3 \cdot 2^3$ **c.** $9 - 2[(3 \cdot 5 - 7^2) \div 2 + 2^2]$

Solution:

a. $\begin{aligned} 36 \div 4 - 6 \cdot 2^2 &= 36 \div 4 - 6 \cdot 4 & &\text{Exponents.} \\ &= 9 - 24 & &\text{Divide and multiply, left to right.} \\ &= -15 & &\text{Subtract.} \end{aligned}$

b. $\begin{aligned} 2(3^2 - 1) - 3 \cdot 2^3 &= 2(9 - 1) - 3 \cdot 8 & &\text{Exponents.} \\ &= 2(8) - 3 \cdot 8 & &\text{Subtract inside the parentheses.} \\ &= 16 - 24 & &\text{Multiply.} \\ &= -8 & &\text{Subtract (or add algebraically).} \end{aligned}$

c. $\begin{aligned} 9 - 2[(3 \cdot 5 - 7^2) \div 2 + 2^2] &= 9 - 2[(3 \cdot 5 - 49) \div 2 + 4] & &\text{Exponents.} \\ &= 9 - 2[(15 - 49) \div 2 + 4] & &\text{Multiply inside the parentheses.} \\ &= 9 - 2[(-34) \div 2 + 4] & &\text{Subtract inside the parentheses.} \\ &= 9 - 2[-17 + 4] & &\text{Divide inside the brackets.} \\ &= 9 - 2[-13] & &\text{Add inside the brackets.} \\ &= 9 + 26 & &\text{Multiply.} \\ &= 35 & &\text{Add.} \end{aligned}$

● ●

Practice Problems

1. List all the prime numbers less than 12.

2. Find the prime factorization of 75.

3. Find the prime factorization of 117.

Find the value of each expression by using the rules for order of operations.

4. $14 \div 2 + 2 \cdot 6 + 30 \div 3$

5. $3[8 + 2(1 - 10)] - 15 \div 5$

Answers to Practice Problems: **1.** $2, 3, 5, 7, 11$ **2.** $75 = 3 \cdot 5^2$ **3.** $3^2 \cdot 13$ **4.** 29 **5.** -33

1.5 Exercises

Rewrite each of the following integers in a base and exponent form without using an exponent of 1.

1. 25   **2.** 36   **3.** 49   **4.** 81   **5.** 121   **6.** 169   **7.** 64

8. 144   **9.** 8   **10.** 27   **11.** 125   **12.** 32   **13.** 243   **14.** 9

15. 100   **16.** 1000

Find the value of each of the following expressions.

17. 8^2 **18.** 1^5 **19.** 10^2 **20.** 6^3 **21.** 7^3 **22.** 9^2

23. 10^4 **24.** 5^4 **25.** 13^0 **26.** $(-16)^0$ **27.** $(-15)^2$ **28.** 20^2

29. 30^3 **30.** 50^2 **31.** $(-10)^4$ **32.** $(-5)^3$ **33.** $(-6)^2$ **34.** -6^2

35. -2^6

Use a calculator to find the value of each of the following expressions.

36. 21^4 **37.** 5^6 **38.** 43^5 **39.** 2^{18} **40.** 63^0

41. -15^2 **42.** $(-12)^5$ **43.** $(6-8)^3$ **44.** $(20-30)^4$ **45.** $(18-32)^2$

46. List the prime numbers less than 50.

Find the prime factorization of each integer.

47. 70   **48.** 80   **49.** 43   **50.** 59   **51.** 225   **52.** 65

53. 363   **54.** 120   **55.** 90   **56.** 128   **57.** 140   **58.** 340

59. 1000   **60.** 560

Use the rules for order of operations to evaluate each of the following expressions.

61. a. $24 \div 4 \cdot 6$ **b.** $24 \cdot 4 \div 6$ **62. a.** $20 \div 5 \cdot 2$ **b.** $20 \cdot 5 \div 2$

63. $15 \div (-3) \cdot 3 - 10$ **64.** $20 \cdot 2 \div 2^2 + 5(-2)$

65. $3^3 \div (-9) \cdot (4 - 2^2) + 5(-2)$ **66.** $4^2 \div (-8)(-2) + 3(2^2 - 5^2)$

67. $14 \cdot 3 \div (-2) - 6(4)$ **68.** $6(13 - 15)^2 \cdot 8 \div 2^2 + 3(-1)$

69. $-10 + 15 \div (-5) \cdot 3^2 - 10^2$ **70.** $16 \cdot 3 \div (2^2 - 5)$

71. $2 - 5[(-20) \div (-4) \cdot 2 - 40]$ **72.** $9 - 6[(-21) \div 7 \cdot 2 - (-8)]$

73. $(7-10)\left[49\div(-7)+20\cdot3-(-10)\right]$

74. $(9-11)\left[(-10)^2\cdot2+6(-5)^2-10^3+3\cdot5\right]$

75. $8-9\left[(-39)\div(-13)+7(-2)-(-2)^3\right]$

76. $6-20\left[(-15)\div3\cdot5+6\cdot2\div3\right]$

77. $\left|16-20\right|\left[32\div\left|3-5\right|-5^2\right]$

78. $\left|10-30\right|\left[4^2\cdot\left|5-8\right|\div(-2)^3+\left|17-18\right|\right]$

Writing and Thinking About Mathematics

79. Explain, in your own words, why the following expression cannot be evaluated.
$(24-2^4)+6(3-5)\div(3^2-9)$

Hawkes Learning Systems: Introductory Algebra

Factoring Positive Integers
Factoring Integers
Order of Operations

1.6	# Properties of Real Numbers

Objectives

After completing this section, you will be able to:

1. Complete statements by using the properties of real numbers.

2. Name the real number properties that justify given statements.

Recognizing Types of Numbers

In mathematics, it is important to be able to recognize and name various types of numbers. The conditions stated or implied in certain word problems allow for only positive solutions or only integer solutions. For example, the height of a person or the length of a piece of string cannot be a negative number. The count of students in a class cannot be a negative integer. With these ideas in mind, we review some of the various types of numbers already discussed.

Integers

$0, 4, -6, 14$, and -35 are all integers.

Rational Numbers

$4, -7, 9.5, -16.8, \dfrac{3}{5}, 9\dfrac{3}{10}$, and $6.45454545\ldots$ are all rational numbers.

In decimal form, each number would be a terminating decimal or an infinite repeating decimal. Remember that all integers are also rational numbers.

Irrational Numbers

$\sqrt{3}, \sqrt{5}, -\sqrt{10}, \pi$, and $\dfrac{2}{\sqrt{7}}$ are all irrational numbers.

In decimal form, each number would be an infinite nonrepeating decimal.

All of the examples shown here are real numbers.

Example 1: Types of Numbers ● ● ● ● ● ● ● ● ● ● ● ● ● ● ● ●

Given the set of numbers $\{-43.2, -\sqrt{19}, -2, -\frac{3}{7}, 0, 0.58, \pi, \sqrt[3]{62}, 6, 13.5555...\}$, tell which numbers are

a. integers   b. rational numbers   c. irrational numbers   d. real numbers.

Solution:

a.   The integers are $-2, 0$, and 6.

b.   The rational numbers are $-43.2, -2, -\frac{3}{7}, 0, 0.58, 6$, and $13.5555...$.

c.   The irrational numbers are $-\sqrt{19}, \pi$, and $\sqrt[3]{62}$.

d.   All of the numbers are real numbers.

● ●

In Chapter 1, we have discussed real numbers and the order of real numbers on a real number line. However, the discussions of operations have been restricted to integers only. All the rules for operating with integers, positives and negatives, apply to all real numbers, including decimals and fractions. Operations with decimals and fractions will be covered in Chapter 2. As we work with all types of real numbers and algebraic expressions, we will need to understand that certain properties are true for some operations and not for others. For example,

the order of the numbers in addition *does not* change the result:

$$17 + 5 = 5 + 17 = 22$$

but, the order of the numbers in subtraction *does* change the result:

$$17 - 5 = 12 \quad \text{but} \quad 5 - 17 = -12$$

We say that addition is commutative while subtraction is not commutative. The various properties of real numbers under the operations of addition and multiplication are summarized here. These properties are used throughout algebra and mathematics in developing formulas and general concepts.

Properties of Addition

For any real numbers a, b, and c,

Examples

Commutative Property of Addition:

$$a + b = b + a$$

$3 + 6 = 6 + 3$

Associative Property of Addition:

$$(a + b) + c = a + (b + c)$$

$(2 + 5) + 4 = 2 + (5 + 4)$

Additive Identity:

$$a + 0 = a$$

$20 + 0 = 20$

Additive Inverse:

$$a + (-a) = 0$$

$10 + (-10) = 0$

$-8 + (+8) = 0$ *[Note that +8 = -(-8).]*

Properties of Multiplication

For any real numbers a, b, and c,

Examples

Commutative Property of Multiplication:

$$a \cdot b = b \cdot a$$

$4 \cdot 9 = 9 \cdot 4$

Associative Property of Multiplication:

$$(a \cdot b) \cdot c = a \cdot (b \cdot c)$$

$(6 \cdot 2) \cdot 7 = 6 \cdot (2 \cdot 7)$

Multiplicative Identity:

$$a \cdot 1 = a$$

$-2 \cdot 1 = -2$

Zero Factor Law:

$$a \cdot 0 = 0 \cdot a = 0$$

$-5 \cdot 0 = 0 \cdot (-5) = 0$

Multiplicative Inverse:

$$a \cdot \frac{1}{a} = 1 \quad (a \neq 0)$$

$3 \cdot \frac{1}{3} = 1$

(Fractions will be discussed in Chapter 2)

Distributive Property

Distributive Property of Multiplication over Addition:

$$a(b + c) = a \cdot b + a \cdot c$$

$3(x + 5) = 3 \cdot x + 3 \cdot 5$

NOTES

The number 0 is called the **additive identity element** or simply the **additive identity**. Likewise, the number 1 is called the **multiplicative identity element** or simply the **multiplicative identity**.

In illustrating the distributive property, we can write $5(x - 6) = 5x - 30$ where -30 is the product of 5 and -6. Remember that the raised dot is optional when indicating multiplication between two variables or a number and a variable. As was stated in Section 1.4, the number is called the **coefficient** of the variable. Thus, 5 is the coefficient of x in the expression $5x$.

Example 2: State the Property • • • • • • • • • • • • • • • •

State the name of each property being illustrated.

a. $(-7) + 13 = 13 + (-7)$

    Solution: Commutative Property of Addition

b. $8 + (9 + 1) = (8 + 9) + 1$

    Solution: Associative Property of Addition

c. $(-25) \cdot 1 = -25$

    Solution: Multiplicative Identity

d. $3(x + y) = 3x + 3y$

    Solution: Distributive Property

e. $4(3 \cdot 2) = (4 \cdot 3) \cdot 2$

    Solution: Associative Property of Multiplication

In each of the following equations, state the property illustrated and show that the statement is true for the value given for the variable by substituting the value in the equation and evaluating.

f. $x + 14 = 14 + x;\ x = -4$

    Solution: The commutative property of addition is illustrated.
$$(-4) + 14 = 14 + (-4) = 10$$

g. $2x + 5 = 5 + 2x;\ x = 10$

> **Solution:** The commutative property of addition is illustrated.
> $$2 \cdot 10 + 5 = 5 + 2 \cdot 10 = 25$$

h. $12(\,y + 3\,) = 12y + 36;\ y = -2$

> **Solution:** The distributive property is illustrated.
> $$12(\,-2 + 3\,) = 12(\,1\,) = 12 \ \text{and}\ 12(\,-2\,) + 36 = -24 + 36 = 12$$

●●●●●●●●●●●●●●●●●●●●●●●●●●●●●●●●●●●●●

Practice Problems

Determine the property being illustrated.

1. $(\,-2 \cdot 5\,) \cdot 2 = -2 \cdot (\,5 \cdot 2\,)$ **2.** $15 \cdot 0 = 0 \cdot 15 = 0$

3. $2 + 7 = 7 + 2$ **4.** $2(\,y + 5\,) = 2 \cdot y + 2 \cdot 5$

1.6 Exercises

1. Given the set of numbers $\{-3.56,\ -\sqrt{8},\ -\dfrac{5}{8},\ -1,\ 0,\ 0.7,\ \pi,\ 4,\ \dfrac{13}{2},\ \sqrt[3]{25}\ \}$, tell which numbers are

    a. integers   **b.** rational numbers   **c.** irrational numbers   **d.** real numbers

2. Given the set of numbers $\{-123.21,\ -\sqrt{118},\ -\dfrac{32}{11},\ -2\dfrac{1}{3},\ 0.9,\ \dfrac{\pi}{2},\ \dfrac{17}{10},\ 5,\ \dfrac{\sqrt{107}}{5},$
$9.343434...\ \}$, tell which numbers are

    a. integers   **b.** rational numbers   **c.** irrational numbers   **d.** real numbers

Complete the expressions in Exercises 3 – 17 using the given property.

3. $7 + 3 =$ ____	Commutative property of addition
4. $(\,6 \cdot 9\,) \cdot 3 =$ ____	Associative property of multiplication
5. $19 \cdot 4 =$ ____	Commutative property of multiplication
6. $18 + 5 =$ ____	Commutative property of addition
7. $6(\,5 + 8\,) =$ ____	Distributive property
8. $16 + (\,9 + 11\,) =$ ____	Associative property of addition

Answers to Practice Problems: 1. Associative Property of Multiplication **2.** Zero Factor Law
3. Commutative Property of Addition **4.** Distributive Property

9. $2 \cdot (3x) =$ _____   Associative property of multiplication

10. $3(x + 5) =$ _____   Distributive property

11. $3 + (x + 7) =$ _____   Associative property of addition

12. $9(x + 5) =$ _____   Distributive property

13. $6 \cdot 0 =$ _____   Zero factor law

14. $6 \cdot 1 =$ _____   Multiplicative identity

15. $0 + (x + 7) =$ _____   Additive identity

16. $0 \cdot (-13) =$ _____   Zero factor law

17. $2(x - 12) =$ _____   Distributive property

Name the property of real numbers illustrated.

18. $5 + 16 = 16 + 5$ **19.** $5 \cdot 16 = 16 \cdot 5$ **20.** $32 \cdot 1 = 32$ **21.** $32 + 0 = 32$

22. $5 + (3 + 1) = (5 + 3) + 1$ **23.** $5 + (3 + 1) = (3 + 1) + 5$

24. $5(x + 7) = (x + 7) \cdot 5$ **25.** $13(y + 2) = (y + 2) \cdot 13$

26. $13(y + 2) = 13y + 26$ **27.** $6(2 \cdot 9) = (2 \cdot 9) \cdot 6$

In each of the following equations, state the property illustrated and show that the statement is true for the value of $x = 4, y = -2,$ or $z = 3$ by substituting the corresponding value in the equation and evaluating.

28. $6 \cdot x = x \cdot 6$ **29.** $19 + z = z + 19$

30. $8 + (5 + y) = (8 + 5) + y$ **31.** $(2 \cdot 7) \cdot x = 2 \cdot (7 \cdot x)$

32. $5(x + 18) = 5x + 90$ **33.** $(2z + 14) + 3 = 2z + (14 + 3)$

34. $(6 \cdot y) \cdot 9 = 6 \cdot (y \cdot 9)$ **35.** $11 \cdot x = x \cdot 11$

36. $z + (-34) = -34 + z$ **37.** $3(y + 15) = 3y + 45$

38. $2(3 + x) = 2(x + 3)$ **39.** $(y + 2)(y - 4) = (y - 4)(y + 2)$

40. $5 + (x - 15) = (x - 15) + 5$ **41.** $z + (4 + x) = (4 + x) + z$

42. $(x + y) + z = x + (y + z)$

In Exercises 43 – 47, first evaluate each expression by using the rules for order of operations and then use the distributive property to evaluate the same expression. The value must be the same.

43. $6(3 + 8)$ **44.** $7(8 - 5)$ **45.** $10(2 - 9)$ **46.** $13(5 + 3)$

47. $5(14 - 16)$

Chapter 1 Index of Key Ideas and Terms

Types of Numbers

Counting numbers (or natural numbers)   page 2

$N = \{\,1, 2, 3, 4, 5, 6, 7, 8, 9, 10, 11, \dots\,\}$

Whole numbers   page 2

$W = \{\,0, 1, 2, 3, 4, 5, 6, \dots\,\}$

Integers   page 4

Integers: $\{\,\dots, -3, -2, -1, 0, 1, 2, 3, \dots\,\}$

Positive integers: $\{\,1, 2, 3, 4, 5, \dots\,\}$

Negative integers: $\{\,\dots, -4, -3, -2, -1\,\}$

The integer 0 is neither positive nor negative.

Rational numbers   page 5

A **rational number** is a number that can be written in the form $\dfrac{a}{b}$, where a and b are integers and $b \neq 0$.

OR,

A **rational number** is a number that can be written in decimal form as a terminating decimal or as an infinite repeating decimal.

Irrational numbers   page 5

Irrational numbers are numbers that can be written as infinite nonrepeating decimals.

Real numbers   page 5

All rational and irrational numbers are classified as **real numbers.**

Diagram of Types of Numbers   page 6

Inequality Symbols   pages 7 - 8

Read from left to right:

< "is less than"

> "is greater than"

≤ "is less than or equal to"

≥ "is greater than or equal to"

Note: Inequality symbols may be read from left to right or from right to left.

51

Absolute Value

pages 9 - 11

The **absolute value** of a real number is its distance from 0.
Symbolically,

$$|a| = a \text{ if } a \text{ is a positive number or } 0.$$
$$|a| = -a \text{ if } a \text{ is a negative number.}$$

Addition with Integers

pages 16 - 19

1. To add two integers with like signs, add their absolute values and use the common sign.
2. To add two integers with unlike signs, subtract their absolute values (the smaller from the larger) and use the sign of the number with the larger absolute value.

Additive Inverse

page 22

The opposite of an integer is called its **additive inverse**.
Symbolically, for any integer a, $a + (-a) = 0$.

Subtraction with Integers

pages 22 - 25

To subtract an integer, add its opposite.
Symbolically, for integers a and b, $a - b = a + (-b)$.

Multiplication with Integers

pages 29 - 30

If a and b are positive integers, then

1. The product of two positive integers is positive: $a \cdot b = ab$.
2. The product of two negative integers is positive: $(-a)(-b) = ab$.
3. The product of a positive integer and a negative integer is negative: $a(-b) = -ab$.
4. The product of 0 and any integer is 0: $a \cdot 0 = 0$ and $(-a) \cdot 0 = 0$.

Division with Integers

pages 31 - 32

If a and b are positive integers,

1. The quotient of two positive integers is positive: $\dfrac{a}{b} = +\dfrac{a}{b}$.

2. The quotient of two negative integers is positive: $\dfrac{-a}{-b} = +\dfrac{a}{b}$.

3. The quotient of a positive integer and a negative integer is negative:

$$\frac{-a}{b} = -\frac{a}{b} \quad \text{and} \quad \frac{a}{-b} = -\frac{a}{b}.$$

Division by 0 is undefined.   page 31

Exponents   pages 36 - 38

In general, for any whole number n and integer a,

1. $\underbrace{a \cdot a \cdot a \cdot a \cdot \ldots \cdot a}_{n \text{ factors}} = a^n$

2. $a^1 = a$

3. $a^0 = 1$ (for $a \neq 0$)

(The expression 0^0 is undefined.)

Prime Numbers   page 39

A **prime number** is a whole number (other than 0 or 1) that has exactly two different factors, itself and 1.

A whole number (other than 0 or 1) that is not prime is **composite**.

Tests for Divisibility   page 39

An integer is divisible

By 2: if the units digit of an integer is 0, 2, 4, 6, or 8.

By 3: if the sum of the digits of an integer is divisible by 3.

By 5: if the units digit is 0 or 5.

By 6: if it is divisible by both 2 and 3.

By 9: if the sum of the digits is divisible by 9.

By 10: if the units digit is 0.

Rules for Order of Operations   pages 40 - 42

1. Simplify within grouping symbols, such as parentheses ( ), brackets [ ], and braces { }. (Starting with the innermost grouping and working out.)

2. Find any powers indicated by exponents.

3. Moving from **left to right**, perform any multiplications or divisions **in the order they appear**.

4. Moving from **left to right**, perform any additions or subtractions **in the order they appear**.

Properties of Real Numbers

Properties of Addition   page 47

For any real numbers a, b, and c,

Commutative Property of Addition:

$a + b = b + a$

Associative Property of Addition:

$(a + b) + c = a + (b + c)$

Additive Identity:

$a + 0 = a$

Additive Inverse:

$a + (-a) = 0$

Properties of Multiplication   page 47

For any real numbers a, b, and c,

Commutative Property of Multiplication:

$a \cdot b = b \cdot a$

Associative Property of Multiplication:

$(a \cdot b) \cdot c = a \cdot (b \cdot c)$

Multiplicative Identity:

$a \cdot 1 = a$

Zero Factor Law:

$a \cdot 0 = 0 \cdot a = 0$

Multiplicative Inverse

$a \cdot \dfrac{1}{a} = 1 \quad (a \neq 0)$

Distributive Property of Multiplication over Addition: page 47

$a(b + c) = a \cdot b + a \cdot c$

Chapter 1 Review

For a review of the topics and problems from Chapter 1, look at the following lessons from *Hawkes Learning Systems: Introductory Algebra*

Name That Real Number
Introduction to Absolute Values
Addition with Integers
Subtraction with Integers
Multiplication and Division with Integers
Factoring Positive Integers
Factoring Integers
Order of Operations
Properties of Real Numbers

Chapter 1 Test

1. Given the set of numbers $\{ -10, -\pi, -\frac{3}{4}, -\sqrt{5}, 0, \frac{7}{6}, 3\frac{4}{9}, 7.121212... \}$, tell which numbers are

    a. integers   **b.** rational numbers   **c.** irrational numbers   **d.** real numbers

2. **a.** True or False; All integers are rational numbers. (Explain your answer.)
    b. True or False; All rational numbers are integers. (Explain your answer.)

3. Explain why the expressions -10^2 and $(-10)^2$ have different values.

4. Fill in the blanks with the proper symbol: $<$, $>$, or $=$.
    a. -4 ____ -2 **b.** $-(-2)$ ____ 0 **c.** $|-8|$ ____ $|8|$

5. Graph the following set of numbers on a real number line:
$$\left\{ -2, -0.4, |-1|, \frac{7}{3}, 3.1 \right\}$$

6. On a real number line, graph the set of all integers less than or equal to 2.

7. What integers satisfy the equation $|y| = 7$?

8. List the prime numbers less than 50.

List the integers that satisfy the inequalities in Exercises 9 and 10, then graph the integers on a real number line.

9. $|x| < 3$ **10.** $|x| \geq 9$

In Exercises 11 – 18, perform the indicated operations.

11. $-15 + 45 + (-17)$ **12.** $17 - 107$

13. $13 - 16 + 5 - 20$ **14.** $-28 - (-47)$

15. $9 - (-6) - 14$ **16.** $(-27)(-6)$

17. $\dfrac{-35}{-5}$ **18.** $(42)(-6)(3)(0)$

19. Write 64 in a base and exponent form without using 1 as an exponent.

20. Use the rules for order of operations to evaluate:
    a. $16 \div 4^2 + 2(12 - 5^2)$
    b. $15 - 60[7(3^2 - 3 \cdot 3) + (-2)(6)] - 10^3$
    c. $|16 - 20| [4^2 - (13 + 2^3) - |-6|]$

21. Find the prime factorization for
    a. 80   **b.** 175

22. Name each property illustrated.
    a. $7 \cdot 1 = 7$ **b.** $x + 3 = 3 + x$ **c.** $(9 + 1) + 2 = 9 + (1 + 2)$
    d. $-5 \cdot 0 = 0$ **e.** $6(x + 2) = 6x + 12$ **f.** $2(x + 5) = (x + 5) \cdot 2$

23. From the sum of -14 and -16, subtract the product of -6 and -5.

24. If x is a variable that represents any integer, state the definition of $|x|$.

25. Use a calculator to find the value of each of the following expressions.
    a. $(-5 - 6)^7$ **b.** $32 \div 2^3 \cdot 5 - 16 + 4(7 - 8)$

26. In a 5-day week, the Dow Jones stock market average showed a gain of 32 points, a gain of 140 points, a loss of 30 points, a loss of 53 points, and a gain of 63 points. What was the net change in the stock market for the week? If the Dow started the week at 11,670 points, what was the average at the end of the week?

Fractions, Decimals, and Algebraic Expressions

Did You Know?

Almost all of mathematics uses the language of sets to simplify notation and to help in understanding concepts. The theory of sets is one of the few branches of mathematics that was initially developed almost completely by one person, Georg Cantor.

Georg Cantor was born in St. Petersburg, Russia, but spent his adult life in Germany, first as a student at the University of Berlin and later as a professor at the University of Halle. Cantor's research and development of ideas concerning sets was met by ridicule and public attacks on his character and work. In 1885, Cantor suffered the first of a series of mental breakdowns, probably caused by the attacks on his work by other mathematicians, most notably a former teacher, Leopold Kronecker. It is suggested that Kronecker kept Cantor from becoming a professor at the prestigious University of Berlin.

We are not used to seeing intolerance toward new ideas among the scientific community, and, as you study mathematics, you will probably have difficulty imagining how anyone could have felt hostile toward or threatened by Cantor's ideas. However, his idea of a set with an **infinite** number of elements was thought of as revolutionary by the mathematical establishment.

The problem is that infinite sets have the following curious property: a part may be numerically equal to a whole. For example,

$$N = \{\, 1, 2, 3, 4, \ 5, \ldots, n, \ldots \}$$
$$\updownarrow \updownarrow \updownarrow \updownarrow \ \updownarrow \qquad \updownarrow$$
$$E = \{\, 2, 4, 6, 8, 10, \ldots, 2n, \ldots \}$$

Sets N and E are numerically equal because set E can be put in one-to-one correspondence (matched or counted) with set N. So there are as many even numbers as there are natural numbers! This is contrary to common sense and, in Cantor's time, contrary to usual mathematical assumptions.

"The essence of mathematics lies in its freedom."

Georg Cantor (1845 - 1918)

A ny Olympic athlete will tell you that practice is the key to success. Chapter 2 allows you to practice the basic arithmetic skills you already have with fractions, decimals, and percents. At the same time, you will find an integrated introduction to elementary algebra concepts using variables, formulas, and positive and negative numbers. Your success in this course may depend on how much you "practice" the basic skills represented in this chapter. Good luck and try your best. Keep a positive attitude.

2.1   Multiplying and Dividing Fractions

Objectives

After completing this section, you will be able to:

1. Reduce fractions to lowest terms.

2. Write fractions as equivalent fractions with specified denominators.

3. Multiply and divide fractions.

The definition of **rational number**, as stated in Chapter 1, is repeated here for convenience and easy reference.

Rational Number

A *rational number* is a number that can be written in the form $\dfrac{a}{b}$ where a and b are integers and b ≠ 0.

OR

A *rational number* is a number that can be written in decimal form as a terminating decimal or as an infinite repeating decimal.

(Note that some fractions cannot be written with integers in the numerator and denominator. For example, $\dfrac{\pi}{3}$ is a fraction but also an infinite nonrepeating decimal and, therefore, an irrational number. See Chapter 9 for more details.)

For now, we will deal with the **fractional** form, $\dfrac{a}{b}$, of rational numbers and we will use the term **rational number** and **fraction** to mean the same thing. The decimal forms of rational numbers and their relationship to fractions will be discussed in Section 2.3.

Examples of rational numbers (fractions):

$$\frac{1}{2}, \ \frac{3}{10}, \ \frac{-11}{7}, \ -\frac{13}{2}, \ -10, \text{ and } 15.$$

Note that every integer is also a rational number because integers can be written in fraction form with a denominator of 1. Thus,

$$0 = \frac{0}{1}, \ 1 = \frac{1}{1}, \ 2 = \frac{2}{1}, \ 3 = \frac{3}{1}, \text{ and so on.}$$

Also,

$$-1 = \frac{-1}{1}, \ -2 = \frac{-2}{1}, \ -3 = \frac{-3}{1}, \ -4 = \frac{-4}{1}, \text{ and so on.}$$

In general, fractions can be used to indicate:
  1. Equal parts of a whole, or
  2. Division

Example 1: Fractions ●

a. $\frac{1}{2}$ can mean 1 of 2 equal parts

If you read $\frac{1}{2}$ of a book, then you can view this as having read one of two equal parts of the book. If the book has 50 pages, then you read 25 pages, since, as we will see, $\frac{25}{50} = \frac{1}{2}$.

b. $\frac{-45}{9}$ can mean to divide: $(-45) \div 9$.

● ●

Before actually operating with fractions, we need to understand the use and placement of negative signs in fractions. Consider the following three results involving fractions, negative signs, and the meaning of division:

$$-\frac{16}{8} = -2, \ \frac{-16}{8} = -2, \text{ and } \frac{16}{-8} = -2$$

Thus, the three different placements of a negative sign give the same results, as noted in the general statement or rule on the next page.

Rules for the Placement of Negative Signs in Fractions

If a and b are real numbers, and $b \neq 0$, then

$$-\frac{a}{b} = \frac{-a}{b} = \frac{a}{-b}.$$

For example:

$$-\frac{3}{4b} = \frac{-3}{4b} = \frac{3}{-4b}$$

Example 2: Negative Signs

a. $-\dfrac{12}{4} = \dfrac{-12}{4} = \dfrac{12}{-4} = -3$

b. $-\dfrac{1}{5} = \dfrac{-1}{5} = \dfrac{1}{-5}$

Multiplication

Multiplication

To **multiply** two fractions, multiply the numerators and multiply the denominators.

$$\frac{a}{b} \cdot \frac{c}{d} = \frac{a \cdot c}{b \cdot d}$$

For example:

$$\frac{2}{3} \cdot \frac{7}{5} = \frac{2 \cdot 7}{3 \cdot 5} = \frac{14}{15}$$

Remember that the number 1 is called the **multiplicative identity** since the product of 1 with any number is that number. That is,

$$\frac{a}{b} \cdot 1 = \frac{a}{b}.$$

Thus, if $k \neq 0$, we have

$$\frac{a}{b} = \frac{a}{b} \cdot 1 = \frac{a}{b} \cdot \frac{k}{k} = \frac{a \cdot k}{b \cdot k}$$

This relationship is called the **Fundamental Principle of Fractions**.

The Fundamental Principle of Fractions

$$\frac{a}{b} = \frac{a \cdot k}{b \cdot k}, \text{ where } k \neq 0$$

We can use the Fundamental Principle to build a fraction to **higher terms** (find an equal fraction with a larger denominator) or reduce to **lower terms** (find an equal fraction with a smaller denominator).

To reduce a fraction, factor both the numerator and denominator, then use the Fundamental Principle to "divide out" any common factors. If the numerator and the denominator have no common prime factors, the fraction has been **reduced to lowest terms**. Finding the prime factorization of the numerator and denominator before reducing, while not necessary, will help guarantee a fraction is in lowest terms.

Example 3: Fundamental Principal of Fractions • • • • • • • •

a. Raise $\frac{3}{7}$ to higher terms with a denominator of 28.

Solution: Use $k = 4$ since $7 \cdot 4 = 28$.

$$\frac{3}{7} = \frac{3 \cdot 4}{7 \cdot 4} = \frac{12}{28}$$

b. Raise $-\frac{5}{8}$ to higher terms with a denominator of $16a$.

Solution: Use $k = 2a$ since $8 \cdot 2a = 16a$.

$$-\frac{5}{8} = -\frac{5 \cdot 2a}{8 \cdot 2a} = -\frac{10a}{16a}$$

Animation:
Reduction of Proper
Fractions, Reduction
of Positive and
Negative Fractions

c. Reduce $\frac{-12}{20}$ to lowest terms by using prime factorizations.

Solution: $\quad \dfrac{-12}{20} = \dfrac{-1 \cdot 2 \cdot 2 \cdot 3}{2 \cdot 2 \cdot 5} = -1 \cdot \dfrac{2}{2} \cdot \dfrac{2}{2} \cdot \dfrac{3}{5} = -1 \cdot 1 \cdot 1 \cdot \dfrac{3}{5} = -\dfrac{3}{5}$

$\quad$ or $\dfrac{-12}{20} = \dfrac{-1 \cdot 4 \cdot 3}{4 \cdot 5} = -1 \cdot \dfrac{4}{4} \cdot \dfrac{3}{5} = -1 \cdot 1 \cdot \dfrac{3}{5} = -\dfrac{3}{5}$

$\quad$ or $\dfrac{-12}{20} = \dfrac{-1 \cdot \cancel{2} \cdot \cancel{2} \cdot 3}{\cancel{2} \cdot \cancel{2} \cdot 5} = -\dfrac{3}{5}$

Continued on next page...

d. Find the product $\dfrac{15ac}{28b^2} \cdot \dfrac{4bc}{9a^3}$ in lowest terms. (Do not find the product directly. Factor and reduce as you multiply.)

Solution: $\dfrac{15ac}{28b^2} \cdot \dfrac{4bc}{9a^3} = \dfrac{\cancel{3} \cdot 5 \cdot \cancel{a} \cdot c \cdot \cancel{4} \cdot \cancel{b} \cdot c}{\cancel{4} \cdot 7 \cdot \cancel{b} \cdot b \cdot \cancel{3} \cdot 3 \cdot \cancel{a} \cdot a \cdot a} = \dfrac{5c^2}{21a^2b}$

e. Find the product $\dfrac{4a}{12b} \cdot \dfrac{3b}{7a}$ in lowest terms. (Note that the number 1 is implied to be a factor even if it is not written.)

Solution: $\dfrac{4a}{12b} \cdot \dfrac{3b}{7a} = \dfrac{\cancel{4} \cdot \cancel{a} \cdot \cancel{3} \cdot \cancel{b} \cdot 1}{\cancel{4} \cdot \cancel{3} \cdot \cancel{b} \cdot 7 \cdot \cancel{a}} = \dfrac{1}{7}$

Here we write the factor 1 because all other factors have been "divided out."

We could write $\dfrac{4a}{12b} \cdot \dfrac{3b}{7a} = \dfrac{4 \cdot a \cdot 3 \cdot b}{4 \cdot 3 \cdot b \cdot 7 \cdot a}$

$$= \dfrac{4}{4} \cdot \dfrac{3}{3} \cdot \dfrac{a}{a} \cdot \dfrac{b}{b} \cdot \dfrac{1}{7}$$

$$= 1 \cdot 1 \cdot 1 \cdot 1 \cdot \dfrac{1}{7} = \dfrac{1}{7}$$

● ●

Finding the product of two fractions can be thought of as finding one fractional part **of** the other. Thus, to find a fraction **of** a number means to multiply the fraction and the number.

Example 4: Fraction of a Number ● ● ● ● ● ● ● ● ● ● ● ● ● ●

Find $\dfrac{2}{3}$ of $\dfrac{5}{7}$.

Solution: $\dfrac{2}{3} \cdot \dfrac{5}{7} = \dfrac{2 \cdot 5}{3 \cdot 7} = \dfrac{10}{21}$

● ●

Division

We will now see that division with fractions is accomplished by multiplication. That is, if we know how to multiply fractions, we automatically know how to divide them.

Reciprocal

> If $a \neq 0$ and $b \neq 0$, the **reciprocal** of $\dfrac{a}{b}$ is $\dfrac{b}{a}$, and $\dfrac{a}{b} \cdot \dfrac{b}{a} = 1$.

Consider the division problem $\dfrac{2}{3} \div \dfrac{5}{6}$. The indicated division can be written in the form of a complex fraction as follows:

$$\frac{2}{3} \div \frac{5}{6} = \frac{\dfrac{2}{3}}{\dfrac{5}{6}} \qquad \text{Write the division in fraction form.}$$

$$= \frac{\dfrac{2}{3} \cdot \dfrac{6}{5}}{\dfrac{5}{6} \cdot \dfrac{6}{5}} \qquad \begin{array}{l}\text{Use the reciprocal of the denominator and multiply by} \\[4pt] \text{1 in the form } \dfrac{\dfrac{6}{5}}{\dfrac{6}{5}}.\end{array}$$

$$= \frac{\dfrac{2}{3} \cdot \dfrac{6}{5}}{\dfrac{\cancel{5}}{\cancel{6}} \cdot \dfrac{\cancel{6}}{\cancel{5}}} \qquad \text{Simplify.}$$

$$= \frac{\dfrac{2}{3} \cdot \dfrac{6}{5}}{1} \qquad \leftarrow \frac{5}{6} \cdot \frac{6}{5} = 1$$

$$= \frac{2}{3} \cdot \frac{6}{5}$$

Thus, we have

$$\frac{2}{3} \div \frac{5}{6} = \frac{2}{3} \cdot \frac{6}{5} = \frac{2 \cdot 2 \cdot \cancel{3}}{\cancel{3} \cdot 5} = \frac{4}{5}$$

This example and the related discussion lead to the following definition.

Division

> To **divide** by a nonzero fraction, multiply by its reciprocal:
>
> $$\frac{a}{b} \div \frac{c}{d} = \frac{a}{b} \cdot \frac{d}{c}$$

Example 5: Division

a. $\left(-\dfrac{3}{4}\right) \div \left(-\dfrac{2}{5}\right)$

Solution: $\left(-\dfrac{3}{4}\right) \div \left(-\dfrac{2}{5}\right) = \left(-\dfrac{3}{4}\right) \cdot \left(-\dfrac{5}{2}\right) = \dfrac{15}{8}$

Note that, just as with integers, the product of two negative fractions is positive.

In algebra, $\dfrac{15}{8}$, an **improper fraction** (a fraction with the numerator greater than the denominator), is preferred to the mixed number $1\dfrac{7}{8}$.

Improper fractions are perfectly acceptable as long as they are reduced, meaning the numerator and denominator have no common prime factors. We will discuss mixed numbers in more detail in the next section.

b. $\dfrac{26}{35} \div \dfrac{39}{20}$

Solution: $\dfrac{26}{35} \div \dfrac{39}{20} = \dfrac{26}{35} \cdot \dfrac{20}{39} = \dfrac{2 \cdot \cancel{13} \cdot 2 \cdot 2 \cdot \cancel{5}}{\cancel{5} \cdot 7 \cdot 3 \cdot \cancel{13}} = \dfrac{8}{21}$

Note carefully that we factored and reduced before multiplying the numerator and denominator. It would not be wise to multiply first because we would then have to factor two large numbers. For example,

$$\dfrac{26}{35} \div \dfrac{39}{20} = \dfrac{26}{35} \cdot \dfrac{20}{39} = \dfrac{520}{1365}$$

and now we have to factor 520 and 1365. However, in the problem these were already factored since $520 = 26 \cdot 20$ and $1365 = 35 \cdot 39$.

c. $\dfrac{-21a^2}{5b} \div 3a$

Solution: $\dfrac{-21a^2}{5b} \div 3a = \dfrac{-21a^2}{5b} \cdot \dfrac{1}{3a} = \dfrac{-1 \cdot \cancel{3} \cdot 7 \cdot \cancel{a} \cdot a}{5 \cdot \cancel{3} \cdot \cancel{a} \cdot b} = \dfrac{-7a}{5b} \left(\text{or } -\dfrac{7a}{5b} \right)$

Note: The reciprocal of $3a$ is $\dfrac{1}{3a}$ since $3a = \dfrac{3a}{1}$.

If the product of two numbers is known and one of the numbers is also known, then the other number can be found by dividing the product by the known number. For example, with whole numbers, suppose the product of two numbers is 36 and one of the numbers is 9. What is the other number? Since $\frac{36}{9} = 4$, the other number is 4.

Example 6: Division

If the product of $\frac{3}{8}$ with another number is $-\frac{5}{16}$, what is the other number?

Solution: Divide the product by the given number.

$$-\frac{5}{16} \div \frac{3}{8} = -\frac{5}{16} \cdot \frac{8}{3} = -\frac{5 \cdot \cancel{8}}{2 \cdot \cancel{8} \cdot 3} = -\frac{5}{6}$$

The other number is $-\frac{5}{6}$.

Note that we could at least anticipate that the other number would be negative since the product is negative and the given number is positive.

2.1 Exercises

In Exercises 1 – 4, supply the missing numbers so that each fraction will be raised to higher terms as indicated.

1. $\frac{5}{6} = \frac{5}{6} \cdot \frac{?}{?} = \frac{?}{48}$

2. $\frac{3}{13} = \frac{3}{13} \cdot \frac{?}{?} = \frac{?}{52}$

3. $\frac{0}{9} = \frac{0}{9} \cdot \frac{?}{?} = \frac{?}{63b}$

4. $\frac{-7}{24} = \frac{-7}{24} \cdot \frac{?}{?} = \frac{?}{72x}$

Reduce each fraction to lowest terms in Exercises 5 – 16.

5. $\frac{18}{45}$

6. $\frac{35}{63}$

7. $\frac{150xy}{350y}$

8. $\frac{60a}{75a}$

9. $\frac{-12a^2b}{-100a}$

10. $\frac{6x^2}{-51x}$

11. $\frac{-30y}{45y^2}$

12. $\frac{66ab^2}{88ab}$

13. $\frac{-28}{56x^2}$

14. $\frac{34x^2}{-51x}$

15. $\frac{-12y^3}{35y^2}$

16. $\frac{8x^3}{15y^3}$

17. Find $\dfrac{1}{2}$ of $\dfrac{3}{4}$.   **18.** Find $\dfrac{2}{7}$ of $\dfrac{5}{7}$.   **19.** Find $\dfrac{7}{8}$ of 40.   **20.** Find $\dfrac{1}{3}$ of $\dfrac{1}{3}$.

In Exercises 21 – 39, multiply or divide as indicated and reduce each answer to lowest terms.

21. $\dfrac{-3}{8} \cdot \dfrac{4}{9}$

22. $\dfrac{4}{5} \cdot \dfrac{-3}{7}$

23. $\dfrac{9}{10x} \div \dfrac{10}{9x}$

24. $\dfrac{4}{5a} \div \dfrac{1}{5a}$

25. $\dfrac{-16x}{7} \cdot \dfrac{49}{64x}$

26. $\dfrac{26b}{51a} \cdot \dfrac{4a}{-39}$

27. $\dfrac{15}{4} \cdot \dfrac{5}{6} \cdot \dfrac{16}{9}$

28. $\dfrac{9a}{15} \cdot \dfrac{10}{3b} \cdot \dfrac{1}{5a^2}$

29. $\dfrac{-36x}{52xy} \cdot \dfrac{-26}{33x} \cdot \dfrac{22x}{9}$

30. $\dfrac{-9x}{40y} \cdot \dfrac{35x}{15} \cdot \dfrac{65y}{10}$

31. $\dfrac{-3}{2} \cdot \dfrac{7}{44} \cdot \dfrac{17}{1} \cdot \dfrac{22}{34}$

32. $\dfrac{21}{30} \div (-7)$

33. $\dfrac{4}{13} \div 0$

34. $0 \div \dfrac{7}{3x}$

35. $\dfrac{92}{7a} \div \dfrac{46a}{77}$

36. $\dfrac{45a}{30b^2} \cdot \dfrac{12ab}{18b^3} \div \dfrac{10ac}{9c^2}$

37. $\dfrac{15p}{3q^2} \cdot \dfrac{2pq}{28p^3} \div \dfrac{14p^2q^3}{10q^2}$

38. $\dfrac{35m}{21n^2} \div \dfrac{40mn^2}{27n^3} \cdot \dfrac{25m^2n}{9m^4}$

39. $\dfrac{72a^3}{4b^2} \div \dfrac{36ab^2}{16b^4} \div \dfrac{20a^2c^3}{8c^3}$

40. Multiply the quotient of $\dfrac{19}{2}$ and $\dfrac{13}{4}$ by $\dfrac{13}{2}$.

41. Divide the product of $\dfrac{5}{8}$ and $\dfrac{9}{10}$ by the product of $\dfrac{9}{10}$ and $\dfrac{4}{5}$.

42. Find the product of $\dfrac{11}{4}$ with the quotient of $\dfrac{8}{5}$ and $\dfrac{11}{6}$.

43. A glass is 6 inches tall. If the glass is $\dfrac{1}{3}$ full of milk, what is the height of the milk in the glass?

44. A study showed that $\dfrac{7}{10}$ of the students in an elementary school were over 4 feet tall. If the school had an enrollment of 500 students, how many were over 4 feet tall? How many were less than or equal to 4 feet tall?

45. A bus is carrying 60 passengers. This is $\dfrac{5}{6}$ of the capacity of the bus.

a. Is the capacity of the bus more or less than 60?

b. If you were to multiply 60 by $\dfrac{5}{6}$, would the product be more or less than 60?

c. What is the capacity of the bus?

46. The continent of Africa covers approximately 11,707,000 square miles. This is $\frac{1}{5}$ of the land area in the world. What is the approximate total land area in the world?

47. If you have $20 and you spend $7 on a glass of milk and a piece of pie, what fraction of your money did you spend? What fraction of your money do you still have?

48. In a class of 40 students, 5 received a grade of A. What fraction of the class did not receive an A?

49. Suppose that a ball is dropped from a height of 30 ft. and that each bounce reaches to $\frac{3}{8}$ of the previous height. How high will the ball bounce on the second bounce?

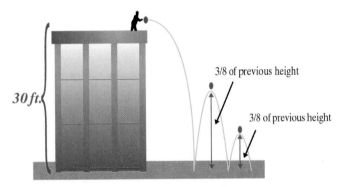

50. An airplane is carrying 180 passengers. This is $\frac{9}{10}$ of its capacity.

   a. Is the capacity more or less than 180?
   b. What is the capacity of the airplane?

51. The product of $\frac{3}{5}$ with another number is $\frac{-5}{8}$. What is the other number?

52. Valley Community College has 4000 students. Of these, $\frac{1}{8}$ are full-time students. Of the full-time students, $\frac{3}{10}$ are in favor of having a soccer team. Of the students that are not full-time, $\frac{4}{5}$ are in favor of having a soccer team. How many students in the school are not in favor of having a soccer team?

Writing and Thinking About Mathematics

53. Explain, in your own words, why 0 does not have a reciprocal.

Hawkes Learning Systems: Introductory Algebra

Reduction of Proper Fractions
Reduction of Improper Fractions
Reduction of Positive and Negative Fractions
Multiplying and Dividing Fractions

2.2 Adding and Subtracting Fractions

After completing this section, you will be able to:

1. Add and subtract fractions with like denominators.

2. Find the least common multiple (LCM) of two or more numbers.

3. Add and subtract fractions with unlike denominators.

4. Evaluate fractional expressions by using the rules for order of operations.

Addition and Subtraction

Finding the **sum** of two or more fractions with the same denominator is similar to adding whole numbers of some particular item. For example, the sum of 5 apples and 6 apples is 11 apples. Similarly, the sum of 5 seventeenths and 6 seventeenths is 11 seventeenths, or

$$\frac{5}{17} + \frac{6}{17} = \frac{11}{17}.$$

The following definition formally explains the above example.

Adding Two Fractions With Like Denominators

To **add** two fractions $\frac{a}{b}$ and $\frac{c}{b}$ with common denominator b, add the numerators a and c and use the common denominator.

$$\frac{a}{b} + \frac{c}{b} = \frac{a+c}{b}$$

A formal proof of this relationship involves the distributive property and the fact that

$$\frac{a}{b} = a \cdot \frac{1}{b}.$$

Proof:

$$\frac{a}{b}+\frac{c}{b}=a\cdot\frac{1}{b}+c\cdot\frac{1}{b}$$

$$=(a+c)\frac{1}{b} \qquad\qquad \text{Distributive property}$$

$$=\frac{a+c}{b}$$

Example 1: Like Denominators • • • • • • • • • • • • • • •

a. $\dfrac{3}{8}+\dfrac{4}{8}$

Solution: $\dfrac{3}{8}+\dfrac{4}{8}=\dfrac{3+4}{8}=\dfrac{7}{8}$

b. $\dfrac{9}{10}+\dfrac{3}{10}$

Solution: $\dfrac{9}{10}+\dfrac{3}{10}=\dfrac{9+3}{10}=\dfrac{12}{10}=\dfrac{\cancel{2}\cdot 6}{\cancel{2}\cdot 5}=\dfrac{6}{5}$

c. $\dfrac{2}{15}+\dfrac{3}{15}+\dfrac{1}{15}+\dfrac{6}{15}$

Solution: $\dfrac{2}{15}+\dfrac{3}{15}+\dfrac{1}{15}+\dfrac{6}{15}=\dfrac{2+3+1+6}{15}=\dfrac{12}{15}=\dfrac{\cancel{3}\cdot 4}{\cancel{3}\cdot 5}=\dfrac{4}{5}$

d. $\dfrac{3}{8x}+\dfrac{7}{8x}+\dfrac{5}{8x}$

Solution: $\dfrac{3}{8x}+\dfrac{7}{8x}+\dfrac{5}{8x}=\dfrac{3+7+5}{8x}=\dfrac{15}{8x}$

• •

NOTES

In Example 1b above, the fraction $\dfrac{6}{5}$ is called an **improper fraction** because the numerator is larger than the denominator. Such unfortunate terminology implies that there is something wrong with improper fractions. This is not the case, and improper fractions are used throughout the study of mathematics. Improper fractions can be changed to mixed numbers, with a whole number and a fraction part. Divide the numerator by the denominator and use the remainder as the numerator of the new fraction. For example,

$$\frac{6}{5}=1\frac{1}{5}, \quad \frac{11}{7}=1\frac{4}{7}, \quad \text{and} \quad \frac{35}{6}=5\frac{5}{6}.$$

The decision of whether to leave an answer in the form of an improper fraction or as a mixed number is optional. Generally, in algebra we will keep the improper fraction form. However, in the case of an application where units of measurement are involved, a mixed number form may be preferred. That is, we would probably write $\frac{3}{2}$ feet as $1\frac{1}{2}$ feet.

To find the sum of fractions with different denominators, we need the concepts of **multiples** and **least common multiple**. **Multiples** of a number are the products of that number with the counting numbers.

Counting numbers:	1,	2,	3,	4,	5,	6,	7,	8,	9,	...
Multiples of 6:	6,	12,	18,	⟨24⟩,	30,	36,	42,	⟨48⟩	54,	...
Multiples of 8:	8,	16,	⟨24⟩,	32,	40,	⟨48⟩,	56,	64,	⟨72⟩,	...

For the multiples of 6 and 8, the common multiples are 24, 48, 72, 96, 120, . . .

The smallest of these, **24**, is called the **least common multiple (LCM)**. Note that the LCM is **not** $6 \cdot 8 = 48$. In this case, the LCM is 24 and it is smaller than the product of 6 and 8. We can use **prime factorizations** to find the LCM of two or more numbers using the following steps.

To Find the LCM

1. *List the prime factorization of each number.*

2. *List the prime factors that appear in any one of the prime factorizations.*

3. *Find the product of these primes using each prime the greatest number of times that it appears in any one prime factorization.*

   Note: *In terms of exponents, the LCM is the product of the highest power of each of the prime factors.*

Example 2: LCM ●

a. Find the LCM of the numbers 27, 15, and 60.

Solution:
$$27 = 9 \cdot 3 = 3 \cdot 3 \cdot 3 = 3^3$$
$$15 = 3 \cdot 5$$
$$60 = 10 \cdot 6 = 2 \cdot 5 \cdot 2 \cdot 3 = 2^2 \cdot 3 \cdot 5$$

$$\left.\right\} \quad \text{LCM} = 2^2 \cdot 3^3 \cdot 5 = 540$$

Continued on next page...

b. Find the LCM for $4x$, x^2y, $6x^2$, and $18y^3$.
(**Hint:** Treat each variable as a prime factor.)

Solution:

$$\left.\begin{array}{l} 4x = 2^2 \cdot x \\ x^2y = x^2 \cdot y \\ 6x^2 = 2 \cdot 3 \cdot x^2 \\ 18y^3 = 2 \cdot 3^2 \cdot y^3 \end{array}\right\} \quad \text{LCM} = 2^2 \cdot 3^2 \cdot x^2 \cdot y^3 = 36x^2y^3$$

Adding Fractions with Different Denominators

1. Find the LCM of the denominators.

2. Change each fraction to an equal fraction with the LCM as the denominator.

3. Add the new fractions.

The LCM of the denominators of fractions is called the **least common denominator**, or **LCD**.

Example 3: Unlike Denominators

a. $\dfrac{1}{4} + \dfrac{3}{8} + \dfrac{3}{10}$

Solution:

$$\left.\begin{array}{l} 4 = 2^2 \\ 8 = 2^3 \\ 10 = 2 \cdot 5 \end{array}\right\} \quad \text{LCM= LCD} = 2^3 \cdot 5 = 40$$

To get the common denominator in each fraction, we see

$$40 = 4 \cdot 10 = 8 \cdot 5 = 10 \cdot 4$$

$$\frac{1}{4} + \frac{3}{8} + \frac{3}{10} = \left(\frac{1}{4} \cdot \frac{10}{10}\right) + \left(\frac{3}{8} \cdot \frac{5}{5}\right) + \left(\frac{3}{10} \cdot \frac{4}{4}\right)$$

Multiply each fraction by 1 in the form $\dfrac{k}{k}$.

$$= \frac{10}{40} + \frac{15}{40} + \frac{12}{40}$$

Each fraction has the same denominator.

$$= \frac{37}{40}$$

Add the fractions.

b. $\dfrac{5}{21a}+\dfrac{5}{28a}$

$\qquad$ **Solution:** $\left.\begin{array}{l}21a = 3\cdot 7\cdot a \\[4pt] 28a = 2^2\cdot 7\cdot a\end{array}\right\}$ $\text{LCM} = \text{LCD} = 2^2\cdot 3\cdot 7\cdot a = 84a = 21a\cdot 4 = 28a\cdot 3$

In each fraction, the numerator and denominator are multiplied by the same number to get 84a as the denominator.

$$\dfrac{5}{21a}+\dfrac{5}{28a}=\left(\dfrac{5}{21a}\cdot\dfrac{4}{4}\right)+\left(\dfrac{5}{28a}\cdot\dfrac{3}{3}\right)$$

$$=\dfrac{20}{84a}+\dfrac{15}{84a}=\dfrac{35}{84a}$$

$$=\dfrac{\cancel{7}\cdot 5}{\cancel{7}\cdot 12a}=\dfrac{5}{12a}$$

Subtracting Fractions

The **difference** of two fractions with a common denominator is found by subtracting the numerators and using the common denominator.

$$\dfrac{a}{b}-\dfrac{c}{b}=\dfrac{a-c}{b}$$

Just as with addition, if the two fractions do not have the same denominator, find equal fractions with the least common denominator (LCD).

Example 4: Subtraction

a. $\dfrac{1}{8a}-\dfrac{5}{8a}$

$\qquad$ **Solution:** $\dfrac{1}{8a}-\dfrac{5}{8a}=\dfrac{1-5}{8a}=\dfrac{-4}{8a}=\dfrac{-1\cdot\cancel{4}}{\cancel{4}\cdot 2\cdot a}=\dfrac{-1}{2a}$ $\qquad\left(\text{or }-\dfrac{1}{2a}\right)$

b. $\dfrac{1}{45}-\dfrac{1}{72}$

$\qquad$ **Solution:** $\left.\begin{array}{l}45 = 3^2\cdot 5 \\[4pt] 72 = 2^3\cdot 3^2\end{array}\right\}$ $\text{LCM} = \text{LCD} = 2^3\cdot 3^2\cdot 5 = 360 = 45\cdot 8 = 72\cdot 5$

$$\dfrac{1}{45}-\dfrac{1}{72}=\left(\dfrac{1}{45}\cdot\dfrac{8}{8}\right)-\left(\dfrac{1}{72}\cdot\dfrac{5}{5}\right)=\dfrac{8}{360}-\dfrac{5}{360}=\dfrac{3}{360}=\dfrac{\cancel{3}\cdot 1}{\cancel{3}\cdot 120}=\dfrac{1}{120}$$

Order of Operations

To evaluate an expression such as $\dfrac{1}{2} + \dfrac{3}{8} \div \dfrac{3}{4}$, we use the same rules for order of operations discussed in Section 1.5. These rules apply to all types of numbers. They are restated here for convenience and easy reference.

Rules for Order of Operations

1. *Simplify within grouping symbols, such as parentheses ( ), brackets [ ], and braces { }, working from the inner most grouping outward.*

2. *Find any powers indicated by exponents.*

3. *Moving from **left to right**, perform any multiplications or divisions in the order they appear.*

4. *Moving from **left to right**, perform any additions or subtractions in the order they appear.*

Example 5: Order of Operations

a. $\dfrac{1}{2} + \dfrac{3}{8} \div \dfrac{3}{4}$

Solution: $\dfrac{1}{2} + \dfrac{3}{8} \div \dfrac{3}{4} = \dfrac{1}{2} + \dfrac{3}{8} \cdot \dfrac{4}{3}$ Divide (Multiply by the reciprocal).

$= \dfrac{1}{2} + \dfrac{\cancel{3} \cdot \cancel{4} \cdot 1}{2 \cdot \cancel{4} \cdot \cancel{3}}$ Reduce.

$= \dfrac{1}{2} + \dfrac{1}{2}$ Add.

$= \dfrac{1+1}{2}$

$= \dfrac{2}{2}$ Reduce.

$= 1$

Continued on next page...

b. $\dfrac{2}{7} \cdot \dfrac{14}{3} + \dfrac{5}{8} \cdot \dfrac{2}{5}$

Solution: $\dfrac{2}{7} \cdot \dfrac{14}{3} + \dfrac{5}{8} \cdot \dfrac{2}{5} = \dfrac{2 \cdot 14}{7 \cdot 3} + \dfrac{5 \cdot 2}{8 \cdot 5}$ Multiply.

$$= \dfrac{2 \cdot \cancel{7} \cdot 2}{\cancel{7} \cdot 3} + \dfrac{\cancel{5} \cdot \cancel{2} \cdot 1}{\cancel{2} \cdot 4 \cdot \cancel{5}}$$ Reduce.

$$= \dfrac{4}{3} + \dfrac{1}{4}$$

$$= \left(\dfrac{4}{3} \cdot \dfrac{4}{4} \right) + \left(\dfrac{1}{4} \cdot \dfrac{3}{3} \right)$$ Common denominator is 12.

$$= \dfrac{16}{12} + \dfrac{3}{12} = \dfrac{19}{12}$$ Add.

c. Divide the sum of $\dfrac{5}{8}$ and $\dfrac{3}{4}$ by $\dfrac{3}{2}$.

Solution: First find the sum of $\dfrac{5}{8}$ and $\dfrac{3}{4}$.

$$\dfrac{5}{8} + \dfrac{3}{4} = \dfrac{5}{8} + \left(\dfrac{3}{4} \cdot \dfrac{2}{2} \right) = \dfrac{5}{8} + \dfrac{6}{8} = \dfrac{11}{8}$$ LCD is 8.

Now divide the sum by $\dfrac{3}{2}$.

$$\dfrac{11}{8} \div \dfrac{3}{2} = \dfrac{11}{8} \cdot \dfrac{2}{3} = \dfrac{11 \cdot \cancel{2}}{\cancel{2} \cdot 4 \cdot 3} = \dfrac{11}{12}$$

The answer is $\dfrac{11}{12}$.

d. $\dfrac{3}{x} + \dfrac{2}{5}$

Solution: In this case the LCD $= x \cdot 5 = 5x$

$$\dfrac{3}{x} + \dfrac{2}{5} = \left(\dfrac{3}{x} \cdot \dfrac{5}{5} \right) + \left(\dfrac{2}{5} \cdot \dfrac{x}{x} \right)$$

$$= \dfrac{15}{5x} + \dfrac{2x}{5x}$$

$$= \dfrac{15 + 2x}{5x}$$ This factor cannot be reduced.

• •

Perform the indicated operations and reduce all answers to lowest terms.

1. $-5 \cdot \left(\dfrac{3}{10} \right)^2$

2. $\dfrac{3}{4} \div \dfrac{4}{3}$

3. $\dfrac{7}{3} \div \dfrac{4}{5} \cdot \dfrac{3}{20}$

4. $\dfrac{1}{2a} - \dfrac{3}{2a}$

5. $\dfrac{5}{24} + \dfrac{7}{36}$

6. $\dfrac{1}{3} \div \dfrac{1}{2} + \dfrac{1}{5} \cdot \dfrac{5}{3}$

2.2 Exercises

Find (a) the factors of each number and (b) the first six multiples of each of the numbers in Exercises 1 – 4.

1. 6

2. 12

3. 15

4. 18

Find the LCM for each set of numbers or algebraic expressions in Exercises 5 – 10.

5. 24, 15, 10

6. 49, 25, 35

7. $8x, 10y, 20xy$

8. $20xz, 24xy, 32yz$

9. $14x^2, 21xy, 35x^2y^2$

10. $60x, 105x^2y, 120xy$

Perform the indicated operations in Exercises 11 – 40. Reduce each answer to lowest terms.

11. $\dfrac{2}{9} + \dfrac{5}{9}$

12. $\dfrac{2}{7} + \dfrac{8}{7} + \dfrac{4}{7}$

13. $\dfrac{14}{23} - \dfrac{9}{23}$

14. $\dfrac{7}{15} - \dfrac{2}{15}$

15. $\dfrac{11}{12} - \dfrac{13}{12}$

16. $\dfrac{5}{16} - \dfrac{9}{16}$

17. $\dfrac{5}{9} + \dfrac{7}{9}$

18. $\dfrac{11}{15a} - \dfrac{7}{15a} - \dfrac{2}{15a}$

19. $\dfrac{-19}{25x} - \dfrac{7}{25x} - \dfrac{2}{25x}$

20. $\dfrac{5}{6} + \dfrac{1}{8}$

21. $\dfrac{11}{12} + \dfrac{7}{15}$

22. $\dfrac{27}{40} - \dfrac{5}{8}$

23. $\dfrac{2}{7} - \dfrac{1}{6}$

24. $\dfrac{5}{24} + \dfrac{7}{18} - \dfrac{3}{8}$

25. $\dfrac{8}{35} + \dfrac{6}{10} - \dfrac{5}{14}$

26. $\dfrac{8}{9} - \dfrac{5}{12} - \dfrac{1}{4}$

27. $\dfrac{2}{3x} + \dfrac{1}{4x} - \dfrac{1}{6x}$

28. $\dfrac{7}{10y} - \dfrac{1}{5y} + \dfrac{1}{3y}$

29. $\dfrac{11}{24x} - \dfrac{7}{18x} + \dfrac{5}{36x}$

30. $\dfrac{7}{6b} - \dfrac{9}{10b} + \dfrac{4}{15b}$

31. $\dfrac{1}{28a} + \dfrac{8}{21a} - \dfrac{5}{12a}$

32. $\dfrac{5}{y} - \dfrac{1}{3}$

33. $\dfrac{6}{y} + \dfrac{1}{2}$

34. $\dfrac{x}{4} + \dfrac{1}{5}$

35. $\dfrac{x}{14} - \dfrac{1}{7}$

36. $\dfrac{y}{3} - \dfrac{1}{4} - \dfrac{1}{6}$

37. $\dfrac{a}{5} - \dfrac{1}{3} - \dfrac{1}{2}$

38. $\dfrac{a}{2} + \dfrac{1}{6} + \dfrac{2}{3}$

39. $\dfrac{b}{4} + \dfrac{1}{2} + \dfrac{3}{5}$

40. $\dfrac{1}{7} - \dfrac{10}{x}$

Answers to Practice Problems: 1. $-\dfrac{9}{20}$ **2.** $\dfrac{9}{16}$ **3.** $\dfrac{7}{16}$ **4.** $-\dfrac{1}{a}$ **5.** $\dfrac{29}{72}$ **6.** 1

Simplify the expressions in Exercises 41 – 50 using the rules for order of operations.

41. $\dfrac{3}{8} \cdot \dfrac{4}{5} + \dfrac{1}{15}$

42. $\dfrac{1}{3} \div \dfrac{1}{2} - \dfrac{5}{6} \cdot \dfrac{3}{4}$

43. $\left(\dfrac{5}{6}\right)^2 \div \dfrac{5}{12} - \dfrac{3}{8}$

44. $\left(\dfrac{2}{5}\right)^2 \cdot \dfrac{3}{8} + \dfrac{1}{5} \cdot \dfrac{3}{4}$

45. $\dfrac{3}{4a} \div \dfrac{3}{16a} - \dfrac{2b}{3} \cdot \dfrac{3}{4b}$

46. $\dfrac{3}{4x} - \dfrac{5}{6} \div \dfrac{5x}{8} - \dfrac{1}{3x}$

47. $\dfrac{-5}{7y} - \dfrac{1}{2y} \cdot \dfrac{2}{3} - \dfrac{1}{21y}$

48. $\left(\dfrac{1}{5x} - \dfrac{2}{3x}\right) \div \left(\dfrac{5}{7y} - \dfrac{1}{3y}\right)$

49. $\left(\dfrac{7}{10x} + \dfrac{3}{5x}\right) \div \left(\dfrac{1}{2x} + \dfrac{3}{7x}\right)$

50. $\left(\dfrac{1}{5a} - \dfrac{2}{3a}\right) \div \left(\dfrac{7}{7b} - \dfrac{1}{3b}\right)$

51. The California lottery is based on choosing any six of the integers from 1 to 51. The probability of winning the lottery can be found by multiplying the fractions

$$\dfrac{6}{51} \cdot \dfrac{5}{50} \cdot \dfrac{4}{49} \cdot \dfrac{3}{48} \cdot \dfrac{2}{47} \cdot \dfrac{1}{46}$$

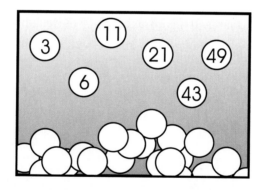

Multiply and reduce the product of these fractions to find the probability of winning the lottery in the form of a fraction with numerator 1.

52. The product of $\dfrac{9}{16}$ and $\dfrac{13}{7}$ is added to the quotient of $\dfrac{10}{3}$ and $\dfrac{7}{4}$. What is the sum?

53. Find the quotient if the sum of $\dfrac{4}{5}$ and $\dfrac{11}{15}$ is divided by the difference between $\dfrac{7}{12}$ and $\dfrac{4}{15}$.

54. In the scale for the blue print of the first floor of a house, 1 in. represents 48 ft. If on the drawing the first floor is $\frac{7}{8}$ in. wide by $\frac{5}{4}$ in. long, what will be the actual length and width of the first floor?

Scale: 1 in. = 48 ft.

dining room
porch
?
kitchen   up   living room
porch
FIRST FLOOR
?

55. The seventeenth hole at the local golf course is a par 4 hole. Ralph drove his ball 258 yards. If this distance was $\frac{3}{4}$ the length of the hole, how long is the seventeenth hole?

56. A watch has a rectangular-shaped display screen that is $\frac{5}{8}$ inch by $\frac{1}{2}$ inch. The display screen has a border of metal that is $\frac{1}{5}$ inch thick. What are the dimensions (length and width) of the watch (including the metal border)?

57. Delia's income is $2700 a month and she plans to budget $\frac{1}{3}$ of her income for rent and $\frac{1}{10}$ of her income for food.

   a. What fraction of her income does she plan to spend each month on these two items?

   b. What amount of money does she plan to spend each month on these two items?

58. The tennis club has 400 members, and they are considering putting in a new tennis court. The cost of the new court is going to involve an assessment of $250 for each member. Of the six-tenths of the members who live near the club, $\frac{4}{5}$ are in favor of the assesment. However, of the other members who live further away, only $\frac{3}{10}$ are in favor of the assessment.

a. If a vote were taken today, would more than one-half of the members vote for or against the new court?

b. By how many votes would the question pass or fail if more than one-half of the members must vote in favor for the question to pass?

Hawkes Learning Systems: Introductory Algebra

Adding and Subtracting Fractions

2.3   Decimal Numbers and Fractions

After completing this section, you will be able to:

1. Understand place value.

2. Read and write decimal numbers.

3. Round off decimal numbers.

4. Operate with decimal numbers.

5. Change fractions to decimal form.

6. Change decimals to fraction form.

Decimal numbers are used in most of the daily calculations we make, such as balancing a checkbook, measuring the distance we drive, and weighing a package to be mailed. Sometimes fractions are used and need to be changed to decimal form, as in the price of a stock of $3\frac{3}{4}$ to $3.75 or $1\frac{1}{2}$ cups of flour to 1.5 cups. Even to be able to understand and use a calculator, we must be able to understand and operate with decimal numbers. In fact, the stock markets have just recently done away with fractions in quoting the prices of stocks and have begun to use decimal numbers in stock pricing.

Place Value

A **decimal number** (or simply a **decimal**) is a fraction that has a power of ten in its denominator. Thus,

$$\frac{3}{10}, \frac{51}{100}, \frac{24}{10}, \frac{17}{1}, \text{ and } \frac{3}{1000}$$

are all decimal numbers. Decimal numbers can be written with a decimal point and a place value system that indicates whole numbers to the left of the decimal point and fractions less than 1 to the right of the decimal point. The values of each place are indicated in Figure 2.1.

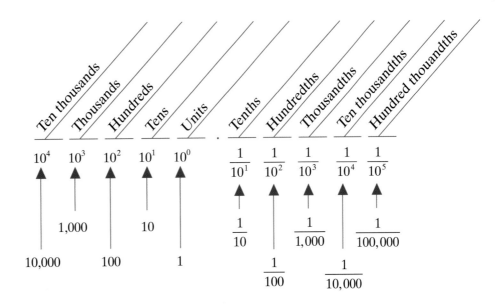

Figure 2.1

Reading and Writing Decimal Numbers

In reading a fraction such as $\dfrac{231}{1000}$, we read the numerator as a whole number ("two hundred thirty-one") and then attach the name of the denominator ("thousand**ths**"). Note that **ths** (or **th**) is used to indicate the fraction. **And** is used to indicate the decimal point. For example:

$\dfrac{231}{1000}$ is read "two hundred thirty-one thousandths."

$3\dfrac{17}{100}$ is read "three **and** seventeen hundredths."

Example 1: Writing Decimal Numbers ● ● ● ● ● ● ● ● ● ● ● ●

Write each of the following numbers in decimal notation and in words.

a. $64\dfrac{5}{10}$ **b.** $7\dfrac{9}{1000}$

Solution: **a.** 64.5 in decimal notation
        "sixty-four and five tenths" in words

b. 7.009 in decimal notation
        "seven and nine thousandths" in words

● ●

Rounding Off Decimal Numbers

Human made measuring devices such as rulers, yard sticks, meter sticks, clocks, and surveying instruments give only approximate measurements. For example, a carpenter may wish to cut a board of length 3.5 feet. As he cuts the board, the actual length cut will be approximately 3.5 feet, maybe 3.47 feet or 3.56 feet.

In this text we will use the following rule for rounding off decimal numbers. (Note that there are other rules for rounding off. For example, one rule is to round off to the nearest even digit.)

Rule for Rounding Off Decimal Numbers

1. *Look at the digit immediately to the right of the place of desired accuracy.*

2. *If this digit is 5 or greater, increase the digit in the desired place of accuracy by 1 and drop all the remaining digits to the right. (If a 9 is to be made larger, then the next digit to the left will also be increased by 1.)*

3. *If this digit is less than 5, leave the digit in the desired place of accuracy unchanged and drop all digits to the right.*

Example 2: Rounding •

Round off each decimal number to the indicated place.

Animation:
Decimals and
Rounding

a. 43.8476 (nearest thousandth)   **b.** 17.6035 (nearest hundredth)
c. 246.95 (nearest tenth)

Solution: **a.** 43.8476   7 is in the thousandths position. Look at the next digit to the right, 6. Since 6 is larger than 5, change 7 to 8 and drop the 6.

Thus, 43.8476 rounds off to 43.848 to the nearest thousandth.

b. 17.6035   0 is in the hundredths position. Look at the next digit to the right, 3. Since 3 is smaller than 5, leave the 0 and drop both the 3 and 5.

Thus, 17.6035 rounds off to 17.60 to the nearest hundredth. Note that the 0 is left in the answer because it indicates the place of rounding.

c. 246.95   9 is in the tenths position. Look at the next digit to the right, 5, and change the 9 to 10. But this also affects the 6.

Thus, 246.95 rounds off to 247.0 to the nearest tenth.

Operating with Decimal Numbers

To add or subtract decimal numbers, line up the decimal points, one under the other, then add or subtract as with whole numbers. Place the decimal point in the answer in line with the other decimal points. This technique guarantees that digits having the same place value will be added to or subtracted from each other.

The rules for operating with positive and negative numbers are the same regardless of the type of number. That is, with positive and negative decimal numbers or fractions, we follow the same rules we discussed in Chapter 1 for operating with positive and negative integers.

Example 3: Addition and Subtraction

a. Add: 37.498 + 5.63 + 42.781

   Solution: 37.498
                      5.630
                    +42.781
                     85.909

b. Subtract: 13.99 − 26.872

   Solution: Find the difference and then attach the negative sign.
                      26.872
                    −13.990
                      12.882
             Thus, 13.99 − 26.872 = −12.882

To find the product of two decimal numbers, multiply as with whole numbers. Then place the decimal point so that the number of digits to its right is equal to the sum of the number of digits to the right of the decimal points in the numbers being multiplied.

Example 4: Multiplication •••••••••••••••••••••••

Find the products.

a.   4.78
       × 0.3

b.   −16.4
       × 0.517

Solution:

4.78	2 digits to the right
× 0.3	1 digit to the right
1.434	3 digits to the right

Solution:

$$\begin{array}{r} -16.4 \\ \times\ 0.517 \\ \hline 1148 \\ 1640 \\ 82000 \\ \hline -8.4788 \end{array}$$

•••••••••••••••••••••••••••••••••••

To find the quotient of two decimal numbers, move the decimal point in the divisor to the right to get a whole number. Move the decimal point in the dividend the same number of places. Divide as with whole numbers.

Example 5: Division •••••••••••••••••••••••

Find the quotient: $3.2\overline{)51.52}$

Solution: $3.2\overline{)51.52}$

$32.\overline{)515.2}$ Move decimal points one place

$32\overline{)515.2}^{\ \ \ \cdot}$ Decimal point in quotient

$32\overline{)515.2}^{\ \ 16.1}$ Divide

$$\begin{array}{r} 32\ \ \\ \hline 195\ \\ 192\ \\ \hline 32 \\ 32 \\ \hline 0 \end{array}$$

•••••••••••••••••••••••••••••••••••

Fractions and Decimals

Fractions can be written in decimal form by dividing the numerator by the denominator. The result will be either

    1. a terminating decimal or
    2. an infinite repeating decimal.

In both cases the number is classified as a **rational number**.

Irrational numbers can also be written in decimal form, but they fall into a third category,

    3. infinite nonrepeating decimals.

For example,

$$\frac{3}{4} = 0.75 \qquad \text{is a terminating decimal.}$$

$$-\frac{1}{3} = -0.33333... = -0.\overline{3} \quad \text{is an infinite repeating decimal. (Note that a bar can be placed over the repeating digits to indicate an infinite repeating decimal.)}$$

$$\pi = 3.1415926535... \quad \text{is an infinite nonrepeating decimal.}$$

Example 6: Fraction to Decimal ● ● ● ● ● ● ● ● ● ● ● ● ● ●

Change $3\frac{7}{8}$ to decimal form.

Solution: $3\frac{7}{8} = \frac{31}{8}$ so we divide. With a calculator $\frac{31}{8} = 3.875$.

        Or, we can write

$$3\frac{7}{8} = 3 + \frac{7}{8} = 3 + 0.875 = 3.875$$

Continued on next page...

Using long division, we get the same terminating decimal:

$$
\begin{array}{r}
3.875 \\
8\overline{)31.000} \\
\underline{24} \\
70 \\
\underline{64} \\
60 \\
\underline{56} \\
40 \\
\underline{40} \\
0
\end{array}
$$

0 remainder means a terminating decimal.

● ●

To change a decimal to fraction form, just write the fraction part of the decimal in the numerator and use the position (place value) of the last digit on the right as the denominator. Reduce the fraction.

Example 7: Decimal to Fraction ● ● ● ● ● ● ● ● ● ● ● ● ● ● ●

a. Write the decimal number 0.56 in fraction form, reduced.

Solution: $0.56 = \dfrac{56}{100} = \dfrac{4 \cdot 14}{4 \cdot 25} = \dfrac{14}{25}$

● ●

Example 8: Calculator Examples ● ● ● ● ● ● ● ● ● ● ● ● ●

Use your calculator to perform the indicated operations.

a. $8.6321 + 7.5476 + 2.143 + 17.8293$

Solution: 36.1520

b. $(14.763)(0.47)(321.6)$

Solution: 2231.456976

c. Change the fraction $\dfrac{1}{7}$ to decimal form as an infinite repeating decimal. Then round off the decimal to ten thousandths (four decimal places).

Solution: $\dfrac{1}{7} = 1 \div 7 = 0.142857142857... = 0.\overline{142857}$ as an infinite repeating decimal.

(Remember that a bar can be placed over the repeating pattern of digits to indicate an infinite repeating decimal.)

$\dfrac{1}{7} = 0.1429$ Accurate to four decimal places

● ●

The TI-83 Plus calculator has commands that will change numbers in fraction form into decimal form and numbers in decimal form into fraction form. The two commands are found by pressing (MATH) and noting choices

1:>Frac and **2:>Dec**

Choice **1:>Frac** will change a decimal number into fraction form. Note that if the decimal number is greater than 1, this form will be an improper fraction and not a mixed number. Choice **2:>Dec** will change a fraction into decimal form. Choice **2:>Dec** is not as useful as the first choice because the calculator automatically treats fractions as indicating division and gives the division in decimal form. Example 9 illustrates these two situations.

Example 9: Calculator Examples ● ● ● ● ● ● ● ● ● ● ● ● ● ●

Use a TI-83 Plus calculator (or other graphing calculator) to perform the following:

a. Change the decimal number 0.82 into fraction form.

b. Change the fraction $\dfrac{5}{8}$ into decimal form.

c. Find the sum $\dfrac{3}{5} + \dfrac{3}{20}$ in fraction form.

Solution: a. Proceed as follows:

Enter the numbers 0.82 (or just .82).

Press (MATH).

Select **>Frac** by pressing (ENTER).

Press (ENTER) again.

Continued on next page...

The screen will display

Note that the fraction is reduced.

b. Proceed as follows:

Enter the fraction $\dfrac{5}{8}$ by entering $5 \div 8$.

Press .

The calculator will automatically give the answer in decimal form and the screen will appear as follows:

c. Proceed as follows:

Enter the sum of the fractions as $(3 \div 5) + (3 \div 20)$.

Press MATH.

Press ENTER.

Press ENTER.

The display on the screen will appear as follows:

Note that the fraction is reduced.

2.3 Exercises

Write each of the numbers in Exercises 1 – 6 in decimal form and in words.

1. $\dfrac{86}{100}$ **2.** $\dfrac{75}{1000}$ **3.** $5\dfrac{1}{10}$ **4.** $7\dfrac{3}{100}$ **5.** $-18\dfrac{6}{100}$ **6.** $-21\dfrac{19}{1000}$

Write each of the numbers in Exercises 7 – 12 in decimal notation.

7. eighty-seven thousandths
8. two and thirty-five hundredths
9. negative five and fourteen hundredths
10. negative twenty and forty-five ten thousandths
11. seven and twenty-one ten thousandths
12. one hundred seventy-three and one hundred forty-six thousandths

In Exercises 13 – 18, round off each number to the place indicated.

13. 91.076 (nearest tenth) **14.** 0.053 (nearest tenth)
15. 0.495 (nearest hundredth) **16.** 6.0072 (nearest hundredth)
17. 67.0572 (nearest thousandth) **18.** 75.4579 (nearest thousandth)

Add or subtract as indicated in Exercises 19 – 26.

19. $243.7 + 65.22 + 8.31$ **20.** $29.51 + 17.2 + 72.4$ **21.** $420.43 - 156.92$
22. $87.0 - 66.31$ **23.** $65.13 - 44.81 - 75.9$ **24.** $-84.1 + 17.63 - 12.98$
25. $-147.0 + 79.6 - 85.43$ **26.** $6.49 + 103.81 - 59.62$

Find the indicated products in Exercises 27 – 30.

27. $(60.4)(1.8)$ **28.** $(21.6)(0.83)$ **29.** $(1.23)(-6.2)$ **30.** $(-5.83)(-0.24)$

Find the indicated quotients in Exercises 31 and 32.

31. $5.1\overline{)77.01}$ **32.** $0.023\overline{)0.4922}$

Use long division to change the fractions to decimals in Exercises 33 – 36.

33. $\dfrac{3}{8}$ **34.** $\dfrac{4}{5}$ **35.** $\dfrac{1}{20}$ **36.** $\dfrac{3}{50}$

In Exercises 37 – 42, use a calculator to find the decimal form of each fraction. If the decimal is non-terminating, write it using the bar notation over the repeating pattern of digits.

37. $\dfrac{7}{16}$ **38.** $\dfrac{5}{18}$ **39.** $-\dfrac{6}{11}$ **40.** $-\dfrac{1}{22}$ **41.** $\dfrac{32}{27}$ **42.** $\dfrac{140}{3}$

Use a calculator to find the answers to Exercises 43 – 48 and then to change each of the answers into fraction form.

43. $0.57 + 0.73$

44. $1.58 - 2.36$

45. $-2.78 - 1.93 + 10.002$

46. $13.175 - 16.32 - 20.784$

47. $\dfrac{7}{8} + \dfrac{3}{4} + \dfrac{1}{10}$

48. $\dfrac{9}{100} - \dfrac{17}{100} - \dfrac{9}{10} + \dfrac{2}{3}$

Writing and Thinking About Mathematics

49. Suppose that a fraction between 0 and 1, such as $\dfrac{1}{4}$ or $\dfrac{5}{6}$ is multiplied by some other number.

    a. Discuss the situations (and give examples) in which the product will be less than the other number.

    b. Discuss the situations (and give examples) in which the product will be equal to the other number.

    c. Discuss the situations (and give examples) in which the product will be more than the other number.

Hawkes Learning Systems: Introductory Algebra

Decimals and Rounding

2.4   Change in Value and Average

Objectives

After completing this section, you will be able to:

1. Solve word problems that require the use of integers.

Change in Value

Subtraction can be used to find the change in value between two readings of measures such as temperatures, distances, and altitudes. To calculate the change between two values, including direction (negative for down, positive for up) use the following rule:

Change In Value

First find the end value and then subtract the beginning value.

(Change in Value) = (End Value) − (Beginning Value)

Example 1: Change in Value ● ● ● ● ● ● ● ● ● ● ● ● ● ● ● ●

a. On a winter day, the temperature dropped from 35° F at noon to 6° below zero (−6° F) at 7 p.m. What was the change in temperature?

Solution: end temperature − beginning temperature = change in temperature

$$-6° \quad - \quad 35° \quad =$$
$$-6° \quad + \quad (-35°) \quad = \quad -41°$$

The change in temperature was −41°. (This means that the temperature *dropped* by 41°.)

b. A jet pilot flew her plane from an altitude of 30,000 ft. to an altitude of 12,000 ft. What was the change in altitude?

Continued on next page...

Solution: end altitude – beginning altitude = change in altitude

$$12{,}000 \quad - \quad 30{,}000 \quad = \quad -18{,}000 \text{ ft.}$$

(This means that the plane *descended* 18,000 feet.)

● ●

The **net change** in a measure is the algebraic sum of several signed numbers. Example 2 illustrates how positive and negative numbers can be used to find the net change of weight (gain or loss) over a period of time.

Example 2: Net Change

Sue weighed 130 lbs. when she started to diet. The first week she lost 7 lbs., the second week she gained 2 lbs., and the third week she lost 5 lbs. What was her weight after 3 weeks of dieting?

Solution:
$$
\begin{aligned}
130 + (-7) + (+2) + (-5) &= 123 + (+2) + (-5) \\
&= 125 + (-5) \\
&= 120 \text{ lbs.}
\end{aligned}
$$

● ●

Average

You may already be familiar with the concept of the **average** of a set of numbers. The average is also called the **arithmetic average** or **mean**. Your grade in this course may be based on the average of your exam scores. Newspapers and magazines report average income, average sales, average attendance at sporting events, and so on. The mean of a set of numbers is particularly important in the study of statistics. For example, scientists might be interested in the mean IQ of the students attending a certain university or the mean height of students in the fourth grade.

Average

*The **average** (or **mean**) of a set of numbers is the value found by adding the numbers in the set and then dividing the sum by the number of numbers in the set.*

Example 3: Average •

a. At noon on five consecutive days in Aspen, Colorado the temperatures were −5°, 7°, 6°, −7°, and 14° (in degrees Fahrenheit). (Negative numbers represent temperatures below zero.) Find the average of these noon-day temperatures.

Solution: First, add the five temperatures.

$$-5 + 7 + 6 + (-7) + 14 = 15$$

Now, divide the sum, 15, by the number of temperatures, 5.

$$\frac{15}{5} = 3$$

The average noon temperature was 3° F.

b. In a placement exam for mathematics, a group of ten students had the following scores: 3 students scored 75, 2 students scored 80, 1 student scored 82, 3 students scored 85, and 1 student scored 88. What was the mean score for this group of students?

Solution: To find the total of all the scores, we multiply and then add. This is more efficient than adding all ten scores.

$75 \cdot 3 = 225$

$80 \cdot 2 = 160$

$82 \cdot 1 = 82$

$85 \cdot 3 = 255$ Multiply.

$88 \cdot 1 = 88$

$225 + 160 + 82 + 255 + 88 = 810$ Add.

$810 \div 10 = 81$ Divide by the number of scores.

The mean score on the placement test for this group of students was 81.

Continued on next page...

c. The following speeds (in miles per hour) of fifteen cars were recorded at a certain point on a freeway.

$$70 \quad 75 \quad 65 \quad 60 \quad 61$$
$$64 \quad 68 \quad 72 \quad 59 \quad 68$$
$$82 \quad 76 \quad 70 \quad 68 \quad 50$$

Find the average speed of these cars. (One car received a speeding ticket.)

Solution: Using a calculator, the sum of the speeds is 1008 mph.
Dividing by 15 gives the average speed:

$$1008 \div 15 = 67.2 \text{ mph}$$

Practice Problems

1. What should be added to −8 to get a sum of 20?

2. At noon, the temperature was 40° F. At 3:00 p.m., the temperature was 32° F. What was the change in temperature?

3. Find the average of the integers −7, 13, −20, and −30.

2.4 Exercises

Write each problem as a sum or difference, then simplify.

1. 23 lbs. lost, 13 lbs. gained, 6 lbs. lost

2. $24 earned, $17 spent, $2 earned

3. 4° rise, 3° rise, 9° drop

4. $10 withdrawal, $25 deposit, $18 deposit, $9 withdrawal

5. $47 earned, $22 spent, $8 earned, $45 spent

6. $20 won, $42 lost, $58 won, $11 lost

7. 14° rise, 6° drop, 11° rise, 15° drop

8. $53 withdrawal, $8 withdrawal, $48 deposit, $17 withdrawal

9. $187 profit, $241 loss, $82 profit, $26 profit

10. Snap-O Mousetrap stock rose $3 Monday, rose $6 Tuesday, and dropped $7 Wednesday. What was the net change in price of the stock?

Answers to Practice Problems: 1. 28 **2.** −8° F **3.** −11

11. In the first quarter of a recent football game, Fumbles A. Lott carried the ball six times with the following results: a gain of 6 yd., a gain of 3 yd., a loss of 4 yd., a gain of 2 yd., no gain or loss, and a loss of 3 yd. What was his net yardage for the first quarter?

12. Beginning at 7° above zero, the temperature rose 4°, then dropped 2°, and then dropped 6°. Find the final temperature.

13. Starting at the third floor, an elevator went down 1 floor, up 3 floors, up 7 floors, and then down 4 floors. Find the final location of the elevator.

14. Bill lost 2 lbs. the first week of his diet, lost 6 lbs. the second week, gained 1 lb. the third week, lost 4 lbs. the fourth week, and gained 3 lbs. the fifth week. What was the total loss or gain? If Bill weighed 223 lbs. at the time he began his diet, what was his weight after 5 weeks of dieting?

15. Jeff works Friday, Saturday, and Sunday in a restaurant. His salary is $10 a night plus tips. Friday night he received $65 in tips but spent $9 for food. Saturday he bought a new shirt for $22, spent $19 for food and received $72 in tips. Sunday he spent $15 for food and received $48 in tips. How much money did he have left after three days of work?

16. What should be added to −5 to get a sum of 17?

17. What should be added to −10 to get a sum of −23?

18. What should be added to 39 to get a sum of 16?

19. What should be added to 24 to get a sum of 13?

20. What should be added to −37 to get a sum of −54?

21. What should be added to 15 to get a sum of −18?

22. What should be added to −12 to get a sum of 15?

23. From the sum of −4 and −17, subtract the sum of −12 and 6.

24. From the sum of 11 and −13, subtract the sum of 19 and −8.

25. Find the average of the integers −36, −18, −10, and −40.

26. Find the mean of the integers −11, −15, 3, 17, and −9.

27. Mr. Chung received a bank statement indicating that he was overdrawn by $63. How much must he deposit to bring his balance to $157?

28. Ms. Martinez knew that the balance in her checking account was $450. She made deposits of $300 and $750 and wrote checks for $85, $620, and $35. What was her new balance?

29. At 2 p.m. the temperature was 76°F. At 8 p.m. the temperature was 58°F. What was the change in temperature? What was the average change in temperature per hour?

30. The temperature at 5:00 a.m. was 8° below zero; at noon, the temperature was 27°F. What was the change in temperature? What was the average change in temperature per hour?

31. Lotsa-Flavor Chewing Gum stock opened on Monday at $47 per share and closed Friday at $39 per share. Find the change in price of the stock.

32. The famous French mathematician René Descartes lived from 1596 to 1650. How long did he live?

33. If you travel from the top of Mt. Whitney, elevation 14,495 ft., to the floor of Death Valley, elevation 282 ft. below sea level, what is the change in elevation?

34. A submarine submerged 280 ft. below the surface of the sea fired a rocket that reached an altitude of 30,000 ft. What was the change in altitude of the rocket?

35. The great English mathematician and scientist Isaac Newton was born in 1642 and died in 1727. How old was he when he died?

36. Find the mean of the set of integers −10, 15, 16, −17, −34, and −42.

37. Find the mean of the set of integers −72, −100, −54, 82, and −96.

38. The temperature readings for 20 days at 3 p.m. at a local ski resort were recorded as follows:

$$24° \quad 11° \quad −5° \quad 14° \quad 15° \quad 5° \quad −6° \quad 13° \quad −2° \quad −8°$$
$$−10° \quad 32° \quad 31° \quad −7° \quad −9° \quad 4° \quad −18° \quad −9° \quad 5° \quad 20°$$

What was the average of the recorded temperatures for these 20 days?

39. On an exam in history, a class of twenty-one students had the following test scores: 4 scored 65, 3 scored 70, 6 scored 78, 2 scored 82, 1 scored 85, 3 scored 91, and 2 scored 95. What was the mean score (to the nearest hundredth) on this test for the class?

40. Fifteen students scored the following scores on an exam in accounting: 1 scored 67, 4 scored 73, 3 scored 77, 2 scored 80, 3 scored 88, and 2 scored 93. What was the average score for these students?

41. Twenty business executives made the following numbers of telephone calls during one week. Find the mean number of calls (to the nearest tenth) made by these executives.

$$20 \quad 16 \quad 14 \quad 11 \quad 51 \quad 18 \quad 16 \quad 42 \quad 49 \quad 12$$
$$40 \quad 36 \quad 28 \quad 52 \quad 25 \quad 18 \quad 22 \quad 33 \quad 9 \quad 19$$

42. The blood calcium level (in milligrams per deciliter) for 20 patients was reported as follows:

$$8.2 \quad 10.2 \quad 9.3 \quad 8.5 \quad 7.3 \quad 9.4 \quad 11.1 \quad 10.0 \quad 8.5 \quad 9.9$$
$$9.7 \quad 9.6 \quad 8.3 \quad 9.8 \quad 9.1 \quad 8.6 \quad 10.2 \quad 9.4 \quad 9.1 \quad 9.2$$

Find the mean blood calcium level (to the nearest tenth) for these patients.

43. Thirty students made the following scores on the final exam in an algebra course. Find the mean score (to the nearest tenth) on the final exam.

$$82 \quad 93 \quad 80 \quad 65 \quad 52 \quad 48 \quad 78 \quad 69 \quad 75 \quad 75$$
$$98 \quad 95 \quad 63 \quad 76 \quad 78 \quad 80 \quad 95 \quad 70 \quad 73 \quad 81$$
$$82 \quad 89 \quad 71 \quad 66 \quad 67 \quad 44 \quad 74 \quad 85 \quad 85 \quad 92$$

The frequency of a number is simply a count of how many times that number appears. In statistics, data is commonly given in the table form of a frequency distribution as illustrated in Exercises 44 and 45. To find the mean, multiply each number by its frequency, add these products, and divide the sum by the sum of the frequencies.

44. The heights of twenty-two men were recorded in the following frequency distribution. Find the mean height (to the nearest tenth of an inch) for these men.

Height (in inches)	Frequency
68	2
69	3
72	8
73	5
74	2
75	1
78	1

45. The students in a psychology class were asked the number of books that they had read in the last month. The following frequency distribution indicates the results. Find the mean number of books read (to the nearest tenth) by these students.

Number of Books	Frequency
0	3
1	2
2	6
3	4
4	2
5	1

46. The bar graph shows the approximate amounts of time per week spent watching TV for six groups (by age and sex) of people 18 years of age and older. What is the average amount of time per week people over the age of 18 spend watching TV? (Assume each group has the same number of people.)

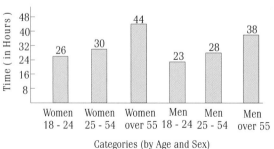

Weekly TV Viewing by Age and Sex

Categories (by Age and Sex)

47. The following pictograph shows the number of phone calls received by a 1 hour radio talk show in Seattle during one week. (Not all calls actually get on the air.) What was the mean number of calls per show received that week?

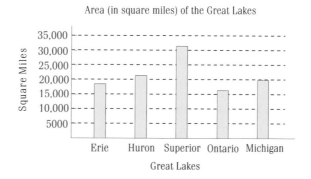

48. The following bar graph shows the area of each of the five Great Lakes. Lake Superior with an area of about 32,000 square miles is the world's largest fresh water lake. What is the mean size of these lakes?

Area (in square miles) of the Great Lakes

Hawkes Learning Systems: Introductory Algebra

Applications (Integers)

2.5   Simplifying and Evaluating Algebraic Expressions

After completing this section, you will be able to:

1. Simplify algebraic expressions by combining like terms.

2. Evaluate expressions for given values of the variables.

Simplifying Algebraic Expressions

A single number is called a **constant**. Any constant or variable or the indicated product and/or quotient of constants and powers of variables is called a **term**. Examples of terms are

$$16, \ 3x, \ -5.2, \ 1.3xy, \ -5x^2, \ 14a^2b^2, \ \text{and} \ -\frac{x}{y^2}.$$

A number written next to a variable (as in $10x$) or a variable written next to another variable (as in xy) indicates multiplication. In the term $5x^2$, the constant 5 is called the **numerical coefficient** of x^2 (or simply the **coefficient** of x^2).

If there is not a number written next to a variable, the coefficient is understood to be 1. For example,

$$x = 1 \cdot x, \ a^3 = 1 \cdot a^3, \ xy = 1 \cdot xy$$

If a "−" sign is next to a variable, the coefficient is understood to be −1. For example,

$$-x = -1 \cdot x, \ -y^5 = -1 \cdot y^5, \ \text{and} \ -mn^2 = -1 \cdot mn^2.$$

Like terms (or similar terms) are terms that are constants or terms that contain the same variables (if any) raised to the same powers.

Like Terms

-6, 1.84, 145, $\dfrac{3}{4}$ are like terms because each term is a constant.

$-3a$, $15a$, $2.6a$, $\dfrac{2}{3}a$ are like terms because each term contains the same variable a, raised to the same power, 1. (Remember that $a = a^1$.)

$5xy^2$ and $-3.2xy^2$ are like terms because each term contains the same two variables, x and y, with x first-degree in both terms and y second-degree in both terms.

Unlike Terms

$8x$ and $-9x^2$ are unlike terms (**not** like terms) because the variable x is not of the same power in both terms.

Example 1: Like Terms •

a. From the following list of terms, pick out the like terms:

$$-7,\ 2x,\ 4.1,\ -x,\ 3x^2y,\ 5x,\ -6x^2y,\ \text{and}\ 0$$

Solution: $-7, 4.1,$ and 0 are like terms. All are constants.
$2x, -x,$ and $5x$ are like terms.
$3x^2y$ and $-6x^2y$ are like terms.

• •

Algebraic expressions (expressions that indicate operations with terms) such as

$$9x + 10y,\ 5n - n + 3m^2,\ 2x^2 - 8x + 10,\ \text{and}\ \frac{5x + 3x}{4} - 8.2x^2$$

are not terms. However, we would like to simplify expressions that contain like terms. That is, we want to combine like terms. For example, by combining like terms we can write

$$9x + 6x = 15x \ \text{and}\ 6.3n + 2n - n = 7.3n$$

Like terms can be combined by applying the distributive property. Recall that the distributive property states that

$$a(b + c) = ab + ac$$
$$\text{or}\quad ab + ac = a(b + c)$$
$$\text{or}\quad ba + ca = (b + c)a$$

This last form is particularly useful when b and c are numerical coefficients. For example,

$$3x + 5x = (3 + 5)x \qquad \text{By the distributive property}$$
$$= 8x \qquad \text{Add the coefficients.}$$

$$3x^2 - 5x^2 = (3 - 5)x^2 \qquad \text{By the distributive property}$$
$$= -2x^2 \qquad \text{Add the coefficients algebraically.}$$

In the first expression, the like terms $3x$ and $5x$ are combined and the result is $8x$. In the second expression, the like terms $3x^2$ and $-5x^2$ are combined and the result is $-2x^2$.

Example 2: Combine Like Terms ● ● ● ● ● ● ● ● ● ● ● ● ● ● ●

Combine like terms whenever possible.

a. $8x + 10x$

   Solution: $8x + 10x = (8 + 10)x = 18x$ By the distributive property

b. $6.5y - 2.3y$

   Solution: $6.5y - 2.3y = (6.5 - 2.3)y = 4.2y$

c. $4(n - 7) + 5(n + 1)$

   Solution: $4(n - 7) + 5(n + 1) = 4n - 28 + 5n + 5$ Use the distributive property twice.
   $$= 4n + 5n - 28 + 5$$
   $$= 9n - 23 \qquad \text{Combine like terms.}$$

d. $2x^2 + 3a + x^2 - a$

   Solution: $2x^2 + 3a + x^2 - a = 2x^2 + x^2 + 3a - a$
   $$= (2 + 1)x^2 + (3 - 1)a \qquad \textbf{Note: } +x^2 = +1x^2 \text{ and}$$
   $$-a = -1a$$
   $$= 3x^2 + 2a$$

e. $\dfrac{x+3x}{2}+5x$

>   **Solution:** A fraction bar is a symbol of inclusion, like parentheses. So combine like terms in the numerator first.

$$\frac{x+3x}{2}+5x = \frac{4x}{2}+5x$$
$$= \frac{4}{2}\cdot x+5x$$
$$= 2x+5x$$
$$= 7x$$

● ●

Evaluating Algebraic Expressions

In most cases, if an expression is to be evaluated, like terms should be combined first and then the resulting expression evaluated by following the rules for order of operations.

To Evaluate an Algebraic Expression

1. Combine like terms, if possible.

2. Substitute the values given for any variables.

3. Follow the rules for order of operations.

Remember to use parentheses around negative numbers when substituting.
Without parentheses, an evaluation can be dramatically changed and lead to wrong answers, particularly when even exponents are involved. These ideas were discussed in Chapter 1 and we summarize with the exponent 2 as follows:

In general, except for $x = 0$,

    1. $-x^2$ **is negative** $[\,-6^2 = -1\cdot 6^2 = -36\,]$

    2. $(-x)^2$ **is positive** $[\,(-6)^2 = (-6)(-6) = 36\,]$

    3. $-x^2 \neq (-x)^2$ $[\,-36 \neq 36\,]$

Example 3: Evaluate Algebraic Expressions • • • • • • • • •

a.   Evaluate x^2 for $x = 3$ and for $x = -4$.

Solution: For $x = 3$, $x^2 = 3^2 = 9$
For $x = -4$, $x^2 = (-4)^2 = 16$

b.   Evaluate $-x^2$ for $x = 3$ and for $x = -4$.

Solution: For $x = 3$, $-x^2 = -3^2 = -1 \cdot 9 = -9$
For $x = -4$, $-x^2 = -(-4)^2 = -1(16) = -16$

Simplify each expression below by combining like terms; then, evaluate the resulting expression using the given values for the variables. Remember to apply the rules for order of operations.

c. $2x + 5 + 7x$; $x = -3$

Solution: Simplify first:
$$2x + 5 + 7x = 2x + 7x + 5$$
$$= 9x + 5$$
Now evaluate:
$$9x + 5 = 9(-3) + 5$$
$$= -27 + 5$$
$$= -22$$

d. $3ab - 4ab + 6a - a$; $a = 2, b = -1$

Solution: Simplify first:
$$3ab - 4ab + 6a - a = -ab + 5a$$
Now evaluate:
$$-ab + 5a = -1(2)(-1) + 5(2) \qquad \textbf{Note: } -ab = -1ab$$
$$= 2 + 10$$
$$= 12$$

e. $\dfrac{5x + 3x}{4} + 2(x + 1)$; $x = 5$

Solution: Simplify first:
$$\frac{5x + 3x}{4} + 2(x + 1) = \frac{8x}{4} + 2x + 2$$
$$= 2x + 2x + 2$$
$$= 4x + 2$$

Continued on next page...

Now evaluate:

$$4x + 2 = 4 \cdot 5 + 2$$
$$= 20 + 2$$
$$= 22$$

f. $\dfrac{3}{4}x + \dfrac{3}{8}x; \ x = -4$

Solution: Simplify first: $\dfrac{3}{4}x + \dfrac{3}{8}x = \left(\dfrac{3}{4} + \dfrac{3}{8}\right)x = \left(\dfrac{6}{8} + \dfrac{3}{8}\right)x = \dfrac{9}{8}x$

Now evaluate:

$$\dfrac{9}{8}x = \dfrac{9}{\overset{}{\underset{2}{\cancel{8}}}} \cdot (\overset{-1}{\cancel{-4}}) = -\dfrac{9}{2} \quad \text{(or } -4.5 \text{ in decimal form)}$$

● ●

NOTES

Important note about expressions involving fractions in algebra

An expression with a fraction as the coefficient can be written in two different forms. For example, for all values of x, $\dfrac{9}{8}x$ is the same as $\dfrac{9x}{8}$. To see this fact,

note that we can write $\dfrac{9}{8}x = \dfrac{9}{8} \cdot \dfrac{x}{1} = \dfrac{9x}{8}$. However, algebraically, $\dfrac{9}{8x}$ (with the

variable in the denominator) is *not* the same as $\dfrac{9}{8}x$.

Similarly,

$$\dfrac{5}{6}a = \dfrac{5a}{6} \quad \text{and} \quad \dfrac{5}{6a} \neq \dfrac{5}{6}a.$$

Practice Problems

Simplify the following expressions by combining like terms.

1. $-2x - 5x$ **2.** $12y + 6 - y + 10$

3. $5(x - 1) + 4x$ **4.** $2b^2 - a + b^2 + a$

Simplify the expression, then evaluate the resulting expression if $x = 3$ and $y = -2$.

5. $2(x + 3y) + 4(x - y)$

Answers to Practice Problems: 1. $-7x$ **2.** $11y + 16$ **3.** $9x - 5$ **4.** $3b^2$ **5.** $6x + 2y; 14$

2.5 Exercises

In Exercises 1 – 6, pick out the like terms in each list of terms.

1. $-5, \dfrac{1}{6}, 7x, 8, 9x, 3y$

2. $-2x^2, -13x^3, 5x^2, 14x^2, 10x^3$

3. $5xy, -x^2, -6xy, 3x^2y, 5x^2y, 2x^2$ **4.** $3ab^2, -ab^2, 8ab, 9a^2b, -10a^2b, ab, 12a^2$

5. $24, 8.3, 1.5xyz, -1.4xyz, -6, xyz, 5xy^2z, 2xyz^2$

6. $-35y, 1.62, -y^2, -y, 3y^2, \dfrac{1}{2}, 75y, 2.5y^2$

Find the value of each numerical expression in Exercises 7 – 10.

7. $(-8)^2$ **8.** -8^2 **9.** -11^2 **10.** $(-6)^2$

Simplify by combining like terms in each of the expressions in Exercises 11 – 42.

11. $8x + 7x$ **12.** $3y + 8y$ **13.** $5x - 2x$ **14.** $7x + (-3x)$

15. $-n - n$ **16.** $-x - x$ **17.** $6y^2 - y^2$ **18.** $16z^2 - 5z^2$

19. $23x^2 - 11x^2$ **20.** $18x^3 + 7x^3$ **21.** $4x + 2 + 3x$ **22.** $3x - 1 + x$

23. $2x - 3y - x$ **24.** $x + y + x - 2y$ **25.** $2x^2 + 5y + 6x^2 - 2y$

26. $4a + 2a - 3b - a$ **27.** $3(n + 1) - n$ **28.** $2(n - 4) + n + 1$

29. $5(a - b) + 2a - 3b$ **30.** $4a - 3b + 2(a + 2b)$

31. $3(2x + y) + 2(x - y)$ **32.** $4(x + 5y) + 3(2x - 7y)$

33. $2x + 3x^2 - 3x - x^2$ **34.** $2y^2 + 4y - y^2 - 3y$

35. $2(n^2 - 3n) + 4(-n^2 + 2n)$ **36.** $3n^2 + 2n - 5 - n^2 + n - 4$

37. $3x^2 + 4xy - 5xy + y^2$ **38.** $2x^2 - 5xy + 11xy + 3$ **39.** $\dfrac{x + 5x}{6} + x$

40. $2y - \dfrac{2y + 3y}{5}$ **41.** $y - \dfrac{2y + 4y}{3}$ **42.** $z - \dfrac{3z + 5z}{4}$

In Exercises 43 – 60, first (a) simplify the expressions and then (b) evaluate the simplified expression at $x = 4, y = 3, a = -2,$ and $b = -1$.

43. $5x + 4 - 2x$ **44.** $7x - 17 - x$ **45.** $8.3x^2 - 5.7x^2 + x^2$

46. $2.4(x+1)+1.3(x-1)$ **47.** $\dfrac{7}{8}x+\dfrac{3}{4}x+\dfrac{1}{2}x$ **48.** $\dfrac{7y}{10}-\dfrac{y}{5}$

49. $\dfrac{a}{6}+\dfrac{a}{2}-\dfrac{2a}{3}$ **50.** $-\dfrac{3}{20}x-\dfrac{3}{10}x-\dfrac{2}{5}x$ **51.** $x-10-3x+2$

52. $3(y-1)+2(y+2)$ **53.** $4(y+3)+5(y-2)$ **54.** $-5(x+y)+2(x-y)$

55. $6a+5a-a+13$ **56.** $3a^2-a^2+4a-5$ **57.** $5ab+b^2-2ab+b^3$

58. $5a+ab^2-2ab^2+3a$ **59.** $\dfrac{3a+5a}{-2}+12a$ **60.** $\dfrac{-4b-2b}{-3}+\dfrac{2b+5b}{7}$

Simplify each of the following expressions.

61. a. $3a+3a$ **b.** $3a\cdot3a$ **c.** $\dfrac{1}{3a}+\dfrac{1}{3a}$ **d.** $\dfrac{1}{3a}\cdot\dfrac{1}{3a}$

 e. $\dfrac{1}{3a}\div\dfrac{1}{3a}$

62. a. $5x+5x$ **b.** $5x\cdot5x$ **c.** $\dfrac{1}{5x}+\dfrac{1}{5x}$ **d.** $\dfrac{1}{5x}\cdot\dfrac{1}{5x}$

 e. $\dfrac{1}{5x}\div\dfrac{1}{5x}$

63. a. $-7ab-7ab$ **b.** $(-7ab)\cdot(-7ab)$ **c.** $\dfrac{1}{7ab}+\dfrac{1}{7ab}$ **d.** $\dfrac{1}{7ab}\cdot\dfrac{1}{7ab}$

 e. $\dfrac{1}{7ab}\div\dfrac{1}{7ab}$

64. a. $8y-8y$ **b.** $8y\cdot(-8y)$ **c.** $\dfrac{1}{8y}-\dfrac{1}{8y}$ **d.** $\dfrac{1}{8y}\cdot\left(-\dfrac{1}{8y}\right)$

 e. $-\dfrac{1}{8y}\div\dfrac{1}{8y}$

65. a. $1.5a+1.5a$ **b.** $1.5a\cdot1.5a$ **c.** $\dfrac{1}{1.5a}+\dfrac{1}{1.5a}$ **d.** $\dfrac{1}{1.5a}\cdot\dfrac{1}{1.5a}$

 e. $\dfrac{1}{1.5a}\div\left(-\dfrac{1}{1.5a}\right)$

66. a. $\dfrac{4x}{3}+\dfrac{4x}{3}$ **b.** $\dfrac{4x}{3}\cdot\dfrac{4x}{3}$ **c.** $\dfrac{4x}{3}-\dfrac{4x}{3}$ **d.** $\dfrac{4x}{3}\cdot\dfrac{3}{4x}$

 e $\dfrac{4x}{3}\div\dfrac{3}{4x}$

Hawkes Learning Systems: Introductory Algebra

Variables and Algebraic Expressions
Simplifying Expressions
Simplifying Expressions with Parentheses
Evaluating Algebraic Expressions

2.6 Translating English Phrases and Algebraic Expressions

After completing this section, you will be able to:

1. Write the meanings of algebraic expressions in words.

2. Write algebraic expressions for word phrases.

Translating English Phrases into Algebraic Expressions

Algebra is a language of mathematicians, and to understand mathematics, you must understand the language. We want to be able to change English phrases into their "algebraic" equivalents and vice versa. So, if a problem is stated in English, we can translate the phrases into algebraic symbols and proceed to solve the problem according to the rules developed for algebra.

The following examples illustrate how certain key words can be translated into algebraic symbols.

Example 1: Key Words • • • • • • • • • • • • • • • • • • •

English Phrase	**Algebraic Expression**
a. 3 **multiplied by** the number represented by x the **product** of 3 and x 3 **times** x	$3x$
b. 3 **added to** a number the **sum** of z and 3 z **plus** 3 3 **more than** z	$z + 3$
c. 2 **times** the quantity found by **adding** a number to 1 **twice** the **sum** of x and 1 the **product** of 2 with the **sum** of x and 1	$2(x + 1)$ *Continued on next page ...*

d. **twice** x **plus** 1

the **sum** of **twice** x and 1

2 **times** x **increased** by 1

1 **more than** the **product** of 2 and a number

$$2x + 1$$

e. the **difference** between 5 **times** a number and 3

3 **less than** the **product** of a number and 5

5 **times** a number **minus** 3

3 **subtracted from** $5n$

5 **multiplied by** a number **less** 3

$$5n - 3$$

● ●

NOTES

In Example 1b, the phrase "the sum of z and 3" was translated as $z + 3$. If the expression had been translated as $3 + z$, there would have been no mathematical error because addition is commutative. That is, $z + 3 = 3 + z$. However, in e., the phrase "3 less than the product of a number and 5" must be translated as it was because subtraction is **not** commutative. Thus,

"3 less than 5 times a number" means $5n - 3$
while "5 times a number less than 3" means $3 - 5n$.

Therefore, be very careful when writing and/or interpreting expressions indicating subtraction. Be sure that the subtraction is in the order indicated by the wording in the problem. The same is true with expressions involving division.

Certain words, such as those in boldface in the previous examples, are the keys to the operations. Learn to look for these words and those from the following list.

__Addition__	__Subtraction__	__Multiplication__	__Division__
add	subtract (from)	multiply	divide
sum	difference	product	quotient
plus	minus	times	
more than	less than	twice	
increased by	decreased by	of (with fractions	
	less	and percent)	

The words **quotient** and **difference** deserve special mention because their use implies that the numbers given are to be operated on in the order given. That is, division and subtraction are done with the values in the same order that they are given in the problem. For example:

the quotient of y and 5 $\longrightarrow$ $\dfrac{y}{5}$

the quotient of 5 and y $\longrightarrow$ $\dfrac{5}{y}$

the difference between 6 and x $\longrightarrow$ $6 - x$

the difference between x and 6 $\longrightarrow$ $x - 6$

If we did not have these agreements concerning subtraction and division, then the phrases just illustrated might have more than one interpretation and be considered **ambiguous**.

An **ambiguous phrase** is one whose meaning is not clear or for which there may be two or more interpretations. This is a common occurrence in ordinary everyday language, and misunderstandings occur frequently. Imagine the difficulties diplomats have in communicating ideas from one language to another trying to avoid ambiguities. Even the order of subjects, verbs, and adjectives may not be the same from one language to another. Translating grammatical phrases in any language into mathematical expressions is quite similar. To avoid ambiguous phrases in mathematics, we try to be precise in the use of terminology, to be careful with grammatical construction, and to follow the rules for order of operations.

Translating Algebraic Expressions into English Phrases

Consider the three expressions to be translated into English:

$$7(n + 1), \quad 6(n - 3), \quad \text{and} \quad 7n + 1$$

In the first two expressions, we indicate the parentheses with a phrase such as "the quantity" or "the sum of" or "the difference between." Without the parentheses, we agree that the operations are to be indicated in the order given. Thus,

$7(n + 1)$ can be translated as "seven times the sum of a number and 1"

$6(n - 3)$ can be translated as "six times the difference between a number and 3"

while $7n + 1$ can be translated as "seven times a number plus 1."

Example 2: Algebraic Expression to Phrase

Write an English phrase that indicates the meaning of each algebraic expression.

a. $5x$ **b.** $2n + 8$ **c.** $3(a - 2)$

Solution:

Algebraic Expression	Possible English Phrase
a. $5x$	The product of a number and 5
b. $2n + 8$	Twice a number increased by 8
c. $3(a - 2)$	Three times the difference between a number and 2

Example 3: Phrase to Algebraic Expression

Change each phrase into an equivalent algebraic expression.

a. The quotient of a number and –4
b. 6 less than 5 times a number
c. Twice the sum of 3 and a number

Solution:

Phrase	Algebraic Expression
a. The quotient of a number and –4	$\dfrac{x}{-4}$
b. 6 less than 5 times a number	$5y - 6$
c. Twice the sum of 3 and a number	$2(3 + n)$

Practice Problems

Change the following phrases to algebraic expressions.

1. 7 less than a number

2. The quotient of y and 5

3. 14 more than 3 times a number

Change the following algebraic expressions into English phrases.

4. $10 - x$ *5. $2(y - 3)$*   *6. $5n + 3n$*

Answers to Practice Problems: **1.** $x - 7$ **2.** $\dfrac{y}{5}$ **3.** $3y + 14$ **4.** 10 decreased by a number **5.** twice the difference between a number and 3 **6.** 5 times a number plus 3 times the same number.

2.6 Exercises

Translate each of the expressions in Exercises 1 – 12 into an equivalent English phrase. (There may be more than one correct translation.)

1. $4x$ **2.** $x + 6$ **3.** $2x + 1$ **4.** $4x - 7$

5. $7x - 5.3$ **6.** $3.2(x + 2.5)$ **7.** $-2(x - 8)$ **8.** $10(x + 4)$

9. $5(2x + 3)$ **10.** $3(4x - 5)$ **11.** $6(x - 1)$ **12.** $9(x + 3)$

Write each pair of expressions in words in Exercises 13 – 16. Notice the differences between the expressions and the corresponding English phrases.

13. $3x + 7$; $3(x + 7)$ **14.** $4x - 1$; $4(x - 1)$ **15.** $7x - 3$; $7(x - 3)$

16. $5(x + 6)$; $5x + 6$

Write the algebraic expression described by each of the word phrases in Exercises 17 – 40. Choose your own variable.

17. 6 added to a number

18. 7 more than a number

19. 4 less than a number

20. A number decreased by 13

21. 5 less than 3 times a number

22. The difference between twice a number and 10

23. The difference between x and 3, all divided by 7

24. 9 times the sum of a number and 2

25. 3 times the difference between a number and 8

26. 13 less than the product of 4 with the sum of a number and 1

27. 5 subtracted from three times a number

28. The sum of twice a number and four times the number

29. 8 minus twice a number

30. The sum of a number and 9 times the number

31. 4 more than the product of 8 with the difference between a number and 6

32. Twenty decreased by 4.8 times a number

33. The difference between three times a number and five times the same number

34. Eight more than the product of 3 times the sum of a number and 6

35. Six less than twice the difference between a number and 7

36. Four less than 3 times the difference between 7 and a number

37. Nine more than twice the sum of 17 and a number

38. **a.** 5 less than 3 times a number
    b. 5 less 3 times a number

39. **a.** 6 less than a number
    b. 6 less a number

40. **a.** 20 less than a number
    b. A number less than 20

Write the algebraic expression described by each of the word phrases in Exercises 41 – 50.

41. The cost of *x* pounds of candy at $4.95 a pound

42. The annual interest on *x* dollars if the rate is 11% per year

43. The number of days in *t* weeks and 3 days

44. The number of minutes in *h* hours and 20 minutes

45. The points scored by a football team on *t* touchdowns (6 points each) and 1 field goal (3 points)

46. The cost of renting a car for one day and driving *m* miles if the rate is $20 per day plus 15 cents per mile

47. The cost of purchasing a fishing rod and reel if the rod costs *x* dollars and the reel costs $8 more than twice the cost of the rod

48. A sales person's weekly salary if he receives $250 as his base plus 9% of the weekly sales of *x* dollars

49. The selling price of an item that costs *c* dollars if the markup is 20% of the cost

50. The perimeter of a rectangle if the width is *w* centimeters and the length is 3 cm less than twice the width

3 cm less than twice the width

Hawkes Learning Systems: Introductory Algebra

 Translating Phrases into Algebraic Expressions

Chapter 2 Index of Key Ideas and Terms

Rational Number   page 58

A **rational number** is a number that can be written in

the form $\dfrac{a}{b}$ where a and b are integers and $b \neq 0$

OR,

A **rational number** is a number that can be written in decimal form as a terminating decimal or as an infinite repeating decimal.

Fraction   pages 58 - 59

In general, fractions can be used to indicate:
1. Equal parts of a whole, or
2. Division

Placement of Negative Signs in Fractions   page 60

If a and b are real numbers, and $b \neq 0$, then

$$-\frac{a}{b} = \frac{-a}{b} = \frac{a}{-b}$$

Multiplication with Fractions   page 60

$\dfrac{a}{b} \cdot \dfrac{c}{d} = \dfrac{a \cdot c}{b \cdot d}$ where $b \neq 0$ and $d \neq 0$

The Fundamental Principle of Fractions   page 61

$\dfrac{a}{b} = \dfrac{a \cdot k}{b \cdot k}$ where $k \neq 0$ and $b \neq 0$

Reciprocal   page 63

If $a \neq 0$ and $b \neq 0$, the **reciprocal** of $\dfrac{a}{b}$ is $\dfrac{b}{a}$, and $\dfrac{a}{b} \cdot \dfrac{b}{a} = 1$

Division with Fractions   page 63

To divide by a nonzero fraction, multiply by its reciprocal:

$$\frac{a}{b} \div \frac{c}{d} = \frac{a}{b} \cdot \frac{d}{c} \text{ where } b \neq 0, c \neq 0, \text{ and } d \neq 0$$

Addition of Fractions with Like Denominators page 69

To add two fractions with a common denominator
add the numerators and use the common denominator.

$$\frac{a}{b} + \frac{c}{b} = \frac{a+c}{b}$$

Least Common Multiple page 71

To find the LCM of a set of integers
1. List the prime factorization of each number.
2. List the prime factors that appear in any one of the
   prime factorizations.
3. Find the product of these primes using each prime
   the greatest number of times that it appears in any one
   prime factorization.

Addition of Fractions with Different Denominators page 72

To add two fractions with different denominators
1. Find the LCM of the denominators (called the LCD).
2. Change each fraction to an equal fraction with the LCD.
3. Add the new fractions.
(**Note**: Treat variables as prime factors.)

Subtraction with Fractions page 73

To subtract two fractions with a common denominator,
subtract the numerators and use the common denominator.

$$\frac{a}{b} - \frac{c}{b} = \frac{a-c}{b}$$

To subtract two fractions with different denominators, find the LCD,
change each fraction to an equal fraction with the LCD, and subtract.

Order of Operations with Fractions page 74

Decimal Numbers

Fractions and Decimals

Chapter 2 Review

For a review of the topics and problems from Chapter 2, look at the
following lessons from *Hawkes Learning Systems: Introductory Algebra*

Reduction of Proper Fractions
Reduction of Improper Fractions
Reduction of Positive and Negative Fractions
Multiplying and Dividing Fractions
Adding and Subtracting Fractions
Decimals and Rounding
Applications (Integers)
Variables and Algebraic Expressions
Simplifying Expressions
Simplifying Expressions with Parentheses
Evaluating Algebraic Expressions
Translating Phrases into Algebraic Expressions

Chapter 2 Test

Supply the missing numbers so that each fraction will be raised to higher terms as indicated.

1. $\dfrac{5}{9} = \dfrac{5}{9} \cdot \dfrac{?}{?} = \dfrac{?}{36}$

2. $\dfrac{2}{3a} = \dfrac{2}{3a} \cdot \dfrac{?}{?} = \dfrac{?}{12a^2}$

Reduce each fraction to lowest terms.

3. $\dfrac{-20y^2}{15y}$

4. $\dfrac{77ab}{121a^3}$

5. Find $\dfrac{3}{5}$ of 160.

6. Find the LCM of $10xy^2$, $18x^2y$, and $15xy$.

Perform the indicated operations and reduce each answer to lowest terms.

7. $\dfrac{-15x}{7} \cdot \dfrac{42}{20x}$

8. $\dfrac{7}{12} \cdot \dfrac{9}{28} \div \dfrac{5}{16}$

9. $\dfrac{4}{15} + \dfrac{1}{3} + \dfrac{3}{10}$

10. $\dfrac{7}{2y} - \dfrac{8}{2y} - \dfrac{3}{2y}$

11. $\dfrac{7}{8} - \dfrac{1}{3} \div \dfrac{5}{6} + \dfrac{1}{4}$

12. $\dfrac{3}{n} + \dfrac{1}{5}$

13. Write the decimal number 17.003 in words.

14. Round off 83.1462 to the nearest hundredth.

Perform the indicated operations.

15. $28.63 + 7.9$

16. $19.1 - 35.6$

17. $(5.61)(3.2)$

18. $27.27 \div 20.2$

19. $2x + 3 - 7x + 4$

20. $5.6y - y - 2.3y + 2.1y$

21. $2a^2 + 3a + 5a^2 - 8a + 10$

(a) Simplify each of the following expressions by combining like terms, and then (b) evaluate each simplified expression for $x = -2$ and $y = 3$.

22. $\dfrac{2}{3}y + \dfrac{1}{4}y - \dfrac{1}{8}y$

23. $2(x - 8)$

Translate each English phrase into an equivalent algebraic expression.

24. 3 less than the product of a number and 6

25. Twice the sum of the number and 5

26. 4 less than twice the sum of a number and 5

27. Find the mean of the set of integers $\{-20, -5, -3, 8, -18, -16\}$.

28. Use a calculator to find the mean (to the nearest tenth) of the data reported in the following frequency distribution.

Scores on History Exam	Frequency
60	2
72	3
75	5
80	6
88	4
90	2
95	1

29. Multiply the sum of $\dfrac{3}{4}$ and $\dfrac{9}{10}$ by the quotient of $\dfrac{2}{3}$ and $-\dfrac{8}{15}$. What is the product?

30. The Veri-Soft computer company stock opened on Monday at $16.50 per share and on the next five days rose $1.50, fell $2.10, rose $3.25, fell $1.75, and fell $1.60 per share. What was the net change in the price of the stock? What was the price of the stock at the closing on Friday?

31. The sale price of a jacket is $60. This is $\dfrac{3}{4}$ of the original price.

a. Is the original price more or less than $75?

b. If you multiply $60 by $\dfrac{3}{4}$, will the product be more or less than $60?

c. What was the original price?

32. You know that the gas tank in your car holds 20 gallons of gas and the gauge reads $\dfrac{1}{4}$ full.

a. How many gallons will be needed to fill the tank?

b. If the price of gas is $1.40 per gallon and you have $20, do you have enough cash to fill the tank? How much extra do you have or how much are you short?

Cumulative Review: Chapters 1 − 2

Perform the indicated operations in Exercises 1 − 8. Reduce each answer to lowest terms.

1. $18 + 7 - 8 + 3$ **2.** $9 - 16 - 11 + 4$ **3.** $26 - 17$ **4.** $18 - 33$

5. $(14)(-9)$ **6.** $(-27)(-17)$ **7.** $273 \div 7$ **8.** $744 \div (-6)$

9. Find the average of each set of numbers:
    a. $17, 14, 20$
    b. $-18, -22, -27, -37$

10. Find the powers:
    a. 3^4
    b. $(-8)^3$

11. Write each expression in exponential form:
    a. $2 \cdot 2 \cdot 2 \cdot a \cdot a \cdot b$
    b. $3 \cdot 3 \cdot 5 \cdot 5 \cdot a \cdot b \cdot b \cdot b$

12. Find the prime factorization of each number:
    a. 60
    b. 153

Find the LCM for each set of numbers or expressions in Exercises 13 and 14.

13. $18, 27, 36$ **14.** $12x, 9xy, 24xy^2$

Name the property of real numbers illustrated in Exercises 15 − 20.

15. $(3x)y = 3(xy)$ **16.** $x \cdot 5 = 5 \cdot x$ **17.** $4(3x + 1) = 12x + 4$

18. $x + 17 = 17 + x$ **19.** $(y + 11) + 4 = y + (11 + 4)$ **20.** $6 \cdot 1 = 6$

21. List the numbers that satisfy the equation $|x| = 2.3$.

22. Graph the integers on a real number line that satisfy the inequality $|x| < 2.3$.

Simplify Exercises 23 − 26 by using the rules for order of operations.

23. $3 \cdot 2^3 - 5$ **24.** $(18 + 2 \cdot 3) \div 4 \cdot 3$

25. $(13 \cdot 5 - 5) \div 2 \cdot 3$ **26.** $4[6 - (2 \cdot 5 - 2 \cdot 9) - 11]$

Perform the indicated operations in Exercises 27 − 34. Reduce each answer to lowest terms.

27. $\dfrac{3}{8} + \dfrac{7}{10}$ **28.** $\dfrac{5}{12} - \dfrac{4}{9}$ **29.** $\dfrac{4}{5y} + \dfrac{2}{3y}$ **30.** $\dfrac{7}{4a} - \dfrac{9}{6a}$

31. $\dfrac{3}{x} + \dfrac{1}{6}$ **32.** $\dfrac{11}{9} \div \dfrac{22}{15}$ **33.** $\dfrac{7b}{15} \div \dfrac{2b}{6}$ **34.** $\dfrac{8x}{21} \cdot \dfrac{7}{12x}$

Simplify the expressions in Exercises 35 – 40 by using the rules for order of operations. Reduce all answers to lowest terms.

35. $\dfrac{1}{2} + \dfrac{3}{8} \div \dfrac{3}{4} - \dfrac{5}{6}$

36. $\left(\dfrac{3}{4} - \dfrac{5}{6}\right) \cdot \left(\dfrac{6}{5} + \dfrac{2}{3}\right)$

37. $\dfrac{7}{18} + \dfrac{5}{24} - \left(\dfrac{3}{8}\right)^2$

38. $\left(\dfrac{3}{5}\right)^2 \div \dfrac{8}{5} \cdot \dfrac{8}{3}$

39. $\dfrac{1}{3} + \dfrac{2}{5} \cdot \dfrac{5}{8} \div \dfrac{3}{2} - \dfrac{1}{2}$

40. $\left(\dfrac{1}{3} - \dfrac{3}{4}\right) \div \left(\dfrac{2}{3} + \dfrac{2}{7}\right)$

Perform the indicated operations in Exercises 41 – 45.

41. $15.8 + 9.1 - 7.63$

42. $19.31 - 34.62$

43. $24.3 + 6.81 - 16.51$

44. $(25.3)(2.3)$

45. $(14.7)(0.36)$

In Exercises 46 – 49, first (a) simplify each expression, and then (b) evaluate the expression for $x = 5$ and $y = -2$.

46. $|x + y| + |x - y|$

47. $3|x| + 5|y| - 2(x - y)$

48. $3x + 2 - x$

49. $3y^2 - 5y + 7 - y^2 + 2y - 10 + y^3$

50. Translate each English phrase into an algebraic expression:
   a. Twice a number decreased by 3
   b. The quotient of a number and 6 increased by 3 times the number
   c. Thirteen less than twice the sum of a number and 10

51. Translate each algebraic expression into an English phrase:
   a. $6 + 4n$
   b. $18(n - 5)$

52. If the difference between $\dfrac{1}{2}$ and $\dfrac{11}{18}$ is divided by the sum of $\dfrac{2}{3}$ and $\dfrac{7}{15}$, what is the quotient?

53. From the sum of -2 and -11, subtract the product of -4 and 2.

54. Lucia is making punch for a party. She is making 5 gallons. The recipe calls for $\dfrac{2}{3}$ cup of sugar per gallon of punch. How many cups of sugar does she need?

55. Find the mean (to the nearest tenth) of the data in the following frequency distribution.

Clocked Speeds of Bicycles

(to nearest mph)	Frequency
19	4
20	5
22	7
25	3
28	1

56. There are 4000 registered voters in Greenville, and $\frac{3}{8}$ of these voters are registered Republicans. A survey indicates that $\frac{2}{3}$ of the registered Republicans are in favor of Bond Measure XX and $\frac{3}{5}$ of the other registered voters are in favor of this measure.

a. How many of the registered Republicans are in favor of Measure XX?
b. How many of the registered voters are in favor of Measure XX?

Perform the indicated operations in Exercises 57 – 59.

57. $8.6x + 2.3x - 5x$ **58.** $\frac{1}{7}a + \frac{1}{14}a - \frac{1}{21}a$ **59.** $\frac{3x}{5} + \frac{2x}{5} - \frac{x}{15}$

Simplify each of the following expressions.

60. **a.** $5x + 5x$ **b.** $5x \cdot 5x$ **c.** $\frac{1}{5x} + \frac{1}{5x}$ **d.** $\frac{1}{5x} \cdot \frac{1}{5x}$ **e.** $\frac{1}{5x} \div \frac{1}{5x}$

61. **a.** $6y - 6y$ **b.** $6y(-6y)$ **c.** $\frac{1}{6y} + \frac{1}{6y}$ **d.** $\frac{1}{6y}\left(-\frac{1}{6y}\right)$ **e.** $\frac{1}{6y} \div \left(-\frac{1}{6y}\right)$

Solving Equations and Inequalities

Did You Know?

Traditionally, algebra has been defined to mean generalized arithmetic where letters represent numbers. For example, $3 + 3 + 3 + 3 = 4 \cdot 3$ is a special case of the more general algebraic statement that $x + x + x + x = 4 \cdot x$. The name algebra comes from an Arabic word, al-jabr, which means "to restore."

A notorious Moslem ruler, Harun al-Rashid, the caliph made famous in *Tales of the Arabian Nights,* and his son Al-Mamun brought to their Baghdad court many renowned Moslem scholars. One of these scholars was the mathematician Mohammed ibn-Musa al-Khowarizmi, who wrote a text (c. A.D. 800) entitled *Ihm al-jabar w' al-muqabal.* The text included instructions for solving equations by adding terms to both sides of the equation, thus "restoring" equality. The abbreviated title of Al-Khowarizmi's text, *Al-jabar,* became our word for equation solving and operations on letters standing for numbers, **algebra**.

Al-Khowarizmi's algebra was brought to western Europe through Moorish Spain in a Latin translation done by Robert of Chester (c. A.D. 1140). Al-Khowarizmi's name may have sounded familiar to you. It eventually was translated as "algorithm," which came to mean any series of steps used to solve a problem. Thus we speak of the division algorithm used to divide one number by another. Al-Khowarizmi is known as the father of algebra just as Euclid is known as the father of geometry.

One of the hallmarks of algebra is its use of specialized notation that enables complicated statements to be expressed using compact notation. Al-Khowarizmi's algebra used only a few symbols with most statements written out in words. He did not even use symbols for numbers! The development of algebraic symbols and notation occurred over the next 1000 years.

"Algebra is generous. She often gives more than is asked of her."

Jean le Rond d' Alembert (1717? – 1783)

$\mathbf{N}$ow you are ready to learn how to solve equations. You will also find a variety of useful formulas and use your equation-solving skills in manipulating these formulas. Of course, solving equations and working with formulas are useful skills for solving word problems. Remember that one of the major goals of mathematics is to develop the skills that will allow you, with reasoning, to solve a wide variety of problems you may see in your lifetime.

3.1   Solving Linear Equations: $x + b = c$ and $ax = c$

Objectives

After completing this section, you will be able to:

1. *Solve linear equations of the form $x + b = c$.*

2. *Solve linear equations of the form $ax = c$.*

3. *Solve applications involving linear equations.*

Many practical applications involve setting up an equation (or several equations) involving an unknown quantity. The objective is to find the value of this unknown so that the question asked in the application is answered. This means that before we can solve meaningful word problems, one of the basic skills needed is solving equations. In this section, we will discuss solving linear equations in the following two forms:

$$x + b = c \qquad \text{and} \qquad ax = c.$$

Then, in the following section, we will combine the techniques learned here and discuss solving linear equations in general of the form:

$$ax + b = c.$$

An **equation** is a statement that two algebraic expressions are equal. That is, both expressions represent the same number. If an equation contains a variable, the value (or set of values) that gives a true statement when substituted for the variable is called the **solution** (or **solution set**) to the equation. The process of finding the solution is called **solving the equation**.

Linear Equation in x

*If a, b, and c are constants and $a \neq 0$, then a **linear equation in x** (or **first-degree equation in x**) is an equation that can be written in the form*

$$ax + b = c.$$

Note: *The term **first-degree** is sometimes used because the variable x is understood to have an exponent of 1. That is, $x = x^1$. Also, variables other than x, such as y or z, may be used.*

All of the following equations are linear equations in one variable:

$$2x + 3 = 7, \qquad 5y + 6 = -13, \qquad z + 5 = 3z - 5.$$

Solving Equations of the Form $x + b = c$

To begin, we need the **Addition Principle of Equality**.

Addition Principle of Equality

If the same algebraic expression is added to both sides of an equation, the new equation has the same solutions as the original equation. Symbolically, if A, B and C are algebraic expressions, then the equations

$$A = B$$

and

$$A + C = B + C$$

have the same solutions.

Equations with the same solutions are said to be **equivalent**.

The objective of solving linear (or first-degree) equations is to get the variable by itself on one side of the equation and any constants on the other side. The following procedure will accomplish this.

Procedure for Solving an Equation that Simplifies to the Form $x + b = c$

1. *Combine any like terms on each side of the equation.*
2. *Use the Addition Principle of Equality and add the opposite of a constant term or a variable term (or both) to both sides. The objective is to isolate the variable on one side of the equation with a coefficient of +1.*
3. *Check your answer by substituting it into the original equation.*

Every linear equation has exactly one solution. This means that once we have found a solution, there is no need to search for another solution.

Example 1: Solving $x + b = c$ ●●●●●●●●●●●●●●●●●

Solve each of the following linear equations:

a. $y - 8 = -2$

    Solution: $y - 8 + 8 = -2 + 8$ Add 8, the opposite of -8, to both sides.

$$y = 6$$ Simplify.

    Check: $y - 8 = -2$

$$6 - 8 \overset{?}{=} -2$$ Substitute $y = 6$.

$$-2 = -2$$ True statement.

b. $x - \dfrac{2}{5} = \dfrac{3}{10}$

    Solution: $x - \dfrac{2}{5} = \dfrac{3}{10}$

$$x - \frac{2}{5} + \frac{2}{5} = \frac{3}{10} + \frac{2}{5}$$ Add $\dfrac{2}{5}$, the opposite of $-\dfrac{2}{5}$ to both sides.

$$x = \frac{3}{10} + \frac{4}{10}$$ Simplify. (The common denominator is 10.)

$$x = \frac{7}{10}$$ Simplify.

    Check: $x - \dfrac{2}{5} = \dfrac{3}{10}$

$$\frac{7}{10} - \frac{2}{5} \overset{?}{=} \frac{3}{10}$$ Substitute $x = \dfrac{7}{10}$.

$$\frac{7}{10} - \frac{4}{10} \overset{?}{=} \frac{3}{10}$$ Simplify.

$$\frac{3}{10} = \frac{3}{10}$$ True statement.

As Examples c and d illustrate, the Addition Principle may be applied by adding variable expressions as well as constants to both sides of an equation. Remember that the objective is to isolate the variable on one side of the equation, either the right side or the left side, with a coefficient of +1.

Continued on next page...

c. $4x + 1 - x = 2x - 13 + 5$

Solution: $4x + 1 - x = 2x - 13 + 5$

$3x + 1 = 2x - 8$ Combine like terms on each side.

$3x + 1 + (-1) = 2x - 8 + (-1)$ Add -1, the opposite of $+1$, to both sides.

$3x + 1 - 1 = 2x - 8 - 1$

$3x = 2x - 9$ Simplify.

$3x + (-2x) = 2x - 9 + (-2x)$ Add $-2x$, the opposite of $+2x$, to both sides.

$x = -9$ Simplify.

Check: $4x + 1 - x = 2x - 13 + 5$

$4(-9) + 1 - (-9) \overset{?}{=} 2(-9) - 13 + 5$ Substitute $x = -9$.

$-36 + 1 + 9 \overset{?}{=} -18 - 13 + 5$ Simplify.

$-26 = -26$ True statement.

d. $6y + 2.5 = 7y - 3.6$

Solution: $6y + 2.5 = 7y - 3.6$

$6y + 2.5 + 3.6 = 7y - 3.6 + 3.6$ Add 3.6, the opposite of -3.6, to both sides.

$6y + 6.1 = 7y$ Simplify.

$6y + 6.1 - 6y = 7y - 6y$ Add $-6y$, the opposite of $6y$, to both sides.

$6.1 = y$ Simplify.

Check: $6y + 2.5 = 7y - 3.6$

$6(6.1) + 2.5 \overset{?}{=} 7(6.1) - 3.6$ Substitute $y = 6.1$.

$36.6 + 2.5 \overset{?}{=} 42.7 - 3.6$ Simplify.

$39.1 = 39.1$ True statement.

●●●●●●●●●●●●●●●●●●●●●●●●●●●●●●●●

Solving Equations of the Form $ax = c$

To solve equations of the form $ax = c$, where $a \neq 0$, we can use the idea of the reciprocal of the coefficient a. For example, as we studied in Section 2.1, the reciprocal

of $\dfrac{3}{4}$ is $\dfrac{4}{3}$ and $\dfrac{3}{4} \cdot \dfrac{4}{3} = 1$. Also, we need the **Multiplication Principle of Equality**.

Multiplication Principle of Equality

If both sides of an equation are multiplied by (or divided by) the same nonzero constant, the new equation has the same solutions as the original equation. Symbolically, if A and B are algebraic expressions and C is any nonzero constant, then the equations

$$A = B$$
$$\text{and} \quad AC = BC \qquad \text{where } C \neq 0$$
$$\text{and} \quad \frac{A}{C} = \frac{B}{C} \qquad \text{where } C \neq 0$$

have the same solutions. We say that the equations are equivalent.

Remember that the objective is to get the variable by itself on one side of the equation. That is, we want the variable to have +1 as its coefficient. The following procedure will accomplish this.

Procedure for Solving an Equation That Simplifies to the Form $ax=c$

1. *Combine any like terms on each side of the equation.*
2. *Use the Multiplication Principle of Equality and multiply both sides of the equation by the reciprocal of the coefficient of the variable. (**Note:** This is the same as dividing both sides of the equation by the coefficient.) Thus, the coefficient of the variable will become +1.*
3. *Check your answer by substituting it into the original equation.*

● ● ● ● ● ● ● ● ● ● ● ● ● ● ● ● ● ●

Solve the following equation.

a. $5x = 20$

Solution: $\qquad 5x = 20$

$$\frac{1}{5} \cdot (5x) = \frac{1}{5} \cdot 20 \quad \text{Multiply by } \frac{1}{5}, \text{ the reciprocal of 5.}$$

$$\left(\frac{1}{5} \cdot 5\right)x = \frac{1}{5} \cdot \frac{20}{1} \quad \text{Use the associative property of multiplication.}$$

$$1x = 4 \qquad \text{Simplify.}$$

$$x = 4$$

Check: $\qquad 5x = 20$

$$5 \cdot 4 \overset{?}{=} 20 \qquad \text{Substitute } x = 4.$$

$$20 = 20 \qquad \text{True statement.}$$

Multiplying by the reciprocal of the coefficient is the same as **dividing** by the coefficient itself. So, we can multiply both sides by $\frac{1}{5}$, as we did, or we can divide both sides by 5. In either case, the coefficient of x becomes +1.

$$5x = 20$$

$$\frac{5x}{5} = \frac{20}{5} \qquad \text{Divide both sides by 5.}$$

$$x = 4 \qquad \text{Simplify.}$$

b. $1.1x + 0.2x = 12.2 - 3.1$

Solution: When decimal coefficients or constants are involved, you might want to use a calculator to perform some of the arithmetic.

$$1.1x + 0.2x = 12.2 - 3.1$$

$$1.3x = 9.1 \qquad \text{Combine like terms.}$$

$$\frac{1.3x}{1.3} = \frac{9.1}{1.3} \qquad \text{Use a calculator or pencil and paper to divide.}$$

$$x = 7.0$$

Check: $\qquad 1.1x + 0.2x = 12.2 - 3.1$

$$1.1(\,7\,) + 0.2(\,7\,) \overset{?}{=} 12.2 - 3.1 \qquad \text{Substitute } x = 7.$$

$$7.7 + 1.4 \overset{?}{=} 9.1 \qquad \text{Simplify.}$$

$$9.1 = 9.1 \qquad \text{True statement.}$$

c. $\dfrac{4x}{5} = \dfrac{3}{10}$ (This could be written $\dfrac{4}{5}x = \dfrac{3}{10}$ since $\dfrac{4}{5}x$ is the same as $\dfrac{4x}{5}$.)

Solution: $\dfrac{4x}{5} = \dfrac{3}{10}$

$\dfrac{5}{4} \cdot \dfrac{4}{5}x = \dfrac{5}{4} \cdot \dfrac{3}{10}$ Multiply both sides by $\dfrac{5}{4}$.

$1x = \dfrac{1 \cdot \cancel{5}}{4} \cdot \dfrac{3}{2 \cdot \cancel{5}}$ Simplify.

$x = \dfrac{3}{8}$

Check: $\dfrac{4x}{5} = \dfrac{3}{10}$

$\dfrac{4}{5} \cdot \dfrac{3}{8} \overset{?}{=} \dfrac{3}{10}$ Substitute $x = \dfrac{3}{8}$.

$\dfrac{3}{10} = \dfrac{3}{10}$ True statement.

d. $-x = 4$

Solution: $-x = 4$

$-1x = 4$ -1 is the coefficient of x.

$\dfrac{-1x}{-1} = \dfrac{4}{-1}$ Divide by -1 so that the coefficient is $+1$.

$x = -4$

Check: $-x = 4$

$-(-4) \overset{?}{=} 4$ Substitute $x = -4$.

$4 = 4$ True statement.

● ●

Problem Solving

George Pòlya (1877 – 1985), a famous professor at Stanford University, studied the process of discovery learning. Among his many accomplishments, he developed the following four-step process as an approach to problem solving:

1. Understand the problem.
2. Devise a plan.
3. Carry out the plan.
4. Look back over the results.

For a complete discussion of these ideas, see *How To Solve It by Pòlya* (Princeton University Press, 2nd edition, 1957). The following quote by Pòlya illustrates his sense of humor and his understanding of students' dilemmas: "The traditional mathematics professor of the popular legend is absent-minded. He usually appears in public with a lost umbrella in each hand. He prefers to face a blackboard and to turn his back on the class. He writes *a*, he says *b*, he means *c*, but it should be *d*. Some of his sayings are handed down from generation to generation."

There are a variety of types of applications discussed throughout this text and subsequent courses in mathematics, and you will find these four steps helpful as guidelines for understanding and solving all of them. Applying the necessary skills to solve exercises, such as combining like terms or solving equations, is not the same as accumulating the knowledge to solve problems. **Problem solving can involve careful reading, reflection, and some original or independent thought**.

Basic Steps for Solving Applications

1. Understand the problem. For example,

    a. Read the problem carefully, maybe several times.

    b. Understand all the words.

    c. If it helps, restate the problem in your own words.

    d. Be sure that there is enough information.

2. Devise a plan. For example,

    a. Guess, estimate, or make a list of possibilities.

    b. Draw a picture or diagram.

    c. Represent the unknown quantity with a variable and form an equation.

3. Carry out the plan. For example,

    a. Try all the possibilities you have listed.

    b. Study your picture or diagram for insight into the solution.

    c. Solve any equation that you may have set up.

4. Look back over the results. For example,

    a. Can you see an easier way to solve the problem?

    b. Does your solution actually work? Does it make sense in terms of the wording of the problem? Is it reasonable?

    c. If there is an equation, check your answer in the equation.

NOTES

You may find that many of the applications in this section can be solved by "reasoning," and there is nothing wrong with that approach. Reasoning is a fundamental part of all of mathematics. However, keep in mind that the algebraic techniques you are learning are important. They also involve reasoning and will prove very useful in solving more complicated problems in later sections and in later courses.

Example 3: Applications

a. When a $2.13 tax was added to the price of a skirt, the total bill was $37.63. What was the price of the skirt?

Solution: Let x = price of the skirt
Now, use the relationship: price of skirt + tax = total bill
Set up the equation:

$$x + 2.13 = 37.63$$

Solve the equation:

$x + 2.13 - 2.13 = 37.63 - 2.13$ Use the Addition
Principle by adding
-2.13 to both sides.
         $x = 35.50$ Simplify both sides.

The price of the skirt was $35.50.

b. The original price of a VCR was reduced by $45.50. The sale price was $215.90. What was the original price?

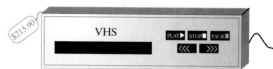

Solution: Let y = price of the VCR.
Now, use the relationship: price of VCR − reduction = sale price
Set up the equation:
     $y - 45.50 = 215.90$
Solve the equation:
         $y - 45.50 + 45.50 = 215.90 + 45.50$ Use the Addition Principle by adding 45.50 to both sides.

               $y = 261.40$ Simplify both sides.

The original price of the VCR was $261.40.

Continued on next page...

c. An exam is given with 15 problems and the students are allowed $1\frac{1}{2}$ hours to take it. How many minutes are allotted for each problem?

Solution: Here we let x = number of minutes allotted per problem. Then the product $15x$ will equal the total time for the exam. Since the time allowed is given in hours, we make the change

$$1\frac{1}{2} \text{ hours } = 90 \text{ minutes.}$$

Then the equation relating the problems and the time is

$$15x = 90$$
$$\frac{15x}{15} = \frac{90}{15}$$
$$x = 6$$

Each problem is allotted 6 minutes.

Practice Problems

Solve the following equations.

1. $-16 = x + 5$ **2.** $6y - 1.5 = 7y$ **3.** $4x = -20$

4. $\frac{3}{5}y = 33$ **5.** $1.7z + 2.4z = 8.2$ **6.** $-x = -8$

7. $3x = 10$ **8.** $5x - 4x + 1.6 = -2.7$

Answers to Practice Problems: 1. $-21 = x$ **2.** $-1.5 = y$ **3.** $x = -5$ **4.** $y = 55$ **5.** $z = 2$ **6.** $x = 8$ **7.** $x = \frac{10}{3}$ **8.** $x = -4.3$

3.1 Exercises

Solve each of the following equations in Exercises 1 – 25. (Remember that the objective is to isolate the variable on one side of the equation with a coefficient of 1.)

1. $y - 5 = 1$

2. $x + 14 = 23$

3. $n + 3 = -7$

4. $x + 3.6 = 2.4$

5. $2y = 10.4$

6. $-8x = 24$

7. $x + 8x = 12 + 15$

8. $10x - 2x = 36 - 100$

9. $13x - 12x - 4 = 21$

10. $-2x - 9 + 5x - 2 = 4x$

11. $7a - 6a + 17 = 17$

12. $3n - n + 14 - 2 - 12 = 3n$

13. $\dfrac{2}{3} = 5x - 4x + \dfrac{1}{6}$

14. $10x - 9x - \dfrac{1}{2} = -\dfrac{9}{10}$

15. $\dfrac{3x}{4} = 15$

16. $\dfrac{2y}{5} = 4$

17. $\dfrac{5}{2}x + 2x = 15 - 33$

18. $\dfrac{5}{3}x + \dfrac{2}{3}x = 21 + 7$

19. $6.4y + 8.8 = 2.0y$

20. $8x = 8.2x + 22.6$

21. $2.9a = 5a - 2.1$

22. $1.5x = 6.5x + 4.6 - 30.1$

23. $-\dfrac{1}{8}x - \dfrac{1}{4}x = \dfrac{5}{8} - \dfrac{1}{4}$

24. $\dfrac{5}{6}y + \dfrac{2}{3} = \dfrac{1}{2}y$

25. $5a + 12a - 6.2 = 8a + 8a$

Problems Related to Geometry

The perimeter, P, of a triangle is equal to the sum of the lengths of the sides, $a, b,$ and c. ($P = a + b + c$)

26. One side of a triangle is 11.5 inches long. A second side is 8.75 inches long. If the perimeter is 24 in., find the length of the third side.

27. The perimeter of a triangle is 71 meters. If the sides are related as shown in the following figure, find the length of each side.

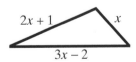

28. Two sides of a triangle measure 43.2 cm and 26.1 cm. Find the length of the third side if the perimeter is 98 cm.

29. The perimeter of a square is 4 times the length of a side. ($P = 4s$) Find the length of a side if the perimeter of a square is $64\frac{1}{2}$ meters.

side

30. The area of a rectangle is found by multiplying the length times the width. ($A = lw$) Find the width of a rectangle with area 75 m² and length 10 m.

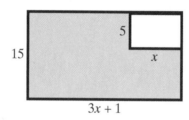

Area = length · width   width

length

31. Find the value of x in the figure if the area of the shaded portion is 95 inches².

5

15

x

$3x + 1$

32. The volume of a rectangular box of cereal is the product of its length, width, and height. ($V = lwh$) Find the width of cereal box with volume 1260 cm³ if its length is 14 cm and its height is 20 cm.

volume $= 1260\text{cm}^3$

14

CEREAL

20

33. The area of a triangle is found by multiplying $\frac{1}{2}$ times the base times the height. $(A = \frac{1}{2}bh)$ Find the length of the base of a triangle with area 180 in.^{2} and height 10 in.

$$A = \frac{1}{2}bh$$

height

base

34. A sail is in the shape of a triangle. Find the height of the sail if its base is 20 ft. and its area is 300 ft^2.

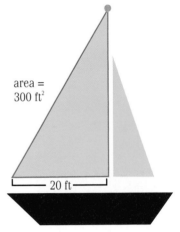

area = 300 ft^2

20 ft

The sum of the measures of the three angles of a triangle is 180°. $(\alpha + \beta + \gamma = 180°)$ (**Note:** α, β, and γ are the Greek lowercase letters alpha, beta, and gamma, respectively.)

35. A triangle with two equal angles is called an **isosceles** triangle. If two angles of an isosceles triangle both measure 45°, is the triangle a **right** triangle? (Does the third angle measure 90°?) Draw a sketch of such a triangle.

36. If all three angles of a triangle measure less than 90°, then the triangle is called an **acute** triangle. Is a triangle with two angles of measure 55° and 20° an acute triangle? What is the measure of the third angle? (**Note:** If a triangle is not acute and not right, then the measure of one of the angles is more than 90°, and it is called an **obtuse** triangle.)

Problems Related to Business

Simple interest, I, is the product of the money invested or principal, P, the annual interest rate, r, and the time, t. ($I = Prt$) The time used in calculating simple interest is one year or less. Use 12 months in a year and 365 days in a year. (If the time is more than one year, then interest is calculated on the interest and a formula for compound interest is used.)

37. The interest earned on an investment of $3240 at 10% was $81. What was the time that the money was invested?

38. A savings account pays an annual interest rate of 5%. How much must be invested to make $400 in interest in 9 months?

39. If you have a credit card debt of $2000 and the simple interest for one month is $40, what is the interest rate that you are paying? (**Note**: Credit card companies usually charge compound interest, particularly on overdue payments.) If you continue to pay only the interest each month for 5 years and then pay off the $2000, how much will you have paid for the $2000 credit card debt?

The profit, P, is equal to the revenue, R, minus the cost, C. ($P = R - C$)

40. Find the revenue (income) of a company that shows a profit of $3.2 million and costs of $1.8 million.

41. A suit cost the store $675 and the owner said that he wants to make a profit of $180. What price should he mark on the suit?

42. An automobile is advertised for a price of $22,500, but when the customer went to buy the car with options that she wanted, the total price came to $25,300 (plus tax). What was the cost of the options?

43. A computer was advertised on sale for $1560. This was a reduction of $130 from the original price. What was the original price?

The distance traveled, d, is equal to the product of the rate, r and the time, t. ($d = rt$)

44. How long will a truck driver take to travel 350 miles if he averages 50 miles per hour?

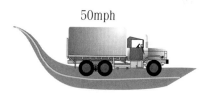

45. How long will it take a train traveling at 40 miles per hour to go 140 miles?

Calculator Problems

Use a calculator to solve Exercises 46 – 50.

46. $y + 32.861 = -17.892$

47. $17.61x + 27.059 = 9.845 + 16.61x$

48. $14.38y - 8.65 + 9.73y = 17.437 + 23.11y$

49. $2.637x = 648.702$

50. $-0.3057y = 316.7052$

Hawkes Learning Systems: Introductory Algebra

Solving Linear Equations Using Addition and Subtraction
Applications of Linear Equations: Addition and Subtraction
Solving Linear Equations Using Multiplication and Division
Applications of Linear Equations: Multiplication and Division

3.2

Solving Linear Equations: $ax + b = c$

After completing this section, you will be able to:

1. *Solve equations of the form $ax + b = c$.*

2. *Supply the reasons for each step in solving an equation.*

Now we are ready to solve equations of the general form $ax + b = c$. That is, we are ready to apply both the Addition Principle and the Multiplication Principle in solving one equation. The initial equation may be of any one of several forms with variables on both sides, parentheses, fractions, or decimals. Our goal is to make the process as simple as possible.

General Procedure for Solving Linear Equations

1. *Simplify each side of the equation by removing any grouping symbols and combining like terms. (This may also involve multiplying both sides by the LCM of the denominators or by a power of 10 to remove decimals. In this manner we will be dealing only with integer constants and coefficients.)*

2. *Use the Addition Principle of Equality to add the opposites of constants and/or variables so that variables are on one side and constants are on the other side.*

3. *Use the Multiplication Principle of Equality to multiply both sides by the reciprocal of the coefficient (or to divide by the coefficient).*

4. *Check your answer by substituting it into the original equation.*

Study each of the following examples carefully. Remember that **the objective is to get the variable on one side by itself with a coefficient of 1**. Note that the equations are written one under the other and the equal signs are aligned vertically.

$$(3n + 4) + n = 24 \qquad \text{Their sum is 24.}$$
$$4n + 4 = 24$$
$$4n + 4 - 4 = 24 - 4$$
$$4n = 20$$
$$\frac{4n}{4} = \frac{20}{4}$$
$$n = 5$$
$$3n + 4 = 19$$

The two integers are 5 and 19.

d. Joe pays $300 per month to rent an apartment. If this is $\frac{2}{5}$ of his monthly income, what is his monthly income?

Solution: Let x = Joe's monthly income.

$$\frac{2}{5}x = 300$$
$$\frac{5}{2} \cdot \frac{2}{5}x = \frac{5}{2} \cdot \frac{300}{1}$$
$$x = 750$$

Joe's monthly income is $750.

Consecutive Integers

Remember that the set of **integers** consists of the whole numbers and their opposites:

$$\{ \dots, -4, -3, -2, -1, 0, 1, 2, 3, 4, \dots \}$$

Even integers are integers that are divisible by 2. The even integers are

$$E = \{ \dots, -6, -4, -2, 0, 2, 4, 6, \dots \}$$

Odd integers are integers that are not even. If an odd integer is divided by 2 the remainder will be 1. The odd integers are

$$O = \{ \dots, -5, -3, -1, 1, 3, 5, \dots \}$$

In this discussion we will be dealing only with integers. Therefore, if you get a result that has a fraction or decimal number (not an integer), you will know that an error has been made and you should correct some part of your work.

The following terms and the ways of representing the integers must be understood before attempting the problems.

Consecutive Integers

*Integers are **consecutive** if each is 1 more than the previous integer. Three consecutive integers can be represented as*

$$n, \ n+1, \ and \ \ n+2$$

Consecutive Odd Integers

*Odd integers are **consecutive** if each is 2 more than the previous odd integer. Three consecutive odd integers can be represented as*

$$n, \ n+2, \ and \ n+4$$

*where n is an **odd** integer.*

Consecutive Even Integers

*Even integers are **consecutive** if each is 2 more than the previous even integer. Three consecutive even integers can be represented as*

$$n, \ n+2, \ and \ n+4$$

*where n is an **even** integer.*

Note that consecutive even and consecutive odd integers are represented in the same way:

$$n, \ n+2, \ and \ \ n+4$$

The value of the first integer, n, determines whether the remaining integers are odd or even. For example,

<table>
<tr><td><u>***n* is odd**</u></td><td></td><td><u>***n* is even**</u></td></tr>
<tr><td>If $n = 11$</td><td>**or**</td><td>If $n = 36$</td></tr>
<tr><td>then $n + 2 = 13$</td><td></td><td>then $n + 2 = 38$</td></tr>
<tr><td>and $n + 4 = 15$</td><td></td><td>and $n + 4 = 40$</td></tr>
</table>

Example 2: Consecutive Integers ● ● ● ● ● ● ● ● ● ● ● ● ● ● ●

a. Find three consecutive integers such that the sum of the first and third is 76 less than three times the second.

Solution: Let n = the first integer
$n + 1$ = the second integer
$n + 2$ = the third integer

Set up and solve the related equation.

$$
\begin{aligned}
n + (n + 2) &= 3(n + 1) - 76 \\
2n + 2 &= 3n + 3 - 76 \\
2n + 2 &= 3n - 73 \\
2n + 2 + 73 - 2n &= 3n - 73 + 73 - 2n \\
75 &= n \\
76 &= n + 1 \\
77 &= n + 2
\end{aligned}
$$

The three consecutive integers are 75, 76, and 77.

Check: $75 + 77 = 152$ and $3(76) - 76 = 228 - 76 = 152$

b. Three consecutive odd integers are such that their sum is −3. What are the integers?

Solution: Let n = the first odd integer
$n + 2$ = the second odd integer
$n + 4$ = the third odd integer

Set up and solve the related equation.

$$
\begin{aligned}
n + (n + 2) + (n + 4) &= -3 \\
3n + 6 &= -3 \\
3n &= -9 \\
n &= -3 \\
n + 2 &= -1 \\
n + 4 &= 1
\end{aligned}
$$

The three consecutive odd integers are −3, −1, and 1.

Check: $(-3) + (-1) + (1) = -3$

Continued on next page...

c. Find three consecutive even integers such that three times the first is 10 more than the sum of the second and third.

Solution: Let n = the first even integer
$n + 2$ = the second even integer
$n + 4$ = the third even integer

Set up and solve the related equation.

$$3n = (n + 2) + (n + 4) + 10$$
$$3n = 2n + 16$$
$$3n - 2n = 2n + 16 - 2n$$
$$n = 16$$
$$n + 2 = 18$$
$$n + 4 = 20$$

The three even integers are 16, 18, and 20.
(Checking shows that $3 \cdot 16$ is 10 more than $18 + 20$.)

• •

3.3 Exercises

Read each problem carefully, translate the various phrases into algebraic expressions, set up an equation, and solve the equation.

1. Five less than a number is equal to 13 decreased by the number. Find the number.

2. Three less than twice a number is equal to the number. What is the number?

3. Thirty-six is 4 more than twice a certain number. Find the number.

4. Fifteen decreased by twice a number is 27. Find the number.

5. Seven times a certain number is equal to the sum of twice the number and 35. What is the number?

6. The difference between twice a number and 3 is equal to 6 decreased by the number. Find the number.

7. Fourteen more than three times a number is equal to 6 decreased by the number. Find the number.

8. Two added to the quotient of a number and 7 is equal to –3. What is the number?

9. The quotient of twice a number and 5 is equal to the number increased by 6. What is the number?

10. Three times the sum of a number and 4 is equal to –9. Find the number.

11. Four times the difference between a number and 5 is equal to the number increased by 4. What is the number?

12. When 17 is added to six times a number, the result is equal to 1 plus twice the number. What is the number?

13. If the sum of twice a number and 5 is divided by 11, the result is equal to the difference between 4 and the number. Find the number.

14. Twice a number increased by three times the number is equal to 4 times the sum of the number and 3. Find the number.

15. If 21 is subtracted from a number and the result is 8 times the number, what is the number?

16. Twice the difference between a number and 10 is equal to 6 times the number plus 14. What is the number?

17. The perimeter (distance around) of a triangle is 24 centimeters. If two sides are equal and the length of the third side is 4 centimeters, what is the length of each of the other two sides?

18. The length of a wire is bent to form a triangle with two sides equal. If the wire is 30 inches long and the two equal sides are each 9 inches long, what is the length of the third side?

19. Two sides of a triangle have the same length and the third side is 3 cm less than the sum of the other two sides. If the perimeter of the triangle is 45 cm, what are the lengths of the three sides?

20. The perimeter of a rectangular shaped swimming pool is 172 meters. If the width is 16 meters less than the length, how long are the width and length of the pool?

21. A classic car is now selling for $1500 more than three times its original price. If the selling price is now $12,000, what was the car's original price?

22. A real estate agent says that the current value of a home is $90,000 more than twice its value when it was new. If the current value is $310,000, what was the value of the home when it was new? The home is 25 years old.

23. The sum of two consecutive odd integers is 60. What are the integers?

24. Find three consecutive integers whose sum is 69.

25. Find two consecutive integers such that twice the first plus three times the second equals 83.

26. Find three consecutive even integers such that the first plus twice the second is 54 less than four times the third.

27. Find three consecutive even integers such that if the first is subtracted from the sum of the second and third, the result is 66.

28. Find three consecutive odd integers such that 4 times the first is 44 more than the sum of the second and third.

29. Find three consecutive even integers such that their sum is 168 more than the second.

30. Find four consecutive integers whose sum is 90.

31. The perimeter of a rectangular parking lot is 410 yards. If the width is 75 yards, what is the length of the parking lot? The concrete in the parking lot is 4 inches thick.

32. Twenty-four plus a number is equal to twice the number plus three times the same number. What is the number?

33. The width of a rectangular-shaped room is $\dfrac{2}{3}$ of the length. If the room is 18 ft. wide, find the length.

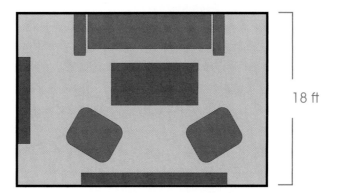

34. Joe Johnson decided to buy a lot and build a house on the lot. He knew that the cost of constructing the house was going to be $25,000 more than the cost of the lot. He told a friend that the total cost was going to be $275,000. As a test to see if his friend remembered the algebra they had together in school, he challenged his friend to calculate what he paid for the lot and what he was going to pay for the house. What was the cost of the lot and the cost of the house?

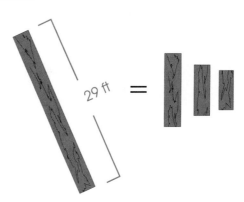

35. A 29 foot board is cut into three pieces at a sawmill. The second piece is 2 feet longer than the first and the third piece is 4 feet longer than the second. What is the length of each of the three pieces?

36. Lucinda bought two boxes of golf balls. She gave the pro-shop clerk a 50-dollar bill and received $10.50 in change. What was the cost of one box of golf balls? (Tax was included.)

37. A mathematics student bought a graphing calculator and a textbook for a course in statistics. If the text cost $49.50 more than the calculator, and the total cost for both was $125.74, what was the cost of each item?

38. The three sides of a triangle are x, $3x - 1$, and $2x + 5$ (as shown in the figure). If the perimeter of the triangle is 64 inches, what is the length of each side?

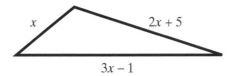

39. Find three consecutive odd integers such that the sum of twice the first and three times the second is 7 more than twice the third.

40. Find three consecutive even integers such that the sum of three times the first and twice the third is twenty less than six times the second.

Make up your own word problem that might use the given equation in its solution. Be creative! Then solve the equation and check to see that the answer is reasonable.

41. $n + (n + 1) = 33$ **42.** $3(n + 1) = n + 53$ **43.** $n + (n + 4) = 3(n + 2)$

44. $x - \dfrac{5}{6}x = \dfrac{5}{3}$ **45.** $\dfrac{x}{2} + \dfrac{1}{3} = \dfrac{3x}{4}$

Hawkes Learning Systems: Introductory Algebra

Applications: Number Problems and Consecutive Integers

Applications: Percent Problems (Discount, Taxes, Commission, Profit, and Others)

After completing this section, you will be able to:

1. *Change decimals to percents.*

2. *Change percents to decimals.*

3. *Find percents of numbers.*

4. *Find fractional parts of numbers.*

5. *Solve word problems involving decimals, fractions, and percents.*

Our daily lives are filled with decimal numbers and percents: stock market reports, batting averages, won-lost records, salary raises, measures of pollution, taxes, interest on savings, home loans, discounts on clothes and cars, and on and on. Since decimal numbers and percents play such a prominent role in everyday life, we need to understand how to operate with them and apply them correctly in a variety of practical situations.

Review of Percent

The word **percent** comes from the Latin *per centum*, meaning "per hundred." So, **percent means hundredths**. Thus, 72%, $\dfrac{72}{100}$, and 0.72 all have the same meaning.

Similarly,

$$\frac{65}{100} = 0.65 = 65\% \quad \text{and} \quad \frac{124}{100} = 1.24 = 124\%$$

The following basic steps are used to change decimals and fractions to percents and vice versa. You may be familiar with these steps from your previous work in arithmetic, but some review should be helpful.

To Change a Decimal to a Percent

Step 1: *Move the decimal point two places to the right. (This is the same as multiplying by 100.)*

Step 2: *Add the % symbol.*

To Change a Fraction (or Mixed Number) to a Percent

Step 1: *Change the fraction to a decimal. (Divide the numerator by the denominator. If the number is a mixed number, first write it as an improper fraction then divide.)*

Step 2: *Change the decimal to a percent.*

Example 1: Decimals to Percents ● ● ● ● ● ● ● ● ● ● ● ● ● ●

Change each of the following numbers to percents.

a. 0.036 **b.** 0.25 **c.** $\dfrac{5}{8}$ **d.** $2\dfrac{1}{3}$

Solution: **a.** 0.036 = 3.6%   Move the decimal point two places to the right and add the % sign.

b. 0.25 = 25%   Move the decimal point two places to the right and add the % symbol.

c. $\dfrac{5}{8} = 0.625 = 62.5\%$ Change the fraction to decimal form by dividing 5 by 8, move the decimal point two places to the right, and add the % sign.

d. $2\dfrac{1}{3} = 2.33\dfrac{1}{3} = 233\dfrac{1}{3}\%$ The fraction $\dfrac{1}{3}$ is not a terminating decimal.

● ●

To Change a Percent to a Decimal

> ***Step 1:*** *Move the decimal point two places to the left.*
>
> ***Step 2:*** *Delete the % symbol.*

To Change a Percent to a Fraction (or Mixed Number)

> ***Step 1:*** *Write the percent as a fraction with denominator 100 and delete the % symbol.*
>
> ***Step 2:*** *Reduce the fraction (or change it to a mixed number if you prefer with the fraction part reduced).*

Example 2: Percents to Decimals, Fractions, or Mixed Numbers

Change each of the following percents as indicated.

a. 47.5% to decimal form   **b.** 56% to fraction form reduced

c. 125% to mixed number form

Solution: **a.** $47.5\% = 0.475$ Move the decimal point two places to the left and delete the % sign.

b. $56\% = \dfrac{56}{100} = \dfrac{\cancel{4} \cdot 14}{\cancel{4} \cdot 25} = \dfrac{14}{25}$ Write the percent with denominator 100, delete the % symbol, and reduce.

c. $125\% = 1.25 = 1\dfrac{25}{100} = 1\dfrac{1}{4}$ Move the decimal point two places to the left, delete the % sign, change the decimal form to mixed number form and reduce.

The Basic Formula $R \cdot B = A$

Now, consider the statement

"15% of 80 is 12."

This statement has three numbers in it. In general, solving a percent problem involves knowing two of these numbers and trying to find the third. That is, **there are three basic types of percent problems**. We can break down the sentence in the following way.

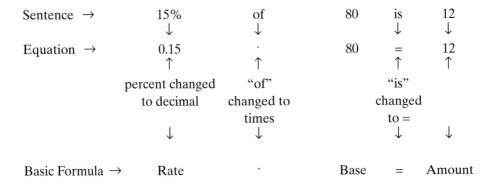

The terms that we have just discussed are explained in detail in the following box.

The Basic Formula $R \cdot B = A$

$R = $ **RATE** *or percent (as a decimal or fraction)*

$B = $ **BASE** *(number we are finding the percent of)*

$A = $ **AMOUNT** *or percentage (a part of the base)*

"of " means to multiply.

"is" means equal (=).

The relationship among R, B, and A is given in the formula

$$R \cdot B = A \quad (or \quad A = R \cdot B)$$

Even though there are just three basic types of percent problems, many people have difficulty deciding whether to multiply or divide in a particular problem. Using the formula $R \cdot B = A$ helps to avoid these difficulties. If the values of any two of the quantities in the formula are known, they can be substituted in the formula and then the missing number can be found by solving the equation. **The process of solving the equation for the unknown quantity determines whether multiplication or division is needed**.

The following examples illustrate how to substitute into the formula and how to solve the resulting equations.

Example 3: Percent of a Number

a. What is 72% of 800?

> **Solution:** $R = 0.72$ and $B = 800$ and A is unknown.
>
> $$R \cdot B = A$$
> $$\downarrow \quad \downarrow \quad \downarrow$$
> $$0.72 \cdot 800 = A \qquad \text{Here, simply multiply to find } A$$
> $$576 = A$$
>
> So, **576** is 72% of 800.

b. 57% of what number is 163.191?

> **Solution:** $R = 0.57$ and B is unknown and A is 163.191.
>
> $$R \cdot B = A$$
> $$\downarrow \quad \downarrow \quad \downarrow$$
> $$0.57 \cdot B = 163.191$$
> $$\frac{0.57 \cdot B}{0.57} = \frac{163.191}{0.57} \qquad \text{Now, divide both sides by 0.57 to find } B.$$
> $$B = 286.3$$
>
> So, 57% of **286.3** is 163.191.

Continued on next page...

c. What percent of 180 is 45?

Solution: R is the unknown and B is 180 and A is 45.

$$R \cdot B = A$$
$$\downarrow \quad \downarrow \quad \downarrow$$
$$R \cdot 180 = 45$$
$$\frac{R \cdot 180}{180} = \frac{45}{180} \qquad \text{Now, divide both sides by 180 to find } R.$$
$$R = 0.25$$
$$R = 25\% \qquad \text{Change the decimal to percent form.}$$

So, **25**% of 180 is 45.

● ●

Applications with Percent

To sell goods that have been in stock for some time or simply to attract new customers, retailers and manufacturers sometimes offer a **discount**, a reduction in the selling price usually stated as a percent of the original price. The new, reduced price is called the **sale price**.

Sales tax is a tax charged on goods sold by retailers, and it is assessed by states and cities for income to operate various services. The **rate of sales tax** (a percent) varies by location.

Example 4: Applications ● ● ● ● ● ● ● ● ● ● ● ● ● ● ● ● ● ●

A bicycle was purchased at a discount of 25% of its original price of $1600. If 6% sales tax was added to the purchase price, what was the total paid for the bicycle?

Solution: There are two ways to approach this problem. One way is to find the discount and then subtract this amount from $1600. Another way is to subtract 25% from 100% to get 75% and then find 75% of $1600. In either case, the sales tax must be calculated on the sale price and added to the sale price.

Continued on next page...

Here, we will calculate the discount and then subtract from the original price.

$$R \quad \cdot \quad B \quad = \quad A$$

$$\downarrow \qquad \downarrow \qquad \downarrow$$

$$0.25 \cdot 1600 = 400 \qquad \text{Discount}$$

Original Price – Discount = $1600 – $400 = $1200   Sale Price

Sales Tax = 6% of Sale Price = $0.06 \cdot \$1200 = \$72$

Total Paid = Sale Price + Sales Tax = $1200 + $72 = $1272

The total paid for the bicycle, including sales tax, was $1272.

● ●

A **commission** is a fee paid to an agent or salesperson for a service. Commissions are usually a percent of a negotiated contract (as to a real estate agent) or a percent of sales.

Example 5: Commission   ● ● ● ● ● ● ● ● ● ● ● ● ● ● ● ● ● ●

A saleswoman earns a salary of $1200 a month plus a commission of 8% on whatever she sells after she has sold $8000 in furniture. What did she earn the month she sold $25,000 worth of furniture?

Solution: First subtract $8000 from $25,000 to find the amount on which the commission is based.

$25,000 – $8000 = $17,000

Now, find the amount of the commission.

$$R \quad \cdot \quad B \quad = \quad A$$

$$\downarrow \qquad \downarrow \qquad \downarrow$$

$0.08 \cdot \$17,000 = \$1360$ commission

Now, add the commission to her salary to find what she earned that month.

$1200 + $1360 = $2560 earned for the month

● ●

Percent of profit for an investment is the ratio (a fraction) that compares the money made to the money invested. If you make two investments of different amounts of money, then the amount of money you make on each investment is not a fair comparison. In comparing such investments, the investment with the greater percent of profit is considered the better investment. **To find the percent of profit, form the fraction (or ratio) of profit divided by the investment and change the fraction to a percent**.

Example 6: Percent of Profit

Calculate the percent of profit for both a and b and tell which is the better investment.
a. $300 profit on an investment of $2400
b. $500 profit on an investment of $5000

Solution: Set up ratios and find the corresponding percents.

$$\textbf{a.} \quad \frac{\$300 \text{ profit}}{\$2400 \text{ invested}} = \frac{300 \cdot 1}{300 \cdot 8} = \frac{1}{8} = 0.125 = 12.5\% \qquad \text{percent of profit}$$

$$\textbf{b.} \quad \frac{\$500 \text{ profit}}{\$5000 \text{ invested}} = \frac{500 \cdot 1}{500 \cdot 10} = \frac{1}{10} = 0.1 = 10\% \qquad \text{percent of profit}$$

Clearly, $500 is more than $300, but 12.5% is greater than 10% and investment **a.** is the better investment.

Practice Problems

1. *Change 0.3 to a percent.*

2. *Change 6.4% to a decimal.*

3. *Change $\dfrac{1}{5}$ to a percent.*

4. *Find 12% of 200.*

5. *Find the percent of profit if $400 is made on an investment of $1500.*

Answers to Practice Problems: 1. 30% **2.** 0.064 **3.** 20% **4.** 24 **5.** 26.67%

3.4 Exercises

Change each of the following numbers to percent form.

1. 0.91   **2.** 0.625   **3.** 1.37   **4.** 0.0075

5. $\dfrac{3}{8}$ **6.** $\dfrac{87}{100}$ **7.** $1\dfrac{1}{2}$ **8.** $\dfrac{2}{3}$

Change each of the following percents to decimal form.

9. 69%   **10.** 7.5%   **11.** 11.3%   **12.** 162%
13. 0.5%   **14.** 235%   **15.** 82%   **16.** 31.4%

Change each of the following percents to fraction form (reduced).

17. 35%   **18.** 72%   **19.** 130%   **20.** 40%

21. Heather's monthly income is $2500. Using the given circle graph answer the following questions.
a. What percentage of her income does Heather spend on rent each month?
b. What percentage of her income is spent on food and entertaiment each month?
c. What percentage of her income does Heather save each month?

22. There were an estimated 20,832 earthquakes worldwide in 1999. Using the given graph answer the following questions. (Round answers to the nearest tenth of a percent.)
a. What percentage of earthquakes in 1999 were less than a magnitude of 4.0?
b. What percentage of earthquakes were of magnitude 3.0 - 3.9?
c. What percentage of earthquakes were 4.0 - 5.9?

Monthly Expenses

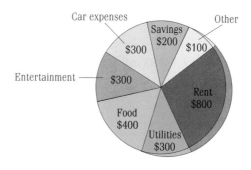

Magnitude of Earthquakes in 1999

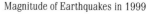

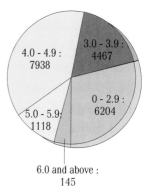

23. In 1998, the US adult population was estimated as 197.4 million. (Round each answer to the nearest tenth.)
a. What percentage of the US adult population was married in 1998?
b. What percentage of the adult population was either widowed or divorced?

24. In July, Daddy Fat's Record Shop sold 3900 CD's.
a. R&B and Rap made up what percentage of their sales?
b. What percentage of their sales was not Rock nor Country CD's?

Marital Status of the Population, 1998 (Data is in millions)

Daddy Fat's CD Sales for July

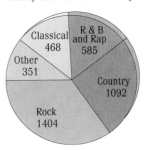

Find the missing rate, base, or amount.

25. 81% of 76 is _____.

26. 102% of 87 is _____.

27. _____% of 150 is 60.

28. _____ % of 160 is 240.

29. 3% of _____ is 65.4

30. 20% of _____ is 45.

31. What percent of 32 is 40?

32. 1250 is 50% of what number?

33. 100 is 125% of what number?

34. Find 24% of 244.

35. What is the percent of profit if
    a. $400 is made on an investment of $5000?
    b. $350 is made on an investment of $3500?
    c. Which was the better investment?

36. What is the percent of profit if
    a. $400 is made on an investment of $2000?
    b. $510 is made on an investment of $3000?
    c. Which was the better investment?

37. Carlos bought 100 shares of CISCO stock at $45 per share. One year later he sold the stock for $5490.
    a. What was the amount of his profit?
    b. What was his percent of profit?

38. Patrick took his parents to dinner and the bill was $70 plus 6% tax. If he left a tip of 15% of the total (food plus tax), what was the total cost of the dinner?

39. A calculator salesman's monthly income is $1500 salary plus a commission of 7% of his sales over $5000. What was his income the month that he sold $11,470 in calculators?

40. A computer programmer was told that he would be given a bonus of 5% of any money his programs could save the company. How much would he have to save the company to earn a bonus of $1000?

41. The property tax on a home was $2550. What was the tax rate if the home was valued at $170,000?

42. In one season a basketball player missed 14% of her free throws. How many free throws did she make if she attempted 150 free throws?

43. A student missed 6 problems on a statistics exam and received a grade of 85%. If all the problems were of equal value, how many problems were on the test?

44. At a department store white sale, sheets and pillowcases were discounted 25%. Sheets were originally marked $12.00 and pillowcases were originally marked $5.00.
a. What was the sale price of the sheets?
b. What was the sale price of the pillowcases?

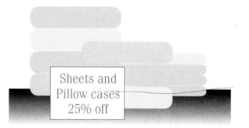

Sheets and
Pillow cases
25% off

45. The Golf Pro Shop had a set of golf clubs that were marked on sale for $560. This was a discount of 20% off the original selling price.
a. What was the original selling price?
b. If the clubs cost the shop $420, what was the percent of profit?

46. If sales tax is figured at 7.25%, how much tax will be added to the total purchase price of three textbooks, priced at $35.00, $55.00, and $70.00? What will be the total amount paid for the books?

47. The dealer's discount to the buyer of a new motorcycle was $499.40. The manufacturer gave a rebate of $1000 on this particular model.

    a. What was the original price if the dealer discount was 7% of the original price?

    b. What would a customer pay for the motorcycle if taxes were 5% of the final selling price and license fees were $642.60?

48. Data cartridges for computer backup were on sale for $27.00 (a package of two cartridges). This price was a discount of 10% off the original price.

    a. Was the original price more than $27 or less than $27?

    b. What was the original price?

49. A man weighed 200 pounds. He lost 20 pounds in 3 months. Then he gained back 20 pounds 2 months later.

    a. What percent of his weight did he lose in the first 3 months?

    b. What percent of his weight did he gain back?

    c. The loss and the gain are the same amount, but the two percents are different. Explain.

50. A car dealer bought a used car for $2500. He marked up the price so that he would make a profit of 25% of his cost.

    a. What was the selling price?

    b. If the customer paid 8% of the selling price in taxes and fees, what was the total cost of the car to the customer? The car was 6 years old.

Writing and Thinking About Mathematics

51. A man and his wife wanted to sell their house and contacted a realtor. The realtor said that the price for the services (listing, advertising, selling, and paperwork) was 6% of the selling price. The realtor asked the couple how much cash they wanted after the realtor fee was deducted from the selling price. They said $141,000 and that therefore, the asking price should be 106% of $141,000 or a total of $149,460. The realtor said that this was the wrong figure and that the percent used should be 94% and not 106%.

    a. Explain why 106% is the wrong percent and $149,460 is not the right selling price.

    b. Explain why and how 94% is the correct percent and just what the selling price should be for the couple to receive $141,000 after the realtor's fee is deducted.

Collaborative Learning Exercise

With the class separated into teams of two to four students, each team is to analyze the following problem and decide how to answer the related questions. Then each team leader is to present the team's answers and related ideas to the class for general discussion.

52. Sam works at a picture framing store and gets a salary of $500 per month plus a commission of 3% on sales he makes over $2000. Maria works at the same store, but she has decided to work on a straight commission of 8% of her sales.

    a. At what amount of sales will Sam and Maria make the same amount of money?

    b. Up to that point, who would be making more?

    c. After that point, who would be making more? Explain briefly. (If you were offered a job at that store, which method of payment would you choose?)

Hawkes Learning Systems: Introductory Algebra

Percents and Applications

3.5   Working with Formulas

Objectives

After completing this section, you will be able to:

1. *Evaluate formulas for given values of the variables.*

2. *Solve formulas for specified variables in terms of the other variables.*

3. *Use formulas to solve a variety of applications.*

Formulas are general rules or principles stated mathematically. There are many formulas in such fields of study as business, economics, medicine, physics, and chemistry as well as mathematics. Some of these formulas and their meanings are shown here.

> **NOTES**
>
> **SPECIAL COMMENT:** Be sure to use the letters just as they are given in the formulas. In mathematics, there is little or no flexibility between capital and small letters as they are used in formulas. In general capital letters have special meanings that are different from corresponding small letters. For example, capital **A** may mean the area of a triangle and small **a** may mean the length of one side, two completely different ideas.

Formula	Meaning
1. $I = Prt$	The simple interest (I) earned by investing money is equal to the product of the principal (P) times the rate of interest (r) times the time (t) in one year or less. (**Note:** If more than one year is involved then the interest is compounded and another formula is used.)
2. $C = \dfrac{5}{9}(F - 32)$	Temperature in degrees Celsius (C) equals $\dfrac{5}{9}$ times the difference between the Fahrenheit temperature (F) and 32.
3. $d = rt$	The distance traveled (d) equals the product of the rate of speed (r) and the time (t).
4. $P = 2l + 2w$	The perimeter (P) of a rectangle is equal to twice the length (l) plus twice the width (w).

5. $L = 2\pi rh$ The lateral surface area, L, (top and bottom not included) of a cylinder is equal to 2π times the radius (r) of the base times the height (h).

6. $F = ma$ In physics, the force, F, acting on an object is equal to its mass (m) times its acceleration (a).

7. $\alpha + \beta + \gamma = 180$ The sum of the angles (α, β, and γ) of a triangle is $180°$. (**Note**: α, β, and γ are the Greek lowercase letters alpha, beta, and gamma, respectively.)

Evaluating Formulas

If you know values for all but one variable in a formula, you can substitute those values and find the value of the unknown variable by using the techniques for solving equations discussed in this chapter. This section presents a variety of formulas from real-life situations, with complete descriptions of the meanings of these formulas. Working with these formulas will help you become familiar with a wide range of applications of algebra and provide practice in solving equations.

Example 1: Evaluating Formulas ● ● ● ● ● ● ● ● ● ● ● ● ● ● ●

a.   A **note** is a loan for a period of 1 year or less, and the interest earned (or paid) is called **simple interest**. A note involves only one payment at the end of the term of the note and includes both principal and interest. The formula for calculating simple interest is:

$$I = Prt$$

where I = **Interest** (earned or paid)
P = **Principal** (the amount invested or borrowed)
r = **rate** of interest (stated as an annual or yearly rate)
t = **time** (one year or part of a year)

Note: The rate of interest is usually given in percent form and converted to decimal or fraction form for calculations. For the purpose of calculations, we will use 360 days in one year and 30 days in a month. Before the use of computers, this was common practice in business and banking.

Maribel loaned $5000 to a friend for 6 months at an interest rate of 8%. How much will her friend pay her at the end of the 6 months?

Continued on next page...

Solution: Here, $P = \$5000$,

$$r = 8\% = 0.08,$$

$$t = 6 \text{ months} = \frac{6}{12} \text{ year} = \frac{1}{2} \text{ year}$$

Find the interest by substituting in the formula $I = Prt$ and evaluating.

$$I = 5000 \cdot \underset{0.04}{\cancel{0.08}} \cdot \frac{1}{\cancel{2}}$$

$$= 5000 \cdot 0.04$$

$$= \$200.00$$

The interest is \$200 and the amount to be paid at the end of 6 months is

Principal + Interest = \$5000 + \$200 = \$5200

b. Given the formula $C = \frac{5}{9}(F - 32)$, first find C if $F = 212°$ and then find F if $C = 20°$.

Solution: $F = 212°$, so substitute 212 for F in the formula.

$$C = \frac{5}{9}(212 - 32)$$

$$= \frac{5}{9}(180)$$

$$= 100$$

That is, $212°$ F is the same as $100°$ C. Water will boil at $212°$ F at sea level. This means that if the temperature is measured in degrees Celsius instead of degrees Fahrenheit, water will boil at $100°$ C at sea level.

$C = 20°$, so substitute 20 for C in the formula.

$$20 = \frac{5}{9}(F - 32) \qquad \text{Now solve for } F.$$

$$\frac{9}{5} \cdot 20 = \frac{9}{5} \cdot \frac{5}{9}(F - 32) \qquad \text{Multiply both sides by } \frac{9}{5}.$$

$$36 = F - 32 \qquad \text{Simplify.}$$

$$68 = F \qquad \text{Add 32 to both sides.}$$

That is, a temperature of $20°$ C is the same as a comfortable spring day temperature of $68°$ F.

Continued on next page...

c. The lifting force, F, exerted on an airplane wing is found by multiplying some constant, k, by the area, A, of the wing's surface and by the square of the plane's velocity, v. The formula is $F = kAv^2$. Find the force on a plane's wing of area 120 ft.^{2} if k is $\dfrac{4}{3}$ and the plane is traveling 80 miles per hour as it takes off.

Solution: We know that $k = \dfrac{4}{3}$, $A = 120$, and $v = 80$. Substitution gives

$$F = \frac{4}{3} \cdot 120 \cdot 80^2$$

$$= \frac{4}{3} \cdot 120 \cdot 6400$$

$$= 160 \cdot 6400$$

$$F = 1{,}024{,}000 \text{ lbs} \quad \text{(The force is measured in pounds.)}$$

● ●

Solving Formulas for Different Variables

We say that the formula $d = rt$ is "solved for" d in terms of r and t. Similarly, the formula $A = \dfrac{1}{2}bh$ is solved for A in terms of b and h, and the formula $P = S - C$ (profit is equal to selling price minus cost) is solved for P in terms of S and C. Many times we want to use a certain formula in another form. We want the formula "solved for" some variable other than the one given in terms of the remaining variables. **Treat the variables just as you would constants in solving linear equations.** Study the following examples carefully.

Example 2: Solving for Different Variables ● ● ● ● ● ● ● ● ● ●

a. Given $d = rt$, solve for t in terms of d and r. We want to represent the time in terms of distance and rate. We will use this concept later in word problems.

Solution:	$d = rt$	Treat r and d as if they were constants.
	$\dfrac{d}{r} = \dfrac{rt}{r}$	Divide both sides by r.
	$\dfrac{d}{r} = t$	Simplify.

Continued on next page...

b. Given $P = a + b + c$, solve for a in terms of $P, b,$ and c. This would be a convenient form for the case in which we know the perimeter and two sides of a triangle and want to find the third side.

Solution:

$P = a + b + c$	Treat P, b, and c as if they were constants.
$P - b - c = a + b + c - b - c$	Add $-b - c$ to both sides.
$P - b - c = a$	Simplify.

c. Given $C = \dfrac{5}{9}(F - 32)$ as in Example 1a, solve for F in terms of C. This would give a formula for finding Fahrenheit temperature given a Celsius temperature value.

Solution:

$C = \dfrac{5}{9}(F - 32)$	Treat C as a constant.
$\dfrac{9}{5} \cdot C = \dfrac{9}{5} \cdot \dfrac{5}{9}(F - 32)$	Multiply both sides by $\dfrac{9}{5}$.
$\dfrac{9}{5}C = F - 32$	Simplify.
$\dfrac{9}{5}C + 32 = F$	Add 32 to both sides.

Thus,

$$F = \frac{9}{5}C + 32 \text{ is solved for } F \text{ and } C = \frac{5}{9}(F - 32) \text{ is solved for } C.$$

These are two forms of the same formula.

d. Given the equation $2x + 4y = 10$, **(i.)** solve first for x in terms of y, and then **(ii.)** solve for y in terms of x. This equation is typical of the algebraic equations that we will discuss in Chapter 7.

Solution: i. Solving for x yields

$2x + 4y = 10$	Treat $4y$ as a constant.
$2x + 4y - 4y = 10 - 4y$	Subtract $4y$ from both sides. (This is the same as adding $-4y$.)
$2x = 10 - 4y$	
$\dfrac{2x}{2} = \dfrac{10 - 4y}{2}$	Divide both sides by 2.
$x = \dfrac{10}{2} - \dfrac{4y}{2}$	Simplify.
$x = 5 - 2y$	

Continued on next page...

ii. Solving for y yields

$$2x + 4y = 10 \qquad \text{Treat } 2x \text{ as a constant.}$$

$$2x + 4y - 2x = 10 - 2x \qquad \text{Subtract } 2x \text{ from both sides.}$$

$$4y = 10 - 2x \qquad \text{Simplify.}$$

$$\frac{4y}{4} = \frac{10 - 2x}{4} \qquad \text{Divide both sides by 4.}$$

$$y = \frac{10}{4} - \frac{2x}{4} \qquad \text{Simplify.}$$

$$y = \frac{5}{2} - \frac{x}{2}$$

or we can write

$$y = \frac{5 - x}{2} \ \text{ or } \ y = -\frac{1}{2}x + \frac{5}{2} \qquad \text{All forms are correct.}$$

e. Given $3x - y = 15$, solve for y in terms of x.

   Solution: Solving for y gives

$$3x - y = 15$$

$$-y = 15 - 3x \qquad \text{Subtract } 3x \text{ from both sides.}$$

$$-1(-y) = -1(15 - 3x) \qquad \text{Multiply both sides by } -1 \text{ (or divide both sides by } -1).$$

$$y = -15 + 3x \qquad \text{Simplify using the distributive property.}$$

f. Given $V = \dfrac{k}{P}$, solve for P in terms of V and k.

   Solution:
$$V = \frac{k}{P}$$

$$V \cdot P = k \qquad \text{Multiply both sides by } P.$$

$$\frac{V \cdot P}{V} = \frac{k}{V} \qquad \text{Divide both sides by } V.$$

$$P = \frac{k}{V}$$

Practice Problems

1. $2x - y = 5$; solve for y.

2. $2x - y = 5$; solve for x.

3. $A = \dfrac{1}{2}bh$; solve for h.

4. $L = 2\pi rh$; solve for r.

5. $P = 2l + 2w$; solve for w.

Answers to Practice Problems: **1.** $y = 2x - 5$ **2.** $x = \dfrac{y+5}{2}$ **3.** $h = \dfrac{2A}{b}$ **4.** $r = \dfrac{L}{2\pi h}$ **5.** $w = \dfrac{P - 2l}{2}$

3.5 Exercises

Simple Interest

For Exercises 1 – 5, refer to Example 1(a) for information concerning simple interest and the related formula $I = Prt$.

1. You want to borrow $4000 at 12% for only 90 days. How much interest would you pay?

2. A savings account of $3500 is left for 9 months and draws simple interest at a rate of 7%.
   a. How much interest is earned?
   b. What is the balance in the account at the end of the 9 months?

3. For how many days must you leave $1000 in a savings account at 5.5% to earn $11.00 in interest?

4. What is the rate of interest charged if a 3-month loan of $2500 is paid off with $2562.50?

5. What principal would you need to invest to earn $450 in simple interest in 6 months if the rate of interest was 9%?

In Exercises 6 – 20, read the descriptive information carefully and then substitute the values given in the problem for the corresponding variables in the formulas. Evaluate the resulting expression for the unknown variable.

Velocity

If an object is shot upward with an initial velocity v_0 feet per second, the velocity, v, in feet per second is given by the formula $v = v_0 - 32t$, where t is time in seconds.

6. Find the velocity at the end of 3 seconds if the initial velocity is 144 feet per second.

7. Find the initial velocity of an object if the velocity after 4 seconds is 48 feet per second.

8. An object projected upward with an initial velocity of 106 feet per second has a velocity of 42 feet per second. How many seconds have passed?

Medicine

In nursing, one procedure for determining the dosage for a child is

$$\text{child's dosage} = \frac{\text{age of child in years}}{\text{age of child} + 12} \cdot \text{adult dosage}$$

9. If the adult dosage of a drug is 20 milliliters, how much should a 3-year-old child receive?

10. If the adult dosage of a drug is 340 milligrams, how much should a 5-year-old child receive?

Investment

The amount of money due from investing P dollars is given by the formula $A = P + Prt$, where r is the rate expressed as a decimal and t is the time in years.

11. Find the amount due if $1000 is invested at 6% for 2 years.

12. How long will it take an investment of $600, at an annual rate of 5%, to be worth $750?

Carpentry

The number, N, of rafters in a roof or studs in a wall can be found by the formula $N = \dfrac{l}{d} + 1$, where l is the length of the roof or wall and d is the center-to-center distance from one rafter or stud to the next. Note that l and d must be in the same units.

13. How many rafters will be needed to build a roof 26 ft. long if they are placed 2 ft. on center?

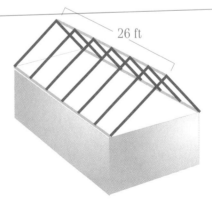

26 ft

14. A wall has studs placed 16 in. on center. If the wall is 20 ft. long, how many studs are in the wall?

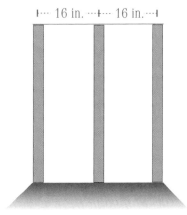

15. How long is a wall if it requires 22 studs placed 16 in. on center?

Cost

The total cost, C, of producing x items can be found by the formula $C = ax + k$, where a is the cost per item and k is the fixed costs (rent, utilities, and so on).

16. Find the cost of producing 30 items if each costs $15 and the fixed costs are $580.

17. It costs $1097.50 to produce 80 dolls per week. If each doll costs $9.50 to produce, find the fixed costs.

18. It costs a company $3.60 to produce a calculator. Last week the total costs were $1308. If the fixed costs are $480 weekly, how many calculators were produced last week?

Depreciation

Many items decrease in value as time passes. This decrease in value is called **depreciation**. One type of depreciation is called **linear depreciation**. The value, V, of an item after t years is given by $V = C - Crt$, where C is the original cost and r is the rate of depreciation expressed as a decimal.

19. If you buy a car for $6000 and depreciate it linearly at a rate of 10% per year, what will be its value after 6 years?

20. A contractor buys a 4 year-old piece of heavy equipment valued at $20,000. If the original cost of this equipment was $25,000, find the rate of depreciation.

Solve for the indicated variable in Exercises 21 – 55.

21. $P = a + b + c$; solve for b.

22. $P = 3s$; solve for s.

23. $F = ma$; solve for m.

24. $C = \pi d$; solve for d.

25. $A = lw$; solve for w.

26. $P = R - C$; solve for C.

27. $R = np$; solve for n.

28. $v = k + gt$; solve for k.

29. $I = A - P$; solve for P.

30. $L = 2\pi rh$; solve for h.

31. $A = \dfrac{m+n}{2}$; solve for m.

32. $P = a + 2b$; solve for a.

33. $I = Prt$; solve for t.

34. $R = \dfrac{E}{I}$; solve for E.

35. $P = a + 2b$; solve for b.

36. $c^2 = a^2 + b^2$; solve for b^2.

37. $\alpha + \beta + \gamma = 180$; solve for β.

38. $y = mx + b$; solve for x.

39. $V = lwh$; solve for h.

40. $v = -gt + v_0$; solve for t.

41. $A = \dfrac{1}{2}bh$; solve for b.

42. $R = \dfrac{E}{I}$; solve for I.

43. $A = \pi r^2$; solve for π.

44. $A = \dfrac{R}{2L}$; solve for L.

45. $K = \dfrac{mv^2}{2g}$; solve for g.

46. $x + 4y = 4$; solve for y.

47. $2x + 3y = 6$; solve for y.

48. $3x - y = 14$; solve for y.

49. $5x + 2y = 11$; solve for x.

50. $-2x + 2y = 5$; solve for x.

51. $A = \dfrac{1}{2}h(b+c)$; solve for b.

52. $A = \dfrac{1}{2}h(b+c)$; solve for h.

53. $R = \dfrac{3(x-12)}{8}$; solve for x.

54. $-2x - 5 = -3(x + y)$; solve for x.

55. $3y - 2 = x + 4y + 10$; solve for y.

Make up a formula for each of the following situations.

56. Each ticket for a concert costs $\$t$ per person and parking costs \$9.00. What is the total cost per car, C, if there are n people in a car?

57. ABC Car Rental charges \$25 per day plus \$0.12 per mile. What would you pay per day for renting a car from ABC if you were to drive the car x miles in one day?

58. Top-Of-The-Line computer company knows that the cost (labor and materials) of producing a computer is \$325 per computer per week and the fixed overhead costs (lighting, rent, etc.) are \$5400 per week. What are the company's weekly costs of producing n computers per week?

59. If T-O-T-L (see Exercise 58) sells its computers for \$683 each, what is its profit per week if it sells the same number, n, that it produces? (Remember that profit is equal to revenue minus costs, or $P = R - C$.)

Writing and Thinking About Mathematics

60. The formula $z = \dfrac{x - \bar{x}}{s}$ is used extensively in statistics. In this formula, x represents one of a set of numbers, $\bar{x}$ represents the average (or mean) of those numbers in the set, and s represents a value called the standard deviation of the numbers. (The standard deviation is a positive number and is a measure of how "spread out" the numbers are.) The values for z are called z-scores, and they measure the number of standard deviation units a number x is from the mean $\bar{x}$.

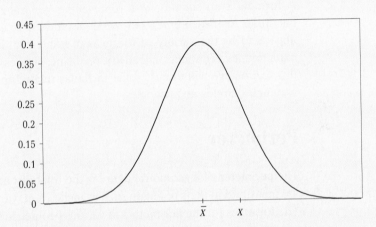

a. If $\bar{x} = 70$, what will be the z-score for $x = 70$? Does this z-score depend on the value of s? Explain.

b. For what values of x will the corresponding z-scores be negative?

c. Calculate your z-score on each of the last two scores in this class. (Your instructor will give you the mean and standard deviation for each test.) What do these scores tell you about your performance on the two exams?

61. Suppose that, for a particular set of exam scores, $\bar{x} = 72$ and $s = 6$. Find the z-score that corresponds to score of

    a. 78   **b.** 66   **c.** 81   **d.** 60

Hawkes Learning Systems: Introductory Algebra

Working with Formulas

3.6   Formulas in Geometry

After completing this section, you will be able to:

1. Recognize and use appropriate geometric formulas for computations.

A **formula** is an equation that represents a general relationship between two or more quantities or measurements. Several variables may appear in a formula. And, as was illustrated by the variety of formulas presented in Section 3.5, formulas are useful in mathematics, economics, chemistry, medicine, physics, and many other fields of study. In this section, we will discuss some formulas related to simple geometric figures such as rectangles, circles, and triangles.

Perimeter

The **perimeter** of a geometric figure is the total distance around the figure. Perimeters are measured in units of length such as inches, feet, yards, miles, centimeters, and meters. The formulas for the perimeters of various geometric figures are shown in Figure 3.1.

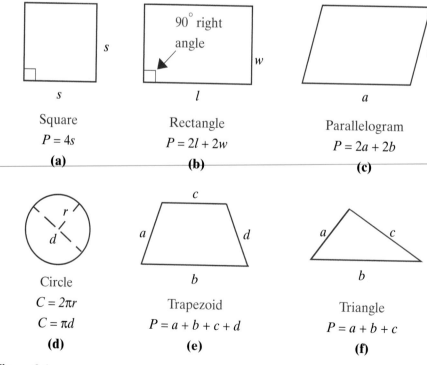

Square
$P = 4s$
(a)

Rectangle
$P = 2l + 2w$
(b)

Parallelogram
$P = 2a + 2b$
(c)

Circle
$C = 2\pi r$
$C = \pi d$
(d)

Trapezoid
$P = a + b + c + d$
(e)

Triangle
$P = a + b + c$
(f)

Figure 3.1

Terminology and Notation Related to Circles

Term	Notation	Definition
1. Circumference	*C*	*The perimeter of a circle is called its* **circumference**.
2. Radius	*r*	*The distance from the center of a circle to any point on the circle is called its* **radius**.
3. Diameter	*d*	*The distance from one point on a circle to another point on the circle measured through its center is called its* **diameter**. *(**Note**: The diameter is twice the radius:* **d = 2r**.*)*
4. Pi	*π*	*π is a Greek letter used for the constant 3.1415926535... (π is an irrational number.)*

For the calculations involving π in this text we will round off to two decimal places and use π = 3.14. Remember, however, that π is an irrational number and is an infinite nonrepeating decimal. Historically, π was discovered by mathematicians trying to find a relationship between the circumference and diameter of any circle. The result is that π is the ratio of the circumference to the diameter of any circle, a rather amazing fact.

Example 1: Perimeter ● ● ● ● ● ● ● ● ● ● ● ● ● ● ● ● ●

a. Find the perimeter of a rectangle with a length of 10 feet and a width of 8 feet.

Solution: **i.** Sketch the figure.

ii.

$$P = 2l + 2w$$

$$P = 2 \cdot 10 + 2 \cdot 8$$

$$P = 20 + 16 = 36$$

The perimeter is 36 feet.

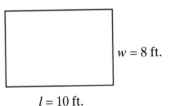

$w = 8$ ft.

$l = 10$ ft.

Continued on next page...

b. Find the circumference of a circle with a diameter of 3 centimeters.

Solution 1:   **i.** Sketch the figure.
ii. $C = \pi d$
$C = 3.14(\,3\,)$
$\quad = 9.42$ cm

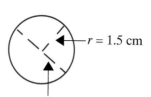

Solution 2:   **i.** Sketch the figure.
ii. $C = 2\pi r$
$C = 2(\,3.14\,)(\,1.5\,)$
$\quad = 9.42$ cm

Note that the result is the same with either formula. The circumference is approximately 9.42 centimeters.

c. Find the perimeter of the triangle with sides labeled as in the figure.

Solution: **i.** Sketch the figure
ii. $P = a + b + c$
$P = 3 + 6.2 + 8.1$
$\quad = 17.3$ in.

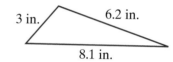

The perimeter is 17.3 inches.

●●●●●●●●●●●●●●●●●●●●●●●●●●●●●●●●●●●●

Area

Area is a measure of the interior, or enclosure, of a surface. Area is measured in square units such as square feet, square inches, square meters, or square miles. For example, the area enclosed by a square of 1 inch on each side is 1 sq. in. (or 1 in.²) [Figure 3.2(a)], while the area enclosed by a square of 1 centimeter on each side is 1 sq. cm (or 1 cm²)[Figure 3.2(b)].

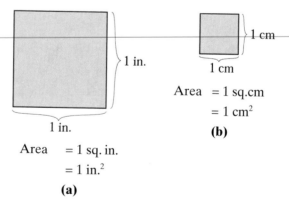

Figure 3.2

The formulas for finding the areas of several geometric figures are shown in Figure 3.3.

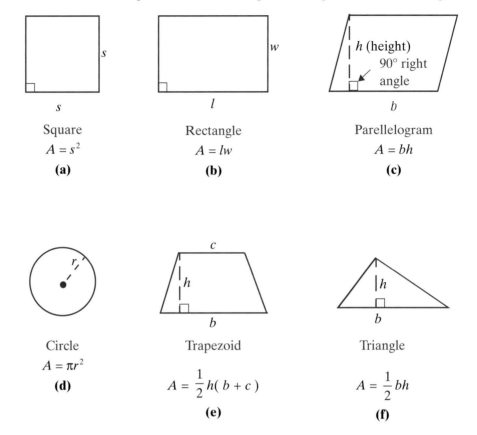

Square
$A = s^2$
(a)

Rectangle
$A = lw$
(b)

Parellelogram
$A = bh$
(c)

Circle
$A = \pi r^2$
(d)

Trapezoid
$A = \dfrac{1}{2} h(b + c)$
(e)

Triangle
$A = \dfrac{1}{2} bh$
(f)

Figure 3.3

Example 2: Area

a. Find the area of a triangle with a height of 3 centimeters and a base of 4 centimeters.

Solution: i. Sketch the figure.

ii. $A = \dfrac{1}{2} \cdot b \cdot h$

$A = \dfrac{1}{2} \cdot 4 \cdot 3$

$= \dfrac{1}{2} \cdot 12 = 6$

The area is 6 square centimeters.

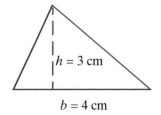

Continued on next page...

b. Find the area of a circle with a radius of 6 inches. (Use $\pi = 3.14$.)

Solution: i. Sketch the figure.

ii. $A = \pi r^2$

$A = 3.14(6)^2$

$= 3.14(36)$

$= 113.04$

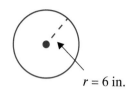

$r = 6$ in.

The area is approximately 113.04 square inches.

● ●

Volume

Volume is a measure of the space enclosed by a three-dimensional figure. Volume is measured in cubic units, such as cubic inches, cubic centimeters, and cubic feet. For example, the volume enclosed by a cube of 1 inch on each edge is 1 cu. in. (or 1 in.³) [Figure 3.4(a)], while the volume enclosed by a cube of 1 centimeter on each edge is 1 cu. cm (or 1 cm³) [Figure 3.4(b)].

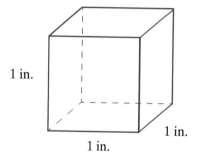

1 in.

1 in.

1 in.

Volume = 1 cu. in.

= 1 in.³

(a)

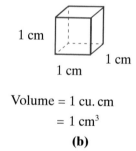

1 cm

1 cm

1 cm

Volume = 1 cu. cm

= 1 cm³

(b)

Figure 3.4

The formulas for finding the volume of several geometric figures are shown in Figure 3.5.

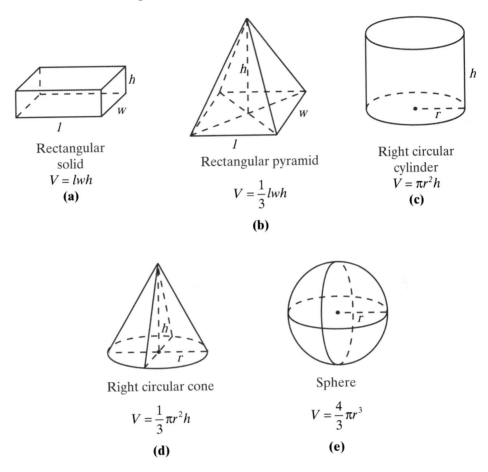

Rectangular
solid
$V = lwh$
(a)

Rectangular pyramid

$V = \dfrac{1}{3}lwh$

(b)

Right circular
cylinder
$V = \pi r^2 h$
(c)

Right circular cone

$V = \dfrac{1}{3}\pi r^2 h$

(d)

Sphere

$V = \dfrac{4}{3}\pi r^3$

(e)

Figure 3.5

Example 3: Volume

a. Find the volume of a right circular cylinder with a radius of 2 centimeters and a height of 5 centimeters.

Solution: **i.** Sketch the figure.

ii. $V = \pi r^2 h$

$= 3.14 \left(2^2 \right) \cdot 5$

$= 3.14 (4) \cdot 5$

$= 3.14 (20)$

$= 62.80 \text{ cm}^3$

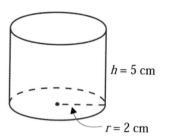

$h = 5$ cm

$r = 2$ cm

The volume is approximately 62.80 cubic centimeters.

b. Find the volume of a sphere with a radius of 4 feet.

Solution: **i.** Sketch the figure.

ii. $V = \dfrac{4}{3} \pi r^3$

$V = \dfrac{4}{3} (3.14) \cdot 4^3$

$= \dfrac{4 (3.14) \cdot 64}{3}$

$= \dfrac{803.84}{3} \text{ ft.}^3$ or about 267.95 ft.^{3}

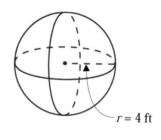

$r = 4$ ft

The volume is approximately 267.95 cubic feet.

Practice Problems

1. Find the area of a square with sides 4 cm long.

2. Find the area of a circle with a diameter 6 in.

3. Find the perimeter of a rectangle 3.5 m long and 1.6 m wide.

Answers to Practice Problems: 1. 16 cm^2 **2.** 28.26 in.^{2} **3.** 10.2 m

3.6 Exercises

In Exercises 1 – 15, select the answer from the right-hand column that correctly matches the statement.

1. The formula for the perimeter of a square _____

a. $V = \dfrac{1}{3}lwh$

2. The formula for the circumference of a circle _____

b. $A = \dfrac{1}{2}h(b+c)$

3. The formula for the perimeter of a triangle _____

c. $A = s^2$

4. The formula for the area of a rectangle _____

d. $A = \dfrac{1}{2}bh$

5. The formula for the area of a square _____

e. $P = 4s$

6. The formula for the area of a trapezoid _____

f. $A = lw$

7. The formula for the area of a triangle _____

g. $P = 2l + 2w$

8. The formula for the area of a parallelogram _____

h. $V = \dfrac{1}{3}\pi r^2 h$

9. The formula for the perimeter of a rectangle _____

i. $V = \dfrac{4}{3}\pi r^3$

10. The formula for the volume of a rectangular pyramid _____

j. $C = 2\pi r$

11. The formula for the volume of a rectangular solid _____

k. $P = a + b + c$

12. The formula for the area of a circle _____

l. $V = lwh$

13. The formula for the volume of a right circular cylinder _____

m. $V = \pi r^2 h$

14. The formula for the volume of a sphere _____

n. $A = \pi r^2$

15. The formula for the volume of a right circular cone _____

o. $A = bh$

Find (a) the perimeter and (b) the area of each figure in Exercises 16 – 21.

16. $P =$ _____ $A =$ _____

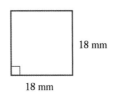

18 mm

18 mm

17. $P =$ _____ $A =$ _____

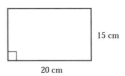

15 cm

20 cm

18. $P =$ _____ $A =$ _____

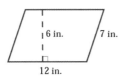

6 in.   7 in.

12 in.

19. $P =$ _____ $A =$ _____

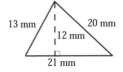

13 mm   20 mm

12 mm

21 mm

20. $P =$ _____ $A =$ _____

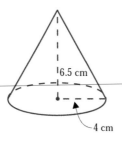

8 cm

5 cm   4 cm   5 cm

14 cm

21. $C =$ _____ $A =$ _____

5 in.

Find the volume of each figure in Exercise 22 – 26. (Use $\pi = 3.14$ and round to the nearest hundredth.)

22. $V =$ _____

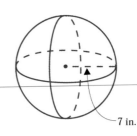

6.5 cm

4 cm

23. $V =$ _____

7 in.

24. $V =$ ____

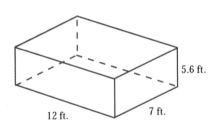

5.6 ft.

12 ft.   7 ft.

25. $V =$ ____

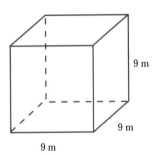

9 m

9 m

9 m

26. $V =$ ____

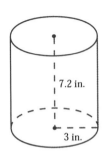

7.2 in.

3 in.

Solve the problems in Exercises 27 – 38. (Use $\pi = 3.14$)

27. What is the perimeter of a rectangle with length of 17 in. and width of 11 in.?
28. Find the area of a square with sides that are 9 ft. long.
29. The base of a triangle is 14 cm and the height is 9 cm. Find the area.
30. Find the circumference of a circle with a radius of 8 m.
31. The radius of the base of a cylindrical tank is 14 ft. If the tank is 10 ft. high, find the volume.
32. The sides of a triangle are 6.2 m, 8.6 m, and 9.4 m. Find the perimeter.
33. What is the volume of a cube with edges of 5 ft.?
34. A rectangular garden plot is 60 ft. long and 42 ft. wide. Find the area.
35. A parallelogram has a base of 20 cm and a height of 13.6 cm. Find the area.
36. A rectangular box is 18 in. long, 10.3 in. wide, and 8 in. high. Find the volume.
37. The diameter of the base of a right circular cone is 15 in. If the height of the cone is 9 in., find the volume.
38. Find the volume of a sphere whose radius is 10 cm. (Round to the nearest hundredth.)

Find the perimeter and area for each figure in Exercises 39 and 40. (Use $\pi = 3.14$)

39. $P = \underline{\hspace{1cm}}$ $A = \underline{\hspace{1cm}}$

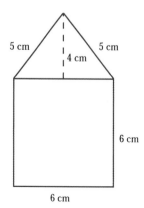

5 cm 5 cm 4 cm 6 cm 6 cm

40. $P = \underline{\hspace{1cm}}$ $A = \underline{\hspace{1cm}}$

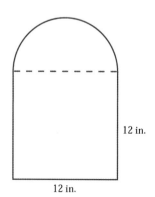

12 in. 12 in.

Calculator Problems

In Exercises 41 – 45, find the answers accurate to two decimal places.

41. Find the area of a rectangle that is 16.54 m long and 12.82 m wide.
42. The radius of a circle is 8.32 in. Find the circumference. (Use $\pi = 3.14$.)
43. Find the area of a trapezoid with bases of 22.36 in. and 17.48 in. and a height of 6.53 in.
44. Find the area of a triangle with a base of 63.52 cm and a height of 41.78 cm.
45. The radius of a sphere is 114.8 cm. Find the volume. (Use $\pi = 3.14$.)

Hawkes Learning Systems: Introductory Algebra

Formulas in Geometry

3.7 Solving Linear Inequalities and Applications

Objectives

After completing this section, you will be able to:

1. Understand intervals of real numbers.

2. Solve linear inequalities.

3. Graph the solutions for linear inequalities.

4. Solve applications by using linear inequalities.

Intervals of Real Numbers

Real numbers, graphs of real numbers on real number lines, and inequalities with the symbols $<, >, \leq$, and $\geq$ were discussed in Section 1.1. In this section, we will expand our work with inequalities to include the concept of **intervals of real numbers**. Remember that dots are used to indicate the graphs of individual real numbers on a real number line.

For example, the set of numbers $\left\{ -\dfrac{3}{2}, -1, 0, 2, \sqrt{10} \right\}$ is graphed in Figure 3.6.

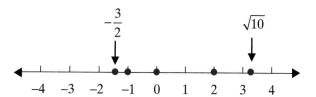

Figure 3.6

Irrational numbers, such as $\pi, \sqrt{2}, -\sqrt{7}$, and $\sqrt{10}$ have been discussed to some extent and will be discussed in more detail in Chapter 9. What is important here is that irrational numbers do indeed correspond to points on a real number line. In fact, we can make the following statement concerning real numbers and points on a line.

One-To-One Correspondence

There is a one-to-one correspondence between the real numbers and the points on a line. That is, each point on a number line corresponds to one real number, and each real number corresponds to one point on a number line.

Now, when an inequality such as $x < 8$ is given, the implication is that x can be **any real number less than 8**. That is, the **solution** to the inequality is the set of all real numbers less than 8. To indicate this solution graphically, an open circle is drawn around 8 on a number line and the line is shaded to the left of 8 as shown in Figure 3.7.

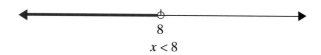

$$8$$
$$x < 8$$

Figure 3.7

Suppose that a and b are two real numbers and that $a < b$. The set of all real numbers between a and b is called an **interval of real numbers**. Intervals of real numbers are commonly used in relationship to ideas such as the length of time to heal a wound, the shelf life of chemicals, the stopping distances for a car traveling at certain speeds, and the distances that a camera can be expected to take clear pictures. As a specific example, a bicyclist may regularly ride from 1 to 3 hours for exercise. This interval of time can be represented in the following way:

$$1 \leq t \leq 3 \quad \text{where } t \text{ represents the time in hours.}$$

Intervals are classified and graphed as indicated in the box that follows. You should keep in mind the following facts as you consider the boxed information.

1. x is understood to represent real numbers.
2. Open dots at endpoints a and b indicate that these points **are not** included in the graph.
3. Solid dots at endpoints a and b indicate that these points **are** included in the graph.

Intervals of Real Numbers

Name	Symbolic Representation	Graph
Open Interval	$a < x < b$	
Closed Interval	$a \leq x \leq b$	
Half - Open Interval	$\begin{cases} a < x \leq b \\ \\ a \leq x < b \end{cases}$	
Open Interval	$\begin{cases} x > a \\ \\ x < a \end{cases}$	
Half - Open Interval	$\begin{cases} x \geq a \\ \\ x \leq a \end{cases}$	

Example 1: Graph •

a. Represent the following graph using algebraic notation, and tell what kind of interval it is.

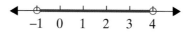

Solution: $-1 < x < 4$ is an open interval.

Continued on next page...

b. Represent the following graph using algebraic notation, and tell what kind of interval it is.

$$2$$

Solution: $x \geq 2$ is a half-open interval.

c. Graph the half-open interval $-3 < x \leq 0$.

Solution:

● ●

Solving Linear Inequalities

Inequalities of the forms

$$ax + b < c \quad \text{and} \quad ax + b \leq c$$
$$ax + b > c \quad \text{and} \quad ax + b \geq c$$
$$c < ax + b < d \quad \text{and} \quad c \leq ax + b \leq d$$

are called **linear inequalities** or **first-degree inequalities**.

The solutions to linear inequalities are intervals of real numbers and the methods for solving linear inequalities are similar to those used to solve linear equations. There is only one important exception:

> **Multiplying or dividing both sides of an inequality by a negative number causes the "sense" of the inequality to be reversed.**

By the sense of the inequality, we mean "less than" or "greater than." Consider the following examples.

We know that $4 < 20$.

Add 5 to both sides	**Multiply both sides by 3**	**Add −10 to both sides**
$4 < 20$	$4 < 20$	$4 < 20$
$4 + 5 \ ? \ 20 + 5$	$3 \cdot 4 \ ? \ 3 \cdot 20$	$4 + (-10) \ ? \ 20 + (-10)$
$9 < 25$	$12 < 60$	$-6 < 10$

In the cases just illustrated involving addition or multiplication by a positive number, the sense of the inequality stayed the same, in these cases, $<$. Now we will see that multiplying or dividing each side by a negative number will reverse the sense of the inequality, from $<$ to $>$, or from $>$ to $<$.

Multiply both sides by −3	**Divide both sides by −2**
$4 < 20$	$4 < 20$
$-3 \cdot 4 \ ? -3 \cdot 20$	$\dfrac{4}{-2} \ ? \ \dfrac{20}{-2}$
$-12 > -60$	$-2 > -10$

In each of these two cases, the sense of the inequality is changed from $<$ to $>$. While two examples do not prove a rule to be true, this particular rule is true and is included in the following rules for solving inequalities.

Rules for Solving Linear (or First Degree) Inequalities

1. *Simplify each side of the inequality by removing any parentheses and combining like terms.*

2. *Add constants and/or variables to or subtract them from both sides of the inequality so that variables are on one side and constants on the other.*

3. *Divide both sides by the coefficient (or multiply them by the reciprocal) of the variable, and **reverse the sense of the inequality if this coefficient is negative**.*

4. *A quick (and generally satisfactory) check is to select any one number in your solution and substitute it into the original inequality.*

As with solving equations, the object of solving an inequality is to find equivalent inequalities of simpler form that have the same solution set. We want the variable with a coefficient of +1 on one side of the inequality and any constants on the other side. One key difference between solving linear equations and solving linear inequalities is that linear equations have only one solution while linear inequalities generally have an infinite number of solutions.

NOTES

Special Note About Reading Inequalities

Inequalities can be read from left to right or from right to left. When reading an inequality with a variable, read the variable first.

For example,

$$4 < x$$ should be read from right to left as "x is greater than 4"

while $x > 4$ should be read from left to right as "x is greater than 4"

$-2 < x < 5.1$ should be read as follows: "x is greater than -2 and x is less than 5.1"

Reading the variable first will help in understanding the inequalities and drawing the correct related graphs.

Example 2: Solving Linear Inequalities • • • • • • • • • • • •

Solve the following linear inequalities and graph the solutions.
a. $5x + 4 \leq -1$

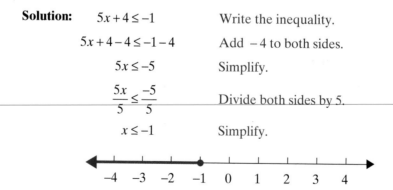

Solution:	$5x + 4 \leq -1$	Write the inequality.
	$5x + 4 - 4 \leq -1 - 4$	Add -4 to both sides.
	$5x \leq -5$	Simplify.
	$\dfrac{5x}{5} \leq \dfrac{-5}{5}$	Divide both sides by 5.
	$x \leq -1$	Simplify.

(**Note:** The shaded dot means that -1 is included.)

Continued on next page...

Check: As a quick check, pick any number less than or equal to –1, say, –4, and substitute it into the original inequality. If a false statement results, then you have made a mistake.

$$5(-4) + 4 \leq -1$$
$$-16 \leq -1 \qquad \text{True statement.}$$

b. $x + 1 > 3 - 2x$

Solution:

$x + 1 > 3 - 2x$	Write the inequality.
$x + 1 + 2x > 3 - 2x + 2x$	Add $2x$ to both sides.
$3x + 1 > 3$	Simplify.
$3x + 1 - 1 > 3 - 1$	Add -1 both sides.
$3x > 2$	Simplify.
$\dfrac{3x}{3} > \dfrac{2}{3}$	Divide both sides by 3.
$x > \dfrac{2}{3}$	Simplify.

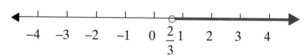

(**Note:** The open circle means that $\dfrac{2}{3}$ is not included in the solution.)

c. $-1 < -2(x + 1) \leq 3$

Solution: Here we are actually solving two inequalities at once:

$$-1 < -2(x + 1) \text{ and } -2(x + 1) \leq 3.$$

We want to find those numbers that satisfy both inequalities.

$-1 < -2(x+1) \leq 3$	Write the expression.
$-1 < -2x - 2 \leq 3$	Use the distributive property.
$-1 + 2 < -2x - 2 + 2 \leq 3 + 2$	Add 2 to all three expressions.
$1 < -2x \leq 5$	Simplify.
$\dfrac{1}{-2} > \dfrac{-2x}{-2} \geq \dfrac{5}{-2}$	Divide each expression by -2 and reverse the sense of each inequality.
$-\dfrac{1}{2} > x \geq -\dfrac{5}{2}$	Simplify.

Continued on next page...

Or, rewriting with the smaller number on the left,

$$-\frac{5}{2} \le x < -\frac{1}{2}$$

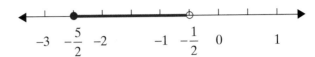

Check: For a quick check, substitute $x = -2$.

$$-1 \overset{?}{<} -2(-2+1) \overset{?}{\le} 3$$

$$-1 \overset{?}{<} -2(-1) \overset{?}{\le} 3$$

$$-1 < 2 \le 3 \qquad \text{True statement.}$$

While a check does not guarantee there are no errors, it is a good indicator.

Applications of Linear Inequalities

We are generally familiar with the use of equations to solve word problems. In the following two examples, we show that solving inequalities can be related to real-world problems.

Example 3: Applications

a. A physics student has grades of 85, 98, 93, and 90 on four examinations. If he must average 90 or better to receive an A for the course, what score can he receive on the final exam and earn an A? (Assume that the final exam counts the same as the other exams.)

Solution: Let x = score on final exam. The average is found by adding the scores and dividing by 5.

$$\frac{85+98+93+90+x}{5} \ge 90$$

$$\frac{366+x}{5} \ge 90$$

$$\cancel{5}\left(\frac{366+x}{\cancel{5}}\right) \ge 5 \cdot 90$$

$$366 + x \ge 450$$

$$x \ge 450 - 366$$

$$x \ge 84$$

Continued on next page...

If the student scores 84 or more on the final exam, he will average 90 or more and receive an A in physics.

b. Ellen is going to buy 30 stamps, some 15-cent and some 34-cent. If she has $6.02, what is the maximum number of 34-cent stamps she can buy?

Solution: Let x = number of 34-cent stamps, then $30 - x$ = number of 15-cent stamps. Ellen cannot spend more than $6.02.

$$0.34x + 0.15(30 - x) \leq 6.02$$
$$0.34x + 4.50 - 0.15x \leq 6.02$$
$$0.19x + 4.50 \leq 6.02$$
$$0.19x \leq 6.02 - 4.50$$
$$0.19x \leq 1.52$$
$$x \leq 8$$

Ellen can buy at most eight 34-cent stamps if she buys a total of 30 stamps.

• •

Practice Problems

Solve the inequalities and graph the solutions.

1. $7 + x < 3$

2. $\dfrac{x}{2} + 1 \geq \dfrac{x}{3}$

3. $-5 \leq 2x + 1 < 9$

Answers to Practice Problems: **1.** $x < -4$ **2.** $x \geq -6$ **3.** $-3 \leq x < 4$

3.7 Exercises

In Exercises 1 – 10, determine whether each of the inequalities is true or false. If the inequality is false, rewrite it in a form that is true using the same numbers or expressions.

1. $3 < -3$

2. $-14 > 4$

3. $|-5| > -5$

4. $|-7| \leq 7$

5. $-3.4 > -3.5$

6. $\dfrac{1}{2} > \dfrac{1}{3}$

7. $-6 \leq 0$

8. $\dfrac{5}{8} \geq \dfrac{7}{16}$

9. $|-4| \leq |4|$

10. $-12.5 < -10.2$

In Exercises 11 – 20, graph the set of real numbers that satisfies each of the inequalities and tell what type of interval it represents.

11. $-2 \leq x < 3$

12. $-4 \leq x \leq -1$

13. $x \leq -4$

14. $x > 6$

15. $-2 < x \leq 1$

16. $x \geq -5$

17. $1 \leq x < \dfrac{10}{3}$

18. $x < \dfrac{13}{4}$

19. $x > \dfrac{1}{3}$

20. $3 \leq x \leq 10$

Solve the inequalities in Exercises 21 – 50 and graph the solutions.

21. $4x + 5 \geq -6$

22. $2x + 3 > -8$

23. $3x + 2 > 2x - 1$

24. $3y + 2 \leq y + 8$

25. $y - 6.1 \leq 4.2 - y$

26. $4x - 2.3 < 6x + 6.5$

27. $2.9x - 5.7 > -3 - 0.1x$

28. $3y - 21.9 \geq 11.04 - 3.1y$

29. $5y + 6 < 2y - 2$

30. $4 - 2x < 5 + x$

31. $4 + x > 1 - x$

32. $x - 6 > 3x + 5$

33. $\dfrac{x}{4} + 1 \leq 5 - \dfrac{x}{4}$

34. $\dfrac{x}{2} - 1 \leq \dfrac{5x}{2} - 3$

35. $\dfrac{x}{3} - 2 > 1 - \dfrac{x}{3}$

36. $\dfrac{5x}{3} + 2 > \dfrac{x}{3} - 1$

37. $-(x + 5) \leq 2x + 4$

38. $-3(2x - 5) \leq 3(x - 1)$

39. $x - (2x + 5) \geq 7 - (4 - x) + 10$

40. $x - 3(4 - x) + 5 \geq -2(3 - 2x) - x$

41. $-5 < 1 - x < 3$

42. $2 < 2x - 4 \leq 8$

43. $1 < 3x - 2 < 4$

44. $-5 < 4x + 1 < 7$

45. $-5 \leq 5x + 3 < 9$

46. $-1 < 3 - 2x < 6$

47. $\dfrac{3}{4} \leq \dfrac{1}{2}x + 6 \leq \dfrac{15}{2}$

48. $\dfrac{1}{3} \leq \dfrac{5}{6}x - 2 < \dfrac{14}{15}$

49. $-\dfrac{1}{2} < -\dfrac{2}{3}x + 5 \leq \dfrac{5}{9}$

50. $0 \leq -\dfrac{5}{8}x + 1 < \dfrac{9}{20}$

51. To receive a B grade, a student must average 80 or more but less than 90. If John received a B in the course and had five grades of 94, 78, 91, 86, and 87 before taking the final exam, what were the possible grades for his final if there were 100 points possible?

1st test	94
2nd test	86
HW avg	78
quiz avg	91
last test	87
final exam	?

52. The range for a C grade is 70 or more but less than 80. Before taking the final exam, Clyde had grades of 59, 68, 76, 84, and 69. If the final exam is counted as two tests, what is the minimum grade he could make on the final to receive a C? If there were 100 points possible, could he receive a B grade if an average of at least 80 points was required?

53. The temperature of a mixture in a chemistry experiment varied from 15° C to 65° C. What is the temperature expressed in degrees Fahrenheit?

$$\left[C = \frac{5}{9}(F - 32) \right]$$

54. The temperature at Braver Lake ranged from a low of 23° F to a high of 59° F. What is the equivalent range of temperatures in degrees Celsius?

$$\left[F = \frac{9}{5}C + 32 \right]$$

55. The sum of the lengths of any two sides of a triangle must be greater than the third side. If a triangle has one side that is 17 cm and a second side that is 1 cm less than twice the third side, what are the possible lengths for the second and third sides?

56. The sum of four times a number and 21 is greater than 45 and less than 73. What are the possible values for the number?

57. In order for Chuck to receive a B in his mathematics class, he must have a total of at least 400 points. If he has scores of 72, 68, 85, and 89, what scores can he make on the final and receive a B? (The maximum possible score on the final is 100.)

58. In Exercise 57, if the final exam counted twice, could Chuck receive a grade of A in the class if it takes 540 points for an A? Assume that the maximum possible score on the final is 100 points.

59. The Concert Hall has 400 seats. For a concert, the admission will be $6.00 for adults and $3.50 for students. If the expense of producing the concert is $1825, what is the least number of adult tickets that must be sold to realize a profit if all seats are taken?

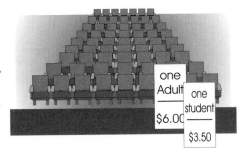

60. The Pep Club is selling candied apples to raise money. The price per apple is $2.50 until Friday, when they will sell for $2.00. If the Pep Club sells 200 apples, what is the minimum number they must sell at $2.50 each in order to raise at least $475.00?

61. Better-Car Rental charges $15 per day plus $0.25 per mile and Best-Car Rental charges $25 per day plus $0.15 per mile. How many miles can be driven in one day with a Better car and still have a bill less than what a Best car would cost?

62. Gary and Susan want to have their wedding reception at a local restaurant and have found two that seem very nice. The Seaside Stop charges $500 plus $13.00 per plate for the guests and the Tall Shrimp charges $800 plus $10.50 per plate. How many guests can they have before the cost at Seaside Stop would become more than or equal to the cost at the Tall Shrimp?

Hawkes Learning Systems: Introductory Algebra

 Solving Linear Inequalities

Chapter 3 Index of Key Ideas and Terms

Linear Equations   page 124

If a, b, and c are constants and $a \neq 0$, then a **linear equation in x** (or **first-degree equation in x**) is an equation that can be written in the form $ax + b = c$.

Addition Principle of Equality   page 125

If the same algebraic expression is added to both sides of an equation, the new equation has the same solutions as the original equation. Symbolically, if A, B, and C are algebraic expressions, then the equations

$$A = B$$
$$\text{and } A + C = B + C$$

have the same solutions.

Equivalent Equations   page 125

Equations with the same solutions are said to be **equivalent**.

Multiplication Principle of Equality   page 128

If both sides of an equation are multiplied by (or divided by) the same nonzero constant, the new equation has the same solutions as the original equation. Symbolically, if A and B are algebraic expressions and C is any nonzero constant, then the equations

$$A = B$$

$$\text{and} \quad AC = BC \qquad \text{where } C \neq 0$$

$$\text{and} \quad \frac{A}{C} = \frac{B}{C} \qquad \text{where } C \neq 0$$

have the same solutions.

Procedure for Solving Applications   pages 131 - 132

George Pòlya's four-step process as an approach to problem solving:

    1. Understand the problem
    2. Devise a plan.
    3. Carry out the plan.
    4. Look back over the results.

General Procedure for Solving Linear Equations   page 140

   1. Simplify each side of the equation by removing any grouping
      symbols and combining like terms. (This may also involve
      multiplying both sides by the LCM of the denominators or by a
      power of 10 to remove decimals. In this manner we will be dealing
      only with **integer** constants and coefficients.)

   2. Use the Addition Principle of Equality to add the opposites of
      constants and/or variables so that variables are on one side and
      constants are on the other side.

   3. Use the Multiplication Principle of Equality to multiply both sides
      by the reciprocal of the coefficient (or to divide by the coefficient).

   4. Check your answer by substituting it into the original equation.

Number Problems   pages 149 - 151

Consecutive Integers   pages 151 - 152

   Consecutive integers are two integers that differ by 1.
   Two consecutive integers can be represented as n and $n + 1$.

Consecutive Even Integers   pages 151 - 152

   Consecutive even integers are two even integers that differ by 2.
   Two consecutive even integers can be represented as n and $n + 2$ where
   n is even.

Consecutive Odd Integers   pages 151 - 153

   Consecutive odd integers are two odd integers that differ by 2.
   Two consecutive odd integers can be represented as n and $n + 2$ where
   n is odd.

Percent   page 159

   To Change a Decimal to a Percent: page 160
   Step 1: Move the decimal point two places to the right.
          (This is the same as multiplying by 100.)
   Step 2: Add the % symbol.

   To Change a Fraction (or Mixed Number) to a Percent: page 160
   Step 1: Change the fraction to a decimal. (Divide the numerator
          by the denominator. If the number is a mixed number,
          first write it as an improper fraction then divide.)
   Step 2: Change the decimal to a percent.

To Change a Percent to a Decimal:

Step 1: Move the decimal point two places to the left.

Step 2: Delete the % symbol.

page 161

To Change a Percent to a Fraction (or Mixed Number):

Step 1: Write the percent as a fraction with denominator 100 and delete the % symbol.

Step 2: Reduce the fraction (or change it to a mixed number if you prefer with the fraction part reduced).

page 161

The Formula for Solving Percent Problems

$R \cdot B = A$ (or $A = R \cdot B$)

R = **RATE** or percent (as a decimal or fraction)

B = **BASE** (number we are finding the percent of)

A = **AMOUNT** or percentage (a part of the base)

"of " means to multiply.

"is" means equal (=).

page 162 - 163

Percent of Profit

To find the **percent of profit**, form the fraction (or ratio) of profit divided by the investment and change the fraction to a percent.

page 166

Formulas

pages 172 - 173

Formulas in Geometry

Perimeter

pages 184 - 189
page 184

Square:	$P = 4s$
Rectangle:	$P = 2l + 2w$
Parallelogram:	$P = 2a + 2b$
Circle:	$C = 2\pi r$ or $C = \pi d$
Trapezoid:	$P = a + b + c + d$
Triangle:	$P = a + b + c$

Area

pages 186 - 187

Square:	$A = s^2$
Rectangle:	$A = lw$
Parallelogram:	$A = bh$
Circle:	$A = \pi r^2$
Trapezoid:	$A = \dfrac{1}{2}h(b+c)$
Triangle:	$A = \dfrac{1}{2}bh$

Volume   pages 188 - 189

 Rectangular Solid: $V = lwh$

 Rectangular Pyramid: $V = \dfrac{1}{3}lwh$

 Right Circular Cylinder: $V = \pi r^2 h$

 Right Circular Cone: $V = \dfrac{1}{3}\pi r^2 h$

 Sphere: $V = \dfrac{4}{3}\pi r^3$

Circles

Terminology and Notation Related to Circles: page 185

Term	Notation	Definition
1. Circumference	C	The perimeter of a circle is called its **circumference**.
2. Radius	r	The distance from the center of a circle to any point on the circle is called its **radius**.
3. Diameter	d	The distance from one point on a circle to another point on the circle measured through its center is called its **diameter**. (**Note:** The diameter is twice the radius: $\boldsymbol{d = 2r}$.)
4. Pi	π	π is a Greek letter used for the constant 3.1415926535... (π is an irrational number.)

Intervals of Real Numbers   pages 195 - 197

There is a one-to-one correspondence between the real numbers and the points on a line. That is, each point on a number line corresponds to one real number, and each real number corresponds to one point on a number line.

Types of Intervals
  Open Interval $a < x < b,\ x > a,\ x < a$
  Closed Interval $a \leq x \leq b$
  Half-Open Interval $a < x \leq b,\ a \leq x < b,\ x \geq a,\ x \leq a$

Linear (or First-Degree) Inequalities

Rules for Solving Linear Inequalities

page 198
page 199

1. Simplify each side of the inequality by removing any parentheses and combining like terms.
2. Add constants and/or variables to or subtract them from both sides of the inequality so that variables are on one side and constants are on the other.
3. Divide both sides by the coefficient (or multiply them by the reciprocal) of the variable and reverse the sense of the inequality if this coefficient is negative.
4. A quick (and generally satisfactory) check is to select any one number in your solution and substitute it into the original inequality.

Chapter 3 Review

For a review of the topics and problems from Chapter 3, look at the following lessons from *Hawkes Learning Systems: Introductory Algebra*

Solving Linear Equations Using Addition and Subtraction
Applications of Linear Equations: Addition and Subtraction
Solving Linear Equations Using Multiplication and Division
Applications of Linear Equations: Multiplication and Division
Solving Linear Equations
Applications: Number Problems and Consecutive Integers
Percents and Applications
Working with Formulas
Formulas in Geometry
Solving Linear Inequalities

Chapter 3 Test

Solve the equations in Exercises 1 – 6.

1. $\frac{5}{3}x + 1 = -4$

2. $4x - 5 - x = 2x + 5 - x$

3. $8(3 + x) = -4(2x - 6)$

4. $\frac{3}{2}x + \frac{1}{2}x = -3 + \frac{3}{4}$

5. $0.7x + 2 = 0.4x + 8$

6. $4(2x - 1) + 3 = 2(x - 4) - 5$

Find the missing number or percent in Exercises 7 and 8.

7. 62% of 180 is _____.

8. 48 is _____% of 150.

9. What simple interest would be earned if $5000 is invested at 9% for 9 months?

10. Which is a better investment, an investment of $6000 that earns a profit of $360 or an investment of $10,000 that earns a profit of $500? Explain your answer in terms of percent.

11. The triangle shown here indicates that the sides can be represented as $x, x + 1,$ and $2x - 1$. What is the length of each side if the perimeter is 12 meters?

In Exercises 12 – 13, solve each formula for the indicated variable.

12. $N = mrt + p$; solve for m.

13. $5x + 3y - 7 = 0$; solve for y.

14. Given the formula $C = \frac{5}{9}(F - 32)$ that relates Fahrenheit and Celsius temperatures, find the Celsius temperature equal to $-4°$ F.

15. Given the equation $3x - 2y = 18$, find the value of x that corresponds to the value of 4 for y.

16. Find (a) the perimeter and (b) the area of the figure shown here.

17. Find the volume of a sphere with diameter 20 cm.

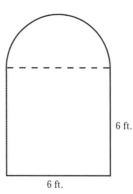

6 ft.

6 ft.

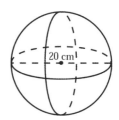

20 cm

Solve each of the inequalities in Exercises 18 – 21 and graph each solution set on a real number line.

18. $3x + 2 \le -8$

19. $-\dfrac{3}{4}x - 6 > 9$

20. $2(x - 7) < 4(2x + 3)$

21. $-2 \le 4x + 7 \le 3$

In Exercises 22 – 26, set up an equation for each word problem and solve.

22. One number is 5 more than twice another. Their sum is −22. Find the numbers.

23. Find two consecutive integers such that twice the first added to three times the second is equal to 83.

24. A rectangle is 37 inches long and has a perimeter of 122 inches.
   a. Find the width of the rectangle.
   b. Find the area of the rectangle.

25. A man's suit was on sale for $547.50.
   a. If this price was a discount of 25% from the original price, what was the original price?
   b. What would the suit cost a buyer if sales tax was 6% and the alterations were $25?

26. Find three consecutive odd integers such that three times the second is equal to 27 more than the sum of the first and the third.

27. **a.** Find the volume of concrete needed for a sidewalk that is 6 feet by 100 feet by 4 inches.

    b. Find the number of cubic yards (to the nearest tenth) in the sidewalk.
(**Note**: 12 in. = 1 ft. and 27 ft^3 = 1 yd.^{3})

28. If a cone has a height of 6 in. and a circular base with a radius of 4 in., what is its volume?

Cumulative Review: Chapters 1 – 3

Determine whether each statement in Exercises 1 – 3 is true or false. If a statement is false, rewrite it in a form that is a true statement. (There may be more than one way to correct a statement.)

1. $|-5| \leq 5$ **2.** $2^3 - 8 > 8 - 3 \cdot 4$ **3.** $\dfrac{3}{4} \geq |-1|$

Exercises 4 and 5: On a number line, graph the integers that satisfy the stated inequality.

4. $|x| \leq 6$ **5.** $|x| = x$

6. Explain, in your own words, why division by 0 is undefined.

7. List the prime numbers less than 50.

8. Find the prime factorization of each integer.
   a. 150   **b.** 125   **c.** 240

Exercises 9 – 13: Use the rules for order of operations to evaluate each expression.

9. $5^2 \cdot 2^3 - 24 \div 2 \cdot 3$ **10.** $-36 \div (-2)^2 + 20 - 2(16 - 17)$

11. $(7 - 10)[49 \div (-7) + 20 \cdot 3 - 5 \cdot 15 - (-10)] + (-1)^0$

12. $\left(\dfrac{2}{3}\right)^2 \div \dfrac{5}{18} - \dfrac{3}{8}$ **13.** $\left(\dfrac{9}{10} + \dfrac{2}{15}\right) \div \left(\dfrac{1}{2} - \dfrac{2}{3}\right)$

14. Use a calculator to find the value of the expression: $(15)^3 - 160 + 13(-5)$

Exercises 15 – 18: (a) Simplify each expression by combining like terms. (b) Evaluate the simplified form of each expression for $x = -2$ and $y = 3$.

15. $-4(y + 3) + 2y$ **16.** $3(x + 4) - 5 + x$ **17.** $2(x^2 + 4x) - (x^2 - 5x)$

18. $\dfrac{3(5x - x)}{4} - 2x - 7$

Exercises 19 – 22: Write an algebraic expression described by each of the phrases.

19. The difference between 9 and twice a number

20. Three times the sum of a number and 10

21. The number of hours in x days and 5 hours

22. The number of points scored by a football team on T touchdowns (6 points each), E extra points (1 point each), and F field goals (3 points each).

Exercises 23 – 30: Solve the following equations.

23. $9x - 8 = 4x - 13$ **24.** $6 - (x - 2) = 5$ **25.** $10 + 4x = 5x - 3(x + 4)$

26. $4.4 + 0.6x = 1.2 - 0.2x$ **27.** $-2(5x + 12) - 5 = -6(3x - 2) - 10$

28. $\dfrac{3}{4}y - \dfrac{1}{2} = \dfrac{5}{8}y + \dfrac{1}{10}$ **29.** $\dfrac{1}{2} + \dfrac{1}{5}y = \dfrac{2}{15}y + 1$ **30.** $-\dfrac{3}{4}y + \dfrac{3}{8}y = \dfrac{1}{6}y$

Exercises 31 – 33: Solve for the indicated variable.

31. $h = vt - 16t^2$; solve for v.   **32.** $A = P + Prt$; solve for r.

33. $5x - 3y = 14$; solve for y.

Exercises 34 – 36: Find the missing number.

34. 75% of _____ is 100.   **35.** 20.24 is _____ % of 92.   **36.** 25.2% of 65 is _____ .

Exercises 37 – 40: Solve each of the inequalities and graph each solution set on a real number line.

37. $2x + 5 - 3 \le 6$ **38.** $-3(7 - 2x) \ge 2 + 3x - 10$

39. $-\dfrac{1}{2}x + \dfrac{7}{8} < \dfrac{2}{3}x + \dfrac{1}{2}$ **40.** $-6.2 \le 2x - 4 \le 10.8$

41. Find (a) the perimeter and (b) area of the figure shown here. (Use $\pi = 3.14$.)

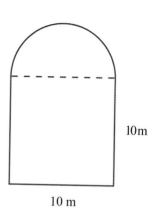

10m

10 m

42. Find the volume of the circular cylinder with dimensions as shown in the figure. (Use $\pi = 3.14$.)

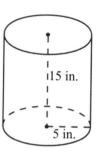

15 in.

5 in.

Exercises 43 – 54: Set up an equation or inequality for each problem and solve for the unknown quantity.

43. Your neighbor has a used (red) pickup for sale and it is perfect for you. However, you have no money. With your new job you would be able to buy it in 6 months, but you know that the truck will certainly be sold by then. Your uncle has agreed to loan you $2000 for 6 months at 8% interest and you have agreed to the terms. How much will you pay your uncle in interest for the 6 months loan?

44. Computers are on sale at a 30% discount. If you know that you will pay 6.5% in sales taxes, what will you pay for a computer that is priced at $680?

45. If twice a certain number is increased by 3, the result is 8 less than three times the number. Find the number.

46. Find three consecutive even integers such that the sum of the first and twice the second is equal to 14 more than twice the third.

47. Find four consecutive integers such that the sum of the first, second, and fourth integers is 2 less than the third.

48. ALPHA Truck Rental charges $35 a day plus $0.55 per mile and BETA Truck Rental charges $45 a day but only $0.25 per mile. For how many miles can you drive an ALPHA truck in one day and keep the cost below the cost of a BETA truck?

49. Jay works at a furniture store for a salary of $800 per month plus a commission of 6% on his sales over $5000. Kay works at the same store but works only on a straight commission of 10%. In one particular month, Jay and Kay sold the same value in furniture and made the same amount of money. What was the dollar value of furniture sold by each? What amount of money did each make for that month?

50. Find the height of a cone that has a volume of 175π in.^{3} and a radius of 7 in.

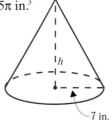

7 in.

51. The range for a grade of B is an average of 75 or more but less than 85 in an English class. If your grades on the first four exams were 73, 65, 77, and 74, what possible grades can you get on the final exam to earn a grade of B? (Assume that 100 is the maximum number of points on the final exam.)

52. What principal would you need to invest to earn $450 in interest in 6 months if the rate of interest was 9%?

53. Which is the better investment: (a) a profit of $800 on an investment of $10,000 or (b) a profit of $1200 on an investment of $15,000? Explain in terms of percents.

54. An artist has a piece of wire 30 inches long and wants to bend it into the shape of a triangle so that the sides are of length $x, 2x + 2,$ and $2x + 3$. This triangle will be a right triangle and the two shortest sides will be perpendicular to each other.
a. What will be the length of each side of the triangle?
b. What will be the area of the triangle?

Graphing Linear Equations in Two Variables

Did You Know?

In Chapter 4, you will be introduced to the idea of a graph of an algebraic equation. A graph is simply a picture of an algebraic relationship. This topic is more formally called **analytic geometry**. It is a combination of algebra (the equation) and geometry (the picture).

The idea of combining algebra and geometry was not thought of until René Descartes wrote his famous *Discourse on the Method of Reasoning* in 1637. The third appendix in this book, "La Geometrie," made Descartes' system of analytic geometry known to the world. In fact, you will find that analytic geometry is sometimes called **Cartesian geometry**.

René Descartes is perhaps better known as a philosopher than as a mathematician; he is often referred to as the father of modern philosophy. His method of reasoning was to apply the same logical structure to philosophy that had been developed in mathematics, especially geometry.

In the Middle Ages, the highest forms of knowledge were believed to be mathematics, philosophy, and theology. Many famous people in history who have reputations as poets, artists, philosophers, and theologians were also creative mathematicians. Almost every royal court had mathematicians whose work reflected glory on the royal sponsor who paid the mathematician for his research and court presence.

Descartes, in fact, died in 1650 after accepting a position at the court of the young warrior-queen, Christina of Sweden. Apparently, the frail French philosopher-mathematician, who spent his mornings in bed doing mathematics, could not stand the climate of Sweden and the hardships imposed by Christina in her demand that Descates tutor her in mathematics each morning at 5 o'clock in an unheated castle library.

"Divide each problem that you examine into as many parts as you can and as you need to solve them more easily."

René Descartes (1596 – 1650)

Thanks to René Descartes (1596 – 1650), the ideas of algebra and geometry can be combined. He developed an entire theory connecting algebraic expressions and equations with points on geometric graphs. In particular, in Chapter 4, you will see that equations with two variables can be related to graphs or "pictures" of straight lines in a plane. By studying the equations in special forms, you will be able to tell the position and basic properties of the corresponding lines before you start to draw the graphs.

4.1   The Cartesian Coordinate System

Objectives

After completing this section, you will be able to:

1. Name ordered pairs corresponding to points on graphs.

2. Graph ordered pairs in the Cartesian coordinate system.

3. Find ordered pairs that satisfy given equations.

René Descartes (1596 – 1650), a famous French mathematician, developed a system for solving geometric problems using algebra. This system is called the **Cartesian coordinate system** in his honor. Descartes based his system on a relationship between points in a plane and **ordered pairs** of real numbers. This section begins by relating algebraic formulas with ordered pairs and then shows how these ideas can be related to geometry.

Equations in Two Variables

Equations such as $d = 60t$, $I = 0.05P$, and $y = 2x + 3$ represent relationships between pairs of variables. For example, in the first equation, if $t = 3$, then $d = 60 \cdot 3 = 180$. With the understanding that t is first and d is second, we can represent $t = 3$ and $d = 180$ in the form of an ordered pair $(3, 180)$. In general, if t is the first number and d is the second number, then solutions to the equation $d = 60t$ can be written in the form of ordered pairs (t, d). Thus, we see that $(180, 3)$ is different from $(3, 180)$. **The order of the numbers in an ordered pair is critical**.

We say that $(3, 180)$ **satisfies the equation** or **is a solution of the equation** $d = 60t$. Similarly, $(100, 5)$ satisfies $I = 0.05P$ where $P = 100$ and $I = 0.05(100) = 5$. Also, $(2, 7)$ satisfies $y = 2x + 3$, where $x = 2$ and $y = 2 \cdot 2 + 3 = 7$.

In an ordered pair such as (x, y), x is called the **first component** (or **first coordinate**) and y is called the **second component** (or **second coordinate**). To find ordered pairs that satisfy an equation such as $y = 2x + 3$, we can choose any value for one variable and then find the corresponding value for the other variable by substituting into the equation.

For example, for the equation $y = 2x + 3$:

Choices For x:	**Substitution:**	**Ordered Pairs:**
$x = 1$	$y = 2 \cdot 1 + 3 = 5$	$(1, 5)$
$x = -2$	$y = 2(-2) + 3 = -1$	$(-2, -1)$
$x = \dfrac{1}{2}$	$y = 2 \cdot \dfrac{1}{2} + 3 = 4$	$\left(\dfrac{1}{2}, 4 \right)$

The ordered pairs $(1, 5)$, $(-2, -1)$, and $\left(\dfrac{1}{2}, 4 \right)$ are just three ordered pairs that satisfy the equation $y = 2x + 3$. There are an infinite number of such ordered pairs. Any real number could have been chosen for x and the corresponding value for y calculated.

Since the equation $y = 2x + 3$ is solved for y, we say that the value of y "depends" on the choice of x. Thus, in an ordered pair of the form (x, y), the second coordinate, y, is called the **dependent variable** and the first coordinate, x, is called the **independent variable**.

In the following tables the first variable, in each case, is the independent variable and the second variable is the dependent variable. Corresponding ordered pairs would be of the form (t, d), (P, I), and (x, y). The choices for the values of the independent variables are arbitrary. There are an infinite number of other values that could have just as easily been chosen.

$d = 60t$		$I = 0.05P$		$y = 2x + 3$	
t	d	P	I	x	y
5	$60 \cdot 5 = 300$	100	$0.05(100) = 5$	-2	$2(-2) + 3 = -1$
10	$60 \cdot 10 = 600$	200	$0.05(200) = 10$	-1	$2(-1) + 3 = 1$
12	$60 \cdot 12 = 720$	500	$0.05(500) = 25$	0	$2(0) + 3 = 3$
15	$60 \cdot 15 = 900$	1000	$0.05(1000) = 50$	3	$2(3) + 3 = 9$

Graphing Ordered Pairs

The Cartesian coordinate system relates algebraic equations and ordered pairs to geometry. In this system, two number lines intersect at right angles and separate the plane into four **quadrants**. The **origin**, designated by the ordered pair $(0, 0)$, is the point of intersection of the two lines. The horizontal number line is called the **horizontal axis** or **x-axis**. The vertical number line is called the **vertical axis** or **y-axis**. Points that lie on either axis are not in any quadrant. They are simply on an axis (Figure 4.1).

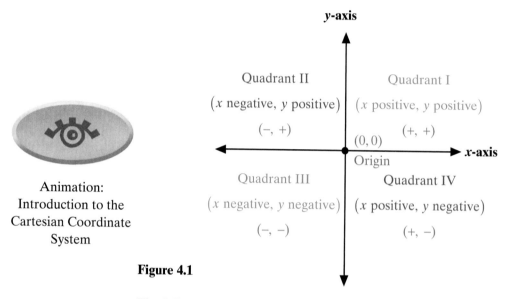

Animation:
Introduction to the
Cartesian Coordinate
System

Figure 4.1

The following important relationship between ordered pairs of real numbers and points in a plane is the cornerstone of the Cartesian coordinate system.

One-to-One Correspondence

There is a one-to-one correspondence between points in a plane and ordered pairs of real numbers.

In other words, for each point there is one and only one corresponding ordered pair of real numbers, and for each ordered pair of real numbers there is one and only one corresponding point.

The **graphs of the points** A (2, 1), B (−2, 3), C (−3, −2), D (1, −2), and E (3, 0) are shown in Figure 4.2. [**Note**: An ordered pair of real numbers and the corresponding point on the graph are frequently used to refer to each other. Thus, the ordered pair (2, 1) and the point (2, 1) are interchangeable ideas.]

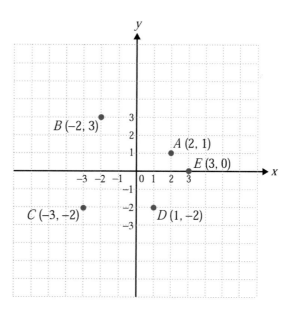

Figure 4.2

Example 1: Graph the Ordered Pairs

Graph the sets of ordered pairs.

a. $\{(-2, 1), (0, 2), (1, 3), (2, -3)\}$
(**Note**: The listing of ordered pairs in each set can be in any order.)

Solution:

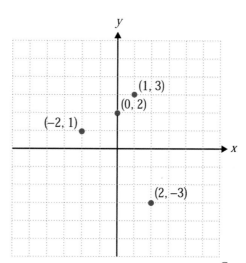

Animation:
Introduction to the
Cartesian Coordinate
System

Continued on next page...

b. $\{(-1, 3), (0, 1), (1, -1), (2, -3), (3, -5)\}$

Solution:

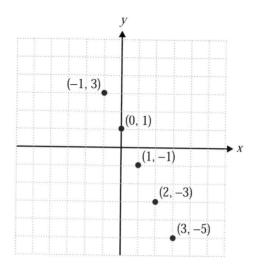

The points (ordered pairs) in Example 1b can be shown to satisfy the equation $y = -2x + 1$. For example, using $x = -1$ in the equation yields,

$$y = -2(-1) + 1 = 2 + 1 = 3$$

and the ordered pair $(-1, 3)$ satisfies the equation. Similarly, letting $y = 1$ gives,

$$1 = -2x + 1$$
$$0 = -2x$$
$$0 = x$$

and the ordered pair $(0, 1)$ satisfies the equation.

We can write all the ordered pairs in Example 1b in table form.

x	$-2x + 1 = y$
-1	$-2(-1) + 1 = 3$
0	$-2(0) + 1 = 1$
1	$-2(1) + 1 = -1$
2	$-2(2) + 1 = -3$
3	$-2(3) + 1 = -5$

Example 2: Ordered Pairs ● ● ● ● ● ● ● ● ● ● ● ● ● ● ● ● ●

a. Determine which, if any, of the ordered pairs $(0, -2)$, $\left(\dfrac{2}{3}, 0\right)$, and $(2, 5)$ satisfy the equation $y = 3x - 2$.

Solution: We will substitute 0, $\dfrac{2}{3}$, and 2 for x and see if the corresponding y-values match those in the given ordered pairs.

$x = 0$: $y = 3(\,0\,) - 2 = -2$ so, $(0, -2)$ satisfies the equation.

$x = \dfrac{2}{3}$: $y = 3\left(\dfrac{2}{3}\right) - 2 = 0$ so, $\left(\dfrac{2}{3}, 0\right)$ satisfies the equation.

$x = 2$: $y = 3(\,2\,) - 2 = 4$ so, $(2, 4)$ satisfies the equation.

The point $(2, 5)$ does not satisfy the equation $y = 3x - 2$ since $y = 4$ when $x = 2$.

b. Determine the missing coordinate in each of the following ordered pairs so that the point will satisfy the equation $2x + 3y = 12$:

$$(0, \quad), (3, \quad), (\quad, 0), (\quad, -2).$$

Solution: For $(0, \quad)$, let $x = 0$:

$$2(\,0\,) + 3y = 12$$
$$3y = 12$$
$$y = 4$$

For $(3, \quad)$, let $x = 3$:

$$2(\,3\,) + 3y = 12$$
$$6 + 3y = 12$$
$$3y = 6$$
$$y = 2$$

The ordered pair is $(0, 4)$.

The ordered pair is $(3, 2)$.

For $(\quad, 0)$, let $y = 0$:

$$2x + 3(\,0\,) = 12$$
$$2x = 12$$
$$x = 6$$

For $(\quad, -2)$, let $y = -2$:

$$2x + 3(\,-2\,) = 12$$
$$2x - 6 = 12$$
$$2x = 18$$
$$x = 9$$

The ordered pair is $(6, 0)$.

The ordered pair is $(9, -2)$.

Continued on next page...

c. Complete the table below so that each ordered pair will satisfy the equation $y = 1 - 2x$.

x	y
0	
	3
$\dfrac{1}{2}$	
5	

Solution: Substituting each given value for x and y into the equation $y = 1 - 2x$ gives the following table of ordered pairs.

x	y
0	1
−1	3
$\dfrac{1}{2}$	0
5	−9

For $x = 0$:
$$y = 1 - 2(0) = 1$$

For $x = \dfrac{1}{2}$:
$$y = 1 - 2\left(\dfrac{1}{2}\right) = 0$$

For $y = 3$:
$$3 = 1 - 2x$$
$$2 = -2x$$
$$-1 = x$$

For $x = 5$:
$$y = 1 - 2(5) = -9$$

● ●

NOTES

Although this discussion is related to ordered pairs of real numbers, most of the examples use ordered pairs of **integers**. This is because ordered pairs of integers are relatively easy to locate on a graph and relatively easy to read from a graph. Ordered pairs with fractions, decimals, or radicals (irrational numbers) must be located by estimating the positions of the points. The precise coordinates intended for such points can be difficult or impossible to read because large dots must be used so the points can be seen. **Even with these difficulties, you should understand that we are discussing ordered pairs of real numbers and that points with fractions, decimals, and radicals as coordinates do exist and should be plotted by estimating their positions.**

Practice Problems

1. *Determine which ordered pairs satisfy the equation $3x + y = 14$.*

   a. $(5, -1)$ **b.** $(4, 2)$ **c.** $(-1, 17)$

2. *Given $3x + y = 5$, find the missing coordinate of each ordered pair so that it will satisfy the equation.*

   a. $(0, \quad)$ **b.** $\left(\dfrac{1}{3}, \quad\right)$ **c.** $(\quad, 2)$

3. *Complete the table so that each ordered pair will satisfy the equation $y = \dfrac{2}{3}x + 1$.*

x	y
0	
	-2
-3	
6	

Answers to Practice Problems: 1. All satisfy the equation. **2. a.** $(0, 5)$ **b.** $\left(\dfrac{1}{3}, 4\right)$ **c.** $(1, 2)$

3.

x	y
0	1
$-\dfrac{9}{2}$	-2
-3	-1
6	5

4.1 Exercises

List the sets of ordered pairs corresponding to the graphs in Exercises 1 – 10. Assume that the grid lines are marked one unit apart. (**Note:** There is no particular order to be followed in listing ordered pairs.)

1.

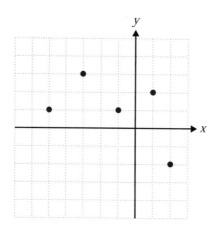

2.

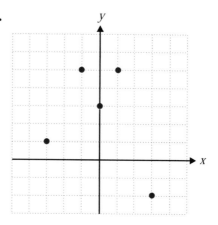

3.

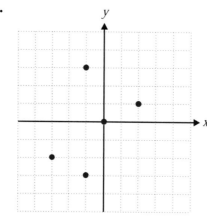

4.

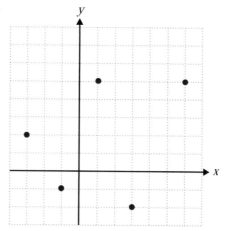

5.

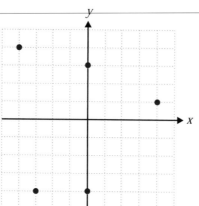

6.

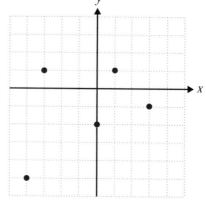

7.

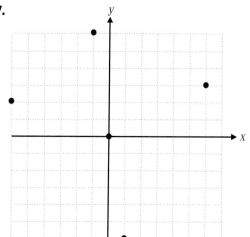

8.

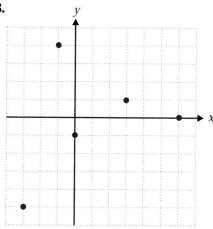

9.

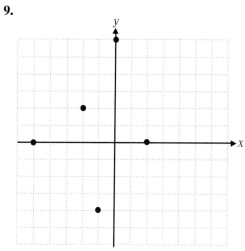

10.

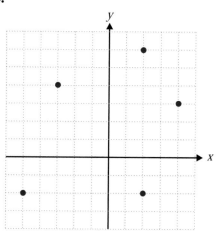

Graph the sets of ordered pairs and label the points in Exercises 11 – 24.

11. $\{(4, -1), (3, 2), (0, 5), (1, -1), (1, 4)\}$

12. $\{(-1, -1), (-3, -2), (1, 3), (0, 0), (2, 5)\}$

13. $\{(1, 2), (0, 2), (-1, 2), (2, 2), (-3, 2)\}$

14. $\{(1, 0), (3, 0), (-2, 1), (-1, 1), (0, 0)\}$

15. $\{(-1, 4), (0, -3), (2, -1), (4, 1)\}$

16. $\{(-1, -1), (0, 1), (1, 3), (2, 5), (3, 10)\}$

17. $\{(4, 1), (0, -3), (1, -2), (2, -1)\}$

18. $\{(0, 1), (1, 0), (2, -1), (3, -2), (4, -3)\}$

19. $\{(1,4),(-1,-2),(0,1),(2,7),(-2,-5)\}$ **20.** $\left\{(1,-3),\left(-4,\dfrac{3}{4}\right),\left(2,-2\dfrac{1}{2}\right),\left(\dfrac{1}{2},4\right)\right\}$

21. $\left\{(0,0),\left(-1,\dfrac{7}{4}\right),\left(3,-\dfrac{1}{2}\right)\right\}$ **22.** $\left\{\left(\dfrac{3}{4},\dfrac{1}{2}\right),\left(2,-\dfrac{5}{4}\right),\left(\dfrac{1}{3},-2\right),\left(-\dfrac{5}{3},2\right)\right\}$

23. $\{(1.6,-2),(3,2.5),(-1,1.5),(0,-2.3)\}$ **24.** $\{(-2,2),(-3,1.6),(3,0.5),(1.4,0)\}$

Determine which of the given ordered pairs satisfy the equation in Exercises 25 – 30.

25. $2x - y = 4$

  a. $(1,1)$
  b. $(2,0)$
  c. $(1,-2)$
  d. $(3,2)$

26. $x + 2y = -1$

  a. $(1,-1)$
  b. $(1,0)$
  c. $(2,1)$
  d. $(3,-2)$

27. $4x + y = 5$

  a. $\left(\dfrac{3}{4},2\right)$
  b. $(4,0)$
  c. $(1,1)$
  d. $(0,3)$

28. $2x - 3y = 7$

  a. $(1,3)$
  b. $\left(\dfrac{1}{2},-2\right)$
  c. $\left(\dfrac{7}{2},0\right)$
  d. $(2,1)$

29. $2x + 5y = 8$

  a. $(4,0)$
  b. $(2,1)$
  c. $(1,1.2)$
  d. $(1.5,1)$

30. $3x + 4y = 10$

  a. $(-2,3)$
  b. $(0,2.5)$
  c. $(4,-2)$
  d. $(1.2,1.6)$

Determine the missing coordinate in each of the ordered pairs so that it will satisfy the equation given in Exercises 31 – 40.

31. $x - y = 4$
  $(0,\ \ ),(2,\ \ ),(\ \ ,0),(\ \ ,-3)$

32. $x + y = 7$
  $(0,\ \ ),(-1,\ \ ),(\ \ ,0),(\ \ ,3)$

33. $x + 2y = 6$
  $(0,\ \ ),(2,\ \ ),(\ \ ,0),(\ \ ,4)$

34. $3x + y = 9$
  $(0,\ \ ),(4,\ \ ),(\ \ ,0),(\ \ ,3)$

35. $4x - y = 8$
  $(0,\ \ ),(1,\ \ ),(\ \ ,0),(\ \ ,-4)$

36. $x - 2y = 2$
  $(0,\ \ ),(4,\ \ ),(\ \ ,0),(\ \ ,3)$

37. $2x + 3y = 6$
  $(0,\ \ ),(-1,\ \ ),(\ \ ,0),(\ \ ,-2)$

38. $5x + 3y = 15$
  $(0,\ \ ),(2,\ \ ),(\ \ ,0),(\ \ ,4)$

39. $3x - 4y = 7$

$(0, \ \ ), (1, \ \ ), (\ \ , 0), \left(\ \ , \dfrac{1}{2} \right)$

40. $2x + 5y = 6$

$(0, \ \ ), \left(\dfrac{1}{2}, \ \ \right), (\ \ , 0), (\ \ , 2)$

Complete the tables in Exercises 41 – 55 so that each ordered pair will satisfy the given equation. Graph the resulting sets of ordered pairs.

41. $y = 3x$

x	y
0	
	−3
−2	
	6

42. $y = -2x$

x	y
0	
	4
3	
	−2

43. $y = 2x - 3$

x	y
0	
	−1
−2	
	$\dfrac{1}{2}$

44. $y = 3x + 5$

x	y
0	
	−1
−2	
	$\dfrac{1}{2}$

45. $y = 7 - 3x$

x	y
0	
	0
−1	
	$\dfrac{1}{3}$

46. $y = 6 - 2x$

x	y
0	
	0
−2	
	$\dfrac{1}{2}$

47. $y = \dfrac{3}{4}x + 2$

x	y
0	
	5
−4	
	$\dfrac{5}{4}$

48. $y = \dfrac{3}{2}x - 1$

x	y
0	
	2
−2	
	$-\dfrac{5}{2}$

49. $3x - 5y = 9$

x	y
0	
	0
−2	
	−1

50. $4x + 3y = 6$

x	y
0	
	0
3	
	-1

51. $5x - 2y = 10$

x	y
0	
	0
-2	
	-1

52. $3x - 2y = 12$

x	y
	0
0	
	-3
2	

53. $x - y = 1.5$

x	y
0	
1.5	
	2.3
	-2.5

54. $2x + 3.2y = 6.4$

x	y
0	
3.2	
	0.8
	-0.2

55. $3x + y = -2.4$

x	y
	0
0	
	0.6
1.6	

56. Given the equation $I = 0.08P$, where I is the interest earned on a principal P at the rate of 8%:

a. Make a table of ordered pairs for the values of P and I if P has the values $1000, $2000, $3000, $4000, and $5000.

b. Graph the points corresponding to the ordered pairs.

P	I
1000	
2000	
3000	
4000	
5000	

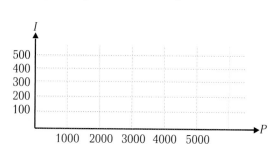

57. Given the equation $F = \dfrac{9}{5}C + 32$ where C is temperature in degrees Celsius and F is the corresponding temperature in degrees Fahrenheit:

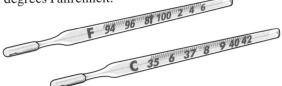

a. Make a table of ordered pairs for the values of C and F if C has the values $-20°$, $-10°, -5°, 0°, 5°, 10°,$ and $15°$.

b. Graph the points corresponding to the ordered pairs.

C	F
−20	
−10	
−5	
0	
5	
10	
15	

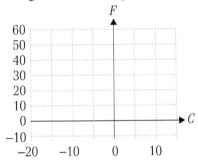

Writing and Thinking About Mathematics

58. In statistics, data is sometimes given in the form of ordered pairs where each ordered pair represents two pieces of information about one person. For example, ordered pairs might represent the height and weight of a person or the person's number of years of education and that person's annual income. The ordered pairs are plotted on a graph and the graph is called a **scatter diagram** (or **scatter plot**). Such scatter diagrams are used to see if there is any pattern to the data and, if there is, then to use the diagram to predict the value for one of the variables if the value of the other is known. For example, if you know that a person's height is 5 ft. 6 in., then his or her weight might be predicted from information indicated in a scatter diagram that has several points of known information about height and weight.

a. The following table of values indicates the number of push-ups and the number of sit-ups that ten students did in a physical education class. Plot these points in a scatter diagram.

Person	#1	#2	#3	#4	#5	#6	#7	#8	#9	#10
x (push-ups)	20	15	25	23	35	30	42	40	25	35
y (sit-ups)	25	20	20	30	32	36	40	45	18	40

Continued on next page...

b. Does there seem to be a pattern in the relationship between push-up and sit-ups? What is this pattern?

c. Using the scatter diagram in Part a, predict the number of sit-ups that a students might be able to do if they have just done each of the following numbers of push-ups: 22, 32, 35, and 45. (**Note:** In each case, there is no one correct answer. The answers are only estimates based on the diagram.)

59. Ask ten friends or fellow students what their height and weight is. Organize the data in table form and then plot the corresponding scatter diagram. Knowing your own height, does the pattern indicated in the scatter diagram seem to predict your weight?

60. Ask ten friends or fellow students what their height and age is. Organize the data in table form and then plot the corresponding scatter diagram. Knowing your own height, does the pattern indicated in the scatter diagram seem to predict your age? Do you think that all scatter diagrams can be used to predict information related to the two variables graphed? Explain.

Hawkes Learning Systems: Introductory Algebra

 Introduction to the Cartesian Coordinate System

4.2 Graphing Linear Equations in Two Variables

Objectives

After completing this section, you will be able to:

1. Graph linear equations by choosing and plotting any two points on the line.

2. Graph linear equations by calculating x- and y-intercepts.

3. Graph horizontal lines.

4. Graph vertical lines.

The Standard Form: $Ax + By = C$

In Section 4.1, we discussed ordered pairs and graphed a few points (ordered pairs) that satisfied particular equations. Now, suppose we want to graph all the points that satisfy the equation $y = 2x + 3$. The fact is, there are an infinite number of such points. In Figure 4.3, we have graphed five points to try to find a pattern.

x	$2x + 3 = y$
0	$2(0) + 3 = 3$
-1	$2(-1) + 3 = 1$
$\dfrac{1}{2}$	$2\left(\dfrac{1}{2}\right) + 3 = 4$
1	$2(1) + 3 = 5$
-2	$2(-2) + 3 = -1$

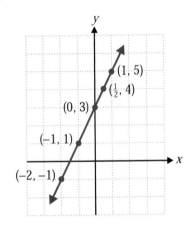

Figure 4.3

The five points in Figure 4.3 appear to lie on a straight line. They in fact do lie on a straight line, and any ordered pair that satisfies the equation $y = 2x + 3$ will also lie on that same line.

What determines whether or not the points that satisfy an equation will lie on a straight line? The points that satisfy any equation of the form

$$Ax + By = C \quad \text{(Standard Form)}$$

will lie on a straight line. The equation is called a **linear equation** and is considered the **standard form** for the equation of a line. We can write the equation $y = 2x + 3$ in the standard form as

$$-2x + y = 3 \quad \text{or} \quad 2x - y = -3.$$

Note that in the standard form $Ax + By = C$, A and B may be positive, negative, or 0, but A and B cannot **both** be 0.

Since we now know that the graph will be a straight line, only two points are necessary to determine the entire graph (two points determine a line). The choice of the two points depends on the choice of any two values of x or any two values of y. A third point is sometimes chosen as insurance against a mistake and to help place the graph of the line in the right position. Remember that there are an infinite number of points on a line, so there are an infinite number of possible choices for x (or y). The only restriction is that the ordered pair must satisfy the equation.

Example 1: Graph by Plotting Points ● ● ● ● ● ● ● ● ● ● ● ● ● ●

a.   Draw the graph of the linear equation $x + 3y = 6$

Solution:

For $x = 0$:
$$0 + 3y = 6$$
$$y = 2$$

For $x = 3$:
$$3 + 3y = 6$$
$$3y = 3$$
$$y = 1$$

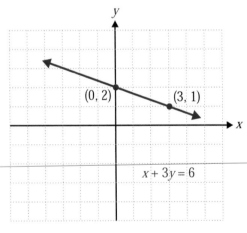

Two points on the graph are $(0, 2)$ and $(3, 1)$. You may have chosen two other values for x and calculated the corresponding y-values. You would still get the same straight line as the graph. For accuracy in graphing, avoid choosing two points close together.

Continued on next page...

b. Draw the graph of the linear equation $2x - 5y = 10$.

Solution:

For $x = 5$:
$$2 \cdot 5 - 5y = 10$$
$$10 - 5y = 10$$
$$-5y = 0$$
$$y = 0$$

For $x = 0$:
$$2 \cdot 0 - 5y = 10$$
$$0 - 5y = 10$$
$$y = -2$$

For $x = -5$:
$$2(-5) - 5y = 10$$
$$-10 - 5y = 10$$
$$-5y = 20$$
$$y = -4$$

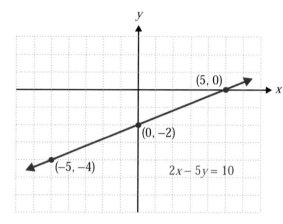

Graphing three points is a good idea, if only to be sure the graph is in the right position and no error has been made in the calculations for the other points.

y-intercept and *x*-intercept

While the choice of the values for x or y can be arbitrary, letting $x = 0$ will locate the point on the graph where the line crosses the y-axis. This point is called the **y-intercept**. The **x-intercept** is the point found by letting $y = 0$. These two points are generally easy to locate and are frequently used as the two points for drawing the graph of a linear equation.

> ### Example 2: Graph Using the *y*- and *x*- Intercepts ● ● ● ● ● ●

Graph the following linear equations by locating the *y*-intercepts and the *x*-intercepts.

a. $3y + 2x = 6$

Solution: $x = 0 \rightarrow y = 2$
$ y = 0 \rightarrow x = 3$

$(0, 2)$ is the *y*-intercept
$(3, 0)$ is the *x*-intercept

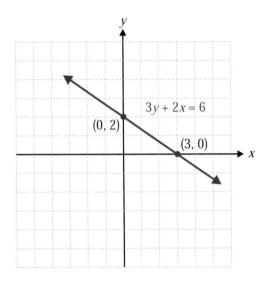

b. $2x - 3y = 12$

Solution: $x = 0 \rightarrow y = -4$
$ y = 0 \rightarrow x = 6$

$(0, -4)$ is the *y*-intercept
$(6, 0)$ is the *x*-intercept

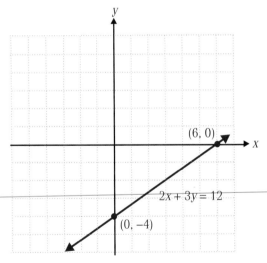

● ●

With the understanding that the *x*-coordinate of the **y-intercept** is always 0, the corresponding *y*-coordinate is also called the **y-intercept**. Thus, in Examples 2a and 2b just given, we will say that the *y*-intercept is 2 and –4, respectively rather than giving the ordered pairs (0, 2) and (0, –4). Similarly the **x-intercept** is 3 in Example 2a, and 6 in Example 2b.

Lines that Contain the Origin

If the line goes through the origin, both the *x*-intercept and *y*-intercept will be 0. In this case, some other point must be used.

> **Example 3: Lines that Contain the Origin** ● ● ● ● ● ● ● ● ● ● ●

Graph the linear equation $y = 3x$.

Solution: Locate two points on the graph.

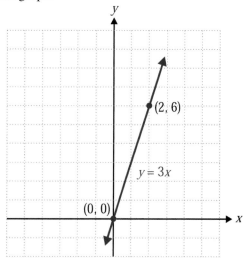

$$x = 0 \;\rightarrow\; y = 0$$
$$x = 2 \;\rightarrow\; y = 6$$

Horizontal and Vertical Lines

Now, consider an equation in the form $0x + By = C$ where the coefficient of *x* is 0. For example,

$$0x + y = 3 \qquad \text{or just} \qquad y = 3.$$

Every value chosen for *x* will be multiplied by 0 and the corresponding *y*-value will be 3. In effect, *x*-values have no influence on the *y*-value. In table form three such points are:

x	y
−2	3
0	3
5	3

Regardless of what value is chosen for *x*, the corresponding *y*-value is 3. Thus, the graph of the equation $y = 3$ is a **horizontal line** (see Figure 4.4).

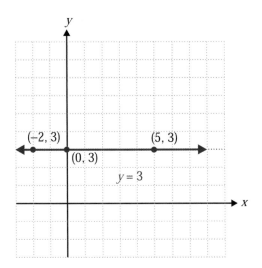

Figure 4.4

Next, consider an equation in the form $Ax + 0y = C$ where the coefficient of y is 0. For example,

$$x + 0y = 2 \qquad \text{or just} \qquad x = 2.$$

Every value chosen for y will be multiplied by 0 and the corresponding x-value will be 2. In effect the y-values have no influence on the value of x. In table form three such points are:

x	y
2	4
2	0
2	−3

Regardless of what is chosen for y, the corresponding x-value is 2. Thus, the graph of the equation $x = 2$ is a **vertical line** (see Figure 4.5).

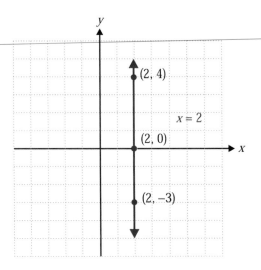

Figure 4.5

We can make the following two statements about horizontal and vertical lines:

1. The graph of any equation of the form $y = b$ is a horizontal line.
2. The graph of any equation of the form $x = a$ is a vertical line.

Example 4: Horizontal and Vertical Lines ● ● ● ● ● ● ● ● ● ● ● ●

Graph each of the following linear equations.

a. $2y = 5$

Solution: Solving for y:

$$2y = 5$$

$$y = \frac{5}{2}$$

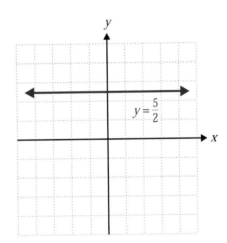

b. $x + 6 = 0$

Solution: Solving for x:
$$x + 6 = 0$$
$$x = -6$$

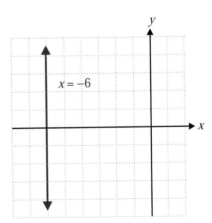

● ●

Using a TI-83 Plus Calculator to Graph Straight Lines

By following the steps outlined here you will be able to see the graphs of all straight lines that are not vertical. The calculator will not graph vertical lines using the methods described here.

To have the calculator graph a nonvertical straight line, you must first solve the equation for y. For example,

given the equation: $2x + y = 3$

solving for y gives: $y = -2x + 3$ (or $y = 3 - 2x$)

Step 1: Press the **MODE** key and set all the highlighted keys as shown in the diagram. If the mode is not as shown, press the down arrow until you reach the desired line and press **ENTER**.

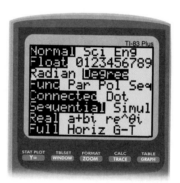

It is particularly important that **Func** is highlighted. This stands for function. Function concepts and notation will be introduced later in this chapter.

Step 2: Press the **ZOOM** key and then press 6 to get the **ZStandard** window for your graphs. This will give scales from −10 to 10 for both the x-axis and the y-axis. If you later decide that you would like some other scales on either axis, press **WINDOW** and set the window values to the appropriate values.

Step 3: Press the Y= key in the upper left corner of the keyboard. Here you will see the following window.

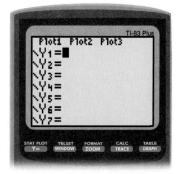

Step 4: Enter the expression for *y* here after **Y₁ =**. The letter *x* is entered by pressing the x,T,θ,*n* key located just below the MODE key. Also note that the negative sign (−) is next to the ENTER key.

The screen should appear as follows:

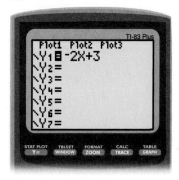

Note that the notation on the screen **Y₁ =, Y₂ =, Y₃ =**, and so forth, allows you to graph several equations at once. We will discuss and use this feature later.

Step 5: Press the GRAPH key in the upper right corner of the keyboard. Your graph should appear as follows:

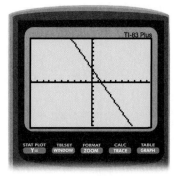

To clear the screen you may press CLEAR or 2nd QUIT.

4.2 Exercises

For Exercises 1 – 6, use your knowledge of x-intercepts and y-intercepts to match each of the following equations with its graph.

1. $4x + 3y = 12$

3. $x + 2y = 8$

5. $x + 4y = 0$

2. $4x - 3y = 12$

4. $-x + 2y = 8$

6. $5x - y = 10$

a.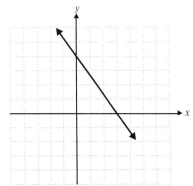

b.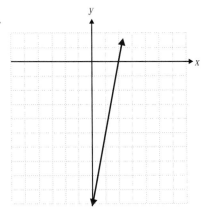

c.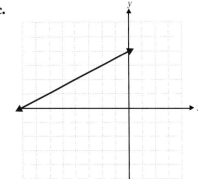

d.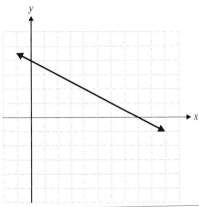

e.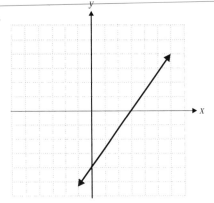

f.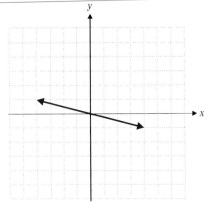

In Exercises 7 – 40, first tell (just by looking at the equation) whether the graph will be a vertical line, a horizontal line, or neither. Then, graph the line corresponding to each linear equation by locating any two (or three) ordered pairs (points) that satisfy the equation.

7. $y = 2x$

8. $y = 3x$

9. $y = -x$

10. $y = -5x$

11. $y = x - 4$

12. $y = x + 3$

13. $y = x + 2$

14. $y = x - 6$

15. $y = 4 - x$

16. $y = 8 - x$

17. $y = 2x - 1$

18. $y = 5 - 2x$

19. $3y = 12$

20. $10 - 2y = 0$

21. $2x - 8 = 0$

22. $9 - 4x = 0$

23. $x - 2y = 4$

24. $x + 3y = 5$

25. $2x + y = 0$

26. $2y = x$

27. $5y = 0$

28. $2y - 3 = 0$

29. $4x = 0$

30. $-\dfrac{3}{4}x - 1 = 0$

31. $2x + 3y = 7$

32. $4x + 3y = 11$

33. $3y - 2x = 4$

34. $3x - 2y = 6$

35. $5x + 2y = 9$

36. $2x - 7y = -14$

37. $4x + 2y = -10$

38. $y = \dfrac{1}{2}x + 1$

39. $y = \dfrac{1}{3}x - 3$

40. $\dfrac{2}{3}x + y = 4$

Graph the linear equations in Exercises 41 – 55 by locating the y-intercept and the x-intercept.

41. $x + y = 4$

42. $x - 2y = 6$

43. $3x - 2y = 6$

44. $3x - 4y = 12$

45. $5x + 2y = 10$

46. $3x + 7y = -21$

47. $2x - y = 9$

48. $4x + y = 7$

49. $x + 3y = 5$

50. $x - 6y = 3$

51. $\dfrac{1}{2}x - y = 4$

52. $\dfrac{2}{3}x - 3y = 4$

53. $\dfrac{1}{2}x - \dfrac{3}{4}y = 6$

54. $5x + 3y = 7$

55. $2x + 3y = 5$

In Exercises 56 – 60, solve each of the equations for y and use your graphing calculator to graph each line.

56. $x + y = 5$

57. $2y + x = 0$

58. $3y - 15 = 0$

59. $3x - y = 4$

60. $5x - 2y = 10$

Writing and Thinking About Mathematics

Each of the following equations is not linear and the corresponding graph is not a straight line. Make a table and find several points, some where x is positive and some where x is negative, to determine the nature of the corresponding graph. After you have plotted your points and analyzed the nature of the graph, enter the expression for y in your graphing calculator to verify that you have a reasonably accurate graph.

61. $y = \dfrac{4}{x}$

62. $y = x^2$

63. $y = -x^2$

64. $y = x^2 - 5$

65. $y = x^3$

Hawkes Learning Systems: Introductory Algebra

Graphing Linear Equations by Plotting Points

4.3

The Slope-Intercept Form: $y = mx + b$

Objectives

After completing this section, you will be able to:

1. *Find the slopes of lines given two points.*

2. *Write the equations of lines given the slopes and y-intercepts.*

3. *Graph lines given the slopes and y-intercepts.*

4. *Find the slopes and y-intercepts of lines.*

5. *Find the slopes of horizontal and vertical lines.*

The Meaning of Slope

A carpenter is given a set of house plans that call for a 5 : 12 roof (or a roof with a **pitch** of 5 : 12) (Figure 4.6). This means that the roof must be constructed so that for every 5 inches of rise (vertical distance), there are 12 inches of run (horizontal distance). That is, the ratio of rise to run is $\dfrac{5}{12}$.

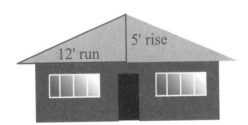

Figure 4.6

Note that this ratio can be in units other than inches such as feet or meters. (See Figure 4.7.)

$$\frac{\text{rise}}{\text{run}} = \frac{5}{12} = \frac{2\frac{1}{2}}{6} = \frac{10}{24}$$

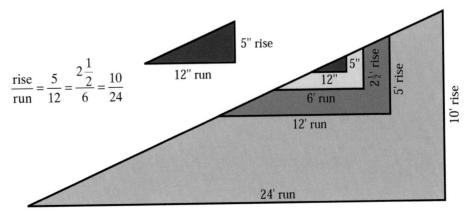

Figure 4.7

What if another roof is to have a pitch of 7 : 12? Would the carpenter then construct the roof so that for every 7 inches of rise, there would be 12 inches of run? Yes. The ratio of rise to run would be $\dfrac{7}{12}$. (See Figure 4.8.)

$$\frac{\text{rise}}{\text{run}} = \frac{7}{12} = \frac{3\frac{1}{2}}{6} = \frac{14}{24}$$

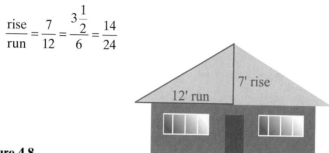

Figure 4.8

Other examples involving the ratio of rise to run are the slope (or incline) of a road and the slope of a ditch. These ideas are very important to drivers, particularly truck drivers in the mountains, and to construction of highways and drainage along the highways. For a straight line, the ratio of rise to run is called the **slope** of the line.

$$\text{slope} = \frac{\text{rise}}{\text{run}}$$

The graph of the linear equation $y = \dfrac{1}{2}x + 1$ is shown in Figure 4.9. What do you think is the slope of the line? Do you think the slope is positive or negative? Do you think the slope might be $\dfrac{1}{2}$?

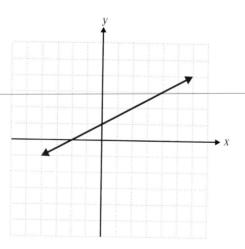

Figure 4.9

The concept of slope also relates to situations that involve ideas other than construction. For example, miles per hour in traveling and gallons of water per day used at your home, can be illustrated graphically and the **rate of change** is the slope of the line in the graph. (See Figure 4.10.)

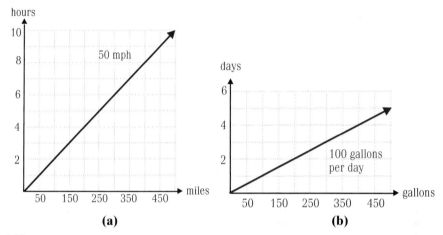

(a)   (b)

Figure 4.10

Calculating the Slope

To calculate the slope of a line:

> **1.** Find **any two** points on the line.
> **2.** Calculate the ratio of rise to run.

Regardless of the two points used on the line, the right triangles formed (as illustrated in Figure 4.7) will be similar triangles. A fact of geometry is that corresponding sides of similar triangles are proportional. That is, the ratio of rise to run, which indicates the slope of the line, is the same for every one of these right triangles.

For example, consider the line $y = 3x - 1$, and find any two points on the line. Two such points can be found by letting $x = -1$ and $x = 3$:

$$
\begin{array}{ll}
\text{For } x = -1: & \text{For } x = 3: \\
y = 3(-1) - 1 & y = 3(3) - 1 \\
y = -3 - 1 & y = 9 - 1 \\
y = -4 & y = 8
\end{array}
$$

Now, using $P_1(-1, -4)$ and $P_2(3, 8)$, the coordinates of P_3 are $(3, -4)$, as shown in Figure 4.11(b). P_3 has the same x-coordinate as P_2 and the same y-coordinate as P_1.

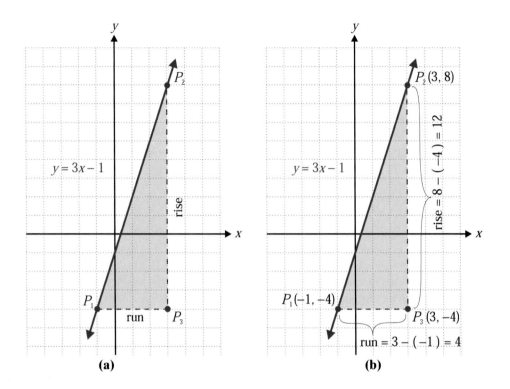

(a)   **(b)**

Figure 4.11

NOTES   In the notation P_1, 1 is called a **subscript** and P_1 is read "P sub 1". Similarly, P_2 is read "P sub 2" and P_3 is read "P sub 3."

For the line $y = 3x - 1$ and using the points $(-1, -4)$ and $(3, 8)$,

$$\text{slope} = \frac{\text{rise}}{\text{run}} = \frac{8 - (-4)}{3 - (-1)} = \frac{12}{4} = 3$$

Now find the slope of the line $y = 3x + 2$. Locate two points on the line and calculate the ratio. For example,

For $x = 0$:   For $x = 1$:

$y = 3\cdot 0 + 2 = 2$ $y = 3\cdot 1 + 2 = 5$

Thus, $P_1(0, 2)$ and $P_2(1, 5)$ are as shown in Figure 4.12. The point $P_3(1, 2)$ is shown to help illustrate the rise and the run.

$$\text{slope} = \frac{\text{rise}}{\text{run}} = \frac{5-2}{1-0} = \frac{3}{1} = 3$$

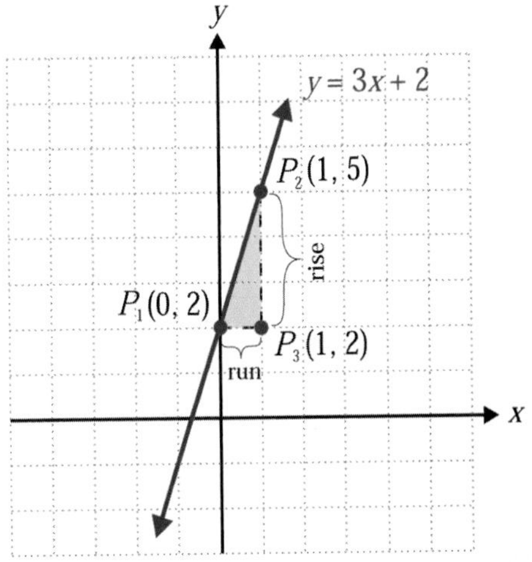

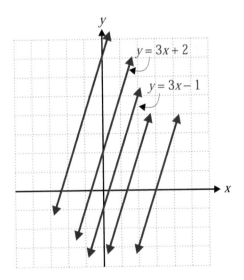

Figure 4.12

Notice that, as we have just illustrated, the two lines $y = 3x - 1$ and $y = 3x + 2$ have the same slope, 3 (See Figure 4.13). This means that the lines are **parallel**. All lines with the same slope are parallel. All the lines in Figure 4.13 have slope 3 and are parallel.

Figure 4.13   **Parallel lines have the same slope.**

By using subscript notation, we can develop a formula for the slope of any line.

Slope

*Let $P_1(x_1, y_1)$ and $P_2(x_2, y_2)$ be two points on a line. Then $P_3(x_2, y_1)$ is at the right angle shown in Figure 4.14, and the **slope** can be calculated as follows.*

$$slope = \frac{rise}{run} = \frac{y_2 - y_1}{x_2 - x_1}$$

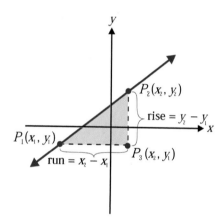

Figure 4.14

Example 1: Slope •

a. Using the formula for slope, find the slope of the line $2x + 3y = 6$.

Solution:

Let $x_1 = 0$:

$$2 \cdot 0 + 3y_1 = 6$$
$$3y_1 = 6$$
$$y_1 = 2$$

Let $x_2 = -3$:

$$2(-3) + 3y_2 = 6$$
$$-6 + 3y_2 = 6$$
$$3y_2 = 12$$
$$y_2 = 4$$

$$(x_1, y_1) = (0, 2) \quad \text{and} \quad (x_2, y_2) = (-3, 4)$$

$$slope = \frac{y_2 - y_1}{x_2 - x_1} = \frac{4 - 2}{-3 - 0} = \frac{2}{-3} = -\frac{2}{3}$$

Continued on next page...

b. Suppose that the order of the points in Example 1 is changed. That is, $(x_1, y_1) = (-3, 4)$ and $(x_2, y_2) = (0, 2)$. Will this make a difference in the slope?

Solution: Slope $= \dfrac{y_2 - y_1}{x_2 - x_1} = \dfrac{2 - 4}{0 - (-3)} = \dfrac{-2}{3} = -\dfrac{2}{3}$

Thus, the slope is the same even if the order of the points is reversed.

● ●

As demonstrated in Examples 1a and 1b, changing the order of the points does not make a difference in the value of the slope. Both the numerator and the denominator change signs so the fraction has the same value. In Example 1a, $\dfrac{2}{-3} = -\dfrac{2}{3}$ and in Example 1b, $\dfrac{-2}{3} = -\dfrac{2}{3}$. The important part of the procedure is that **the coordinates must be subtracted in the same order in both the numerator and the denominator.**

In general,

$$\text{slope} = \frac{y_2 - y_1}{x_2 - x_1} = \frac{y_1 - y_2}{x_1 - x_2}$$

In Example 1a, the negative slope for the line $2x + 3y = 6$ means that the line slants (or slopes) downward to the right (See Figure 4.15).

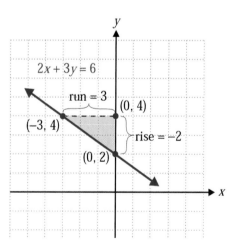

Figure 4.15

Slopes of Horizontal and Vertical Lines

In Section 4.2, we discussed the graphs of horizontal lines ($y = b$) and vertical lines ($x = a$), but the slopes of lines of these types were not discussed.

To find the slope of a horizontal line, such as $y = 3$, find two points on the line and substitute in the slope formula. Note that any two points on the line will have the same y-coordinate; namely, 3. Two such points are $(-2, 3)$ and $(5, 3)$. Using these two points in the formula for slope gives:

$$\text{slope} = \frac{y_2 - y_1}{x_2 - x_1} = \frac{3-3}{5-(-2)} = \frac{0}{7} = 0$$

In fact, the numerator will always be 0 because the y-values will all be 3 regardless of the x-values. (See Figure 4.16.)

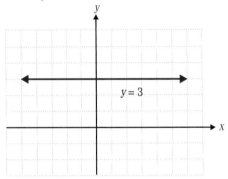

Figure 4.16

To find the slope of a vertical line, such as $x = 4$, find two points on the line and substitute in the slope formula. Now, any two points on the line will have the same x-coordinate; namely, 4. Two such points are $(4, 1)$ and $(4, 6)$. The slope formula gives:

$$\text{slope} = \frac{y_2 - y_1}{x_2 - x_1} = \frac{6-1}{4-4} = \frac{5}{0}, \text{which is } \textbf{undefined}.$$

In fact, the denominator will always be 0 because the x-values will all be 4 regardless of the y-values. (See Figure 4.17.)

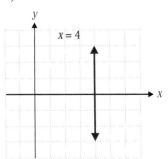

Figure 4.17

The following two general statements are true for horizontal and vertical lines:

1. For horizontal lines (of the form $y = b$), the slope is 0.

2. For vertical lines (of the form $x = a$), the slope is undefined.

Example 2: Vertical and Horizontal Lines ● ● ● ● ● ● ● ● ● ● ●

a. Find the equation and slope of the horizontal line through the point $(-2, 6)$.

Solution: The equation is $y = 6$ and the slope is 0.

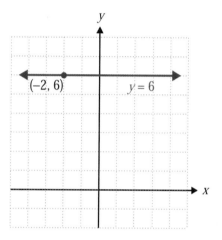

b. Find the equation and slope of the vertical line through the point $(5, 2)$.

Solution: The equation is $x = 5$ and the slope is undefined.

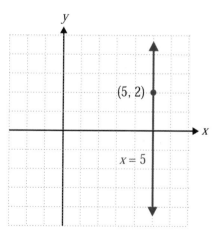

● ●

NOTES

Special Note About Slopes of Lines
All lines that slant downward to the right have negative slopes.
All lines that slant upward to the right have positive slopes.
All horizontal lines have slope 0.
All vertical lines have slope undefined.

The Slope-Intercept Form: $y = mx + b$

In the beginning of this section, we graphed two lines, $y = 3x - 1$ and $y = 3x + 2$. In both cases we found that the slope was 3. In Example 1a, we found that the line $2x + 3y = 6$ has a slope of $-\dfrac{2}{3}$. If we solve this equation for y we get the following results:

$$2x + 3y = 6$$

$$3y = -2x + 6$$

$$y = \frac{-2x}{3} + \frac{6}{3}$$

$$y = -\frac{2}{3}x + 2$$

Observing the relationship between slopes and the equations just discussed, we see that when the equation is solved for y the coefficient of x is the slope in each case. This observation leads to the following discussion which states that if the equation is solved for y, the coefficient of x will always be the slope of the line.

For $y = mx + b$, m is the slope

Statement: *Given an equation in the form $y = mx + b$, then m is the slope.*

Proof: *Suppose that the equation is solved for y and $y = mx + b$. Let (x_1, y_1) and (x_2, y_2) be two points on the line where $x_1 \neq x_2$. Then $y_1 = mx_1 + b$ and $y_2 = mx_2 + b$ so the slope can be calculated as follows:*

$$slope = \frac{y_2 - y_1}{x_2 - x_1} = \frac{(mx_2 + b) - (mx_1 + b)}{x_2 - x_1}$$

$$= \frac{mx_2 + b - mx_1 - b}{x_2 - x_1}$$

$$= \frac{mx_2 - mx_1}{x_2 - x_1}$$

$$= \frac{m(x_2 - x_1)}{x_2 - x_1} = m$$

Therefore, for an equation in the form $y = mx + b$, the slope of the line is m.

For the line $y = mx + b$, the point where $x = 0$ is the point where the line will cross the y-axis. This point is called the **y-intercept**. By letting $x = 0$, we get

$$y = mx + b$$
$$y = m \cdot 0 + b$$
$$y = b$$

Thus, the point $(0, b)$ is the y-intercept. As discussed in Section 4.2, we generally say that b is the y-intercept. These concepts of slope and y-intercept lead to the following definition.

Slope-Intercept Form

> $y = mx + b$ *is called the **slope-intercept form** for the equation of a line.* m *is the **slope** and b is the **y-intercept.**

The following examples show how to find the slope and y-intercept of a line by solving the equation for y and writing the result in the slope-intercept form. With this form, we will see that the lines can be easily graphed.

Example 3: $y = mx + b$ ● ● ● ● ● ● ● ● ● ● ● ● ● ● ● ● ● ●

a. Find the slope and y-intercept of $2x + 3y = 3$.

Solution: $2x + 3y = 3$

$$3y = -2x + 3$$

$$y = \frac{-2x + 3}{3}$$

$$y = -\frac{2}{3}x + 1$$

Therefore, $m = -\dfrac{2}{3}$

and y-intercept $= 1$.

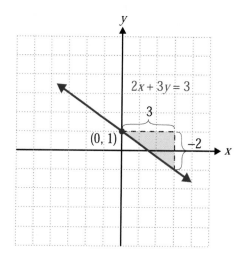

Continued on next page...

To use the slope and *y*-intercept as aids in graphing lines that are not horizontal and not vertical, remember that a line can be graphed if you know two points on the line. Proceed as follows:

1. One point on the line is the *y*-intercept. So, plot this point first. In Example 3a, we have plotted the point $(0, 1)$.
2. To find a second point, use the information given by the slope. Move horizontally (to the right) from the *y*-intercept the number of units indicated by the denominator of the slope. From this position, move vertically the number of units indicated by the numerator of the slope (up if the slope is positive, down if the slope is negative.) In Example 3a, we have moved horizontally 3 units and down 2 units.
3. Draw the line through the two points.

(**Note:** The numbers of units moved in Example 3a need not be exactly 3 and –2, but they must be in the ratio of 3 to –2, such as 6 and –4. That is the ratio must be $\dfrac{\text{rise}}{\text{run}} = -\dfrac{2}{3} = \dfrac{-2}{3} = \dfrac{2}{-3}$.)

b. Find the slope and *y*-intercept of $x - 2y = 6$.

Solution: $x - 2y = 6$

$$-2y = -x + 6$$

$$y = \frac{-x + 6}{-2}$$

$$y = \frac{1}{2}x - 3$$

Therefore, $m = \dfrac{1}{2}$ and $b = -3$.

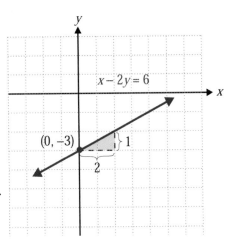

Since $m = +\dfrac{1}{2}$, moving 2 units right and 1 unit up from the *y*-intercept locates another point on the graph. Of course, another point could be located by substituting some value for *x* in the equation. For example, if $x = 4$, then,

$$x - 2y = 6$$

$$4 - 2y = 6$$

$$-2y = 6 - 4$$

$$-2y = 2$$

$$y = -1$$

The point $(4, -1)$ is also on the graph.

Continued on next page...

c. Find the slope and y-intercept of $-4x + 2y = 7$.

Solution:

$$-4x + 2y = 7$$

$$2y = 4x + 7$$

$$y = \frac{4x + 7}{2}$$

$$y = 2x + \frac{7}{2}$$

Therefore, $m = 2$ and $b = \frac{7}{2}$.

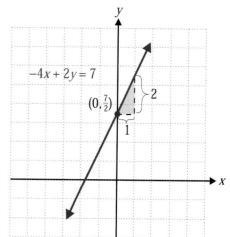

Note that in Examples 3b and 3c, the slopes are positive and the lines slant upward to the right. In Example 3a, the slope is negative and the line slants downward to the right.

Example 4: Find the Equation

Find the equation of the line with y-intercept $(0, -2)$ and slope $m = \frac{3}{4}$.

Solution: In this case, we can substitute directly into the slope-intercept form $y = mx + b$ with $b = -2$ and $m = \frac{3}{4}$.

$$y = \frac{3}{4}x + (-2) \quad \text{or} \quad y = \frac{3}{4}x - 2$$

In standard form, $3x - 4y = 8$. Note that $\frac{3}{4}x - y = 2$ is also in standard form.

For convenience, we multiply by 4 so that the coefficients are integers. Both forms are correct and either form is an acceptable answer.

To quickly graph a horizontal line, any two points with the *y*-value indicated by the equation can be used. To graph a vertical line, any two points with the *x*-value indicated by the equation can be used.

Example 5: Horizontal and Vertical Lines ● ● ● ● ● ● ● ● ● ● ●

a. Graph the horizontal line $y = 2$.

> **Solution:**   Two points with *y*-value 2 are $(-4, 2)$ and $(3, 2)$. The line can be drawn through these two points.

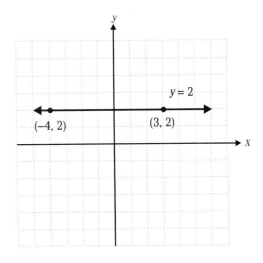

b. Graph the vertical line $x = -3$.

> **Solution:**   Two points with *x*-value -3 are $(-3, 5)$ and $(-3, -1)$. The line can be drawn through these two points.

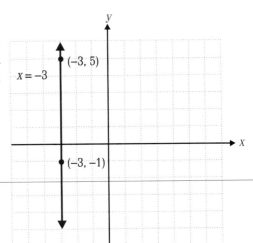

● ●

1. *Find the slope of the line determined by the points (1, 5) and (2, –2).*

2. *Find the equation of the line through the point (0, 3) with slope* $-\dfrac{1}{2}$.

3. *Find the slope, m, and y-intercept, b, for the equation 2x + 3y = 6.*

4. *Write an equation for the horizontal line through the point (–2, 3).*

5. *Write an equation for the vertical line through the point (–1, –2).*

4.3 Exercises

Find the slope of the line determined by each pair of points in Exercises 1 – 10.

1. $(1, 4),\ (2, -1)$ **2.** $(3, 1), (5, 0)$ **3.** $(4, 7), (-3, -1)$

4. $(-6, 2), (1, 3)$ **5.** $(-3, 8), (5, 9)$ **6.** $(0, 0), (-6, -4)$

7. $(4, 2), \left(-1, \dfrac{1}{2}\right)$ **8.** $\left(\dfrac{3}{4}, 2\right), \left(1, \dfrac{3}{2}\right)$ **9.** $\left(\dfrac{7}{2}, 3\right), \left(\dfrac{1}{2}, -\dfrac{3}{4}\right)$

10. $\left(-2, \dfrac{4}{5}\right), \left(\dfrac{3}{2}, \dfrac{1}{10}\right)$

Find the equation and draw the graph of the line passing through the given y-intercept with the given slope in Exercises 11 – 25.

11. $(0, 0),\ m = \dfrac{2}{3}$ **12.** $(0, 1),\ m = \dfrac{1}{5}$ **13.** $(0, -3),\ m = -\dfrac{3}{4}$

14. $(0, -2),\ m = \dfrac{4}{3}$ **15.** $(0, 3),\ m = -\dfrac{5}{3}$ **16.** $(0, 2),\ m = 4$

17. $(0, -1),\ m = 2$ **18.** $(0, 4),\ m = -\dfrac{3}{5}$ **19.** $(0, -5),\ m = -\dfrac{1}{4}$

20. $(0, 5),\ m = -3$ **21.** $(0, -4),\ m = \dfrac{3}{2}$ **22.** $(0, 0),\ m = -\dfrac{1}{3}$

23. $(0, 5),\ m = 1$ **24.** $(0, 6),\ m = -1$ **25.** $(0, -6),\ m = \dfrac{2}{5}$

Answers to Practice Problems: 1. $m = -7$ **2.** $y = -\dfrac{1}{2}x + 3$ **3.** $m = -\dfrac{2}{3}$ and $b = 2$ **4.** $y = 3$ **5.** $x = -1$

Find the slope, m, and the y-intercept, b, for each of the equations in Exercises 26 – 45. Then graph the line.

26. $y = 2x - 3$

27. $y = 3x + 4$

28. $y = 2 - x$

29. $y = -\dfrac{2}{3}x + 1$

30. $x + y = 5$

31. $2x - y = 3$

32. $3x - y = 4$

33. $4x + y = 0$

34. $x - 3y = 9$

35. $x + 4y = -8$

36. $3x + 2y = 6$

37. $2x - 5y = 10$

38. $4x - 3y = -3$

39. $7x + 2y = 4$

40. $6x + 5y = -15$

41. $3x + 8y = -16$

42. $x + 4y = 6$

43. $2x + 3y = 8$

44. $-3x + 6y = 4$

45. $5x - 2y = 3$

In Exercises 46 – 55, find two points on each of the lines and graph the line. State the slope of each line.

46. $y + 5 = 0$

47. $y - 4 = 0$

48. $x = -1$

49. $x - 5 = 0$

50. $2y = -16$

51. $3x + 2 = 0$

52. $5y - 12 = 0$

53. $3y - 5 = 0$

54. $1.5x = -3$

55. $-2x + 6 = 0$

Writing and Thinking About Mathematics

56. The slope of a road is called a "grade." A steep grade is cause for truck drivers to have slow speed limits in mountains. What do you think that a "grade of 12%" means? Draw a picture of a right triangle that would indicate a grade of 12%.

57. Ramps for persons in wheelchairs or otherwise handicapped are now built into most buildings and walkways. (If ramps are not present in a building, then there must be elevators.) What do you think that the slope of a ramp should be for handicapped access? Look in your library or contact your local building permit office to find the recommended slope for such ramps.

Hawkes Learning Systems: Introductory Algebra

 Graphing Linear Equations

4.4   Point-Slope Form: $y - y_1 = m(x - x_1)$

After completing this section, you will be able to:

1. *Graph lines given the slope and one point on each line.*

2. *Write the equations of lines given the slope and one point on each line.*

3. *Write the equations of lines given two points on each line.*

4. *Write linear equations from information given in applied problems.*

Graphing a Line Given a Point and the Slope

Lines represented by equations in the standard form $Ax + By = C$ and in the slope-intercept form $y = mx + b$ have been discussed in Sections 4.2 and 4.3. In Section 4.3, we discussed how to graph a line by using the y-intercept point $(0, b)$, and the slope by moving horizontally and vertically from the y-intercept.

Now, suppose that you are given the slope of a line and a point on the line. This point may or may not be the y-intercept. The graph of the line can be drawn by using the same technique discussed when using the y-intercept and slope. Consider the following example.

Example 1: Graph ●

Graph the line with slope $m = \dfrac{3}{4}$ which passes through the point $(4, 5)$.

Solution: Start from the point $(4, 5)$ and locate another point on the line using the slope as $\dfrac{\text{rise}}{\text{run}} = \dfrac{3}{4}$. Moving 4 units to the right (run) and 3 units up (rise) from $(4, 5)$ will give another point on the line. From $(4, 5)$ you might have moved 4 units left and 3 units down, or 8 units right and then 6 units up. Just move so that the ratio of rise to run is 3 to 4, and you will locate a second point on the graph.

Continued on next page...

(For a negative slope, move either to the right and then down or to the left and then up.)

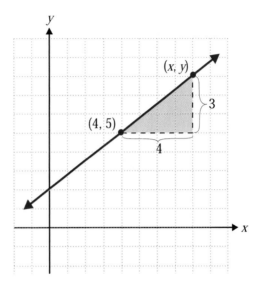

The Point-Slope Form for the Equation of a Line: $y - y_1 = m(x - x_1)$

In Example 1, the graph is drawn given the slope $m = \dfrac{3}{4}$ and a point $(4, 5)$ on the line. How would you find the equation that corresponds to this line? If the slope is $\dfrac{3}{4}$ and (x, y) is to be a point on the line along with $(4, 5)$, then the following discussion will help you determine the equation that corresponds to this line.

$$\text{slope} = \frac{y - 5}{x - 4} \quad \text{and} \quad \text{slope} = \frac{3}{4}$$

Setting these equal to each other gives,

$$\frac{y - 5}{x - 4} = \frac{3}{4}$$

This is the equation we want. To obtain a more common form, we multiply both sides by $x - 4$ and get

$$y - 5 = \frac{3}{4}(x - 4)$$

With this approach, we can say that a line with slope m that passes through a given fixed point (x_1, y_1) can be represented in the form

$$y - y_1 = m(x - x_1)$$

Point-Slope Form

$y - y_1 = m(x - x_1)$ *is called the **point-slope form** for the equation of a line. m is the **slope**, and (x_1, y_1) is a given **point** on the line.*

Example 2: Find the Equation ● ● ● ● ● ● ● ● ● ● ● ● ● ● ●

a. Find an equation of the line that has slope $\dfrac{2}{3}$ and passes through the point $(1, 5)$.

Solution: Substitute into the point-slope form $y - y_1 = m(x - x_1)$.

$$y - 5 = \frac{2}{3}(x - 1)$$

This same equation can be written in the slope-intercept form and in standard form.

$$y - 5 = \frac{2}{3}x - \frac{2}{3}$$

$$y = \frac{2}{3}x - \frac{2}{3} + 5 \qquad\qquad -\frac{2}{3} + 5 = -\frac{2}{3} + \frac{15}{3} = \frac{13}{3}$$

$$y = \frac{2}{3}x + \frac{13}{3} \qquad\qquad \text{Slope-intercept form}$$

Or multiplying each term by 3 gives

$$3y = 2x + 13$$

$$-2x + 3y = 13 \qquad \text{Standard form}$$

$$\text{or}\quad 2x - 3y = -13 \qquad \text{Standard form}$$

b. Find three forms for the equation of the line with slope $-\dfrac{5}{2}$ that contains the point $(6, -1)$.

Solution: Use the point-slope form: $y - y_1 = m(x - x_1)$.

$$y - (-1) = -\frac{5}{2}(x - 6)$$

$$y + 1 = -\frac{5}{2}(x - 6) \qquad \text{Point-slope form}$$

$$y + 1 = -\frac{5}{2}x + 15$$

$$y = -\frac{5}{2}x + 14 \qquad \text{Slope-intercept form}$$

$$2y = -5x + 28 \qquad \text{Multiply by 2.}$$

$$5x + 2y = 28 \qquad \text{Standard form}$$

● ●

You must understand that all three forms for the equation of a line are acceptable and correct. Your answer may be in a different form from that in the back of the text, but it should be algebraically equivalent.

Finding the Equations of Lines Given Two Points

Given two points that lie on a line, the equation of the line can be found by proceeding as follows:

Finding the Equation of a Line

If two points are given,

1. *Use the formula $m = \dfrac{y_2 - y_1}{x_2 - x_1}$ to find the slope.*

2. *Use this slope, m, and either point in the point-slope formula $y - y_1 = m(x - x_1)$.*

The following examples illustrate this common situation.

Example 3: Finding the Equation of a Line ● ● ● ● ● ● ● ● ● ● ●

a. Find an equation of the line passing through the two points $(-1, -3)$ and $(5, -2)$.

Solution: Find m using the formula for slope.

$$m = \frac{y_2 - y_1}{x_2 - x_1} = \frac{-2 - (-3)}{5 - (-1)} = \frac{-2 + 3}{5 + 1} = \frac{1}{6}$$

Using the point-slope form [either point $(-1, -3)$ or point $(5, -2)$ will give the same result] yields

$$y - (-2) = \frac{1}{6}(x - 5) \qquad \text{Using } (5, -2)$$

$$y + 2 = \frac{1}{6}(x - 5) \qquad \text{Point-slope form}$$

$$\text{or} \quad y + 2 = \frac{1}{6}x - \frac{5}{6} \qquad -\frac{5}{6} - 2 = -\frac{5}{6} - \frac{12}{6} = -\frac{17}{6}$$

$$y = \frac{1}{6}x - \frac{17}{6} \qquad \text{Slope-intercept form}$$

$$\text{or} \quad 6y = x - 17$$

$$17 = x - 6y \qquad \text{Standard form } Ax + By = C$$

b. Find an equation of the line that contains the points $(-2, 3)$ and $(5, 1)$.

Solution: $m = \dfrac{y_2 - y_1}{x_2 - x_1} = \dfrac{1 - 3}{5 - (-2)} = \dfrac{-2}{7} = -\dfrac{2}{7}$

Using $(-2, 3)$ in the point-slope form gives

$$y - 3 = -\frac{2}{7}\left[x - (-2)\right] \quad \text{Using } (-2, 3)$$

$$y - 3 = -\frac{2}{7}(x + 2) \qquad \text{Point-slope form}$$

$$y - 3 = -\frac{2}{7}x - \frac{4}{7}$$

$$y = -\frac{2}{7}x - \frac{4}{7} + 3 \qquad -\frac{4}{7} + 3 = -\frac{4}{7} + \frac{21}{7} = \frac{17}{7}$$

$$y = -\frac{2}{7}x + \frac{17}{7} \qquad \text{Slope-intercept form}$$

Continued on next page...

Using $(5, 1)$ in the point-slope form gives

$$y - 1 = -\frac{2}{7}(x - 5) \qquad \text{Point-slope form}$$

$$y - 1 = -\frac{2}{7}x + \frac{10}{7}$$

$$y = -\frac{2}{7}x + \frac{10}{7} + 1$$

$$y = -\frac{2}{7}x + \frac{17}{7} \qquad \text{Slope-intercept form}$$

Thus, we see that using either point yields the same equation. The equation may be written in standard form $2x + 7y = 17$.

● ●

Practice Problems

1. *Write an equation in standard form for the line with slope $\frac{5}{8}$ that contains the point $(2, -2)$.*

2. *Write an equation in standard form for the line passing through the two points $(6, 2)$ and $(-3, 1)$.*

3. *Write an equation in slope-intercept form for the line passing through the point $(-5, 2)$ and parallel to the line $3x + y = 8$. (Remember, parallel lines have the same slope.)*

Answers to Practice Problems: 1. $5x - 8y = 26$ **2.** $x - 9y = -12$ **3.** $y = -3x - 13$

4.4 Exercises

In Exercises 1 – 12, write an equation (in slope-intercept form or standard form) for the line passing through the given point with the given slope.

1. $(0, 3), m = 2$

2. $(1, 4), m = -1$

3. $(-1, 3), m = -\dfrac{2}{5}$

4. $(-2, -5), m = \dfrac{7}{2}$

5. $(-3, 2), m = \dfrac{3}{4}$

6. $(6, -1), m = \dfrac{5}{3}$

7. $(3, -1), m = 0$

8. $(4, 6), m = 0$

9. $(2, -4),$ undefined slope

10. $(-3, 2),$ undefined slope

11. $\left(1, \dfrac{2}{3}\right), m = \dfrac{3}{2}$

12. $\left(\dfrac{1}{2}, -2\right), m = \dfrac{4}{3}$

In Exercises 13 – 24, write an equation (in slope-intercept form or standard form) for the line passing through the given points.

13. $(-2, 3), (1, 2)$

14. $(-5, 1), (2, 0)$

15. $(3, 4), (6, 2)$

16. $(-4, -4), (3, 1)$

17. $(8, 2), (-1, 2)$

18. $\left(\dfrac{1}{2}, -1\right), (3, -1)$

19. $(-5, 2), (1, 4)$

20. $(-2, 6), (3, 1)$

21. $(-2, 4), (-2, 1)$

22. $(3, 0), (3, -3)$

23. $(-4, 2), \left(1, \dfrac{1}{2}\right)$

24. $(0, 2), \left(1, \dfrac{3}{4}\right)$

25. Write an equation for the horizontal line through point $(-2, 5)$.

26. Write an equation for the line parallel to the x-axis through point $(6, -2)$.

27. Write an equation for the line parallel to the y-axis through point $(-1, -3)$.

28. Write an equation for the vertical line through point $(-1, -1)$.

29. Write an equation for the line parallel to the line $3x - y = 4$ through the origin.

30. Write an equation for the line parallel to the line $2x = 5$ through point $(0, 3)$.

31. Write an equation for the line through point $\left(\dfrac{3}{2}, 1\right)$ and parallel to the line $4y - 6 = 0$.

32. Write an equation for the horizontal line through point $(-1, -6)$.

33. Find an equation for the line parallel to $2x - y = 7$ having the same y-intercept as $x - 3y = 6$.

34. Find an equation for the line parallel to $3x - 2y = 4$ having the same y-intercept as $5x + 4y = 12$.

35. The rental rate schedule for a car rental is described as "$30 per day plus 15¢ per mile driven." Write a linear equation for the daily cost of renting a car for x miles ($C = mx + b$).

36. For each job, a print shop charges $5.00 plus $0.05 per page printed.
   a. Write a linear equation for the cost of printing x pages. ($C = mx + b$)
   b. Graph the corresponding part of the line for $x \geq 0$.
   c. Interpret the meaning of the slope of the line.
   d. Explain the fact that the line makes sense only for positive integer values of x.

37. A salesperson's weekly income depends on the amount of her sales. Her weekly income is a salary of $200 plus 9% of her weekly sales.
   a. Write a linear equation for her weekly income for sales of x dollars. ($I = mx + b$)
   b. Graph the corresponding part of the line for $x \geq 0$.
   c. Interpret the meaning of the slope of the line.

38. A manufacturer has determined that the cost, C, of producing x units is given by a linear equation ($C = mx + b$).
   a. If 12 units cost $752 to produce and 15 units cost $800 to produce, find a linear equation for the cost of producing x units. ($C = mx + b$)
   b. What is the slope of the line?
   c. Interpret the meaning of the slope of the line.
   d. Explain the fact that the line makes sense only for positive integer values of x.

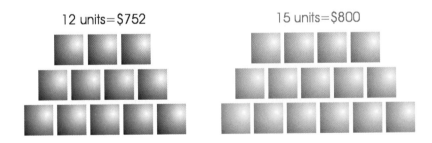

39. A local amusement park found that if the admission was $7, the attendance was about 1000 customers per day. When the price of admission was dropped to $6, attendance increased to about 1200 customers per day.

    a. Write a linear equation for the attendance in terms of the price, p. ($A = mp + b$)

    b. What is the slope of the line?

    c. Interpret the meaning of the slope of the line.

    d. Explain the fact that the line makes sense only for positive values of p.

40. A department store manager has determined that if the price for a necktie is $35, the store will sell 100 neckties per month. However, only 50 neckties per month are sold if the price is raised to $45.

    a. Write a linear equation for the number of neckties sold per month in terms of the price, p. ($N = mp + b$)

    b. What is the slope of the line?

    c. Interpret the meaning of the slope of the line.

    d. Explain the fact that the line makes sense only for positive values of p.

In Exercises 41 – 45, (a) use the given information to find the equation of each line and (b) use a graphing calculator to graph the line.

41. contains the two points $(-1, 2)$ and $(3, 1)$

42. has slope $\dfrac{5}{2}$ and y-intercept $(0, -1)$

43. has slope $-\dfrac{1}{2}$ and passes through the origin

44. contains the two points $(-2, 5)$ and $(3, 5)$

45. contains the point $(-3, -4)$ and has slope $\dfrac{3}{2}$

Hawkes Learning Systems: Introductory Algebra

Finding the Equation of a Line

4.5   Introduction to Functions and Function Notation

After completing this section, you will be able to:

1. *Determine whether a relation is or is not a function.*
2. *Find the domain and range of a relation or a function.*
3. *Use the vertical line test to determine whether a graph is or is not the graph of a function.*
4. *Use function notation.*

The cost of producing computers is a function of the number of computers produced; the rate of your heart beat is a function of your anxiety level; your grade in this class is a function of the time you spend studying. The concept of a function is present in many aspects of our daily lives. In fact, the concept of a function is one of the most useful and important in all of mathematics. With this concept and the related topics, we are able to develop calculus and perform engineering feats such as building bridges, building airplanes, and sending satellites into orbit. Here, we will simply introduce the idea of a function and how to use function notation.

Introduction to Functions

In Section 4.1 we discussed ordered pairs of real numbers and their graphs in the Cartesian coordinate system. Every set of ordered pairs of real numbers can be classified as a **relation**.

Relation

*A **relation** is a set of ordered pairs.*

*The **domain**, **D**, of a relation is the set of all first coordinates in the relation.*

*The **range**, **R**, of a relation is the set of all second coordinates in the relation.*

In graphing relations, the x-axis is sometimes called the **domain axis** and the y-axis is called the **range axis**. In real life applications the variables may not be x and y. For example, the distance, d, an object falls depends on the time, t, that it falls and this relationship can be represented in the form $d = 16t^2$. Thus, in graphing such a relation, the horizontal axis (domain axis) would be labeled the t-axis and the vertical axis (range axis) would be labeled the d-axis.

> ### Example 1: Domain and Range • • • • • • • • • • • • • • • • •

Find the domain and range for each of the following relations.

a. $r = \{(5, 7), (6, 2), (6, 3), (-2, -1)\}$

Solution: $D = \{-2, 5, 6\}$
$R = \{-1, 2, 3, 7\}$

Note that 6 is listed only once in the domain even though it appears in two ordered pairs. The numbers may be listed in any order in both the domain and range. We have listed them in order here for convenience only.

b. $f = \{(-1, 1), (1, 5), (3, 0), (4, 0)\}$

Solution: $D = \{-1, 1, 3, 4\}$
$R = \{0, 1, 5\}$

• •

The relation $f = \{(-1, 1), (1, 5), (3, 0), (4, 0)\}$ in Example 1b meets a particular condition in that each first coordinate has only one corresponding second coordinate. Such a relation is called a **function**. Notice that r in Example 1a is not a function because the first coordinate 6 has more than one corresponding second coordinate. That is, 6 appears as a first coordinate more than once in the listing of the ordered pairs. The fact that 0 appears more than once as a second coordinate in the relation f is not important.

Also, you should be aware of the fact that most functions have an infinite number of ordered pairs and therefore the domains and ranges cannot be listed as illustrated in these examples. These ideas will be studied in greater detail in later courses in mathematics.

Function

*A **function** is a relation in which each domain element has exactly one corresponding range element.*

OR

*A **function** is a relation in which each domain element occurs only once.*

Example 2: Determining Functions ● ● ● ● ● ● ● ● ● ● ● ● ● ●

Determine whether each of the following relations is or is not a function.

a. $r = \{(2,3), (1,6), (2,5), (0,-1), (-3,0)\}$

Solution: r is **not** a function. The first coordinate 2 appears more than once.

b. $s = \{(0,0), (1,1), (2,4), (3,9)\}$

Solution: s is a function. Each first coordinate has only one corresponding second coordinate.

c. $t = \{(1,5), (3,5), (-2,5), (-1,5), (-4,5)\}$

Solution: t is a function. Each first coordinate appears only once. The fact that the second coordinates are all the same has no relevance in the definition of a function.

● ●

The Vertical Line Test

If one point in the graph of a relation is directly above or below another point in the graph, then these points have the same first coordinate (or the same x-coordinate). Such a relation would **not** be a function. Therefore, we can tell whether or not a graph represents a function by using the following **vertical line test**.

Vertical Line Test

If any vertical line intersects a graph of a relation in more than one point, then the relation graphed is not a function.

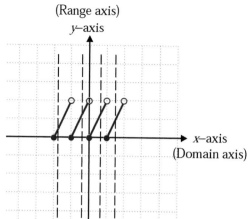

This graph represents a function. Several vertical lines are drawn to illustrate that vertical lines do not intersect the graph in more than one point. Each *x*-value has only one corresponding *y*-value.

(a)

This graph is not a function because the vertical line drawn intersects the graph in more than one point. Thus, for at least one *x*-value, there is more than one corresponding *y*-value.

(b)

Figure 4.18

Example 3: Vertical Line Test ● ● ● ● ● ● ● ● ● ● ● ● ● ●

Use the vertical line test to determine whether each of the following graphs does or does not represent a function.

a.

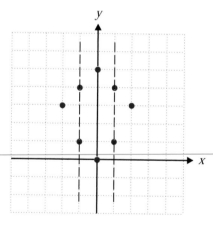

Solution: Not a function. At least one vertical line intersects the graph in more than one point. Listing the ordered pairs shows that several *x*-coordinates are repeated:

$\{(-2, 3), (-1, 1), (-1, 4), (0, 0), (0, 5), (1, 1), (1, 4), (2, 3)\}$

Continued on next page...

b.

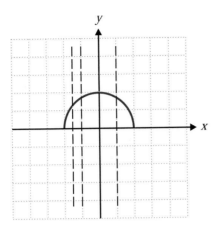

Solution: A function. No vertical line will intersect the graph in more than one point. Several vertical lines are drawn to illustrate this fact. Furthermore, the domain and range can be determined by investigating the graph. We see that the domain and range consist of an infinite number of real values and are intervals of real numbers:

Domain: $-2 \leq x \leq 2$

Range: $0 \leq y \leq 2$

c.

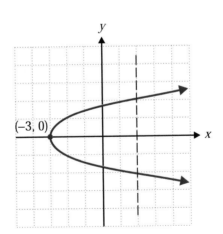

Solution: Not a function. A vertical line intersects the graph in more than one point.

Continued on next page...

d.

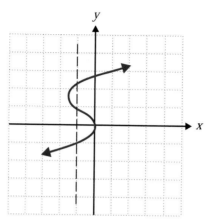

Solution: Not a function. A vertical line intersects the graph in more than one point.

● ●

Function Notation

We have used ordered pair notation (x, y) to represent points on the graphs of relations and functions. As the vertical line test will verify, equations of the form

$$y = mx + b$$

where the equation is solved for y, represent linear functions. The use of x and y in this manner is standard in mathematics and is particularly useful when graphing because we are familiar with the concepts of the x-axis and y-axis.

Another notation, called **function notation**, is more convenient for indicating calculations of values of a function and for indicating operations performed with functions. In function notation,

instead of writing y,   write $f(x)$,   read "f of x."

The letter f is the name of the function. Any letter will do. The letters f, g, h, F, G, and H are commonly used. We have used r, s, and t in previous examples.

Suppose that $y = 3x - 2$ is a given linear equation. Since the equation is solved for y it represents a linear function and we can replace y with the notation $f(x)$ as follows:

$$y = 3x - 2 \quad \text{can be written in the form} \quad f(x) = 3x - 2$$

That is, the notation $f(x)$ represents the y-value for some corresponding x-value.

Now, in function notation, $f(5)$ means to replace x with 5 in the function:

$$f(5) = 3 \cdot 5 - 2 = 15 - 2 = 13$$

Thus, the ordered pair $(5, 13)$ is the same as $(5, f(5))$.

Example 4: Function Evaluation

a. For the function $g(x) = 4x + 3$, find

   i. $g(2)$ 　　　　 **ii.** $g(-3)$ 　　　　 **iii.** $g(0)$

   Solution: 　**i.** $g(2) = 4 \cdot 2 + 3 = 11$
   　　　　　　ii. $g(-3) = 4(-3) + 3 = -9$
   　　　　　iii. $g(0) = 4 \cdot 0 + 3 = 3$

Not all functions are linear functions just as not all graphs are straight lines. The function notation is valid for a wide variety of types of functions. The following examples illustrate the use of function notation with functions other than linear functions.

Example 5: Nonlinear Functions

a. For the function $f(x) = x^2 - 2x + 1$, find
   i. $f(-2)$ 　　　　 **ii.** $f(0)$ 　　　　 **iii.** $f(4)$

   Solution: **i.** $f(-2) = (-2)^2 - 2(-2) + 1 = 4 + 4 + 1 = 9$

   　　　　ii. $f(0) = 0^2 - 2(0) + 1 = 0 - 0 + 1 = 1$

   　　　　iii. $f(4) = 4^2 - 2 \cdot 4 + 1 = 16 - 8 + 1 = 9$

b. For the function $h(x) = 2x^3 - 5x$, find
   i. $h(-1)$ 　　　　 **ii.** $h(3)$ 　　　　 **iii.** $h(-4)$

   Solution: **i.** $h(-1) = 2(-1)^3 - 5(-1) = 2(-1) + 5 = -2 + 5 = 3$

   　　　　ii. $h(3) = 2(3)^3 - 5(3) = 54 - 15 = 39$

   　　　　iii. $h(-4) = 2(-4)^3 - 5(-4) = 2(-64) + 20 = -128 + 20 = -108$

Continued on next page...

c. Given the function $f(x) = x^2 - 10$ with restricted domain $D = \{-1, 0, 2, 3\}$, write the function as a set of ordered pairs.

Solution: For each x-value in D, find the corresponding y-value by substituting into $f(x)$.

$$f(-1) = (-1)^2 - 10 = 1 - 10 = -9$$
$$f(0) = 0^2 - 10 = 0 - 10 = -10$$
$$f(2) = 2^2 - 10 = 4 - 10 = -6$$
$$f(3) = 3^2 - 10 = 9 - 10 = -1$$

So, the function can be written as the following set of ordered pairs:

$$\{(-1, -9), (0, -10), (2, -6), (3, -1)\}$$

● ●

Practice Problems

1. *State the domain and range of the relation { (1, 2), (3, 4), (5, 6) }. Is the relation a function?*

2. *Write the function as a set of ordered pairs given $f(x) = 2x - 5$ and $D = \{-4, 0, 2, 3\}$.*

4.5 Exercises

In Exercises 1 – 10, (a) list the sets of ordered pairs corresponding to the points, (b) give the domain and range, and (c) state whether the relation is or is not a function.

1.

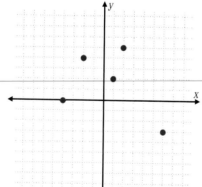

2.

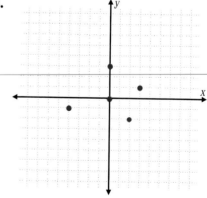

Answers to Practice Problems: 1. $D = \{1, 3, 5\}$ and $R = \{2, 4, 6\}$. It is a function. **2.** $\{(-4, -13), (0, -5), (2, -1), (3, 1)\}$

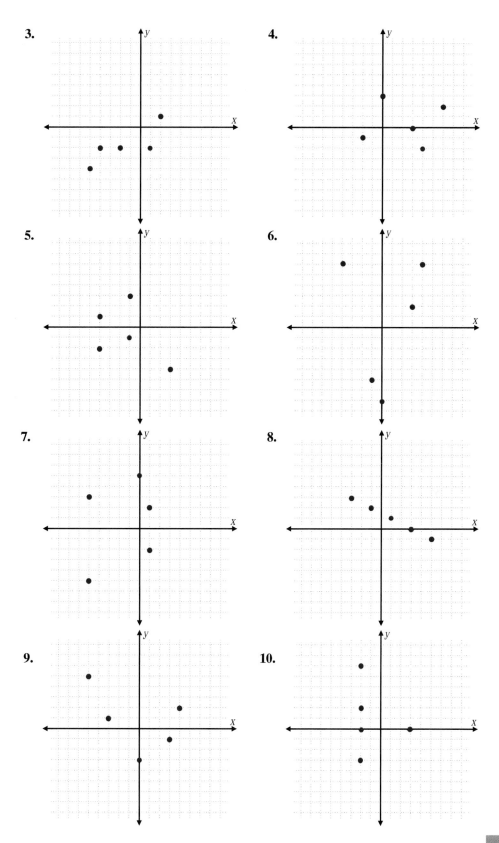

In Exercises 11 – 20, graph each set of ordered pairs and state whether the relation is or is not a function. If the relation is not a function, tell why not.

11. $\{(0,0),(1,6),(4,-2),(-3,5),(2,-1)\}$
12. $\{(1,-5),(2,-3),(-1,-3),(0,2),(4,3)\}$
13. $\{(-4,4),(3,-2),(1,0),(2,-3)\}$
14. $\{(-3,-3),(0,1),(-2,1),(3,1)\}$
15. $\{(0,2),(-1,1),(2,4),(3,5),(-3,5)\}$
16. $\{(-1,4),(-1,2),(-1,0),(-1,6),(-1,-2)\}$
17. $\{(0,0),(-2,5),(2,0.5),(4,-4.6),(5,0.2)\}$
18. $\{(2,-4),(2,-2),(0.5,-1),(0.5,3)\}$
19. $\{(-2,6),(1,-5),(3,1),(-4,3),(-3,2)\}$
20. $\{(5,-1),(6,-1),(-2,0),(5,0),(0.3,0.6)\}$

Use the vertical line test to determine whether or not each of the graphs in Exercises 21 – 32 is a function. If the graph represents a function, state its domain and range.

21.   **22.**

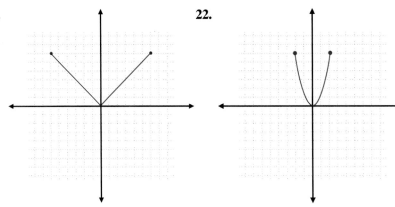

23.   **24.**

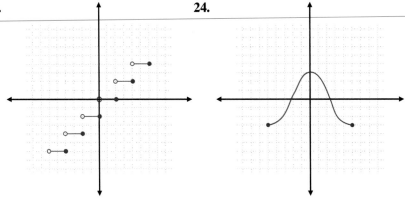

25.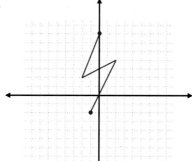

26.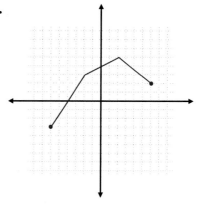

27.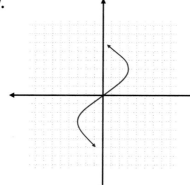

28.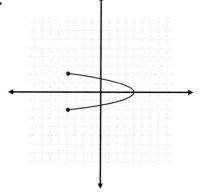

29.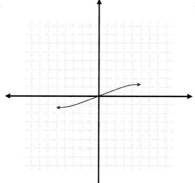

30.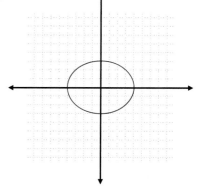

31.   **32.**

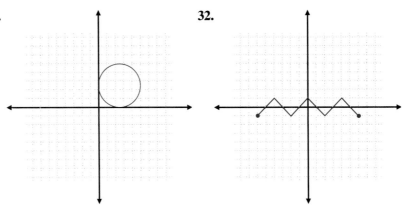

For Exercises 33 – 45, assume that each function has the same domain, $D = \{-2, -1, 0, 1, 5\}$ and find the corresponding functional value (*y*-value) for each *x*-value in the domain.

33. $f(x) = 2x + 1$ **34.** $f(x) = 3x - 5$ **35.** $g(x) = -4x + 2$

36. $g(x) = -x - 7$ **37.** $h(x) = x^2$ **38.** $h(x) = x^3$

39. $F(x) = x^2 - 4x + 4$ **40.** $G(x) = x^2 + 5x + 6$ **41.** $H(x) = x^3 - 8x$

42. $C(x) = 4x^2 + 3x - 7$ **43.** $P(x) = 0.5x^2 - 2x + 3.5$ **44.** $R(x) = 1.5x^2 + 4x - 9$

45. $f(x) = x^3 - 5x^2 + 7x - 2$

Hawkes Learning Systems: Introductory Algebra

Introduction to Functions and Function Notation

Chapter 4 Index of Key Ideas and Terms

There is a **one-to-one correspondence** between points in a
plane and ordered pairs of real numbers.

(A and B cannot both be 0.)

    y-intercept (point where a line crosses the y-axis)
    x-intercept (point where a line crosses the x-axis)

    1. The graph of any equation of the form $y = b$ is a
       horizontal line.
    2. The graph of any equation of the form $x = a$ is a
       vertical line.

$$\text{slope} = \frac{\text{rise}}{\text{run}} = \frac{y_2 - y_1}{x_2 - x_1}$$

Parallel Lines   page 253

    All lines with the same slope are **parallel**.

Slopes of Horizontal and Vertical Lines   pages 256 - 257

    1. For horizontal lines ($y = b$), the slope is 0.
    2. For vertical lines ($x = a$), the slope is undefined.

Slope-Intercept Form   pages 258 - 259

    $y = mx + b$ is called the **slope-intercept form** for the
    equation of a line.
    m is the slope and b is the y-intercept.

Slope Related to Graphs of Lines   page 258

    1. All lines with negative slopes slant downward to the right.
    2. All lines with positive slopes slant upward to the right.
    3. For horizontal lines ($y = b$), the slope is 0.
    4. For vertical lines ($x = a$), the slope is undefined.

Point-Slope Form   pages 267 - 268

    $y - y_1 = m(x - x_1)$ is called the **point-slope form** for the
    equation of a line.
    m is the slope and (x_1, y_1) is a given point on the line.

Finding the Equation of a Line Given Two Points   page 269

    If two points are given,

    1. Use the formula $m = \dfrac{y_2 - y_1}{x_2 - x_1}$ to find the slope.
    2. Use this slope, m, and either point in the point-slope
       formula $y - y_1 = m(x - x_1)$.

Relation   page 275

    A **relation** is a set of ordered pairs.
    The **domain**, D, of a relation is the set of all first coordinates
    in the relation.
    The **range**, R, of a relation is the set of all second coordinates
    in the relation.

Function   pages 276 - 277

    A **function** is a relation in which each domain element has
exactly one corresponding range element.
(or equivalently)
A **function** is a relation in which each domain element occurs
only once.

Vertical Line Test   page 277

    If any vertical line intersects a graph of a relation in
more than one point, then the relation is not a function.

Function Notation: $f(x)$ page 280

    The notation $f(x)$ is read "f of x."
The notation $f(x)$ represents the y-value for some corresponding x-value.

Chapter 4 Review

For a review of the topics and problems from Chapter 4, look at the
following lessons from *Hawkes Learning Systems: Introductory Algebra*

Introduction to the Cartesian Coordinate System
Graphing Linear Equations
Graphing Linear Equations By Plotting Points
Finding the Equation of a Line
Introduction to Functions and Function Notation

Chapter 4 Test

1. Which of the following points lie on the line determined by the equation $x - 4y = 7$?

   a. $(2, -2)$

   b. $(-1, -2)$

   c. $\left(5, -\dfrac{1}{2}\right)$

   d. $(0, 7)$

2. List the ordered pairs corresponding to the points on the graph.

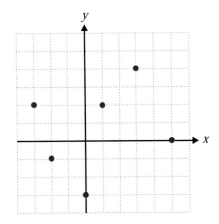

3. Graph the following set of ordered pairs:

   $$\left\{(0, 2), (4, -1), (-3, 2), (-1, -5), \left(2, \dfrac{3}{2}\right)\right\}.$$

4. Graph the line $2x - y = 4$.

5. Graph the equation $3x + 4y = 9$ by locating and labeling the x-intercept and y-intercept.

6. Graph the line passing through the point $(1, -5)$ with the slope $m = \dfrac{5}{2}$.

7. For the line $3x + 5y = -15$, determine the slope, m and the y-intercept, b. Then graph the line.

8. What is the slope of the line that contains the two points $(4, -3)$ and $\left(2, \dfrac{1}{2}\right)$?

9. Write an equation for the line passing through the point $(-3, 2)$ with the slope $m = \dfrac{1}{4}$.

10. Write an equation for the line passing through the points $(3, 1)$ and $(-2, 6)$. What is the slope of this line?

11. Write an equation for the line parallel to the *x*-axis through the point $(3, -2)$. What is the slope of this line?

12. A carpet cleaning machine can be rented for a fixed cost of $9.00 plus $4.00 per hour.
a. Write a linear equation representing the cost, C, in terms of hours used, t.
b. Explain why this equation only makes sense if t is positve.
c. Explain the meaning of the slope of this line.

13. Determine whether the two lines $2x + 3y = 6$ and $y = -\dfrac{2}{3}x + 1$ are or are not parallel. Explain.

14. Write an equation for the line that passes through the point $(1, -6)$ and is parallel to the line $y - 2x = 4$.

15. State the domain and range of the relation $\{(0, 6), (-1, 4), (2, -3), (5, 6)\}$.

16. Given the relation $r = \{(2, -5), (3, -4), (2, 0), (1, -5)\}$. Is this relation a function? Explain.

17. For the function $f(x) = 3x - 4$, find
  a. $f(-1)$ **b.** $f(6)$

18. For the function $h(x) = 3x^2 - 5$, find
  a. $h(-3)$ **b.** $h(2)$

Use the vertical line test to determine whether each graph in Exercises 19 – 22 does or does not represent a function. If the graph represents a function, state its domain and range.

19.

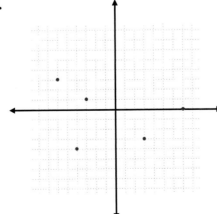

20.

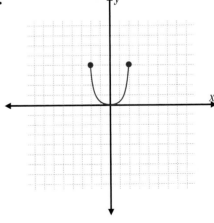

21.

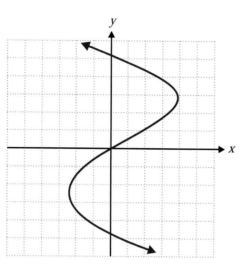

22.

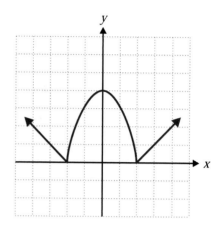

Cumulative Review: Chapters 1 – 4

Use the distributive property to complete the expression in Exercises 1 and 2.

1. $3x + 45 = 3(\quad)$

2. $6x + 16 = 2(\quad)$

Find the LCM for each set of numbers or terms in Exercises 3 and 4.

3. $12, 15, 54$

4. $6a^2, 24ab^3, 30ab, 40a^2b^2$

For each pair of numbers in Exercises 5 – 8, find two factors of the first number whose sum is the second number.

5. $32, 18$

6. $36, -13$

7. $-56, 10$

8. $-48, -8$

In Exercises 9 – 12, determine whether the inequality is true or false. If the inequality is false, rewrite it in a form that is true using the same numbers or expressions.

9. $-15 > 5$

10. $|-6| \geq 6$

11. $\dfrac{7}{8} \leq \dfrac{7}{10}$

12. $-10 < 0$

Solve the equations in Exercises 13 – 16.

13. $7(4 - x) = 3(x + 4)$

14. $-2(5x + 1) + 2x = 4(x + 1)$

15. $\dfrac{2}{3}x + \dfrac{1}{2} = \dfrac{3}{4}$

16. $1.5x - 3.7 = 3.6x + 2.6$

In Exercises 17 – 20, solve for the indicated variable.

17. $C = \pi d$; solve for d.

18. $3x + 5y = 10$, solve for y.

19. $P = 2l + 2w$, solve for w.

20. $V = \dfrac{1}{3}\pi r^2 h$, solve for h.

In Exercises 21 – 24, solve the inequality for x and graph the solution set on a real number line.

21. $5x + 3 \geq 2x - 15$

22. $-16 < 3x + 5 < 17$

23. $\dfrac{1}{2}x + 5 \leq \dfrac{2}{3}x - 7$

24. $-15 < \dfrac{3}{4}x - 9 \leq \dfrac{1}{3}$

On a number line, graph the integers that satisfy each inequality in Exercises 25 and 26.

25. $|x| < 5$ **26.** $|x| > 6$

For each relation represented in Exercises 27 – 30, state the domain and range and state whether or not the relation represents a function.

27.

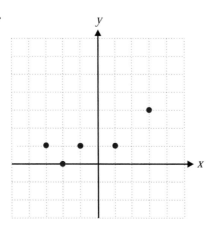

28.

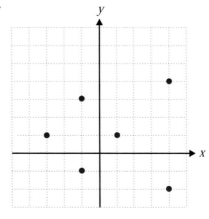

29.

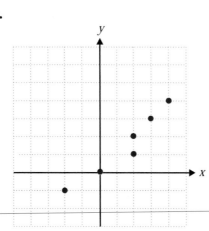

30.

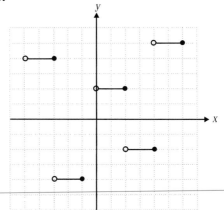

In Exercises 31 and 32, (a) graph each of the relations, (b) state the domain and range, and (c) state whether the relation is or is not a function.

31. $\{(0, 4), (2, 2), (3, 1), (-2, 6), (0, -3)\}$ **32.** $\left\{\left(2, \dfrac{1}{2}\right), (-1, 3), \left(\dfrac{2}{3}, -2\right), (1, 1)\right\}$

Determine which of the given points lie on the line determined by the equations in Exercises 33 and 34.

33. $4x - y = 7$
 a. $(2, 1)$
 b. $(3, 5)$
 c. $(1, -3)$
 d. $(0, 7)$

34. $2x + 5y = 6$
 a. $\left(\dfrac{1}{2}, 1\right)$
 b. $(3, 0)$
 c. $\left(0, \dfrac{6}{5}\right)$
 d. $(-2, 2)$

Graph the linear equations in Exercises 35 – 38.

35. $y = -4x$　　　**36.** $x + 2y = 4$　　　**37.** $x + 4 = 0$　　　**38.** $2y - 6 = 0$

Graph the linear equations in Exercises 39 and 40 by locating the x-intercept and the y-intercept.

39. $2x + y = 4$　　　　　　　　　**40.** $5x - y = 1$

For Exercises 41 and 42, determine the slope, m, and the y-intercept, b, and then graph the line.

41. $y - 2x = 3$　　　　　　　　　**42.** $2x + 5y = 10$

Find the equation of the line determined by each pair of points in Exercises 43 and 44.

43. $(2, 4), (-1, 6)$　　　　　　　　**44.** $(6, -2), (3, 2)$

In Exercises 45 – 48, graph the line and then write an equation (in slope-intercept form or standard form) for it.

45. $(6, -1), m = 0$　　　　　　**46.** $\left(3, \dfrac{5}{2}\right), m = -\dfrac{1}{4}$

47. $(-3, 4),$ slope undefined　　　**48.** $(4, -5), m = \dfrac{3}{2}$

49. Write an equation for the line parallel to $2x + y = 5$ passing through $(1, -2)$.

50. Write an equation for the line parallel to $x - 2y = 5$ having the same y-intercept as $5x + 3y = 9$.

51. For the function $f(x) = -3x + 4$, find **a.** $f(5)$ **b.** $f(-6)$

52. For the function $F(x) = 2x^2 - 5x + 1$, find **a.** $F(2)$ **b.** $F(-5)$

53. Find three consecutive even integers such that the sum of the second and third is equal to three times the first decreased by 14.

54. The length of a rectangle is 9 cm more than its width. The perimeter is 82 cm. Find the dimensions of the rectangle.

55. Twice the difference between a number and 16 is equal to four times the number increased by 6. What is the number?

56. To receive a B grade, a student must average 80 or more but less than 90. If Izumi received a B in the course and had four scores of 93, 77, 90, and 86 before taking the final exam, what were her possible scores for the final? (Assume that the maximum score on the final was 100.)

57. Acme Car Rental charges $0.25 for each mile driven plus $15 a day. Zenith Car Rental charges $50 a day with no extra charge for mileage. How many miles a day can you drive an Acme car and still keep the cost less than a Zenith car?

58. Find three consecutive odd integers such that the sum of the first and three times the second is 28 more than twice the third.

Systems of Linear Equations

Did You Know?

The subject of systems of linear equations received a great deal of attention in nineteenth-century mathematics. However, problems of this type are very old in the history of mathematics. The solution of simultaneous systems of equations was well known in China and Japan. The great Japanese mathematician Seki Shinsuku Kowa (1642-1708) wrote a book on the subject in 1683 that was well in advance of European work on the subject.

Seki Kowa is probably the most distinguished of all Japanese mathematicians. Born into a samurai family, he showed great mathematical talent at an early age. He is credited with the independent development of calculus and is sometimes referred to as "the Newton of Japan." There is a traditional story that Seki Kowa made a journey to the Buddhist shrines at Nara. There, ancient Chinese mathematical manuscripts were preserved, and Seki is supposed to have spent three years learning the contents of the manuscripts that previously no one had been able to understand. As was the custom, much of Seki Kowa's work was done through the solution of very intricate problems. In 1907, the Emperor of Japan presented a posthumous award to the memory of Seki Kowa, who did so much to awaken interest in scientific and mathematical research in Japan.

Native Japanese mathematics (the **wasan**) flourished until the nineteenth century when western mathematics and notation were completely adopted. In the 1940's, solution of large systems of equations became a part of the new branch of mathematics called **operations research.** Operations research is concerned with deciding how to best design and operate man-machine systems, usually under conditions requiring the allocation of scarce resources. Although this new science initially had only military applications, recent applications have been in the areas of business, industry, transportation, meteorology, and ecology. Computers now make it possible to solve extremely large systems of linear equations that are used to model the system being studied. Although computers do much of the work involved in solving systems of equations, it is necessary for you to understand the principles involved by studying small systems involving two unknowns as presented in Chapter 5.

"The advancement and perfection of mathematics are intimately connected with the prosperity of the state."

Napoleon Bonaparte (1769 - 1821)

Many applications involve two (or more) quantities and, by using two (or more) variables, we can form linear equations using the information given. Such a set of equations is called a system, and in this chapter we will develop techniques for solving systems of linear equations.

Graphing systems of equations in two variables is helpful in visualizing the relationship between two equations. However, this approach is somewhat limited since numbers might be quite large or solutions might involve fractions that must be estimated on the graph. Therefore, other algebraic techniques are necessary. We will discuss two of these: substitution and addition. These algebraic techniques are also necessary when we try to solve more than two equations with more than two variables. These ideas and graphing in three dimensions will be left to later courses.

5.1   Systems of Equations: Solutions by Graphing

Objectives

After completing this section, you will be able to:

1. Determine if given points lie on specified lines.

2. Determine if given points lie on both lines in specified systems of equations.

3. Determine, by graphing, if systems of linear equations are consistent, inconsistent, or dependent.

4. Estimate, by graphing, the coordinates of the intersections of consistent systems of linear equations.

Systems of Linear Equations

In Chapter 4, straight lines and their characteristics (slopes and intercepts) were analyzed in terms of linear equations. Many applications, as we will see in Sections 5.4 and 5.5, involve solving pairs of linear equations. Such pairs are said to form a **system of linear equations** (or a set of **simultaneous linear equations**). For example,

$$\begin{cases} x - 2y = 0 \\ 3x + y = 7 \end{cases} \qquad \text{and} \qquad \begin{cases} y = -x + 4 \\ y = 2x + 1 \end{cases}$$

are two systems of linear equations.

The **solution of a system** of linear equations is the set of ordered pairs (or points) that satisfy **both** equations. To determine whether or not a particular ordered pair is a solution to a system, substitute the values for x and y in **both** equations. If the results for both equations are true statements, then the ordered pair is a solution to the system.

Example 1: Solution of a System

Show that $(2, 1)$ is a solution to the system $\begin{cases} x - 2y = 0 \\ 3x + y = 7 \end{cases}$.

Solution: Substitute $x = 2$ and $y = 1$ into **both** equations.
In the first equation:

$$2 - 2(1) \overset{?}{=} 0$$

$$2 - 2 = 0 \quad \text{a true statement}$$

In the second equation:

$$3(2) + 1 \overset{?}{=} 7$$

$$6 + 1 = 7 \quad \text{a true statement}$$

Because $(2, 1)$ satisfies both equations, $(2, 1)$ is a solution to the system.

Example 2: Not a Solution of a System

Show that $(0, 4)$ is not a solution to the system $\begin{cases} y = -x + 4 \\ y = 2x + 1 \end{cases}$.

Solution: Substitute $x = 0$ and $y = 4$ into **both** equations.
In the first equation:

$$4 \overset{?}{=} -(0) + 4$$

$$4 = 0 + 4 \quad \text{a true statement}$$

In the second equation:

$$4 \overset{?}{=} 2(0) + 1$$

$$4 = 0 + 1 \quad \text{a false statement}$$

Because $(0, 4)$ does not satisfy both equations, $(0, 4)$ is **not** a solution to the system.

Solutions by Graphing

In this chapter, we will discuss three methods for solving systems of linear equations. The first method is to **solve by graphing**. In this method, both equations are graphed and the point of intersection (if there is one) is the solution to the system. Geometrically there are three possibilities for the graphs of two lines relative to each other. What do you think that these three possibilities are? The following discussion analyzes the three situations.

A Consistent System:

In Example 2, we showed that $(0, 4)$ is not a solution to the system $\begin{cases} y = -x + 4 \\ y = 2x + 1 \end{cases}$.

However, by graphing both equations as shown in Figure 5.1, we see that the two lines appear to **intersect** at the point $(1, 3)$.

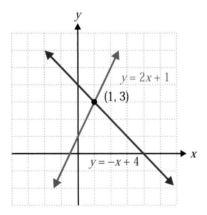

Figure 5.1

We can check that $(1, 3)$ satisfies both equations by substituting $x = 1$ and $y = 3$ into both equations.

$$y = 2x + 1 \qquad\qquad y = -x + 4$$
$$3 \overset{?}{=} 2 \cdot 1 + 1 \qquad\qquad 3 \overset{?}{=} -1 + 4$$
$$3 = 3 \qquad\qquad 3 = 3$$

Since both substitutions give true statements, the lines do intersect at the point $(1, 3)$ and the system is said to be **consistent**.

An Inconsistent System:

Now consider the system $\begin{cases} y = -x + 4 \\ x + y = 2 \end{cases}$. Do the graphs of these lines intersect? If so, where do they intersect? Figure 5.2 indicates that the lines do not intersect. We say that the system is **inconsistent**.

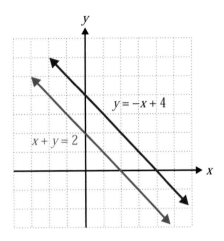

Figure 5.2

The lines do not intersect because they are **parallel**. In this case, we can write both equations in the slope-intercept form and note that they have the same slope, –1, and different y-intercepts.

$$y = -x + 2 \quad \text{with } m = -1 \text{ and } y\text{-intercept } 2$$
$$y = -x + 4 \quad \text{with } m = -1 \text{ and } y\text{-intercept } 4$$

A Dependent System:

For the system $\begin{cases} y = -x + 4 \\ 2y + 2x = 8 \end{cases}$, the graphs of both lines are shown in Figure 5.3. Why is there just one line?

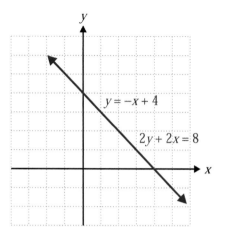

Figure 5.3

There is just one line in Figure 5.3 because both equations represent the same line. The lines not only intersect, they are the same line. That is, they **coincide**. Any point that satisfies one equation will also satisfy the other, and the system has an infinite number of solutions. Putting both equations in the slope-intercept form shows that they are identical.

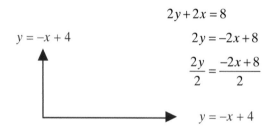

When both equations are simply different forms of the same equation and their graphs are the same line, we say that the system is **dependent**.

When considering a system of linear equations, the system is called a set of **simultaneous equations** to emphasize the idea that the solution of a system is the point that satisfies both equations at the same time, or simultaneously. The following table and graphs summarize the basic ideas and terminology.

Solutions By Graphing

System	Graph	Intersection	Terms
$\begin{cases} y = -x + 4 \\ y = 2x + 1 \end{cases}$		One point: $(1, 3)$	**Consistent**
$\begin{cases} y = -x + 4 \\ x + y = 2 \end{cases}$		No points; lines are parallel.	**Inconsistent**
$\begin{cases} y = -x + 4 \\ 2y + 2x = 8 \end{cases}$		Infinite number of points; lines are the same (they coincide).	**Dependent**

NOTES

The graphing method of solving a system of linear equations shows the geometric relationship between the two lines. However, determining the exact point of intersection on a graph can be difficult or impossible. In particular, when the solution of a system involves fractions or roots, the point of intersection can be only estimated. Algebraic methods, which give exact solutions, will be discussed in Sections 5.2 and 5.3.

Example 3: Solving by Graphing

Determine graphically whether the following systems are (a) consistent, (b) inconsistent, or (c) dependent. If the system is consistent, find (or estimate) the point of intersection.

a. $\begin{cases} y - x = 0 \\ 2x + y = 3 \end{cases}$

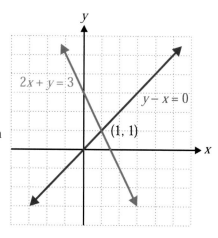

Solution: The system is consistent.
The point of intersection is $(1,1)$.

Check: Substitute $x = 1$ and $y = 1$ into both equations.
For $y - x = 0$, we have $1 - 1 = 0$.
For $2x + y = 3$,
we have $2 \cdot 1 + 1 = 3$.

b. $\begin{cases} x - 2y = 1 \\ x + 3y = 0 \end{cases}$

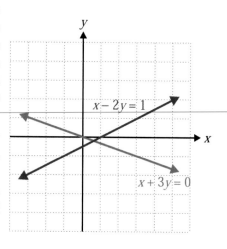

Solution: The system is consistent, but the point of intersection can be only estimated. This example points out the main weakness in solving a system graphically. In such cases, any reasonable estimate will be acceptable. For example, if you estimated $\left(\frac{1}{2}, -\frac{1}{4}\right)$ or $\left(\frac{3}{4}, -\frac{1}{3}\right)$, or some such point, your answer is acceptable. The actual point of intersection is $\left(\frac{3}{5}, -\frac{1}{5}\right)$. We will be able to locate this point precisely using the techniques of the next section.

c. $\begin{cases} y = 3x \\ y - 3x = -4 \end{cases}$

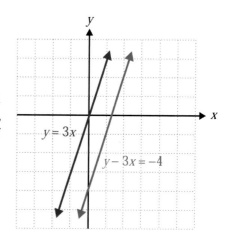

Solution: The system is inconsistent. The lines are parallel with the same slope, 3, and there are no points of intersection.

d. $\begin{cases} x + 2y = 6 \\ y = -\dfrac{1}{2}x + 3 \end{cases}$

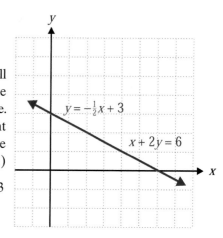

Solution: The system is dependent. All points that lie on one line also lie on the other line. For example, $(4, 1)$ is a point on the line $x + 2y = 6$ since $4 + 2(1) = 6$. The point $(4, 1)$ is also on the line $y = -\dfrac{1}{2}x + 3$ since $1 = -\dfrac{1}{2}(4) + 3$.

● ●

Remember:

1. To use the graphing method, graph the lines as accurately as you can.

2. Be sure to check your solution by substituting back into both of the original equations. (Of course, fractional estimates may not check exactly.)

Using a Graphing Calculator

A graphing calculator can be used to locate (or estimate) the point of intersection of two lines. Consider the system $\begin{cases} 2x + y = 8 \\ x - y = 1 \end{cases}$.

Step 1:   Solve each equation for y. For this system we get

$$\begin{cases} y = -2x + 8 \\ y = x - 1 \end{cases}$$

Step 2:   Press [Y=] and enter the two expressions.

$$Y = -2X + 8 \quad \text{and} \quad Y = X - 1$$

Note: For the leading – sign press the [(−)] key next to [ENTER].

Step 3:   Check the window to be sure that the lines appear. You may press [ZOOM] and then 6 to get the standard window where x values and y-values both go from −10 to +10.

Step 4:   Press the [TRACE] key to estimate the point of intersection (if there is one). The calculator may give only decimal approximations for x and y even though solution values for x and y may be integers. For example, the solution to this system is $(3, 2)$. However, the calculator might show $x = 3.0691489$ and $y = 2.0691489$ as illustrated.

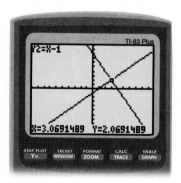

5.1 Exercises

In Exercises 1 – 4, determine which of the given points, if any, lie on both of the lines in the systems of equations by substituting each point into both equations.

1. $\begin{cases} x - y = 6 \\ 2x + y = 0 \end{cases}$

    a. $(1, -2)$
    b. $(4, -2)$
    c. $(2, -4)$
    d. $(-1, 2)$

2. $\begin{cases} x + 3y = 5 \\ 3y = 4 - x \end{cases}$

    a. $(2, 1)$
    b. $(2, -2)$
    c. $(-1, 2)$
    d. $(4, 0)$

3. $\begin{cases} 2x+4y-6=0 \\ 3x+6y-9=0 \end{cases}$

    a. $(1,1)$

    b. $(2,0)$

    c. $\left(0,\dfrac{3}{2}\right)$

    d. $(-1,3)$

4. $\begin{cases} 5x-2y-5=0 \\ 5x=-3y \end{cases}$

    a. $(1,0)$

    b. $\left(\dfrac{3}{5},-1\right)$

    c. $(0,0)$

    d. $(1,4)$

In Exercises 5 – 8, the graphs of the lines represented by each system of equations are given. Determine the solution of the system by looking at the graph. Check your solution by substituting into both equations.

5. $\begin{cases} x+2y=4 \\ x-y=-2 \end{cases}$ **6.** $\begin{cases} x+2y=1 \\ 2x+y=-1 \end{cases}$ **7.** $\begin{cases} 2x-y=6 \\ 3x+y=14 \end{cases}$ **8.** $\begin{cases} x-y=4 \\ 3x-y=6 \end{cases}$

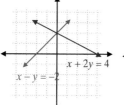 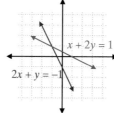 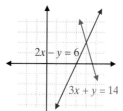 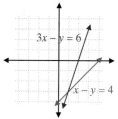

In Exercises 9 – 12, show that each system of equations is inconsistent by determining the slope of each line and the y-intercept. (That is, show that the lines are parallel and do not intersect.)

9. $\begin{cases} 2x+y=3 \\ 4x+2y=5 \end{cases}$ **10.** $\begin{cases} y=\dfrac{1}{2}x+3 \\ x-2y=1 \end{cases}$ **11.** $\begin{cases} 3x-y=8 \\ x-\dfrac{1}{3}y=2 \end{cases}$ **12.** $\begin{cases} 3x-5y=1 \\ 6x-10y=4 \end{cases}$

Graph each of the systems in Exercises 13 – 38 and determine whether the system is (a) consistent, (b) inconsistent, or (c) dependent. If the system is consistent, determine the coordinates of the solution and check your solution.

13. $\begin{cases} 2x-y=4 \\ 3x+y=6 \end{cases}$ **14.** $\begin{cases} x+y-5=0 \\ x-4y=5 \end{cases}$ **15.** $\begin{cases} 3x-y=6 \\ y=3x \end{cases}$

16. $\begin{cases} x-y=5 \\ x=-3 \end{cases}$ **17.** $\begin{cases} x+2y=8 \\ 3x-2y=0 \end{cases}$ **18.** $\begin{cases} 5x-4y=5 \\ 8y=10x-10 \end{cases}$

19. $\begin{cases} 4x - 2y = 10 \\ -6x + 3y = -15 \end{cases}$

20. $\begin{cases} y = 2x + 5 \\ 4x - 2y = 7 \end{cases}$

21. $\begin{cases} \dfrac{1}{2}x + 2y = 7 \\ 2x = 4 - 8y \end{cases}$

22. $\begin{cases} 4x + 3y + 7 = 0 \\ 5x - 2y + 3 = 0 \end{cases}$

23. $\begin{cases} 2x + 3y = 4 \\ 4x - y = 1 \end{cases}$

24. $\begin{cases} 7x - 2y = 1 \\ y = 3 \end{cases}$

25. $\begin{cases} 2x - 5y = 6 \\ y = \dfrac{2}{5}x + 1 \end{cases}$

26. $\begin{cases} y = 4x - 3 \\ x = 2y - 8 \end{cases}$

27. $\begin{cases} y = \dfrac{1}{2}x + 2 \\ x - 2y + 4 = 0 \end{cases}$

28. $\begin{cases} 2x + 3y = 5 \\ 3x - 2y = 1 \end{cases}$

29. $\begin{cases} \dfrac{2}{3}x + y = 2 \\ x - 4y = 3 \end{cases}$

30. $\begin{cases} x + y = 8 \\ 5y = 2x + 5 \end{cases}$

31. $\begin{cases} y = 2x \\ 2x + y = 4 \end{cases}$

32. $\begin{cases} x - y = 4 \\ 2y = 2x - 4 \end{cases}$

33. $\begin{cases} y = x + 1 \\ y + x = -5 \end{cases}$

34. $\begin{cases} x + y = 4 \\ 2x - 3y = 3 \end{cases}$

35. $\begin{cases} 2x + y = 0 \\ 4x + y = -2 \end{cases}$

36. $\begin{cases} 4x + y = 6 \\ 2x + \dfrac{1}{2}y = 3 \end{cases}$

37. $\begin{cases} \dfrac{1}{2}x + \dfrac{1}{3}y = \dfrac{1}{6} \\ \dfrac{1}{4}x + \dfrac{1}{4}y = 0 \end{cases}$

38. $\begin{cases} \dfrac{1}{4}x - y = \dfrac{13}{4} \\ \dfrac{1}{3}x + \dfrac{1}{6}y = -\dfrac{1}{6} \end{cases}$

For Exercises 39 – 42, a word problem is stated with equations given that represent a mathematical model for the problem. Solve the system graphically.

39. The sum of two numbers is 25 and their difference is 15. What are the two numbers?
Let x = one number and y = the other number.

The corresponding modeling system is $\begin{cases} x + y = 25 \\ x - y = 15 \end{cases}$.

40. The perimeter of a rectangle is 50 meters and the length is 5 meters longer than the width. Find the dimensions of the rectangle.
Let x = the length and y = the width.

The corresponding modeling system is $\begin{cases} 2x + 2y = 50 \\ x - y = 5 \end{cases}$.

perimeter = 50m

41. Ten gallons of a salt solution consists of 30% salt. It is the result of mixing a 50% solution with a 25% solution. How many gallons of each of the mixing solutions was used?

Let x = the number of gallons of the 50% solution
and y = the number of gallons of the 25% solution.

The corresponding modeling system is $\begin{cases} x + y = 10 \\ 0.50x + 0.25y = 0.30(10) \end{cases}$.

42. A student bought a calculator and a textbook for a course in algebra. He told his friend that the total cost was $170 (without tax) and that the calculator cost $20 more than twice the cost of the textbook. What was the cost of each item?

Let c = the cost of the calculator
and t = the cost of the textbook.

The corresponding modeling system is $\begin{cases} c + t = 170 \\ c = 2t + 20 \end{cases}$.

In Exercises 43 – 46, use a graphing calculator to estimate the solutions to each of the systems of equations. Remember to solve each equation for y, press ⬭ Y= ⬭ and enter the expressions involving x, and use the ⬭TRACE⬭ key to locate the estimated solutions.

43. $\begin{cases} x + 2y = 9 \\ x - 2y = -7 \end{cases}$

44. $\begin{cases} x - 3y = 0 \\ 2x + y = 7 \end{cases}$

45. $\begin{cases} y = 2 \\ 2x - 3y = -3 \end{cases}$

46. $\begin{cases} 2x - 3y = 0 \\ 3x + 3y = \dfrac{5}{2} \end{cases}$

Hawkes Learning Systems: Introductory Algebra

Solving Systems of Linear Equations by Graphing

5.2   Systems of Equations: Solutions by Substitution

After completing this section, you will be able to:

1. Solve systems of linear equations by substitution.

As we discussed in Section 5.1, solving systems of linear equations by graphing is somewhat limited in accuracy. The graphs must be drawn exactly and even then the points of intersection (if there are any) can be difficult to estimate accurately. In this section, we will develop an algebraic method called the **method of substitution**. This method involves solving one of the equations for one of the variables and then substituting the resulting expression into the other equation. The objective is to get one equation in one variable. For example, consider the system

$$\begin{cases} y = -2x + 5 \\ x + 2y = 1 \end{cases}.$$

How would you substitute? Since the first equation is already solved for y, a reasonable substitution would be to put $-2x + 5$ for y in the second equation. Try this and see what happens.

First equation already solved for y: $y = \boxed{-2x + 5}$

Substitute into second equation: $x + 2y = 1$

$$x + 2(-2x + 5) = 1$$

We now have one equation in only one variable, namely x. The problem has been reduced from one of solving two equations in two variables to solving one equation in one variable. Solve this equation for x. Then find the corresponding y-value by substituting this x-value into **either of the two original equations**.

$$x + 2(-2x + 5) = 1$$
$$x - 4x + 10 = 1$$
$$-3x = -9$$
$$x = 3$$

Substituting $x = 3$ into $y = -2x + 5$ gives

$$y = -2 \cdot 3 + 5$$
$$= -6 + 5$$
$$= -1.$$

Thus, the system is consistent and the solution to the system is $x = 3$ and $y = -1$, or the point $(3, -1)$.

Substitution is not the only algebraic technique for solving a system of linear equations. It does work in all cases but generally is used only when one of the equations is easily solved for one variable.

In the following examples, note how the results in Example 1c indicate that the system is inconsistent and the results in Example 1d indicate that the system is dependent.

Example 1: Solve by Substitution • • • • • • • • • • • • • • •

Solve the following systems of linear equations by using the technique of substitution.

a.
$$\begin{cases} y = \dfrac{5}{6}x + 1 \\ 2x + 6y = 7 \end{cases}$$

Solution: $y = \dfrac{5}{6}x + 1$ is already solved for y. Substituting into the other equation, we find

$$2x + 6y = 7$$
$$2x + 6\left(\frac{5}{6}x + 1\right) = 7$$
$$2x + 5x + 6 = 7$$
$$7x = 1$$
$$x = \frac{1}{7}$$
$$y = \frac{5}{6} \cdot \frac{1}{7} + 1 = \frac{5}{42} + \frac{42}{42} = \frac{47}{42}$$

The system is **consistent** and the solution is $\left(\dfrac{1}{7}, \dfrac{47}{42}\right)$, or $x = \dfrac{1}{7}$ and $y = \dfrac{47}{42}$.

Continued on next page ...

311

b. $\begin{cases} x = -5 \\ y = 2x + 9 \end{cases}$

Solution: In the first equation, $x = -5$ is already solved for x. The equation represents a vertical line. Substituting $x = -5$ into the second equation we have,

$$y = 2x + 9$$

$$y = 2(-5) + 9 = -10 + 9 = -1$$

The system is **consistent** and the solution is $x = -5$ and $y = -1$, or $(-5, -1)$.

c. $\begin{cases} 3x + y = 1 \\ 6x + 2y = 3 \end{cases}$

Solution: Solving for y in the first equation gives

$$3x + y = 1$$

$$y = -3x + 1$$

Substituting for y in the second equation yields

$$6x + 2y = 3$$

$$6x + 2(-3x + 1) = 3$$

$$6x - 6x + 2 = 3$$

$$2 = 3.$$

This last equation, $2 = 3$, is never true. This tells us that the system is **inconsistent**. Graphically, the lines are parallel and there is no point of intersection. The system has no solution.

d. $\begin{cases} x - 2y = 1 \\ 3x - 6y = 3 \end{cases}$

Solution: Solving for x in the first equation gives

$$x - 2y = 1$$

$$x = 2y + 1$$

Substituting for x in the second equation yields

$$3x - 6y = 3$$

$$3(2y + 1) - 6y = 3$$

$$6y + 3 - 6y = 3$$

$$3 = 3.$$

Continued on next page ...

This last equation, 3 = 3, is always true. This tells us that the system is **dependent**. Graphically, the two lines are the same and there are an infinite number of points of intersection. Therefore, the solution consists of all the points that satisfy either of the original equations, $x - 2y = 1$ or $3x - 6y = 3$.

e. $\begin{cases} x + y = 5 \\ 0.2x + 0.3y = 0.9 \end{cases}$

Solution: Solving the first equation for y, we have

$$x + y = 5$$

$$y = 5 - x.$$

Substituting for y in the second equation yields

$$0.2x + 0.3y = 0.9 \qquad \text{Note that you could multiply}$$

$$0.2x + 0.3(5 - x) = 0.9 \qquad \text{each term by 10 to eliminate}$$

$$0.2x + 1.5 - 0.3x = 0.9 \qquad \text{the decimal.}$$

$$-0.1x = -0.6$$

$$\frac{-0.1x}{-0.1} = \frac{-0.6}{-0.1}$$

$$x = 6$$

$$y = 5 - x = 5 - 6 = -1.$$

The system is **consistent** and the solution is $x = 6$ and $y = -1$, or $(6, -1)$.

● ●

Practice Problems

Solve the following systems using the technique of substitution.

1. $\begin{cases} x + y = 3 \\ y = 2x \end{cases}$ **2.** $\begin{cases} y = 3x - 1 \\ 2x + y = 4 \end{cases}$ **3.** $\begin{cases} x + 2y = -1 \\ x - 4y = -4 \end{cases}$

Answer to Practice Problems: 1. $x = 1, y = 2$ **2.** $x = 1, y = 2$ **3.** $x = -2, \; y = \dfrac{1}{2}$

5.2 Exercises

Solve the systems of linear equations in Exercises 1 – 38 by using the technique of substitution. If the system is inconsistent or dependent, say so in your answer.

1. $\begin{cases} x+y=6 \\ y=2x \end{cases}$

2. $\begin{cases} 5x+2y=21 \\ x=y \end{cases}$

3. $\begin{cases} x-7=3y \\ y=2x-4 \end{cases}$

4. $\begin{cases} y=3x+4 \\ 2y=3x+5 \end{cases}$

5. $\begin{cases} x=3y \\ 3y-2x=6 \end{cases}$

6. $\begin{cases} 4x=y \\ 4x-y=7 \end{cases}$

7. $\begin{cases} x-5y+1=0 \\ x=7-3y \end{cases}$

8. $\begin{cases} 2x+5y=15 \\ x=y-3 \end{cases}$

9. $\begin{cases} 7x+y=9 \\ y=4-7x \end{cases}$

10. $\begin{cases} 3y+5x=5 \\ y=3-2x \end{cases}$

11. $\begin{cases} 3x-y=7 \\ x+y=5 \end{cases}$

12. $\begin{cases} 4x-2y=5 \\ y=2x+3 \end{cases}$

13. $\begin{cases} 6x+3y=9 \\ y=3-2x \end{cases}$

14. $\begin{cases} x-y=5 \\ 2x+3y=0 \end{cases}$

15. $\begin{cases} 4x=8 \\ 3x+y=8 \end{cases}$

16. $\begin{cases} x+y=8 \\ 3x+2y=8 \end{cases}$

17. $\begin{cases} y=2x-5 \\ 2x+y=-3 \end{cases}$

18. $\begin{cases} 2x+3y=5 \\ x-6y=0 \end{cases}$

19. $\begin{cases} 2y=5 \\ 3x-4y=-4 \end{cases}$

20. $\begin{cases} x+5y=1 \\ x-3y=5 \end{cases}$

21. $\begin{cases} 3x+8y=-2 \\ x+2y=-1 \end{cases}$

22. $\begin{cases} 4x-4y=9 \\ 3x+y=2 \end{cases}$

23. $\begin{cases} 5x+2y=-10 \\ 7x=4-y \end{cases}$

24. $\begin{cases} x-2y=-4 \\ 3x+y=-5 \end{cases}$

25. $\begin{cases} x+4y=3 \\ 3x-4y=7 \end{cases}$

26. $\begin{cases} 3x-y=-1 \\ 7x-4y=0 \end{cases}$

27. $\begin{cases} x+5y=-1 \\ 2x+7y=1 \end{cases}$

28. $\begin{cases} x+3y=5 \\ 3x+2y=7 \end{cases}$

29. $\begin{cases} 3x-4y-39=0 \\ 2x-y-13=0 \end{cases}$

30. $\begin{cases} \dfrac{x}{3}+\dfrac{y}{5}=1 \\ x+6y=12 \end{cases}$

31. $\begin{cases} \dfrac{x}{5}+\dfrac{y}{4}-3=0 \\ \dfrac{x}{10}-\dfrac{y}{2}+1=0 \end{cases}$

32. $\begin{cases} 6x-y=15 \\ 0.2x+0.5y=2.1 \end{cases}$

33. $\begin{cases} x+2y=3 \\ 0.4x+y=0.6 \end{cases}$

34. $\begin{cases} 0.2x - 0.1y = 0 \\ \qquad\quad y = x + 10 \end{cases}$

35. $\begin{cases} 0.1x - 0.2y = 1.4 \\ \quad\; 3x + y = 14 \end{cases}$

36. $\begin{cases} 3x - 2y = 5 \\ \quad\; y = \dfrac{3}{2}x + 2 \end{cases}$

37. $\begin{cases} \quad\; x = 2y - 7.5 \\ 2x + 4y = -15 \end{cases}$

38. $\begin{cases} \dfrac{1}{2}x + \dfrac{1}{3}y = 4 \\ 3x + 2y = 24 \end{cases}$

For Exercises 39 – 42, a word problem is stated with equations given that represent a mathematical model for the problem. Solve the system by using the method of substitution. (**Note**: These exercises were also given in Section 5.1. Check to see that you arrived at the same answers by both methods.)

39. The sum of two numbers is 25 and their difference is 15. What are the two numbers?

Let x = one number and y = the other number.

The corresponding modeling system is $\begin{cases} x + y = 25 \\ x - y = 15 \end{cases}$.

40. The perimeter of a rectangle is 50 meters and the length is 5 meters longer than the width. Find the dimensions of the rectangle.

Let x = the length and y = the width.

The corresponding modeling system is $\begin{cases} 2x + 2y = 50 \\ \quad\; x - y = 5 \end{cases}$.

perimeter = 50m

41. Ten gallons of a salt solution consists of 30% salt. It is the result of mixing a 50% solution with a 25% solution. How many gallons of each of the mixing solutions was used?

Let x = the number of gallons of the 50% solution and
y = the number of gallons of the 25% solution.

The corresponding modeling system is $\begin{cases} x + y = 10 \\ 0.50x + 0.25y = 0.30(10) \end{cases}$.

42. A student bought a calculator and a textbook for a course in algebra. He told his friend that the total cost was $170 (without tax) and that the calculator cost $20 more than twice the cost of the textbook. What was the cost of each item?

Let c = the cost of the calculator and t = the cost of the textbook.

The corresponding modeling system is $\begin{cases} c + t = 170 \\ c = 2t + 20 \end{cases}$.

Hawkes Learning Systems: Introductory Algebra

Solving Systems of Linear Equations By Substitution

5.3 Systems of Equations: Solutions by Addition

After completing this section, you will be able to:

1. *Solve systems of linear equations by addition.*

2. *Write equations of lines given two points by using the formula $y = mx + b$ and systems of linear equations.*

In Section 5.1, we discussed solving systems of linear equations by graphing, and in Section 5.2 we discussed solving by substitution. We know that solutions by graphing are not necessarily exact and, in some cases, the method of substitution can lead to complicated algebraic steps. That is, solving for x or y can involve several steps and fractions. For example, consider solving the following system by the method of substitution:

$$\begin{cases} 4x - 3y = 1 \\ 3x - 2y = 4 \end{cases}.$$

Solving the first equation for x, we have

$$4x - 3y = 1$$
$$4x = 1 + 3y$$
$$x = \frac{1 + 3y}{4}.$$

Then, substitution gives

$$3\left(\frac{1 + 3y}{4}\right) - 2y = 4.$$

We can go on to finish this process and find the solution. However, the point here is that the technique involves somewhat difficult algebraic manipulations.

Another method, called the **method of addition** (or **method of elimination**), can simplify the process considerably. We begin the discussion with a relatively simple system;

$$\begin{cases} y = 4 - x \\ x = y + 6 \end{cases}.$$

Put both equations in the standard form $Ax + By = C$ and set one equation under the other so that like terms are aligned vertically. Then **add like terms**.

$$
\begin{array}{r}
x + y = 4 \\
\underline{x - y = 6} \\
2x \quad\ = 10
\end{array}
$$

Since $+y$ and $-y$ have opposite coefficients, the y-terms are eliminated, and the resulting equation, $2x = 10$, has only one variable. Just as with the technique of substitution, the solution of the system is reduced to solving one equation in one variable.

$$
\begin{array}{r}
x + y = 4 \\
\underline{x - y = 6} \\
2x \quad\ = 10 \\
x \quad\ = 5
\end{array}
$$

Substitute $x = 5$ into one of the original equations and solve for y.

$$
\begin{array}{r}
x + y = 4 \\
5 + y = 4 \\
y = -1
\end{array}
$$

The solution is $x = 5$ and $y = -1$, or $(5, -1)$.

If the two coefficients of one variable in the system are not opposites, then we multiply each equation by some nonzero constant so that either the two x-coefficients are opposites or the two y-coefficients are opposites. For example, consider the system,

$$
\begin{cases}
4x - 3y = 1 \\
3x - 2y = 4
\end{cases}.
$$

Multiplying each term of the first equation by 2 and each term of the second equation by -3 will result in the y-coefficients being opposites. Or multiplying the terms of the first equation by 3 and the second equation by -4 will give opposite coefficients for the x-terms. Both cases yield the same solution and are illustrated below. The number used to multiply the terms of each equation is in brackets.

Method 1: Eliminate y-terms

$$
\begin{cases}
[2]\ 4x - 3y = 1 \rightarrow & 8x - 6y = \ \ 2 \\
[-3]\ 3x - 2y = 4 \rightarrow & \underline{-9x + 6y = -12}
\end{cases}
$$

$$
\begin{array}{r}
-x \quad\ = -10 \\
x \quad\ = \ \ 10
\end{array}
$$

Substitute $x = 10$ into one of the original equations.

Continued on next page ...

$$4x - 3y = 1$$
$$4 \cdot 10 - 3y = 1$$
$$40 - 3y = 1$$
$$-3y = -39$$
$$y = 13$$

The solution is $x = 10$ and $y = 13$, or $(10, 13)$.

Method 2: Eliminate x-terms

$$\begin{cases} [3] \ \ 4x - 3y = 1 \ \rightarrow \ \ \ 12x - 9y = \ \ 3 \\ [-4] 3x - 2y = 4 \ \rightarrow \ -12x + 8y = -16 \end{cases}$$
$$-y = -13$$
$$y = \ \ 13$$

Substitute $y = 13$ into one of the original equations.

$$4x - 3y = 1$$
$$4x - 3 \cdot 13 = 1$$
$$4x - 39 = 1$$
$$4x = 40$$
$$x = 10$$

The solution is the same: $x = 10$ and $y = 13$, or $(10, 13)$. As illustrated in Method 1 and Method 2, eliminating either the y-terms or the x-terms yields the same results.

What if both the x- and y-terms are eliminated? In this situation, the system is either inconsistent or dependent. Just as with the technique of substitution, the resulting equation involving only constants tells which case is under consideration.

For the system

$$\begin{cases} 2x - y = 6 \\ 4x - 2y = 1 \end{cases}$$

multiplying the terms in the first equation by -2 and then adding gives

$$\begin{cases} [-2] 2x - y = 6 \ \rightarrow \ -4x + 2y = -12 \\ \ \ \ \ \ \ 4x - 2y = 1 \ \rightarrow \ \ \ \ 4x - 2y = \ \ 1 \end{cases}$$
$$0 = -11.$$

Since the equation $0 = -11$ is not true, the system is inconsistent. There is no solution.

For the system

$$\begin{cases} 2x - 2y = 1 \\ 3x - 3y = \dfrac{3}{2} \end{cases}$$

multiplying the terms in the first equation by 3 and the terms in the second equation by –2 and then adding gives,

$$\begin{cases} [3] \ \ 2x - 2y = 1 \ \rightarrow \ \ \ 6x - 6y = \ \ 3 \\ [-2] 3x - 3y = \dfrac{3}{2} \ \rightarrow \ \ -6x + 6y = -3 \end{cases}$$
$$0 = \ \ 0.$$

Since the equation $0 = 0$ is always true, the system is dependent. The solution consists of all points that satisfy the equation $2x - 2y = 1$.

In solving a system by addition, find the constant for multiplying the terms of each equation by trying to get coefficients of like terms to be opposites. There are many possible choices. One approach is to find the least common multiple of the two coefficients already there, and then multiply so that one coefficient will be the LCM and the other coefficient, its opposite.

Thus, for the system,

$$\begin{cases} 4x - 3y = 1 \\ 3x - 2y = 4 \end{cases}$$

in Method 1, the coefficients for the y-terms ended up being –6 and +6. The LCM for 3 and 2 is 6. In Method 2, the coefficients for the x-terms ended up being +12 and –12. The LCM for 4 and 3 is 12.

Example 1: Solve by Addition ● ● ● ● ● ● ● ● ● ● ● ● ● ● ● ● ●

Solve the following systems of equations using the technique of addition.

a. $\begin{cases} 5x + 3y = -3 \\ 2x - 7y = 7 \end{cases}$

    Solution:
$$\begin{cases} [-2] \ 5x + 3y = -3 \rightarrow \ -10x - \ 6y = \ \ 6 \\ [5] \ \ 2x - 7y = 7 \ \rightarrow \ \ \ \ 10x - 35y = 35 \end{cases}$$
$$-41y = 41$$
$$y = -1$$

Continued on next page ...

Substitute $y = -1$ into one of the original equations.

$$5x + 3y = -3$$
$$5x + 3(-1) = -3$$
$$5x - 3 = -3$$
$$5x = 0$$
$$x = 0$$

The solution is $x = 0$ and $y = -1$, or $(0, -1)$.

b. $\begin{cases} y = -2 + 4x \\ 8x - 2y = 4 \end{cases}$

Solution: Rearranging so that both equations are in standard form gives

$$\begin{cases} -4x + y = -2 \\ 8x - 2y = 4 \end{cases}$$

Then
$$\begin{cases} -4x + y = -2 \rightarrow -4x + y = -2 \\ \left[\dfrac{1}{2}\right] \quad 8x - 2y = 4 \rightarrow \underline{\quad 4x - y = 2\quad} \end{cases}$$
$$0 = 0$$

The system is dependent. The solution is the set of all points that satisfy the equation $y = -2 + 4x$ (or the equation $8x - 2y = 4$).

c. $\begin{cases} x + 0.4y = 3.08 \\ 0.1x - y = 0.1 \end{cases}$

Solution:
$$\begin{cases} x + 0.4y = 3.08 \rightarrow 1.0x + 0.4y = 3.08 \\ [-10] \quad 0.1x - y = 0.1 \rightarrow \underline{-1.0x + 10.0y = -1.0} \end{cases}$$
$$10.4y = 2.08$$
$$y = 0.2$$

Substitute $y = 0.2$ into one of the original equations.

$$x + 0.4y = 3.08$$
$$x + 0.4(0.2) = 3.08$$
$$x + 0.08 = 3.08$$
$$x = 3$$

The solution is $x = 3$ and $y = 0.2$, or $(3, 0.2)$.

Continued on next page ...

d. Using the formula $y = mx + b$, find the equation of the line determined by the two points $(3, 5)$ and $(-6, 2)$.

Solution: Write two equations in m and b by substituting the coordinates of the points for x and y.

$$\begin{cases} 5 = 3m + b \rightarrow 5 = 3m + b \\ [-1] \quad 2 = -6m + b \rightarrow \underline{-2 = 6m - b} \end{cases}$$

$$3 = 9m$$

$$\frac{1}{3} = m$$

Substitute $m = \dfrac{1}{3}$ into one of the original equations.

$$5 = 3m + b$$

$$5 = 3 \cdot \frac{1}{3} + b$$

$$5 = 1 + b$$

$$4 = b$$

The equation is $y = \dfrac{1}{3}x + 4$.

• •

Summary of the Method of Addition

1.   *Rewrite (if necessary) both equations in the standard form $Ax + By = C$.*

2.   *Multiply (if necessary) all terms in one (or both) equations so that the coefficients of one of the variables are opposites.*

3.   *Add the like terms of the equations so that one of the variables is eliminated and solve the resulting equation.*

   (If both variables are eliminated and the constant is not 0, the system is inconsitent.)

   (If both variables are eliminated and the constant is 0, the system is dependent.)

4.   *Substitute the solution from Step 3 back into either of the two original equations and solve for the other variable.*

5.   *Check the solutions in both of the original equations.*

Practice Problems

Solve the following systems by using the method of addition.

1. $\begin{cases} 2x + 2y = 4 \\ x - y = -3 \end{cases}$

2. $\begin{cases} 3x + 4y = 12 \\ \dfrac{1}{3}x - 8y = -5 \end{cases}$

3. $\begin{cases} 0.02x + 0.06y = 1.48 \\ 0.03x - 0.02y = 0.02 \end{cases}$

Now that you know three methods for solving a system of linear equations (graphing, substitution, and addition), which method should you use? Consider the following guidelines in making your decision.

Guidelines for Deciding which Method to Use in Solving a System of Linear Equations

1. The graphing method is helpful in "seeing" the geometric relationship between the lines and finding approximate solutions. A calculator can be very helpful here.

2. Both the substitution method and the addition method give exact solutions.

3. The substitution method may be reasonable and efficient if one of the coefficients of one of the variables is 1.

4. In general, the method of addition will prove to be most efficient.

5.3 Exercises

In Exercises 1 – 38, solve the system by using either the substitution method or the addition method (whichever seems better to you). If the system is inconsistent or dependent, say so in your answer.

1. $\begin{cases} 2x - y = 7 \\ x + y = 2 \end{cases}$

2. $\begin{cases} x + 3y = 9 \\ x - 7y = -1 \end{cases}$

3. $\begin{cases} 3x + 2y = 0 \\ 5x - 2y = 8 \end{cases}$

4. $\begin{cases} 4x - y = 7 \\ 4x + y = -3 \end{cases}$

5. $\begin{cases} 2x + 2y = 5 \\ x + y = 3 \end{cases}$

6. $\begin{cases} y = 2x + 14 \\ x = 14 - 3y \end{cases}$

7. $\begin{cases} x = 11 + 2y \\ 2x - 3y = 17 \end{cases}$

8. $\begin{cases} 6x - 3y = 6 \\ y = 2x - 2 \end{cases}$

9. $\begin{cases} x - 2y = 4 \\ y = \dfrac{1}{2}x - 2 \end{cases}$

Answer to Practice Problems: 1. $x = -\dfrac{1}{2},\ y = \dfrac{5}{2}$ **2.** $x = 3,\ y = \dfrac{3}{4}$ **3.** $x = 14,\ y = 20$

10. $\begin{cases} x = 3y + 4 \\ y = 6 - 2x \end{cases}$

11. $\begin{cases} 8x - y = 29 \\ 2x + y = 11 \end{cases}$

12. $\begin{cases} 7x - y = 16 \\ 2y = 2 - 3x \end{cases}$

13. $\begin{cases} 3x + y = -10 \\ 2y - 1 = x \end{cases}$

14. $\begin{cases} 3x + 3y = 18 \\ 4x + 2y = 32 \end{cases}$

15. $\begin{cases} 3x + 2y = 4 \\ x + 5y = -3 \end{cases}$

16. $\begin{cases} x + 2y = 0 \\ 2x = 4y \end{cases}$

17. $\begin{cases} \dfrac{1}{2}x + y = -4 \\ 3x - 4y = 6 \end{cases}$

18. $\begin{cases} x + y = 1 \\ x - \dfrac{1}{3}y = \dfrac{11}{3} \end{cases}$

19. $\begin{cases} 4x + 3y = 2 \\ 3x + 2y = 3 \end{cases}$

20. $\begin{cases} 5x - 2y = 17 \\ 2x - 3y = 9 \end{cases}$

21. $\begin{cases} \dfrac{1}{2}x + 2y = 9 \\ 2x - 3y = 14 \end{cases}$

22. $\begin{cases} 3x + 2y = 14 \\ 7x + 3y = 26 \end{cases}$

23. $\begin{cases} 4x + 3y = 28 \\ 5x + 2y = 35 \end{cases}$

24. $\begin{cases} 2x + 7y = 2 \\ 5x + 3y = -24 \end{cases}$

25. $\begin{cases} 7x - 6y = -1 \\ 5x + 2y = 37 \end{cases}$

26. $\begin{cases} 9x + 2y = -42 \\ 5x - 6y = -2 \end{cases}$

27. $\begin{cases} 7x + 4y = 7 \\ 6x + 7y = 31 \end{cases}$

28. $\begin{cases} 6x - 5y = -40 \\ 8x - 7y = -54 \end{cases}$

29. $\begin{cases} \dfrac{3}{4}x - \dfrac{1}{2}y = 2 \\ \dfrac{1}{3}x - \dfrac{7}{6}y = 1 \end{cases}$

30. $\begin{cases} x + y = 12 \\ 0.05x + 0.25y = 1.6 \end{cases}$

31. $\begin{cases} x + 0.5y = 8 \\ 0.1x + 0.01y = 0.64 \end{cases}$

32. $\begin{cases} 0.5x - 0.3y = 7 \\ 0.3x - 0.4y = 2 \end{cases}$

33. $\begin{cases} 0.6x + 0.5y = 5.9 \\ 0.8x + 0.4y = 6 \end{cases}$

34. $\begin{cases} 2.5x + 1.8y = 7 \\ 3.5x - 2.7y = 4 \end{cases}$

35. $\begin{cases} \dfrac{2}{3}x + \dfrac{1}{2}y = \dfrac{2}{3} \\ 3x + 2y = \dfrac{17}{6} \end{cases}$

36. $\begin{cases} \dfrac{3}{4}x - \dfrac{1}{4}y = \dfrac{3}{8} \\ \dfrac{1}{2}x + \dfrac{1}{2}y = \dfrac{3}{4} \end{cases}$

37. $\begin{cases} \dfrac{1}{6}x - \dfrac{1}{12}y = -\dfrac{13}{6} \\ \dfrac{1}{5}x + \dfrac{1}{4}y = 2 \end{cases}$

38. $\begin{cases} \dfrac{5}{3}x - \dfrac{2}{3}y = -\dfrac{29}{30} \\ 2x + 5y = 0 \end{cases}$

In Exercises 39 – 44, write an equation for the line determined by the two given points using the formula $y = mx + b$ to set up a system of equations with m and b as the unknowns.

39. $(2, 3), (1, -2)$ **40.** $(4, 7), (-3, 2)$ **41.** $(-4, 1), (5, 2)$
42. $(0, 6), (-3, -3)$ **43.** $(3, -4), (7, 7)$ **44.** $(1, -3), (5, -3)$

Calculator Problems

In Exercises 45 – 50, use a graphing calculator to estimate the solutions to each of the systems of equations. Remember to solve each equation for y, press $\boxed{\text{Y=}}$ and enter the expressions involving x, and use the $\boxed{\text{TRACE}}$ key to locate the estimated solutions.

45. $\begin{cases} 0.9x + 1.3y = 1.4 \\ 1.2x - 0.7y = 4.3 \end{cases}$ **46.** $\begin{cases} 1.8x + 2.0y = 4.4 \\ 1.2x + 1.2y = -3.6 \end{cases}$ **47.** $\begin{cases} 1.4x + 3.5y = 7.28 \\ 2.4x - 2.1y = -0.48 \end{cases}$

48. $\begin{cases} 2.2x + 1.5y = 7.69 \\ 4.0x - 0.8y = 7.28 \end{cases}$ **49.** $\begin{cases} 1.3x + 4.1y = 9.294 \\ 0.7x - 1.6y = 4.494 \end{cases}$ **50.** $\begin{cases} 0.09x + 0.17y = 0.6198 \\ 2.10x - 0.90y = 1.6140 \end{cases}$

For Exercises 51 – 54, a word problem is stated with equations given that represent a mathematical model for the problem. Solve the system by using either the method of substitution or the method of addition.

51. Georgia has $10,000 to invest and she is going to put the money into two accounts. One of the accounts will pay 6% interest and the other will pay 10%. How much will she put in each account if she knows that the interest from the 10% account will exceed the interest from the 6% account by $40?
Let x = amount in 10% account and y = amount in 6% account.

Then the system that models the problem is $\begin{cases} x + y = 10,000 \\ 0.10x - 0.06y = 40 \end{cases}$.

52. Money is invested at two rates of interest. One rate is 8% and the other is 6%. If there is $1000 more invested at 8% than at 6%, find the amount invested at each rate if the annual interest from both investments is $640.
Let x = amount invested at 8% and y = amount invested at 6%.

Then the system that models the problem is $\begin{cases} x = y + 1000 \\ 0.08x + 0.06y = 640 \end{cases}$.

53. How many liters each of a 30% acid solution and a 40% acid solution must be used to produce 100 liters of a 36% acid solution?

Let x = amount of 30% solution and y = amount of 40% solution.

Then the system that models the problem is

$$\begin{cases} x + y = 100 \\ 0.30x + 0.40y = 0.36(100) \end{cases}.$$

54. Two cars leave Denver at the same time traveling in opposite directions. One travels at an average speed of 55 mph and the other at 65 mph. In how many hours will they be 420 miles apart?

Let x = time of travel for first car and y = time of travel for second car.

Then the system that models the problem is

$$\begin{cases} x = y \\ 55x + 65y = 420 \end{cases}.$$

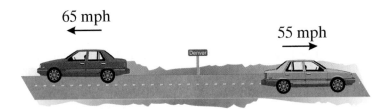

Hawkes Learning Systems: Introductory Algebra

   Solving Systems of Linear Equations by Addition

5.4 Applications: Distance-Rate-Time, Number Problems, Amounts and Costs

Objectives

After completing this section, you will be able to:

1. Solve applied problems related to distance, rate, and time by using systems of linear equations.

2. Solve applied problems related to numbers by using systems of linear equations.

3. Solve applied problems related to amounts and costs by using systems of linear equations.

Systems of equations occur in many practical situations such as supply and demand in business, velocity and acceleration in engineering, money and interest in investments, and mixture in physics and chemistry. Many of the problems in the exercises for Sections 5.1, 5.2, and 5.3 illustrated these ideas and corresponding systems of equations that served to solve the problems were given. The equations were given in those problems to help lead you to understand how the variables and equations are set up in solving problems with systems of equations. As you study the applications in Sections 5.4 and 5.5, you may want to refer to some of those exercises, as well as the examples, as guides in solving the new applications.

In this section, we will study applications related to distance-rate-time concepts ($d = rt$) as well as "fun type" reasoning problems involving such topics as coins and people's ages. Remember that the emphasis in all applications in a course in introductory algebra is to develop your reasoning skills and to teach you how to transfer English phrases into abstract algebraic expressions.

Example 1: Distance-Rate-Time ● ● ● ● ● ● ● ● ● ● ● ● ● ● ● ●

a. A small plane flew 300 miles in 2 hours. Then on the return trip, flying against the wind, it traveled only 200 miles in 2 hours. What were the wind velocity and the speed of the plane? (**Note:** The "speed of the plane" means how fast the plane would be flying with no wind.)

Solution: Let s = speed of plane
and w = wind velocity

Continued on next page ...

	rate	•	time	=	distance
With the wind	$s + w$		2		$2(s + w)$
Against the wind	$s - w$		2		$2(s - w)$

$$\begin{cases} 2(s+w) = 300 \rightarrow 2s + 2w = 300 \\ 2(s-w) = 200 \rightarrow \underline{2s - 2w = 200} \end{cases}$$
$$4s \qquad = 500$$
$$s \qquad = 125$$

Substitute $s = 125$ into one of the original equations.

$$2(125 + w) = 300$$
$$125 + w = 150$$
$$w = 25$$

The speed of the plane was 125 mph, and the wind velocity was 25 mph.

b. Two buses leave Eureka at the same time traveling in opposite directions. One bus travels at 55 mph and the other at 59 mph. How soon will they be 285 miles apart?

Solution: Let x = time of travel for first bus
and y = time of travel for second bus

The system of linear equations is

$$\begin{cases} x = y & \text{Both buses will be traveling the same amount of time.} \\ 55x + 59y = 285 & \text{The sum of the distances each bus travels will be 285.} \end{cases}$$

Substitution gives

$$55x + 59x = 285$$
$$114x = 285$$
$$x = 2.5 \text{ and } y = 2.5$$

Each bus will travel for 2.5 hours and at the end of 2.5 hours they will be 285 miles apart.

Example 2: Number Problem ● ● ● ● ● ● ● ● ● ● ● ● ● ● ● ●

The sum of two numbers is 80 and their difference is 10. What are the two numbers?

Solution: Let x = one number
and y = the other number.

The system of linear equations is

$$\begin{cases} x + y = 80 & \text{The sum is 80.} \\ x - y = 10 & \text{The difference is 10.} \end{cases}$$

Solving the first equation for y and then substituting gives,

$$y = 80 - x$$
$$x - (80 - x) = 10$$
$$x - 80 + x = 10$$
$$2x = 90$$
$$x = 45$$

$$y = 80 - 45 = 35$$

The two numbers are 45 and 35.

Check: $45 + 35 = 80$ and $45 - 35 = 10$

● ●

Example 3: Coins ●

Mike has \$1.05 worth of change in nickels and quarters. If he has twice as many nickels as quarters, how many of each type of coin does he have?

Solution: We use two equations: one relating the number of coins and the other relating the value of the coins. The value of each nickel is 5 cents and the value of each quarter is 25 cents.

Let n = number of nickels
and q = number of quarters.
The system of linear equation is

$$\begin{cases} n = 2q & \text{This equation relates the number of coins.} \\ 0.05n + 0.25q = 1.05 & \text{This equation relates the values of the coins.} \end{cases}$$

Note carefully that in the first equation q is multiplied by 2 because the number of nickels is twice the number of quarters. Therefore, n is bigger. *Continued on next page ...*

The first equation is already solved for n, so substitution into the second equation gives

$$0.05(2q)+0.25q = 1.05$$

$$0.10q+0.25q = 1.05$$

$$10q+25q = 105 \qquad \text{Multiply the equation by 100}$$

$$35q = 105$$

$$q = 3$$

$$n = 2q = 6$$

Mike has 3 quarters and 6 nickels.

● ●

Example 4: Age ●

Pat is 6 years older than her sister Sue. In 3 years she will be twice as old as Sue. How old is each girl now?

Solution: Let P = Pat's age now
and S = Sue's age now
Then the system of linear equations is

$$\begin{cases} P - S = 6 & \text{The difference in their ages is 6.} \\ P + 3 = 2(S+3) & \text{Each age is increased by 3.} \end{cases}$$

Rewrite the second equation in standard form and solve by addition.

$$P + 3 = 2(S+3)$$

$$P + 3 = 2S + 6$$

$$P - 2S = 3$$

$$\begin{cases} P - S = 6 \rightarrow & P - S = 6 \\ [-1]P - 2S = 3 \rightarrow & \underline{-P + 2S = -3} \\ & S = 3 \end{cases}$$

Substitute $S = 3$ into one of the original equations.

$$P - 2(3) = 3$$

$$P - 6 = 3$$

$$P = 9$$

Pat is 9 years old; Sue is 3 years old.

● ●

Example 5: Amounts and Costs • • • • • • • • • • • • • • • •

Three hot dogs and two orders of French fries cost $5.80. Four hot dogs and four orders of fries cost $8.60. What is the cost of a hot dog? What is the cost of an order of fries?

Solution: Let x = cost of one hot dog
and y = cost of one order of fries.
Then the system of linear equations is

$$\begin{cases} 3x+2y=5.80 & \text{Three hot dogs and two orders of French fries cost \$5.80.} \\ 4x+4y=8.60 & \text{Four hot dogs and four orders of fries cost \$8.60.} \end{cases}$$

Both equations are in standard form. Solve using the addition method.

$$\begin{cases} [-2] \ 3x+2y=5.80 \rightarrow -6x-4y=-11.60 \\ \qquad\quad 4x+4y=8.60 \rightarrow \underline{\quad 4x+4y= \quad 8.60 \quad} \end{cases}$$
$$-2x = -3.00$$
$$x = \quad 1.50$$

Substitute $x = 1.50$ into one of the original equations.
$$3(1.50)+2y=5.80$$
$$4.50+2y=5.80$$
$$2y=1.30$$
$$y=0.65$$

One hot dog costs $1.50 and one order of fries costs $0.65.

• •

5.4 Exercises

Solve each problem by setting up a system of two equations in two unknowns and solving the system.

1. The sum of two numbers is 56. Their difference is 10. Find the numbers.

2. The sum of two numbers is 40. The sum of twice the larger and 4 times the smaller is 108. Find the numbers.

3. The sum of two numbers is 36. Three times the smaller plus twice the larger is 87. Find the two numbers.

4. The difference between two numbers is 17. Four times the smaller is equal to 7 more than the larger. What are the numbers?

5. Ken makes a 4-mile motorboat trip downstream in 20 minutes $\left(\frac{1}{3}\text{hr}\right)$. The return trip takes 30 minutes $\left(\frac{1}{2}\text{hr}\right)$. Find the rate of the boat in still water and the rate of the current.

6. Mr. McKelvey finds that flying with the wind he can travel 1188 miles in 6 hours. However, when flying against the wind, he travels only $\frac{2}{3}$ of the distance in the same amount of time. Find the speed of the plane in still air and the wind speed.

7. Randy made a business trip of 190 miles. He averaged 52 mph for the first part of the trip and 56 mph for the second part. If the total trip took $3\frac{1}{2}$ hours, how long did he travel at each rate?

8. Marian drove to a resort 335 miles from her home. She averaged 60 mph for the first part of her trip and 55 mph for the second part. If her total driving time was $5\frac{3}{4}$ hours, how long did she travel at each rate?

9. Mr. Green traveled to a city 200 miles from his home to attend a meeting. Due to car trouble, his average speed returning was 10 mph less than his speed going. If the outbound trip was 4 hours and the return trip was 5 hours, at what rate of speed did he travel to the city?

10. Two trains leave Kansas City at the same time. One train travels east and the other travels west. The speed of the westbound train is 5 mph greater than the speed of the eastbound train. After 6 hours, they are 510 miles a part. Find the rate of each train. Assume the trains travel in a straight line in opposite directions.

11. Steve travels 4 times as fast as Fred. Traveling in opposite directions, they are 105 miles apart after 3 hours. Find their rates of travel.

12. Sue travels 5 mph less than twice as fast as June. Starting at the same point and traveling in the same direction, they are 80 miles apart after 4 hours. Find their speeds.

13. Mary and Linda live 324 miles apart. They start at the same time and travel toward each other. Mary's speed is 8 mph greater than Linda's. If they meet in 3 hours, find their speeds.

14. Two planes leave from points 1860 miles apart at the same time and travel toward each other (at slightly different altitudes, of course). If their rates are 220 mph and 400 mph, how soon will they meet?

15. A jogger runs into the countryside at a rate of 10 mph. He returns along the same route at 6 mph. If the total trip took 1 hour, 36 minutes, how far did he jog?

16. A cyclist traveled to her destination at an average rate of 15 mph. By traveling 3 mph faster, she took 30 minutes less to return. What distance did she travel each way?

17. An airliner's average speed is $3\frac{1}{2}$ times the average speed of a private plane. Two hours after traveling in the same direction they are 580 miles apart. What is the average speed of each plane? (**Hint:** Since they are traveling in the same direction, the distance between them will be the difference of their distances.)

18. Sonja has some nickels and dimes. If she has 30 coins worth a total of $2.00, how many of each type of coin does she have?

19. Louis has a total of 27 coins consisting of quarters and dimes. The total value of the coins is $5.40. How many of each type of coin does he have?

20. Jill is 8 years older than her brother Curt. Four years from now, Jill will be twice as old as Curt. How old is each at the present time?

21. Two years ago, Anna was half as old as Beth. Eight years from now she will be two-thirds as old as Beth. How old are they now?

22. The length of a rectangle is 10 meters more than one-half the width. If the perimeter is 44 meters, what are the length and width?

23. The length of a rectangle is 1 meter less than twice the width. If each side is increased by 4 meters, the perimeter will be 116 meters. Find the length and the width of the original rectangle.

24. The line $y = mx + b$ passes through the two points $(1, 3)$ and $(5, 1)$. Find the equation of the line. (**Hint:** Substitute the values for x and y in the equation and solve the resulting system of equations for m and b.)

25. The line $y = mx + b$ passes through the two points $(-2, -1)$ and $(6, -7)$. Find the equation of the line. (**Hint**: Substitute the values for x and y in the equation and solve the resulting system of equations for m and b.)

26. Your friend challenges you to figure out how many dimes and quarters are in a cash register. He tells you that there are 65 coins and that their value is $11.90. How many dimes and how many quarters are in the register?

27. A bag contains pennies and nickels only. If there are 182 coins in all and their value is $3.90, how many pennies and how many nickels are in the bag?

28. A Christmas charity party sold tickets for $45.00 for adults and $25.00 for children. The total number of tickets sold was 320 and the total for the ticket sales was $13,000. How many adult and how many children's tickets were sold?

29. Tickets for the local high school basketball game were priced at $3.50 for adults and $2.50 for students. If the income for one game was $9550 and the attendance was 3500, how many adults and how many students attended that game?

30. The width of a rectangle is $\dfrac{3}{4}$ of its length. If the perimeter of the rectangle is 140 feet, what are the dimensions of the rectangle?

31. A farmer has 260 meters of fencing to build a rectangular corral. He wants the length to be 3 times as long as the width. What dimensions should he make his corral?

32. Joan went to a book sale on campus and bought paperback books for $0.25 each and hardback books for $1.75 each. If she bought a total of 15 books for $11.25, how many of each type of book did she buy?

33. Admission to the baseball game is $2.00 for general admission and $3.50 for reserved seats. The receipts were $36,250 for 12,500 paid admissions. How many of each ticket, general and reserved, were sold?

34. A men's clothing store sells two styles of sports jackets, one selling for $95 and one selling for $120. Last month, the store sold 40 jackets, with receipts totaling $4250. How many of each style did the store sell?

35. Seventy children and 160 adults attended a movie theater. The total receipts were $620. One adult ticket and 2 children's tickets cost $7. Find the price of each type of ticket.

36. Morton took some old newspapers and aluminum cans to the recycling center. Their total weight was 180 pounds. He received 1.5¢ per pound for the newspapers and 30¢ per pound for the cans. The total received was $14.10. How many pounds of each did Morton have?

37. Frank bought 2 shirts and 1 pair of slacks for a total of $55. If he had bought 1 shirt and 2 pairs of slacks, he would have paid $68. What was the price of each shirt and each pair of slacks?

38. Four hamburgers and three orders of French fries cost $5.15. Three hamburgers and five orders of fries cost $5.10. What would one hamburger and one order of fries cost?

39. A small manufacturer produces two kinds of radios, model X and model Y. Model X takes 4 hours to produce and costs $8 each to make. Model Y takes 3 hours to produce and costs $7 each to make. If the manufacturer decides to allot a total of 58 hours and $126 each week, how many of each model will be produced?

40. A furniture shop refinishes chairs. Employees use two methods to refinish a chair. Method I takes 1 hour and the material costs $3. Method II takes $1\frac{1}{2}$ hours and the material costs $1.50. Last week, they took 36 hours and spent $60 refinishing chairs. How many did they refinish with each method?

Writing and Thinking About Mathematics

41. A two digit number can be written as *ab*, where *a* and *b* are the digits. We do **not** mean that the digits are multiplied, but the value of the number is $10a + b$. For example, the two digit number 34 has a value of $10 \cdot 3 + 4$. Set up and solve a system of equations for the following problem:

The sum of the digits of a two digit number is 13. If the digits are reversed, then the value of the number is increased by 45. What is the number?

Hawkes Learning Systems: Introductory Algebra

Applications: Distance-Rate-Time, Number Problems, Amounts and Costs

| 5.5 | **Applications: Interest and Mixture** |

Objectives

After completing this section, you will be able to:

1. *Solve applied problems related to interest by using systems of linear equations.*

2. *Solve applied problems related to mixture by using systems of linear equations.*

As we have seen throughout Chapter 5, systems of equations occur in many practical applications. In this section we will study two more types of applications: interest (involving money invested) and mixture. These applications can be "wordy" and you will need to study the following examples carefully to understand how to set up the system of equations.

Interest

People in business and banking know several formulas for calculating interest. The formula used depends on the method of payment (monthly or yearly) and the type of interest (simple or compound). Also, penalties for late payments and even penalties for early payments might be involved. In any case, standard notation is to let P represent the principal (amount of money invested or borrowed), r represent the rate of interest (an annual rate), t represent the time, and I represent the interest.

In this section, we will use only the basic formula for simple interest $I = Prt$ with interest calculated on an annual basis. **So, in the special case with $t = 1$, the formula becomes $I = Pr$.**

Example 1: Interest ●

a. James has two investment accounts, one pays 6% interest and the other pays 10% interest. He has $1000 more in the 10% account than he has in the 6% account and the income from the 10% account exceeds the interest from the 6% account by $260 each year. How much does he have in each account?

Solution: Let x = amount invested at 6%
and y = amount invested at 10%.

Continued on next page ...

Then the system of equations is

$$\begin{cases} y - x = 1000 & \text{y is larger than x by \$1000.} \\ 0.10y - 0.06x = 260 & \text{Income from the 10\% account exceeds (is larger} \end{cases}$$

than) income from the 6% account by $260.

We can solve the first equation for y: $y = x + 1000$
Substituting for y in the second equation gives

$$0.10(x + 1000) - 0.06x = 260$$
$$10(x + 1000) - 6x = 26,000 \quad \text{Multiply by 100}$$
$$10x + 10,000 - 6x = 26,000$$
$$4x = 16,000$$
$$x = 4000$$

Substitute $x = 4000$ into one of the original equations.

$$y - 4000 = 1000 \quad \text{or} \quad y = 4000 + 1000 = 5000$$

James has $4000 invested at 6% and $5000 invested at 10%.

b. Lila has $7000 to invest. She decides to separate her funds into two investments. One yields an interest of 7%, and the other, 12%. If she wants an annual income from the investments to be $690, how should she split the money? (**Note:** The higher interest account is considered more risky. Otherwise, she would put all her money in that account.)

Solution: Let x = amount to be invested at 7% and y = amount to be invested at 12%.

Then the system of equations is

$$\begin{cases} x + y = 7000 & \text{The sum of both amounts is \$7000.} \\ 0.07x + 0.12y = 690 & \text{The total interest is \$690.} \end{cases}$$

Both equations are in standard form. Multiplying the first equation by –7 and the second by 100 gives

$$\begin{cases} [-7] & x + y = 7000 & -7x - 7y = -49,000 \\ [100] & 0.07x + 0.12y = 690 & \underline{7x + 12y = 69,000} \end{cases}$$

$$5y = 20,000$$
$$y = 4000$$

Substitute $y = 4000$ into one of the original equations.

$$x + 4000 = 7000$$
$$x = 3000$$

She should invest $3000 at 7% and $4000 at 12%.

Mixture

Problems involving mixtures occur in physics and chemistry and in such places as candy stores or coffee shops. Two or more items of a different percentage of concentration of a chemical such as salt, chlorine, or antifreeze are to be mixed; or two or more types of coffee are to be mixed to form a final mixture that satisfies certain conditions of percentage of concentration.

The basic plan is to write an equation that deals with only one part of the mixture (such as the salt in the mixture). The following examples explain how this can be accomplished.

Example 2: Mixture •

a. How many ounces each of a 10% salt solution and a 15% salt solution must be used to produce 50 ounces of a 12% salt solution?

Solution: Let x = amount of 10% solution
and y = amount of 15% solution.

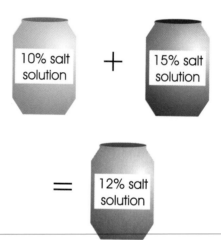

Continued on next page ...

	amount of solution · percent of salt = amount of salt		
10% solution	x	0.10	$0.10x$
15% solution	y	0.15	$0.15y$
12% solution	50	0.12	$0.12(50)$

Then the system of linear equations is

$$\begin{cases} x + y = 50 \\ 0.10x + 0.15y = 0.12(50) \end{cases}$$

The sum of the two amounts must be 50 ounces.

The sum of the amount of salt from each solution equals the total amount of salt in the final solution.

Multiplying the first equation by -10 and the second by 100 gives,

$$\begin{cases} [-10] \quad x + \quad y = 50 \quad \rightarrow -10x - 10y = -500 \\ [100] \quad 0.10x + 0.15y = 0.12(50) \rightarrow \quad \underline{10x + 15y = 600} \end{cases}$$

$$5y = 100$$
$$y = 20$$

Substitute $y = 20$ into one of the original equations.

$$x + 20 = 50$$
$$x = 30$$

Use 30 ounces of the 10% solution and 20 ounces of the 15% solution.

b. How many gallons of a 20% acid solution should be mixed with a 30% acid solution to produce 100 gallons of a 23% solution?

Solution: Let x = amount of 20% solution
and y = amount of 30% solution.

Continued on next page ...

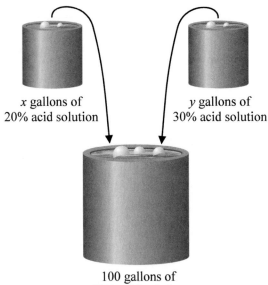

x gallons of
20% acid solution

y gallons of
30% acid solution

100 gallons of
23% acid solution

	amount of solution · percent of acid = amount of acid		
20% solution	x	0.20	$0.20x$
30% solution	y	0.30	$0.30y$
23% solution	100	0.23	$0.23(100)$

Then the system of linear equations is

$$\begin{cases} x + y = 100 \\ 0.20x + 0.30y = 0.23(100) \end{cases}$$

The sum of the two amounts must be 100 gallons.

The sum of the amount of acid from each solution equals the total amount of acid in the final solution.

Multiplying the first equation by –20 and the second by 100 gives,

$$\begin{cases} [-20] \quad x + \quad y = 100 \quad \rightarrow -20x - 20y = -2000 \\ [100] \quad 0.20x + 0.30y = 0.23(100) \rightarrow \quad \underline{20x + 30y = 2300} \end{cases}$$

$$10y = 300$$
$$y = 30$$

Substitute $y = 30$ into one of the original equations.

$$x + 30 = 100$$
$$x = 70$$

Seventy gallons of the 20% solution should be added to 30 gallons of the 30% solution. This will produce 100 gallons of a 23% solution.

5.5 Exercises

1. Carmen invested $9000, part in a 6% passbook account and the rest in a 10% certificate account. If her annual interest was $680, how much did she invest at each rate?

2. Mrs. Brown has $12,000 invested. Part is invested at 6% and the remainder at 8%. If the interest from the 6% investment exceeds the interest from the 8% investment by $230, how much is invested at each rate?

3. Ten thousand dollars is invested, part at 5.5% and part at 6%. The interest from the 5.5% investment exceeds the interest from the 6% investment by $251. How much is invested at each rate?

4. On two investments totaling $9500, Bill lost 3% on one and earned 6% on the other. If his net annual receipts were $282, how much was each investment?

5. Marsha has money in two savings accounts. One rate is 8% and the other is 10%. If she has $200 more in the 10% account, how much is invested at 8% if the total interest is $101?

6. Money is invested at two rates. One rate is 9% and the other is 13%. If there is $700 more invested at 9%, find the amount invested at each rate if the annual interest is $239.

7. Frank has half of his investments in stock paying an 11% dividend and the other half in a debentured stock paying 13% interest. If his total annual interest is $840, how much does he have invested?

8. Betty invested some of her money at 12% interest. She invested $300 more than twice that amount at 10%. How much is invested at each rate if her income is $318 annually?

9. GFA invested some money in a development yielding 24% and $9000 less in a development yielding 18%. If the first investment produces $2820 more per year than the second, how much is invested in each development?

10. Judy invests a certain amount of money at 7% annual interest and three times that amount at 8%. If her annual income is $232.50, how much does she have invested at each rate?

11. Norman has a certain amount of money invested at 5% annual interest and $500 more than twice that amount invested in bonds yielding 7%. His total income from interest is $187. How much does he have invested at each rate?

12. A total of $6000 is invested, part at 8% and the remainder at 12%. How much is invested at each rate if the annual interest is $620?

13. Mr. Brown has $12,000 invested. Part is invested at 9% and the remainder at 11%. If the interest from the 9% investment exceeds the interest from the 11% investment by $380, how much is invested at each rate?

14. Eight thousand dollars is invested, part at 15% and the remainder at 12%. If the annual income from the 15% investment exceeds the income from the 12% investment by $66, how much is invested at each rate?

15. A metallurgist has one alloy containing 20% copper and another containing 70% copper. How many pounds of each alloy must he use to make 50 pounds of a third alloy containing 50% copper?

16. A manufacturer has received an order for 24 tons of a 60% copper alloy. His stock contains only alloys of 80% copper and 50% copper. How much of each will he need to fill the order?

17. A tobacco shop wants 50 ounces of tobacco that is 24% rare Turkish blend. How much each of a 30% Turkish blend and a 20% Turkish blend will be needed?

18. How many liters each of a 40% acid solution and a 55% acid solution must be used to produce 60 liters of a 45% acid solution?

19. A dairy man wants to mix a 35% protein supplement and a standard 15% protein ration to make 1800 pounds of a high-grade 20% protein ration. How many pounds of each should he use?

20. To meet the government's specifications, an alloy must be 65% aluminum. How many pounds each of a 70% aluminum alloy and a 54% aluminum alloy will be needed to produce 640 pounds of the 65% aluminum alloy?

21. A meat market has ground beef that is 40% fat and extra lean ground beef that is only 15% fat. How many pounds of each will be needed to obtain 50 pounds of lean ground beef that is 25% fat?

22. George decides to mix grades of gasoline in his truck. He puts in 8 gallons of regular and 12 gallons of premium for a total cost of $23.80. If premium gasoline costs $0.15 more per gallon than regular, what was the price of each grade of gasoline?

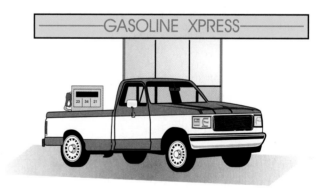

Hawkes Learning Systems: Introductory Algebra

 Applications: Interest and Mixture

Chapter 5 Index of Key Ideas and Terms

System of Linear Equations page 298

A pair of linear equations considered together is called a **system of linear equations** (or a **set of simultaneous equations**).

Solutions by Graphing page 300

The two lines related to a system of equations can be graphed.

a. If the two lines intersect in a single point, the point is the page 300
solution to the system. The system is said to be **consistent**.

b. If the two lines are parallel (do not intersect), the system is page 301
said to be **inconsistent** and there are no solutions.

c. If the two lines coincide (are the same line), the system is said pages 301 - 302
to be **dependent** and there are an infinite number of solutions.

Solutions by Substitution page 310

1. Solve one of the equations for one of the variables.
2. Substitute the resulting expression into the other equation.
3. Solve the new equation for the one variable.
4. Substitute the result back into one of the original equations to find the value of the other variable.

Solutions by Addition pages 317 - 320

1. Write the equations in standard form.
2. Arrange the equations vertically so that like terms are aligned.
3. Multiply one or both of the equations by a constant so that the coefficients of one of the variables are opposites.
4. Add the like terms.
5. Solve the resulting equation.
6. Substitute the results back into one of the original equations to find the value of the other variable.

Guidelines for Deciding which Method to Use page 323
in Solving a System of Linear Equations

1. The graphing method is helpful in "seeing" the geometric relationship between the lines and finding approximate solutions. A calculator can be very helpful here.
2. Both the substitution method and the addition method give exact solutions.
3. The substitution method may be reasonable and efficient if one of the coefficients of one of the variables is 1.
4. In general, the method of addition will prove to be most efficient.

Applications

Chapter 5 Review

For a review of the topics and problems from Chapter 5, look at the following lessons from *Hawkes Learning Systems: Introductory Algebra*

Solving Systems of Linear Equations by Graphing
Solving Systems of Linear Equations By Substitution
Solving Systems of Linear Equations By Addition
Applications: Distance-Rate-Time, Number Problems, Amounts and Costs
Applications: Interest and Mixture

Chapter 5 Test

In Exercises 1 and 2, determine which of the points satisfy the given system of linear equations.

1. $\begin{cases} 3x - 7y = 5 \\ 5x - 2y = -11 \end{cases}$

2. $\begin{cases} x - 2y = 7 \\ 2x - 3y = 5 \end{cases}$

a. $(1, 8)$
b. $(4, 1)$
c. $(-3, -2)$
d. $(13, 10)$

a. $(0, 3)$
b. $(7, 0)$
c. $(1, -1)$
d. $(-11, -9)$

For Exercises 3 – 11 solve as directed. In each exercise state whether the system is (a) consistent, (b) inconsistent, or (c) dependent.

Solve the systems in Exercises 3 and 4 by graphing.

3. $\begin{cases} y = 2 - 5x \\ x - y = 6 \end{cases}$

4. $\begin{cases} x - y = 3 \\ 2x + 3y = 11 \end{cases}$

Solve the systems in Exercises 5 and 6 by using substitution.

5. $\begin{cases} 5x - 2y = 0 \\ y = 3x + 4 \end{cases}$

6. $\begin{cases} x = \dfrac{1}{3}y - 4 \\ 2x + \dfrac{3}{2}y = 5 \end{cases}$

Solve the systems in Exercises 7 and 8 by using the method of additon.

7. $\begin{cases} -2x + 3y = 6 \\ 4x + y = 1 \end{cases}$

8. $\begin{cases} 2x - 4y = 9 \\ 3x + 6y = 8 \end{cases}$

Solve the systems in Exercises 9 – 11 by using any method.

9. $\begin{cases} x + y = 2 \\ y = -2x - 1 \end{cases}$

10. $\begin{cases} 6x + 2y - 8 = 0 \\ y = -3x \end{cases}$

11. $\begin{cases} 7x + 5y = -9 \\ 6x + 2y = 6 \end{cases}$

12. Determine the values of a and b such that the straight line $ax + by = 11$ passes through the two points $(1, -3)$ and $(2, 5)$.

13. Pete's boat can travel 48 miles upstream in 4 hours. The return trip takes 3 hours. Find the speed of the boat in still water and the speed of the current.

14. Eight pencils and two pens cost $2.22. Three pens and four pencils cost $2.69. What is the price of each pen and each pencil?

15. Gary has two investments yielding a total annual interest of $185.60. The amount invested at 8% is $320 less than twice the amount invested at 6%. How much is invested at each rate?

16. A metallurgist needs 2000 pounds of an alloy that is 80% copper. In stock, he has only alloys of 83% copper and 68% copper. How many pounds of each must be used?

17. The perimeter of a rectangle is 60 inches and the length is 4 inches longer than the width. Find the dimensions of the rectangle.

18. Sonia has a bag of coins with only nickels and quarters. She wants you to figure out how many of each type of coin she has and tells you that she has 105 coins and that the value of the coins is $17.25. Tell her that you know algebra and determine how many nickels and how many quarters she has.

Cumulative Review: Chapters 1 – 5

Use the rules for order of operations to evaluate the expressions in Exercises 1 and 2.

1. $-20 + 15 \div (-5) \cdot 2^3 - 11^2$

2. $24 \div 4 \cdot 6 - 36 \cdot 2 \div 3^2$

3. What is 110% of 50?

4. Find $\dfrac{7}{8}$ of 10,000.

Solve each of the equations in Exercises 5 – 8.

5. $2(x - 7) + 14 = -3x + 1$

6. $\dfrac{5x}{6} - \dfrac{2}{3} = \dfrac{x}{2} + \dfrac{1}{4}$

7. $2.3y - 1.6 = 3(1.2y + 2.5)$

8. $\dfrac{a}{5} + \dfrac{a}{7} = \dfrac{a}{2} - \dfrac{5}{14}$

In Exercises 9 and 10, solve the inequality and graph the solution set on a real number line.

9. $3x - 14 \geq 25$

10. $0 \leq 2x + 10 < 1.6$

In Exercises 11 and 12, solve for the indicated variable.

11. $v = k + gt$; solve for t.

12. $3x - 4y = 6$; solve for y.

13. For the equation $2x + 4y = 11$, determine the slope, m, and the y-intercept, b. Then use a graphing calculator to graph the line.

14. Write an equation for the line parallel to the line $x - 3y = -1$ and passing through the point $(1, 2)$. Use a graphing calculator to graph both lines.

15. Determine which of the points lie on both of the lines in the given system of equations.
$$\begin{cases} x - 2y = 6 \\ y = \dfrac{1}{2}x - 3 \end{cases}$$

    a. $(0, 6)$ **b.** $(6, 0)$ **c.** $(2, -2)$ **d.** $\left(3, -\dfrac{3}{2}\right)$

16. Use a graphing calculator to estimate the solution to the following system of linear equations: $\begin{cases} y = 5x - 3 \\ 2x + y = 10 \end{cases}$

For Exercises 17 – 32 solve as directed. In each exercise state whether the system is (a) consistent, (b) inconsistent, or (c) dependent.

Solve the systems in Exercises 17 – 20 by graphing.

17. $\begin{cases} 2x + 3y = 4 \\ 3x - y = 6 \end{cases}$
18. $\begin{cases} 5x + 2y = 3 \\ y = 4 \end{cases}$
19. $\begin{cases} 3x + y = 3 \\ x - 3y = 4 \end{cases}$
20. $\begin{cases} y = 4x - 6 \\ 8x - 2y = -4 \end{cases}$

Solve the systems in Exercises 21 – 24 by using the method of substitution.

21. $\begin{cases} x + y = -4 \\ 2x + 7y = 2 \end{cases}$
22. $\begin{cases} x = 2y \\ y = \dfrac{1}{2}x + 9 \end{cases}$
23. $\begin{cases} 4x + 3y = 8 \\ x + \dfrac{3}{4}y = 2 \end{cases}$
24. $\begin{cases} 2x + y = 0 \\ 7x + 6y = -10 \end{cases}$

Solve the systems in Exercises 25 – 28 using the method of addition.

25. $\begin{cases} 2x + y = 7 \\ 2x - y = 1 \end{cases}$
26. $\begin{cases} 3x - 2y = 9 \\ x - 2y = 11 \end{cases}$
27. $\begin{cases} 2x + 4y = 9 \\ 3x + 6y = 8 \end{cases}$
28. $\begin{cases} x + 5y = 10 \\ y = 2 - \dfrac{1}{5}x \end{cases}$

Solve the systems in Exercises 29 – 32 by using any method.

29. $\begin{cases} 3y - x = 7 \\ x - 2y = -2 \end{cases}$
30. $\begin{cases} x - \dfrac{2}{5}y = \dfrac{4}{5} \\ \dfrac{3}{4}x + \dfrac{3}{4}y = \dfrac{5}{4} \end{cases}$
31. $\begin{cases} x + 3y = 7 \\ 5x - 2y = 1 \end{cases}$
32. $\begin{cases} 3x - 5y = 17 \\ x + 2y = 4 \end{cases}$

In Exercises 33 and 34, write an equation for the line determined by the two given points. Use the formula $y = mx + b$ to set up a system of equations with m and b as the unknowns.

33. $(3, -1), (2, 6)$ **34.** $(-2, 5), (4, -3)$

35. a. Two angles are **complementary** if the sum of their measures is 90°. Find two complementary angles such that one is 10° more than three times the other.

  b. Two angles are **supplementary** if the sum of their measures is 180°. Find two supplementary angles such that one is 30° less than one-half of the other.

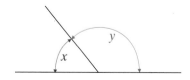

36. Bernice's secretary bought 15 stamps, some 25¢ and some 34¢. If he spent $4.20, how many of each kind did he buy?

37. A boat can travel 24 miles downstream in 2 hours. The return trip takes 3 hours. Find the speed of the boat in still water and the speed of the current.

38. The perimeter of a rectangle is 50 yards and the width is only 3 yards less than the length. What are the dimensions of the rectangle?

39. A company manufactures two kinds of dresses, Model A and Model B. Each Model A dress takes 4 hours to produce and each costs $18. Each Model B takes 2 hours to produce and each costs $7. If during a week there were 52 hours of production and costs of $198, how many of each model were produced?

40. Two trains leave Kansas City at the same time. One train travels east and the other travels west. The speed of the westbound train is 5 mph greater than the speed of the eastbound train. After 6 hours, they are 510 miles apart. Find the rate of each train. (Assume that the trains travel in a straight line in opposite directions.)

41. In a biology experiment, Owen observed that the bacteria count in a culture was approximately 100 at the end of 1 hour. At the end of 3 hours, the bacteria count was about 2000. Write a linear equation describing the bacteria count, N, in terms of the time, t.

42. List the set of ordered pairs corresponding to the points on the graph. Give the domain and range, and indicate if the relation is or is not a function.

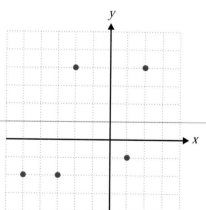

In Exercises 43 and 44, use the vertical line test to determine whether each of the graphs does or does not represent a function.

43.

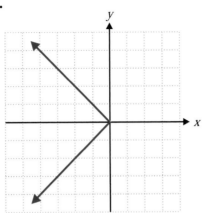

44.

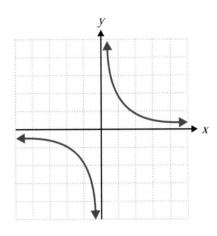

45. Given that $F(x) = 3x - 7$, find each of the following:

   a. $F(-5)$ **b.** $F(10)$ **c.** $F(3.5)$

46. Given that $f(x) = 2x^2 + 3x - 10$, find each of the following:

   a. $f(0)$ **b.** $f(-3)$ **c.** $f(6)$

Exponents and Polynomials

Did You Know?

One of the most difficult problems for students in beginning algebra is to become comfortable with the idea that letters or symbols can be manipulated just like numbers in arithmetic. These symbols may be the cause of "math anxiety." A great deal of publicity has recently been given to the concept that a large number of people suffer from math anxiety, a painful uneasiness caused by mathematical symbols or problem-solving situations. Persons affected by math anxiety find it almost impossible to learn mathematics, or they may be able to learn but be unable to apply their knowledge or do well on tests. Persons suffering from math anxiety often develop math avoidance, so they avoid careers, majors, or classes that will require mathematics courses or skills. The sociologist Lucy Sells has determined that mathematics is a critical filter in the job market. Persons who lack quantitative skills are channeled into high unemployment, low-paying, non-technical areas.

What causes math anxiety? Researchers are investigating the following hypotheses:

1. A lack of skills that leads to a lack of confidence and, therefore, to anxiety;
2. An attitude that mathematics is not useful or significant to society;
3. Career goals that seem to preclude mathematics;
4. A self-concept that differs radically from the stereotype of a mathematician;
5. Perceptions that parents, peers, or teachers have low expectations for the person in mathematics;
6. Social conditioning to avoid mathematics (a particular problem for women).

We hope that you are finding your present experience with algebra successful and that the skills you are acquiring now will enable you to approach mathematical problems with confidence.

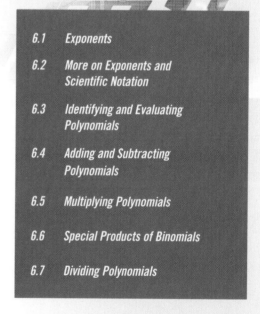

"Mathematics is the queen of the sciences and arithmetic the queen of mathematics."

Karl F. Gauss (1777 – 1855)

In Chapter 6, you will learn the rules of exponents and how exponents can be used to simplify very large and very small numbers. Astronomers and chemists are very familiar with these ideas and the notation used is appropriately called scientific notation. This notation is also used in scientific calculators. (Multiply 5,000,000 by 5,000,000 on your calculator and read the results.)

Polynomials and operations with polynomials (the topics in Chapters 6, 7, and 8) appear at almost every level of mathematics from elementary and intermediate algebra through statistics, calculus, and beyond. Be aware that the knowledge and skills you learn about polynomials will be needed again and again in any mathematics courses you take in the future.

6.1   Exponents

Objectives

After completing this section, you will be able to:

1. Simplify expressions by using properties of integer exponents.

2. Recognize which property of exponents is used to simplify an expression.

The Product Rule

In Section 1.5, an **exponent** was defined as a number that tells how many times a number (called the **base**) is used in multiplication. This definition is limited because it is valid only if the exponents are positive integers. In this section, we will develop four properties of exponents that will help in simplifying algebraic expressions and expand your understanding of exponents to include variable bases, negative exponents, and the exponent 0. (In later courses you will study fractional exponents.)

From Section 1.5, we know that

Exponent

$$6^2 = 6 \cdot 6 = 36$$

Base

and

$$6^3 = 6 \cdot 6 \cdot 6 = 216.$$

Also, the base may be a variable so that

$$x^3 = x \cdot x \cdot x \quad \text{and} \quad x^5 = x \cdot x \cdot x \cdot x \cdot x.$$

Now, to find the products of expressions such as $6^2 \cdot 6^3$ or $x^3 \cdot x^5$ and to simplify these products, we can write down all the factors as follows:

$$6^2 \cdot 6^3 = (6 \cdot 6) \cdot (6 \cdot 6 \cdot 6) = 6^5$$

and

$$x^3 \cdot x^5 = (x \cdot x \cdot x) \cdot (x \cdot x \cdot x \cdot x \cdot x) = x^8.$$

With these examples in mind, what do you think would be a simplified form for the product $3^4 \cdot 3^3$? You were right if you thought 3^7. That is, $3^4 \cdot 3^3 = 3^7$. Notice that in each case, the base stays the same.

The preceding discussion, along with the basic concept of whole-number exponents, leads to the following **Product Rule for Exponents**.

Product Rule for Exponents

If a is a nonzero real number and m and n are integers, then

$$a^m \cdot a^n = a^{m+n}$$

In words, to multiply two powers with the same base, keep the base and add the exponents.

NOTES Remember (see Section 1.5) about the **exponent 1**. If a variable or constant has no exponent written, the exponent is understood to be 1.

For example,
$$y = y^1$$
and
$$7 = 7^1.$$

In general, for any real number a,
$$a = a^1.$$

Example 1: The Product Rule ● ● ● ● ● ● ● ● ● ● ● ● ● ● ● ●

Use the Product Rule for exponents to simplify the following expressions.

a. $x^2 \cdot x^4$

  Solution: $x^2 \cdot x^4 = x^{2+4} = x^6$

b. $y \cdot y^6$

  Solution: $y \cdot y^6 = y^1 \cdot y^6 = y^{1+6} = y^7$

c. $4^2 \cdot 4$

  Solution: $4^2 \cdot 4 = 4^{2+1} = 4^3 = 64$ Note that the base stays 4.

                                                           That is, the bases are not multiplied.

d. $2^3 \cdot 2^2$

  Solution: $2^3 \cdot 2^2 = 2^{3+2} = 2^5 = 32$ Note that the base stays 2.

                                                           That is, the bases are not multiplied.

e. $\left(-2\right)^4 \left(-2\right)^3$

  Solution: $(-2)^4(-2)^3 = (-2)^{4+3} = (-2)^7 = -128$

f. $2y^2 \cdot 3y^9$

                                                           Note that the coefficients are multiplied
  Solution: $2y^2 \cdot 3y^9 = 2 \cdot 3 \cdot y^2 \cdot y^9$ and the exponents are added.

                  $= 6y^{2+9} = 6y^{11}$

g. $(-6xy^2)(-8xy^3)$

  Solution: $(-6xy^2)(-8xy^3) = (-6)(-8)x^{1+1}\,y^{2+3} = 48x^2 y^5$

● ●

The Exponent 0

The Product Rule is stated for *m* and *n* as **integers**. This means that the rule is valid if an exponent is 0 or a negative integer. Therefore, we need to develop an understanding of the meaning of 0 as an exponent and negative integers as exponents.

Study the following patterns of numbers. What do you think are the missing values for $2^0, 3^0$, and 5^0?

$2^5 = 32$	$3^5 = 243$	$5^5 = 3125$
$2^4 = 16$	$3^4 = 81$	$5^4 = 625$
$2^3 = 8$	$3^3 = 27$	$5^3 = 125$
$2^2 = 4$	$3^2 = 9$	$5^2 = 25$
$2^1 = 2$	$3^1 = 3$	$5^1 = 5$
$2^0 = ?$	$3^0 = ?$	$5^0 = ?$

Notice that in the column of powers of 2, each number is $\frac{1}{2}$ of the preceding number. Since $\frac{1}{2} \cdot 2 = 1$, a reasonable guess is that $2^0 = 1$. Similarly, in the column of powers of 3, $\frac{1}{3} \cdot 3 = 1$, so $3^0 = 1$ seems reasonable. Also, $\frac{1}{5} \cdot 5 = 1$, so $5^0 = 1$ fits the pattern. In fact, the missing values are $2^0 = 1, 3^0 = 1$, and $5^0 = 1$.

Another approach to understanding 0 as an exponent involves the Product Rule. Consider the following:

$$2^0 \cdot 2^3 = 2^{0+3} = 2^3 \qquad \text{Using the Product Rule.}$$

and

$$1 \cdot 2^3 = 2^3.$$

So,

$$2^0 \cdot 2^3 = 1 \cdot 2^3 \quad \text{and} \quad 2^0 = 1.$$

Similarly,

$$7^0 \cdot 7^2 = 7^{0+2} = 7^2$$

and

$$1 \cdot 7^2 = 7^2.$$

So,

$$7^0 \cdot 7^2 = 1 \cdot 7^2 \quad \text{and} \quad 7^0 = 1.$$

This discussion leads directly to the **rule for 0 as an exponent**.

The Exponent 0

If a is a nonzero real number, then

$$a^0 = 1$$

The expression 0^0 is undefined.

Example 2: The Exponent 0 ● ● ● ● ● ● ● ● ● ● ● ● ● ● ●

Simplify the following expressions using the rule for 0 as an exponent.

a. 10^0

Solution: $10^0 = 1$

b. $x^0 \cdot x^3$

Solution: $x^0 \cdot x^3 = x^{0+3} = x^3$ *or* $x^0 \cdot x^3 = 1 \cdot x^3 = x^3$

c. $(-6)^0$

Solution: $(-6)^0 = 1$

● ●

The Quotient Rule

Now consider a fraction in which the numerator and denominator are powers with the same base, such as $\dfrac{5^4}{5^2}$ or $\dfrac{x^5}{x^2}$. We can write,

$$\frac{5^4}{5^2} = \frac{\cancel{5} \cdot \cancel{5} \cdot 5 \cdot 5}{\cancel{5} \cdot \cancel{5} \cdot 1} = \frac{5^2}{1} = 25 \quad \text{or} \quad \frac{5^4}{5^2} = 5^{4-2} = 5^2 = 25$$

and

$$\frac{x^5}{x^2} = \frac{\cancel{x} \cdot \cancel{x} \cdot x \cdot x \cdot x}{\cancel{x} \cdot \cancel{x} \cdot 1} = \frac{x^3}{1} = x^3 \quad \text{or} \quad \frac{x^5}{x^2} = x^{5-2} = x^3$$

In fractions, as just illustrated, the exponents can be subtracted. Again, the base remains the same. We now have the following **Quotient Rule for Exponents**.

Quotient Rule for Exponents

If a is a nonzero real number and m and n are integers, then

$$\frac{a^m}{a^n} = a^{m-n}$$

In words, to divide two powers with the same base, keep the base and subtract the exponents. (Subtract the denominator exponent from the numerator exponent.)

Example 3: The Quotient Rule

a. $\dfrac{x^6}{x}$

Solution: $\dfrac{x^6}{x} = x^{6-1} = x^5$

b. $\dfrac{y^8}{y^2}$

Solution: $\dfrac{y^8}{y^2} = y^{8-2} = y^6$

c. $\dfrac{x^2}{x^2}$

Solution: $\dfrac{x^2}{x^2} = x^{2-2} = x^0 = 1$

Note how this example shows another way to justify the idea that $a^0 = 1$. Since the numerator and denominator are the same and not 0, it makes sense that the fraction is equal to 1.

d. $\dfrac{15x^{15}}{3x^3}$

Solution: $\dfrac{15x^{15}}{3x^3} = 5x^{15-3} = 5x^{12}$

Note that the coefficients are divided and the exponents are subtracted.

e. $\dfrac{20x^{10}y^6}{4x^2y^3}$

Solution: $\dfrac{20x^{10}y^6}{4x^2y^3} = 5x^{10-2}y^{6-3} = 5x^8y^3$ Again, note that the coefficients are divided and the exponents are subtracted.

Negative Exponents

Now we will see that the Quotient Rule for exponents leads directly to an understanding of negative exponents. In the fractions in Example 3 (except for 3c), for each base the larger exponent was in the numerator. Therefore, when the exponents were subtracted (the exponent in the denominator was subtracted from the exponent in the numerator), the resulting expression had only positive exponents. But, if the larger exponent is in the denominator, the resulting expression will have a negative exponent. For example,

$$\text{reducing:} \quad \frac{3^3}{3^5} = \frac{\cancel{3} \cdot \cancel{3} \cdot \cancel{3} \cdot 1}{\cancel{3} \cdot \cancel{3} \cdot \cancel{3} \cdot 3 \cdot 3} = \frac{1}{3^2}$$

and using the Quotient Rule: $\quad \dfrac{3^3}{3^5} = 3^{3-5} = 3^{-2}$

Thus, $3^{-2} = \dfrac{1}{3^2}$, and the **Rule for Negative Exponents** follows.

Rule for Negative Exponents

If a is a nonzero real number and n is an integer, then

$$a^{-n} = \frac{1}{a^n}$$

In words, a negative exponent indicates the reciprocal of the expression.

Example 4: Negative Exponents ● ● ● ● ● ● ● ● ● ● ● ● ● ● ● ●

Use the Rule for Negative Exponents to simplify each expression so that it contains only positive exponents.

a. 5^{-1}

Solution: $5^{-1} = \dfrac{1}{5^1} = \dfrac{1}{5}$ Using the Rule for Negative Exponents.

b. x^{-3}

Solution: $x^{-3} = \dfrac{1}{x^3}$ Using the Rule for Negative Exponents.

Continued on next page ...

c. $x^{-9} \cdot x^7$

> **Solution:** Here we use the Product Rule first and then the Rule for Negative Exponents.
>
> $$x^{-9} \cdot x^7 = x^{-9+7} = x^{-2} = \frac{1}{x^2}$$

• •

Each of the expressions in Example 5 is simplified by using the appropriate rules for exponents. Study each example carefully. In each case, the expression is considered simplified if each base appears only once and each base has only positive exponents.

(**Note**: There is nothing wrong with negative exponents. In fact, negative exponents are preferred in later courses. However, so that all answers are the same, in this course we will consider expressions to be simplified if they have only positive exponents.)

Example 5: Combining the Rules of Exponents • • • • • • • •

a. $2^{-5} \cdot 2^8$

> **Solution:** $2^{-5} \cdot 2^8 = 2^{-5+8} = 2^3 = 8$ Using the Product Rule with positive and negative exponents.

b. $\dfrac{x^6}{x^{-1}}$

> **Solution:** $\dfrac{x^6}{x^{-1}} = x^{6-(-1)} = x^{6+1} = x^7$ Using the Quotient Rule with positive and negative exponents.

c. $\dfrac{10^{-5}}{10^{-2}}$

> **Solution:** $\dfrac{10^{-5}}{10^{-2}} = 10^{-5-(-2)} = 10^{-5+2}$ Using the Quotient Rule with positive and negative exponents.
>
> $= 10^{-3} = \dfrac{1}{10^3}$ or $\dfrac{1}{1000}$ Using the Rule for Negative Exponents.

Continued on next page ...

d. $\dfrac{15x^{10} \cdot 2x^2}{3x^{15}}$

$\qquad$ **Solution:** $\qquad \dfrac{15x^{10} \cdot 2x^2}{3x^{15}} = \dfrac{30x^{10+2}}{3x^{15}}$ $\qquad$ Using the Product Rule.

$\qquad\qquad\qquad\qquad\qquad = 10x^{12-15}$ $\qquad$ Using the Quotient Rule.

$\qquad\qquad\qquad\qquad\qquad = 10x^{-3}$

$\qquad\qquad\qquad\qquad\qquad = \dfrac{10}{x^3}$ $\qquad$ Using the Rule for Negative Exponents.

e. $\dfrac{x^6 y^3}{x^2 y^5}$

$\qquad$ **Solution:** $\qquad \dfrac{x^6 y^3}{x^2 y^5} = x^{6-2} y^{3-5} = x^4 y^{-2}$ Using the Quotient Rule with two variables.

$\qquad\qquad\qquad\qquad\qquad = \dfrac{x^4}{y^2}$ $\qquad$ Using the Rule for Negative Exponents.

● ●

NOTES

Special Note about Using the Quotient Rule:
Regardless of the size of the exponents or whether they are positive or negative, the following single subtraction rule can be used with the Quotient Rule.

(numerator exponent – denominator exponent)

This subtraction will always lead to the correct answer.

Summary of the Rules for Exponents

For any nonzero real number a and integers m and n:

1. The Exponent 1: $a = a^1$

2. The Exponent 0: $a^0 = 1$ $\quad (a \neq 0)$

3. The Product Rule: $a^m \cdot a^n = a^{m+n}$

4. The Quotient Rule: $\dfrac{a^m}{a^n} = a^{m-n}$

5. Negative Exponents: $a^{-n} = \dfrac{1}{a^n}$ $\quad (a \neq 0)$

Practice Problems

Simplify each expression.

1. $2^3 \cdot 2^4$

2. $\dfrac{2^3}{2^4}$

3. $\dfrac{x^7 \cdot x^{-3}}{x^{-2}}$

4. $\dfrac{10^{-8} \cdot 10^2}{10^{-7}}$

5. $\dfrac{14x^{-3}y^2}{2x^{-3}y^{-2}}$

6. $\left(9x^4\right)^0$

6.1 Exercises

Simplify each expression and tell which rule (or rules) for exponents you used. The final form of the expressions with variables should contain only positive exponents. Assume that all variables represent nonzero numbers.

1. $3^2 \cdot 3$

2. $7^2 \cdot 7^3$

3. $8^3 \cdot 8^0$

4. 3^{-1}

5. 4^{-2}

6. $\left(-5\right)^{-2}$

7. 6^{-3}

8. $\left(-2\right)^4 \cdot \left(-2\right)^0$

9. $3 \cdot 2^3$

10. $\left(-4\right)^3 \cdot \left(-4\right)^0$

11. $6 \cdot 3^2$

12. $-4 \cdot 5^3$

13. $-2 \cdot 3^3$

14. $3 \cdot 2^{-3}$

15. $4 \cdot 3^{-2}$

16. $-3 \cdot 5^{-2}$

17. $-5 \cdot 2^{-2}$

18. $x^2 \cdot x^3$

19. $x^3 \cdot x$

20. $y^2 \cdot y^0$

21. $y^3 \cdot y^8$

22. x^{-3}

23. y^{-2}

24. $2x^{-1}$

25. $5y^{-4}$

26. $-8y^{-2}$

27. $-10x^{-3}$

28. $5x^6y^{-4}$

29. x^0y^{-2}

30. $3x^0 + y^0$

31. $5y^0 - 3x^0$

32. $\dfrac{7^3}{7}$

33. $\dfrac{9^5}{9^2}$

34. $\dfrac{10^3}{10^4}$

35. $\dfrac{10}{10^5}$

36. $\dfrac{2^3}{2^6}$

37. $\dfrac{x^4}{x^2}$

38. $\dfrac{x^5}{x^3}$

39. $\dfrac{x^3}{x}$

40. $\dfrac{y^6}{y^4}$

41. $\dfrac{x^7}{x^3}$

42. $\dfrac{x^8}{x^3}$

43. $\dfrac{x^{-2}}{x^2}$

44. $\dfrac{x^{-3}}{x}$

45. $\dfrac{x^4}{x^{-2}}$

46. $\dfrac{x^5}{x^{-1}}$

47. $\dfrac{x^{-3}}{x^{-5}}$

48. $\dfrac{x^{-4}}{x^{-1}}$

49. $\dfrac{y^{-2}}{y^{-4}}$

50. $\dfrac{y^3}{y^{-3}}$

Answers to Practice Problems: 1. $2^7 = 128$ **2.** $\dfrac{1}{2}$ **3.** x^6 **4.** 10 **5.** $7y^4$ **6.** 1

51. $3x^3 \cdot x^0$ **52.** $3y \cdot y^4$ **53.** $(5x^2)(2x^2)$ **54.** $(3x^2)(3x)$

55. $(4x^3)(9x^0)$ **56.** $(5x^2)(3x^4)$ **57.** $(-2x^2)(7x^3)$ **58.** $(3y^3)(-6y^2)$

59. $(-4x^5)(3x)$ **60.** $(6y^4)(5y^5)$ **61.** $\dfrac{8y^3}{2y^2}$ **62.** $\dfrac{12x^4}{3x}$

63. $\dfrac{9y^5}{3y^3}$ **64.** $\dfrac{-10x^5}{2x}$ **65.** $\dfrac{-8y^4}{4y^2}$ **66.** $\dfrac{12x^6}{-3x^3}$

67. $\dfrac{21x^4}{-3x^2}$ **68.** $\dfrac{10 \cdot 10^3}{10^{-3}}$ **69.** $\dfrac{10^4 \cdot 10^{-3}}{10^{-2}}$ **70.** $\dfrac{10 \cdot 10^{-1}}{10^2}$

71. $(9x^2)^0$ **72.** $(9x^2y^3)(-2x^3y^4)$ **73.** $(-3xy)(-5x^2y^{-3})$ **74.** $\dfrac{-8x^2y^4}{4x^3y^2}$

75. $\dfrac{-8x^{-2}y^4}{4x^2y^{-2}}$ **76.** $(-2x^{-3}y^5)^0$ **77.** $(3a^2b^4)(4ab^5c)$ **78.** $(-6a^3b^{-4})(4a^{-2}b^8)$

79. $\dfrac{36a^5b^0c}{-9a^{-5}b^{-3}}$ **80.** $\dfrac{25y^6 \cdot 3y^{-2}}{15xy^4}$

Calculator Problems

Use a calculator to evaluate each expression.

81. $(2.16)^0$ **82.** $(-5.06)^2$ **83.** $(1.6)^{-2}$

84. $(6.4)^5 \cdot (2.3)^2$ **85.** $(-14.8)^2 \cdot (21.3)^2$

Hawkes Learning Systems: Introductory Algebra

Simplifying Integer Exponents I

6.2 More on Exponents and Scientific Notation

Objectives

After completing this section, you will be able to:

1. Simplify powers of expressions by using the properties of integer exponents.

2. Write a decimal number in scientific notation.

3. Operate with decimal numbers by using scientific notation.

The summary of the rules for exponents given in Section 6.1 is repeated here for easy reference.

Summary of the Rules for Exponents

For any nonzero real number a and integers m and n:

1. The Exponent 1: $a = a^1$

2. The Exponent 0: $a^0 = 1$ $(a \neq 0)$

3. The Product Rule: $a^m \cdot a^n = a^{m+n}$

4. The Quotient Rule: $\dfrac{a^m}{a^n} = a^{m-n}$

5. Negative Exponents: $a^{-n} = \dfrac{1}{a^n}$ $(a \neq 0)$

Power Rule

Now, consider what happens when a power is raised to a power. For example, to simplify the expressions $(x^2)^3$ and $(2^5)^2$, we can write

$$(x^2)^3 = x^2 \cdot x^2 \cdot x^2 = x^{2+2+2} = x^6$$

and

$$(2^5)^2 = 2^5 \cdot 2^5 = 2^{5+5} = 2^{10}$$

However, this technique can be quite time-consuming when the exponent is large such as in $(3y^3)^{17}$. The **Power Rule for Exponents** gives a convenient way to handle powers raised to powers.

Power Rule for Exponents

If a is a nonzero real number and m and n are integers, then

$$\left(a^m\right)^n = a^{mn}$$

In words, the value of a power raised to a power can be found by multiplying the exponents and keeping the base.

Example 1: Power Rule

Simplify each expression using the Power Rule for Exponents.

a. $\left(x^2\right)^4$

Solution: $\left(x^2\right)^4 = x^{2 \cdot 4} = x^8$

b. $\left(x^5\right)^{-2}$

Solution: $\left(x^5\right)^{-2} = x^{5(-2)} = x^{-10} = \dfrac{1}{x^{10}}$

or $\left(x^5\right)^{-2} = \dfrac{1}{\left(x^5\right)^2} = \dfrac{1}{x^{5 \cdot 2}} = \dfrac{1}{x^{10}}$

c. $\left(y^{-7}\right)^2$

Solution: $\left(y^{-7}\right)^2 = y^{(-7)2} = y^{-14} = \dfrac{1}{y^{14}}$

Rule for Power of a Product

If the base of an exponent is a product, we will see that each factor in the product can be raised to the power indicated by the exponent. For example, $(10x)^3$ indicates that the product of 10 and x is to be raised to the 3rd power and $(-2x^2y)^5$ indicates that the product of -2, x^2, and y is to be raised to the 5th power. We can simplify these expressions as follows:

$$(10x)^3 = 10x \cdot 10x \cdot 10x = 10 \cdot 10 \cdot 10 \cdot x \cdot x \cdot x = 10^3 \cdot x^3 = 1000x^3$$

and

$$\left(-2x^2y\right)^5 = \left(-2x^2y\right)\cdot\left(-2x^2y\right)\cdot\left(-2x^2y\right)\cdot\left(-2x^2y\right)\cdot\left(-2x^2y\right)$$

$$= (-2)(-2)(-2)(-2)(-2)\cdot x^2\cdot x^2\cdot x^2\cdot x^2\cdot x^2\cdot y\cdot y\cdot y\cdot y\cdot y$$

$$= (-2)^5\cdot(x^2)^5\cdot y^5$$

$$= -32x^{10}y^5$$

We can simplify expressions such as these in a much easier fashion by using the following **Power of a Product Rule for Exponents**.

Power of a Product

If a and b are nonzero real numbers and n is an integer then

$$(ab)^n = a^n b^n.$$

In words, a power of a product is found by raising each factor to that power.

Example 2: Rule for Power of a Product ● ● ● ● ● ● ● ● ● ● ● ●

Simplify each expression using the Rule for Power of a Product.

a. $(5x)^2$

    Solution: $(5x)^2 = 5^2\cdot x^2 = 25x^2$

b. $(xy)^3$

    Solution: $(xy)^3 = x^3\cdot y^3 = x^3y^3$

c. $(-7ab)^2$

    Solution: $(-7ab)^2 = (-7)^2 a^2 b^2 = 49a^2 b^2$

d. $(ab)^{-5}$

    Solution: $(ab)^{-5} = a^{-5}\cdot b^{-5} = \dfrac{1}{a^5}\cdot\dfrac{1}{b^5} = \dfrac{1}{a^5 b^5}$

    or, using the Rule of Negative Exponents first and then the Rule for the Power of a Product,

$$(ab)^{-5} = \dfrac{1}{(ab)^5} = \dfrac{1}{a^5 b^5}$$

Continued on next page ...

e. $(x^2y^{-3})^4$

Solution: $(x^2y^{-3})^4 = (x^2)^4 \cdot (y^{-3})^4 = x^8 \cdot y^{-12} = x^8 \cdot \dfrac{1}{y^{12}} = \dfrac{x^8}{y^{12}}$

● ●

NOTES

Special Note about Negative Numbers and Exponents:

In an expression such as $-x^2$, we know that -1 is understood to be the coefficient of x^2. That is,

$$-x^2 = -1 \cdot x^2$$

The same is true for expressions with numbers only such as -7^2. That is,

$$-7^2 = -1 \cdot 7^2 = -1 \cdot 49 = -49.$$

We see that the exponent refers to 7 and **not** to -7. For the exponent to refer to -7 as the base, -7 **must be in parentheses** as follows:

$$(-7)^2 = (-7) \cdot (-7) = +49.$$

As another example,

$$-2^0 = -1 \cdot 2^0 = -1 \cdot 1 = -1 \text{ and } (-2)^0 = 1.$$

Rule for Power of a Quotient

If the base of an exponent is a quotient (in fraction form), we will see that both the numerator and denominator can be raised to the power indicated by the exponent.

For example, in the expression $\left(\dfrac{2}{x}\right)^3$ the quotient (or fraction) $\dfrac{2}{x}$ is raised to the 3rd power.

We can simplify this expression as follows:

$$\left(\frac{2}{x}\right)^3 = \frac{2}{x} \cdot \frac{2}{x} \cdot \frac{2}{x} = \frac{2 \cdot 2 \cdot 2}{x \cdot x \cdot x} = \frac{2^3}{x^3} = \frac{8}{x^3}.$$

Or, we can simplify the expression in a much easier manner by applying the following **Rule for Power of a Quotient**.

Rule for Power of a Quotient

If a and b are nonzero real numbers and n is an integer, then

$$\left(\frac{a}{b}\right)^n = \frac{a^n}{b^n}$$

In words, a power of a quotient (in fraction form) is found by raising both the numerator and the denominator to that power.

Example 3: Rule for the Power of a Quotient • • • • • • • • • •

Simplify each expression using the Rule for the Power of a Quotient.

a. $\left(\dfrac{y}{x}\right)^5$

Solution: $\left(\dfrac{y}{x}\right)^5 = \dfrac{y^5}{x^5}$

b. $\left(\dfrac{2a}{b}\right)^4$

Solution: $\left(\dfrac{2a}{b}\right)^4 = \dfrac{(2a)^4}{b^4} = \dfrac{2^4 a^4}{b^4} = \dfrac{16a^4}{b^4}$

c. $\left(\dfrac{2x}{-3y}\right)^2$

Solution: $\left(\dfrac{2x}{-3y}\right)^2 = \dfrac{(2x)^2}{(-3y)^2} = \dfrac{2^2 x^2}{(-3)^2 y^2} = \dfrac{4x^2}{9y^2}$

• •

Using Combinations of Rules for Exponents

Various combinations of all the rules for exponents are used to simplify the expressions in the following examples. There may be more than one correct sequence of steps to follow. However, no matter how you proceed, if you apply the rules correctly, the answer will be the same in every case.

Example 4: Combination of Rules for Exponents • • • • • • • •

Simplify each expression by using the appropriate rules for exponents.

a. $\left(\dfrac{-2x}{y^2}\right)^3$

Solution: $\left(\dfrac{-2x}{y^2}\right)^3 = \dfrac{(-2x)^3}{(y^2)^3} = \dfrac{(-2)^3 x^3}{y^6} = \dfrac{-8x^3}{y^6}$

Continued on next page ...

b. $\left(\dfrac{3a^5 b}{a^0 b^2}\right)^2$

Solution: $\left(\dfrac{3a^5 b}{a^0 b^2}\right)^2 = \left(3a^{5-0} b^{1-2}\right)^2 = \left(3a^5 b^{-1}\right)^2 = 3^2 \left(a^5\right)^2 \left(b^{-1}\right)^2 = 9a^{10} b^{-2} = \dfrac{9a^{10}}{b^2}$

or

$$\left(\dfrac{3a^5 b}{a^0 b^2}\right)^2 = \dfrac{3^2 (a^5)^2 (b)^2}{(a^0)^2 (b^2)^2} = \dfrac{9a^{10} b^2}{a^0 b^4} = 9a^{10-0} b^{2-4} = 9a^{10} b^{-2} = \dfrac{9a^{10}}{b^2}$$

● ●

Another general approach with fractions involving negative exponents is to note that

$$\left(\dfrac{a}{b}\right)^{-n} = \dfrac{a^{-n}}{b^{-n}} = \dfrac{b^n}{a^n} = \left(\dfrac{b}{a}\right)^n$$

In effect, there are two basic shortcuts with negative exponents and fractions:

1. Taking the reciprocal of a fraction changes the sign of any exponent on the fraction.
2. Moving any term from numerator to denominator or vice versa, changes the sign of the corresponding exponent.

Example 5: Two Approaches   ● ● ● ● ● ● ● ● ● ● ● ● ● ● ●

a. Either of the following approaches can be used to simplify $\left(\dfrac{x^3}{y}\right)^{-4}$.

Solutions: i. $\left(\dfrac{x^3}{y}\right)^{-4} = \left(\dfrac{y}{x^3}\right)^4 = \dfrac{y^4}{\left(x^3\right)^4} = \dfrac{y^4}{x^{12}}$

ii. $\left(\dfrac{x^3}{y}\right)^{-4} = \dfrac{\left(x^3\right)^{-4}}{(y)^{-4}} = \dfrac{x^{-12}}{y^{-4}} = \dfrac{y^4}{x^{12}}$

● ●

Now a complete summary of the Rules for Exponents shows eight rules.

If a and b are nonzero real numbers and m and n are integers:

1. The Exponent 1: $a = a^1$

2. The Exponent 0: $a^0 = 1$

3. The Product Rule: $a^m \cdot a^n = a^{m+n}$

4. The Quotient Rule: $\dfrac{a^m}{a^n} = a^{m-n}$

5. Negative Exponents: $a^{-n} = \dfrac{1}{a^n}$

6. Power Rule: $(ab)^n = a^n b^n$

7. Power of a Product: $(a^m)^n = a^{mn}$

8. Power of a Quotient: $\left(\dfrac{a}{b}\right)^n = \dfrac{a^n}{b^n}$

Scientific Notation

Problems in physics, chemistry, astronomy, and other scientific and technological fields sometimes involve very large or very small numbers. For convenience, these numbers can be written in a form that involves exponents called **scientific notation**. Your calculator also uses this notation.

In scientific notation, all decimal numbers are written as the product (indicted with a cross, ×) of a number between 1 and 10 and some power of 10. **There will always be one digit to the left of the decimal point**.

For example,

$$56,000 = 5.6 \times 10^4 \qquad \text{(Note that 5.6 is between 1 and 10.)}$$
$$872,000 = 8.72 \times 10^5$$
$$0.0049 = 4.9 \times 10^{-3}$$

To understand this notation and how to find the correct power of 10, consider the place value system we use for decimal numbers (see Figure 6.1).

$$\overline{10^7}\ \overline{10^6}\ \overline{10^5}\ \overline{10^4}\ \overline{10^3}\ \overline{10^2}\ \overline{10^1}\ \overline{10^0}\ \cdot\ \overline{10^{-1}}\ \overline{10^{-2}}\ \overline{10^{-3}}\ \overline{10^{-4}}\ \overline{10^{-5}}\ \overline{10^{-6}}\ \overline{10^{-7}}$$

Values of each position in the decimal system using the powers of 10

Figure 6.1

Write 56,000 with the value of each position indicated:

$$56,000 = \frac{5}{10^4}\ \frac{6}{10^3}\ \frac{0}{10^2}\ \frac{0}{10^1}\ \frac{0}{10^0}\ .$$

Now, we see that the digit 5 is in the position corresponding to 10^4. That is, if we multiply 5.6×10^4, that will give the value 56,000. Similarly,

$$872,000 = \frac{8}{10^5}\ \frac{7}{10^4}\ \frac{2}{10^3}\ \frac{0}{10^2}\ \frac{0}{10^1}\ \frac{0}{10^0}\ . = 8.72 \times 10^5$$

since the digit 8 is in the 10^5 place value position. Also,

$$0.0049 = 0. \frac{0}{10^{-1}}\ \frac{0}{10^{-2}}\ \frac{4}{10^{-3}}\ \frac{9}{10^{-4}} = 4.9 \times 10^{-3}$$

since the digit 4 is in the 10^{-3} place value position.

Your calculator will also give large answers in scientific notation. For example, use your calculator to multiply

93,000,000 miles   The distance to the
sun from the earth.

and

250,000 miles   The distance to the
moon from the earth.

Your calculator probably shows an expression like

2.325 13 or 2.325 E 13

This is the calculator's version of the scientific notation 2.325×10^{13}. To see how to arrive at this value using scientific notation and the properties of exponents, we can calculate in the following way:

$$93{,}000{,}000 \cdot 250{,}000 \quad = 9.3 \times 10^7 \cdot 2.5 \times 10^5$$
$$= 9.3 \cdot 2.5 \times 10^7 \cdot 10^5$$
$$= 23.25 \times 10^{7+5}$$
$$= 2.325 \times 10^1 \times 10^{12}$$
$$= 2.325 \times 10^{13}$$

Example 6: Scientific Notation

a. Write 8,720,000 in scientific notation.

Solution: The digit 8 is in the seventh position to the left of the understood decimal point. In powers of 10, this is the 10^6 position. So,

$$8{,}720{,}000 = 8.72 \times 10^6 \qquad \text{8.72 is between 1 and 10.}$$

To check, move the decimal point 6 places to the right and get the original number:

$$8.72 \times 10^6 = 8.720000. = 8{,}720{,}000.$$
$$1\ 2\ 3\ 4\ 5\ 6$$

Note that multiplying by 10^6 is the same as multiplying by 1,000,000.

b. Write 0.000000376 in scientific notation.

Solution: The digit 3 is in the seventh position to the right of the decimal point. In powers of 10, this is the 10^{-7} position. So,

$$0.000000376 = 3.76 \times 10^{-7} \qquad \text{3.76 is between 1 and 10.}$$

To check, move the decimal point 7 places to the left and get the original number:

$$3.76 \times 10^{-7} = 0.0000003.76 = 0.000000376.$$
$$7\ 6\ 5\ 4\ 3\ 2\ 1$$

Note that multiplying by 10^{-7} is the same as dividing by 10,000,000.

Continued on next page ...

c. Multiply 300,000 and 700 using scientific notation and leave your answer in scientific notation. Verify your answer using your calculator.

Solution: $300{,}000 \cdot 700 = 3.0 \times 10^5 \cdot 7.0 \times 10^2$

$$= 3 \cdot 7 \times 10^5 \cdot 10^2$$
$$= 21 \times 10^7$$
$$= 2.1 \times 10^1 \times 10^7$$
$$= 2.1 \times 10^8$$

d. Simplify the expression using scientific notation and leave your answer in scientific notation. Verify your answer using your calculator.

Solution: $\dfrac{0.0042 \cdot 0.0003}{0.21} = \dfrac{4.2 \times 10^{-3} \cdot 3 \times 10^{-4}}{2.1 \times 10^{-1}}$

$$= \frac{4.2 \cdot 3}{2.1} \times \frac{10^{-7}}{10^{-1}}$$
$$= 6 \times 10^{-7+1}$$
$$= 6 \times 10^{-6}$$

e. A light-year is the distance light travels in one year. Find the length of a light-year in scientific notation if light travels 186,000 miles per second.

Solution:
$$60 \text{ sec } = 1 \text{ minute}$$
$$60 \text{ min } = 1 \text{ hour}$$
$$24 \text{ hr. } = 1 \text{ day}$$
$$365 \text{ days } = 1 \text{ year}$$
$$1 \text{ light-year } = 186{,}000 \cdot 60 \cdot 60 \cdot 24 \cdot 365$$
$$= 5{,}865{,}696{,}000{,}000$$
$$= 5.865696 \times 10^{12} \text{ mi.}$$

● ●

Practice Problems

Simplify each expression.

1. $\dfrac{x^{-2}x^5}{x^{-7}}$ **2.** $\left(\dfrac{a^2 b^3}{4}\right)^0$ **3.** $\dfrac{-3^2 \cdot 5}{2 \cdot (-3)^2}$ **4.** $\left(\dfrac{5x}{3b}\right)^{-2}$

5. Write the number in scientific notation: *186,000 miles per second (speed of light in miles per second).*

Answers to Practice Problems: 1. x^{10} **2.** 1 **3.** $\dfrac{-5}{2}$ **4.** $\dfrac{9b^2}{25x^2}$ **5.** 1.86×10^5

6.2 Exercises

Use the rules for exponents to simplify each of the expressions in Exercises 1 – 40. Assume that all variables represent nonzero real numbers.

1. -3^4 **2.** -5^2 **3.** $(-3)^4$ **4.** -20^2 **5.** $(-10)^6$

6. $(-4)^6$ **7.** $(6x^3)^2$ **8.** $(-3x^4)^2$ **9.** $4(-3x^2)^3$ **10.** $7(y^{-2})^4$

11. $-3(7xy^2)^0$ **12.** $5(x^2y^{-1})$

13. $-2(3x^5y^{-2})^{-3}$ **14.** $\left(\dfrac{3x}{y}\right)^3$ **15.** $\left(\dfrac{-4x}{y^2}\right)^2$ **16.** $\left(\dfrac{6m^3}{n^5}\right)^0$

17. $\left(\dfrac{3x^2}{y^3}\right)^2$ **18.** $\left(\dfrac{-2x^2}{y^{-2}}\right)^2$ **19.** $\left(\dfrac{x}{y}\right)^{-2}$ **20.** $\left(\dfrac{2a}{b}\right)^{-1}$

21. $\left(\dfrac{2x}{y^5}\right)^{-2}$ **22.** $\left(\dfrac{3x}{y^{-2}}\right)^{-1}$ **23.** $\left(\dfrac{4a^2}{b^{-3}}\right)^{-3}$ **24.** $\left(\dfrac{-3}{xy^2}\right)^{-3}$

25. $\left(\dfrac{5xy^3}{y}\right)^2$ **26.** $\left(\dfrac{m^2n^3}{mn}\right)^2$ **27.** $\left(\dfrac{2ab^3}{b^2}\right)^4$ **28.** $\left(\dfrac{-7^2x^2y}{y^3}\right)^{-1}$

29. $\left(\dfrac{2ab^4}{b^2}\right)^{-3}$ **30.** $\left(\dfrac{5x^3y}{y^2}\right)^2$ **31.** $\left(\dfrac{2x^2y}{y^3}\right)^{-4}$ **32.** $\left(\dfrac{x^3y^{-1}}{y^2}\right)^2$

33. $\left(\dfrac{2a^2b^{-1}}{b^2}\right)^3$ **34.** $\left(\dfrac{6y^5}{x^2y^{-2}}\right)^2$ **35.** $\left(\dfrac{7x^{-2}y}{xy^{-1}}\right)^2$ **36.** $\dfrac{-5x^6y^7}{3x^{-6}y^2}$

37. $\dfrac{(3x^2y^{-1})^{-2}}{(6x^{-1}y)^{-3}}$ **38.** $\left(\dfrac{2x^{-3}}{5y^{-2}}\right)^2$ **39.** $\dfrac{(4x^{-2})(6x^5)}{(9y)(2y^{-1})}$ **40.** $\dfrac{(5x^2)(3x^{-1})^2}{(25y^3)(6y^{-2})}$

In Exercises 41 – 46, write the numbers in scientific notation.

41. 86,000   **42.** 927,000   **43.** 0.0362

44. 0.0061   **45.** 18,300,000   **46.** 376,000,000

In Exercises 47 – 52, write the numbers in decimal form.

47. 4.2×10^{-2} **48.** 8.35×10^{-3} **49.** 7.56×10^6

50. 6.132×10^{-5} **51.** 8.515×10^8 **52.** 9.374×10^7

In Exercises 53 – 66, first write each of the numbers in scientific notation. Then perform the indicated operations and leave your answer in scientific notation. Use a calculator to verify your answers.

53. $300 \cdot 0.00015$

54. $0.000024 \cdot 40,000$

55. $0.0003 \cdot 0.0000025$

56. $0.00005 \cdot 0.00013$

57. $\dfrac{3900}{0.003}$

58. $\dfrac{4800}{12,000}$

59. $\dfrac{125}{50,000}$

60. $\dfrac{0.0046}{230}$

61. $\dfrac{0.02 \cdot 3900}{0.013}$

62. $\dfrac{0.0084 \cdot 0.003}{0.21 \cdot 60}$

63. $\dfrac{0.005 \cdot 650 \cdot 3.3}{0.0011 \cdot 2500}$

64. $\dfrac{5.4 \cdot 0.003 \cdot 50}{15 \cdot 0.0027 \cdot 200}$

65. $\dfrac{\left(1.4 \times 10^{-2}\right)(922)}{\left(3.5 \times 10^{3}\right)\left(2.0 \times 10^{6}\right)}$

66. $\dfrac{(4300)\left(3.0 \times 10^{2}\right)}{\left(1.5 \times 10^{-3}\right)\left(860 \times 10^{-2}\right)}$

67. The mass of a hydrogen atom is approximately 0.00000000000000000000000167 grams. Write this number in scientific notation.

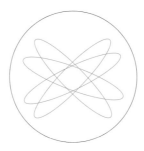

68. The weight of an atom is expressed in atomic weight units (amu), where 1 amu = 0.0000000000000000000000000016605 kilograms. Express the weight in scientific notation.

69. The distance light travels in one year is called a light-year. A light-year is approximately 5,866,000,000,000 miles. Write this number in scientific notation. How far does light travel in 2.3 years?

70. Light travels approximately 3×10^{10} centimeters per second. How many centimeters would this be per minute? per hour? Express your answers in scientific notation.

71. The mass of the earth is about 5,980,000,000,000,000,000,000,000,000 grams. Write this number in scientific notation.

72. An atom of gold weighs approximately 3.25×10^{-22} grams. What would be the weight of 2000 atoms of gold? Express your answer in scientific notation.

73. An ounce of gold contains 5×10^{22} atoms. All the gold ever taken out of the Earth is estimated to be 3.0×10^{31} atoms. How many ounces of gold is this?

74. There are approximately 6×10^{13} cells in an adult human body. Express this number in decimal form.

75. Write the number 1.0×10^{10} in decimal form. About how many years are in 1.0×10^{10} minutes?

Hawkes Learning Systems: Introductory Algebra

Simplifying Integer Exponents II

6.3 Identifying and Evaluating Polynomials

After completing this section, you will be able to:

1. Define a polynomial.

2. Classify a polynomial as a monomial, binomial, trinomial, or a polynomial with more than three terms.

3. Evaluate a polynomial for given values of the variable.

Definition of a Polynomial

A **term** is an expression that involves only multiplication and/or division with constants and/or variables. Remember that a number written next to a variable indicates multiplication, and the number is called the **numerical coefficient** (or **coefficient**) of the variable. For example,

$$3x, \quad -5y^2, \quad 17, \quad \text{and} \quad \frac{x}{y}$$

are all algebraic terms. In the term $3x$, 3 is the coefficient of x. In the term $-5y^2$, -5 is the coefficient of y^2 and 1 is the coefficient of the term $\frac{x}{y}$. A term that consists of only a number, such as 17, is called a **constant** or a **constant term**.

Monomial

> A **monomial in x** is a term of the form
>
> $$kx^n$$
>
> where k is a real number and n is a whole number.
>
> **n** is called the **degree** of the term, and **k** is called the **coefficient**.

A monomial may have more than one variable, and the degree of such a monomial is the sum of the degrees of its variables. For example,

$$4x^2y^3 \text{ is a } 5^{\text{th}} \text{ degree monomial in } x \text{ and } y.$$

However, in this chapter, only monomials of one variable (note that any variable may be used in place of x) will be discussed. Monomials may have fractional or negative coefficients; however, monomials may **not** have fractional or negative exponents. These facts are part of the definition since in the expression kx^n, k (the coefficient) can be any real number, but n (the exponent) must be a whole number.

Expressions that **are not** monomials: $3\sqrt{x}, \ -15x^{\frac{2}{3}}, \ 4a^{-2}$

Expressions that **are** monomials: $17, \ 3x, \ 5y^2, \ \frac{2}{7}a^4$

Since $x^0 = 1$, a nonzero constant can be multiplied by x^0 without changing its value. Thus, we say that a **nonzero constant is a monomial of degree 0.** For example,

$$17 = 17x^0 \qquad \text{and} \qquad -6 = -6x^0$$

which means that the constants 17 and –6 are monomials of degree 0. However, for the special number 0, we can write

$$0 = 0x^2 = 0x^5 = 0x^{13}$$

and we say that **the constant 0 is a monomial of no degree.**

A **polynomial** is a monomial or the algebraic sum of monomials. Examples of polynomials are

$$3x, \ y + 5, \ 4x^2 - 7x + 1, \ \text{and} \ a^{10} + 5a^3 - 2a^2 + 6.$$

In the formal definition of a polynomial, we make use of **subscript notation** for the coefficients. A subscript is a number written to the right and below a letter. Subscripts allow the use of the same letter while indicating different numbers. For example, a_1 (read "a sub 1") and a_2 (read "a sub 2") indicate two different numbers even though the same letter, a, is used.

Polynomial in x

A ***polynomial in x*** *is an expression of the form*

$$a_n x^n + a_{n-1} x^{n-1} + a_{n-2} x^{n-2} + \ldots + a_1 x^1 + a_0$$

where $n, n-1, n-2, \ldots, 1, 0$ are whole numbers
and $a_n, a_{n-1}, \ldots, a_1, a_0$ are real numbers with $a_n \neq 0$.
*n is called **the degree of the polynomial.***
*a_n is called **the leading coefficient of the polynomial.***

As an example,

$$x^3 + 5x^2 - 8x + 14 \text{ is a third-degree polynomial in } x.$$

We can write

$$x^3 + 5x^2 - 8x + 14 = a_3 x^3 + a_2 x^2 + a_1 x + a_0$$

And, since the corresponding coefficients must be equal, we have

$$a_3 = 1 \text{ (The leading coefficient is 1.)}, a_2 = 5, a_1 = -8, \text{ and } a_0 = 14.$$

Similarly,

$$3y^4 - 2y^3 + 4y - 6 \text{ is a fourth-degree polynomial in } y$$

and writing

$$3y^4 - 2y^3 + 4y - 6 = a_4 y^4 + a_3 y^3 + a_2 y^2 + a_1 y + a_0$$

we see that

$$a_4 = 3 \text{ (The leading coefficient is 3.)}, a_3 = -2, a_2 = 0, a_1 = 4, \text{ and } a_0 = -6.$$

In each of these examples, the terms have been written so that the exponents decrease in order from left to right. We say that the terms are written in **descending order**. If the terms increase in order from left to right, then we say that the terms are written in **ascending order**. As a general rule, for consistency and style in operating with polynomials, the polynomials in this chapter will be written in descending order.

Some forms of polynomials are used so frequently that they are given special names. These names are indicated in the following box with examples. Remember that all are also classified as polynomials.

Classification of Polynomials

		Example	
Monomial	*polynomial with one term*	$7a^3$	*Third-degree monomial*
Binomial	*polynomial with two terms*	$3x^2 + 5x$	*Second-degree binomial*
Trinomial	*polynomial with three terms*	$2x^5 + 3x^2 + 1$	*Fifth-degree trinomial*

> **Example 1: Simplify Polynomials** • • • • • • • • • • • • • •

Simplify each of the following polynomials by combining like terms. Write the polynomial in descending order and state the degree and type of the polynomial.

a. $5x^3 + 7x^3$

Solution: $5x^3 + 7x^3 = (5 + 7)x^3 = 12x^3$ Third-degree monomial.

b. $5x^3 + 7x^3 - 2x$

Solution: $5x^3 + 7x^3 - 2x = 12x^3 - 2x$ Third-degree binomial.

c. $\frac{1}{2}y + 3y - \frac{2}{3}y^2 - 7$

Solution: $\frac{1}{2}y + 3y - \frac{2}{3}y^2 - 7 = -\frac{2}{3}y^2 + \frac{7}{2}y - 7$ Second-degree trinomial.

d. $x^2 + 8x - 15 - x^2$

Solution: $x^2 + 8x - 15 - x^2 = 8x - 15$ First-degree binomial.

e. $-3y^4 + 2y^2 + y^{-1}$

Solution: This expression is not a polynomial since y has a negative exponent.

• •

Evaluating a Polynomial

In evaluating any numerical expression, we follow the rules for order of operations discussed in Section 1.5. Similarly, to evaluate a polynomial for a given value of the variable, we substitute the value for the variable wherever it occurs in the polynomial and follow the rules for order of operations.

A polynomial in x can be thought of as a function of x. Thus, using the notation for functions discussed in Section 4.5, we can write $p(x) = $ a polynomial in x. This notation is particularly appropriate when evaluating a polynomial, as illustrated in Example 2.

> **Example 2: Evaluating Polynomials** • • • • • • • • • • • • • •

a. Evaluate $p(x) = 4x^2 + 5x - 15$ for $x = 3$.

Solution: For $p(x) = 4x^2 + 5x - 15$,

$$p(3) = 4 \cdot 3^2 + 5 \cdot 3 - 15 \qquad \text{Substitute 3 for } x.$$
$$= 4 \cdot 9 + 15 - 15$$
$$= 36 + 15 - 15$$
$$= 36$$

Continued on next page ...

b. Evaluate $p(y) = 5y^3 + y^2 - 3y + 8$ for $y = -2$.

Solution: For $p(y) = 5y^3 + y^2 - 3y + 8$,

$$p(-2) = 5(-2)^3 + (-2)^2 - 3(-2) + 8 \quad \text{Note the use of}$$
$$= 5(-8) + 4 - 3(-2) + 8 \qquad \text{parentheses around } -2.$$
$$= -40 + 4 + 6 + 8$$
$$= -22$$

● ●

Practice Problems

Combine like terms and state the degree and type of the polynomial.

1. $8x^3 - 3x^2 - x^3 + 5 + 3x^2$ **2.** $5y^4 + y^4 - 3y^2 + 2y^2 + 4$

3. *For the polynomial* $p(x) = x^2 - 5x - 5$, *find (a)* $p(3)$ *and (b)* $p(-1)$.

6.3 Exercises

In Exercises 1 – 10, identify the expression as a monomial, binomial, trinomial, or not a polynomial.

1. $3x^4$ **2.** $5y^2 - 2y + 1$ **3.** $-2x^{-2}$ **4.** $8x^3 - 7$

5. $14a^7 - 2a - 6$ **6.** $17x^{\frac{2}{3}} + 5x^2$ **7.** $6a^3 + 5a^2 - a^{-3}$ **8.** $-3y^4 + 2y^2 - 9$

9. $\frac{1}{2}x^3 - \frac{2}{5}x$ **10.** $\frac{5}{8}x^5 + \frac{2}{3}x^4$

Simplify the polynomials in Exercises 11 – 30. Write the polynomial in descending order and state the degree and type of the simplified polynomial. For each polynomial, write the values for $a_n, a_{n-1}, a_{n-2}, ..., a_1$, and a_0.

11. $y + 3y$ **12.** $4x^2 - x + x^2$ **13.** $x^3 + 3x^2 - 2x$

14. $3x^2 - 8x + 8x$ **15.** $x^4 - 4x^2 + 2x^2 - x^4$ **16.** $2 - 6y + 5y - 2$

17. $-x^3 + 6x + x^3 - 6x$ **18.** $11x^2 - 3x + 2 - 7x^2$ **19.** $6a^5 + 2a^2 - 7a^3 - 3a^2$

20. $2x^2 - 3x^2 + 2 - 4x^2 - 2 + 5x^2$ **21.** $4y - 8y^2 + 2y^3 + 8y^2$

22. $2x + 9 - x + 1 - 2x$ **23.** $5y^2 + 3 - 2y^2 + 1 - 3y^2$

24. $13x^2 - 6x - 9x^2 - 4x$ **25.** $7x^3 + 3x^2 - 2x + x - 5x^3 + 1$

26. $-3y^5 + 7y - 2y^3 - 5 + 4y^2 + y^2$ **27.** $x^4 + 3x^4 - 2x + 5x - 10 - x^2 + x$

Answers to Practice Problems: **1.** $7x^3 + 5$; third degree binomial **2.** $6y^4 - y^2 + 4$; fourth degree trinomial **3. a.** -11 **b.** 1

28. $a^3 + 2a^2 - 6a + 3a^3 + 2a^2 + 7a + 3$

29. $2x + 4x^2 + 6x + 9x^3$

30. $15y - y^3 + 2y^2 - 10y^2 + 2y - 16$

In Exercises 31 – 40, evaluate the polynomials for the given value of the variable.

31. $p(x) = x^2 + 14x - 3; x = -1$

32. $p(y) = y^3 - 5y^2 + 6y + 2; y = 2$

33. $p(x) = 3x^3 - 9x^2 - 10x - 11; \ x = 3$

34. $p(x) = -5x^2 - 8x + 7; x = -3$

35. $p(x) = 8x^4 + 2x^3 - 6x^2 - 7; x = -2$

36. $p(a) = a^3 + 4a^2 + a + 2; a = -5$

37. $p(a) = 2a^4 + 3a^2 - 8a; a = -1$

38. $p(y) = -4y^3 + 5y^2 + 12y - 1; y = -10$

39. $p(x) = x^5 - x^3 + x - 2; x = 2$

40. $p(x) = 3x^6 - 2x^5 + x^4 - x^3 - 3x^2 + 2x - 1; x = 1$

Writing and Thinking About Mathematics

First-degree polynomials are also called **linear polynomials**, second-degree polynomials are called **quadratic polynomials**, and third-degree polynomials are called **cubic polynomials**. The related functions are called linear functions, quadratic functions, and cubic functions, respectively.

41. Use a graphing calculator to graph the following linear functions.
   a. $p(x) = 2x + 3$ **b.** $p(x) = -3x + 1$ **c.** $p(x) = \frac{1}{2}x$

42. Use a graphing calculator to graph the following quadratic functions.
   a. $p(x) = x^2$ **b.** $p(x) = x^2 + 6x + 9$ **c.** $p(x) = -x^2 + 2$

43. Use a graphing calculator to graph the following cubic functions.
   a. $p(x) = x^3$ **b.** $p(x) = x^3 - 4x$ **c.** $p(x) = x^3 + 2x^2 - 5$

44. Make up a few of your own linear, quadratic, and cubic functions and graph these functions with your calculator. Using the results from Exercises 41, 42, and 43, and your own functions, describe in your own words:
   a. the nature of graphs of linear functions.
   b. the nature of graphs of quadratic functions.
   c. the nature of graphs of cubic functions.

Hawkes Learning Systems: Introductory Algebra

Identifying Polynomials
Evaluating Polynomials

6.4 | Adding and Subtracting Polynomials

After completing this section, you will be able to:

1. Add polynomials.

2. Subtract polynomials.

3. Simplify expressions by removing grouping symbols and combining like terms.

We have discussed identifying polynomials in terms of degree and type. In the remainder of this chapter, we will discuss algebraic operations with polynomials. That is, we will learn to add, subtract, multiply, and divide with polynomials.

Addition with Polynomials

Animation:
Addition and
Subtraction of
Polynomials

The **sum** of two or more polynomials is found by combining like terms. The polynomials may be written horizontally or vertically. For example,

$$(x^2 - 5x + 3) + (2x^2 - 8x - 4) + (3x^3 + x^2 - 5)$$
$$= 3x^3 + (x^2 + 2x^2 + x^2) + (-5x - 8x) + (3 - 4 - 5)$$
$$= 3x^3 + 4x^2 - 13x - 6$$

If the polynomials are written in a vertical format, we write like terms one beneath the other in a column format and add the like terms in each column.

$$
\begin{array}{r}
x^2 - 5x + 3 \\
2x^2 - 8x - 4 \\
3x^3 + x^2 \quad\quad - 5 \\
\hline
3x^3 + 4x^2 - 13x - 6
\end{array}
$$

> **Example 1: Addition with Polynomials** • • • • • • • • • • • • • •

    a. Add as indicated: $(5x^3 - 8x^2 + 12x + 13) + (-2x^2 - 8) + (4x^3 - 5x + 14)$

        Solution: $(5x^3 - 8x^2 + 12x + 13) + (-2x^2 - 8) + (4x^3 - 5x + 14)$

$$= (5x^3 + 4x^3) + (-8x^2 - 2x^2) + (12x - 5x) + (13 - 8 + 14)$$

$$= 9x^3 - 10x^2 + 7x + 19$$

    b. Find the sum: $(x^3 - x^2 + 5x) + (4x^3 + 5x^2 - 8x + 9)$

        Solution:
$$\begin{array}{l} x^3 - \ \ x^2 + 5x \\ \underline{4x^3 + 5x^2 - 8x + 9} \\ 5x^3 + 4x^2 - 3x + 9 \end{array}$$

• •

Subtraction with Polynomials

If a negative sign is written in front of a polynomial in parentheses, the meaning is the opposite of the entire polynomial. The opposite can be found by changing the sign of every term in the polynomial.

$$-(2x^2 + 3x - 7) = -2x^2 - 3x + 7$$

We can also think of the opposite of a polynomial as -1 times the polynomial, then applying the distibutive property as follows:

$$\begin{aligned} -(2x^2 + 3x - 7) &= -1(2x^2 + 3x - 7) \\ &= -1(2x^2) - 1(3x) - 1(-7) \\ &= -2x^2 - 3x + 7 \end{aligned}$$

The result is the same with either approach. So the **difference** between two polynomials can be found by changing the sign of each term of the second polynomial and then combining like terms.

$$\begin{aligned} (5x^2 - 3x - 7) - (2x^2 + 5x - 8) \ \ &= 5x^2 - 3x - 7 - 2x^2 - 5x + 8 \\ &= 5x^2 - 2x^2 - 3x - 5x - 7 + 8 \\ &= 3x^2 - 8x + 1 \end{aligned}$$

If the polynomials are written in a vertical format, one beneath the other, we change the signs of the terms of the polynomial being subtracted and then combine like terms.

Subtract:

$$5x^2 - 3x - 7$$
$$\underline{-(2x^2 + 5x - 8)} \longrightarrow$$

$$5x^2 - 3x - 7$$
$$\underline{-2x^2 - 5x + 8}$$
$$3x^2 - 8x + 1$$

Example 2: Subtraction with Polynomials ● ● ● ● ● ● ● ● ● ●

a. Subtract as indicated: $(9x^4 - 22x^3 + 3x^2 + 10) - (5x^4 - 2x^3 - 5x^2 + x)$

Solution: $(9x^4 - 22x^3 + 3x^2 + 10) - (5x^4 - 2x^3 - 5x^2 + x)$
$$= 9x^4 - 22x^3 + 3x^2 + 10 - 5x^4 + 2x^3 + 5x^2 - x$$
$$= 9x^4 - 5x^4 - 22x^3 + 2x^3 + 3x^2 + 5x^2 - x + 10$$
$$= 4x^4 - 20x^3 + 8x^2 - x + 10$$

b. Find the difference:
$$8x^3 + 5x^2 - 14$$
$$\underline{-(-2x^3 + x^2 + 6x)}$$

Solution:

$$8x^3 + 5x^2 - 14$$
$$\underline{-(-2x^3 + x^2 + 6x)} \longrightarrow$$

$$8x^3 + 5x^2 + 0x - 14$$
$$\underline{2x^3 - x^2 - 6x + 0}$$
$$10x^3 + 4x^2 - 6x - 14$$

Write in 0's for missing powers to help with alignment of like terms.

● ●

If an expression contains more than one pair of grouping (or inclusion) symbols, such as parentheses ( ), brackets [ ], or braces { }, simplify by working to remove the innermost pair of symbols first.

Example 3: Simplify ●

Simplify each of the following expressions.

a. $5x - [\, 2x + 3\,(4 - x) + 1\,] - 9$

Solution: $5x - [\, 2x + 3(4 - x) + 1\,] - 9$
$$= 5x - [\, 2x + 12 - 3x + 1\,] - 9$$
$$= 5x - [\, -x + 13\,] - 9$$
$$= 5x + x - 13 - 9$$
$$= 6x - 22$$

Work with the parentheses first since they are included inside the brackets.

Continued on next page ...

b. $10 - x + 2\,[\,x + 3\,(x-5) + 7\,]$

Solution: $10 - x + 2\,[\,x + 3(x-5) + 7\,]$

$= 10 - x + 2\,[\,x + 3x - 15 + 7\,]$ Work with the parentheses first since they are included inside the brackets.

$= 10 - x + 2\,[\,4x - 8\,]$

$= 10 - x + 8x - 16$

$= 7x - 6$

● ●

Practice Problems

1. *Add:* $(15x + 4) + (3x^2 - 9x - 5)$

2. *Subtract:* $(-5x^3 - 3x + 4) - (3x^3 - x^2 + 4x - 7)$

3. *Simplify:* $2 - [\,3a - (4 - 7a) + 2a\,]$

6.4 Exercises

Find the indicated sum in Exercises 1 – 20.

1. $(2x^2 + 5x - 1) + (x^2 + 2x + 3)$

2. $(x^2 + 2x - 3) + (x^2 + 5)$

3. $(x^2 + 7x - 7) + (x^2 + 4x)$

4. $(x^2 + 3x - 8) + (3x^2 - 2x + 4)$

5. $(2x^2 - x - 1) + (x^2 + x + 1)$

6. $(3x^2 + 5x - 4) + (2x^2 + x - 6)$

7. $(-2x^2 - 3x + 9) + (3x^2 - 2x + 8)$

8. $(x^2 + 6x - 7) + (3x^2 + x - 1)$

9. $(-4x^2 + 2x - 1) + (3x^2 - x + 2) + (x - 8)$

10. $(8x^2 + 5x + 2) + (-3x^2 + 9x - 4)$

11. $(x^2 + 2x - 1) + (3x^2 - x + 2) + (x - 8)$

12. $(x^3 + 2x - 9) + (x^2 - 5x + 2) + (x^3 - 4x^2 + 1)$

13.
$$\begin{array}{r} x^2 + 4x - 4 \\ \underline{-2x^2 + 3x + 1} \end{array}$$

14.
$$\begin{array}{r} x^3 + 3x^2 +\ \ x \\ \underline{-2x^3 - x^2 + 2x - 4} \end{array}$$

15.
$$\begin{array}{r} 7x^3 + 5x^2 +\ \ x - 6 \\ -3x^2 + 4x + 11 \\ \underline{-3x^3 - 2x^2 - 5x +\ 2} \end{array}$$

16.
$$\begin{array}{r} 5x^2 - 3x + 11 \\ \underline{-2x^2 +\ \ x - 6} \end{array}$$

17.
$$\begin{array}{r} x^3 + 3x^2\qquad - 4 \\ 7x^2 + 2x + 1 \\ \underline{x^3 +\ \ x^2 - 6x} \end{array}$$

18.
$$\begin{array}{r} x^3 + 2x^2\qquad - 5 \\ -2x^3\qquad + x - 9 \\ \underline{x^3 - 2x^2\qquad + 14} \end{array}$$

19.
$$\begin{array}{r} 2x^2 + 4x - 3 \\ \underline{3x^2 - 9x + 2} \end{array}$$

20.
$$\begin{array}{r} x^3 + 5x^2 + 7x - 3 \\ 4x^2 + 3x - 9 \\ \underline{4x^3 + 2x^2\qquad - 2} \end{array}$$

Answers to Practice Problems: 1. $3x^2 + 6x - 1$ **2.** $-8x^3 + x^2 - 7x + 11$ **3.** $-12a + 6$

Find the indicated difference in Exercises 21 – 40.

21. $(2x^2 + 4x + 8) - (x^2 + 3x + 2)$ **22.** $(3x^2 + 7x - 6) - (x^2 + 2x + 5)$

23. $(x^4 + 8x^3 - 2x^2 - 5) - (2x^4 + 10x^3 - 2x^2 + 11)$

24. $(-3x^4 + 2x^3 - 7x^2 + 6x + 12) - (x^4 + 9x^3 + 4x^2 + x - 1)$

25. $(x^2 - 9x + 2) - (4x^2 - 3x + 4)$ **26.** $(2x^2 - x - 10) - (-x^2 + 3x - 2)$

27. $(7x^2 + 4x - 9) - (-2x^2 + x - 9)$ **28.** $(6x^2 + 11x + 2) - (4x^2 - 2x - 7)$

29. $(3x^4 - 2x^3 - 8x - 1) - (5x^3 - 3x^2 - 3x - 10)$

30. $(x^5 + 6x^3 - 3x^2 - 5) - (2x^5 + 8x^3 + 5x + 17)$

31. $(9x^2 + 6x - 5) - (13x^2 + 6)$ **32.** $(8x^2 + 9) - (4x^2 - 3x - 2)$

33. $(x^3 + 4x^2 - 7) - (3x^3 + x^2 + 2x + 1)$

34. $\begin{array}{r} 14x^2 - 6x + 9 \\ -\left(8x^2 + x - 9\right) \\ \hline \end{array}$ **35.** $\begin{array}{r} x^3 + 6x^2 - 3 \\ -\left(-x^3 + 2x^2 - 3x + 7\right) \\ \hline \end{array}$ **36.** $\begin{array}{r} 9x^2 - 3x + 2 \\ -\left(4x^2 - 5x - 1\right) \\ \hline \end{array}$

37. $\begin{array}{r} 5x^4 + 8x^2 + 11 \\ -\left(-3x^4 + 2x^2 - 4\right) \\ \hline \end{array}$ **38.** $\begin{array}{r} 11x^2 + 5x - 13 \\ -\left(-3x^2 + 5x + 2\right) \\ \hline \end{array}$ **39.** $\begin{array}{r} 3x^3 + 9x - 17 \\ -\left(x^3 + 5x^2 - 2x - 6\right) \\ \hline \end{array}$

40. $\begin{array}{r} -3x^4 + 10x^3 - 8x^2 - 7x - 6 \\ -\left(2x^4 + x^3 + 5x + 6\right) \\ \hline \end{array}$

Simplify each of the expressions in Exercises 41 – 55, and write the polynomials in order of descending powers.

41. $5x + 2(x - 3) - (3x + 7)$ **42.** $-4(x - 6) - (8x + 2)$

43. $11 + [\, 3x - 2(1 + 5x)\,]$ **44.** $2x + [\, 9x - 4(3x + 2) - 7\,]$

45. $8x - [\, 2x + 4(x - 3) - 5\,]$ **46.** $17 - [\, -3x + 6(2x - 3) + 9\,]$

47. $3x^3 - [\, 5 - 7(x^2 + 2) - 6x^2\,]$ **48.** $10x^3 - [\, 8 - 5(3 - 2x^2) - 7x^2\,]$

49. $(2x^2 + 4) - [\, -8 + 2(7 - 3x^2) + x\,]$ **50.** $-[\, 6x^2 - 3(4 + 2x) + 9\,] - (x^2 + 5)$

51. $2[\, 3x + (x - 8) - (2x + 5)] - (x - 7)$

52. $-3[\, -x + (10 - 3x) - (8 - 3x)] + (2x - 1)$

53. $(x^2 - 1) + x[\, 4 + (3 - x)]$ **54.** $x(x - 5) + [\, 6x - x(4 - x)]$

55. $x(2x + 1) - [\, 5x - x(2x + 3)]$

56. Subtract $2x^2 - 4x$ from $7x^3 + 5x$.

57. Subtract $3x^2 - 4x + 2$ from the sum of $4x^2 + x - 1$ and $6x - 5$.

58. Subtract $-2x^2 + 6x + 12$ from the sum of $2x^2 + 3x - 4$ and $x^2 - 13x + 2$.

59. Add $5x^3 - 8x + 1$ to the difference between $2x^3 + 14x - 3$ and $x^2 + 6x + 5$.

60. Find the sum of $10x - 2(3x + 5)$ and $3(x - 4) + 16$.

Hawkes Learning Systems: Introductory Algebra

 Adding and Subtracting Polynomials

6.5 Multiplying Polynomials

After completing this section, you will be able to:

1. Multiply polynomials.

Up to this point, we have multiplied terms such as $5x^2 \cdot 3x^4 = 15x^6$ by using the Product Rule for Exponents. Also, we have applied the distributive property to expressions such as $5(2x + 3) = 10x + 15$.

Now we will use both of these procedures to multiply polynomials. We will discuss first the product of a monomial with a polynomial of two or more terms; second, the product of two binomials; and third, the product of a binomial with a polynomial of more than two terms.

Multiplying a Polynomial by a Monomial

Using the distributive property $a(b + c) = ab + ac$ with multiplication indicated on the left, we can find the product of a monomial with a polynomial of two or more terms as follows:

$$5x(2x + 3) = 5x \cdot 2x + 5x \cdot 3 = 10x^2 + 15x$$
$$3x^2(4x - 1) = 3x^2 \cdot 4x + 3x^2(-1) = 12x^3 - 3x^2$$
$$-4a^5(a^2 - 8a + 5) = -4a^5 \cdot a^2 - 4a^5(-8a) - 4a^5(5) = -4a^7 + 32a^6 - 20a^5$$

Multiplying Two Polynomials

Now suppose that we want to multiply two binomials, say, $(x + 3)(x + 7)$. We will apply the distributive property in the following way with multiplication indicated on the right of the parentheses.

Compare $(x + 3)(x + 7)$ to

$$(a + b)c = ac + bc.$$

Think of $(x + 7)$ as taking the place of c. Thus,

$$(a + b)\, c \quad = \quad ac \quad + \quad bc$$

takes the form $\quad (x + 3)(x + 7) \quad = \quad x(x + 7) \quad + \quad 3(x + 7)$

Completing the products on the right, using the distributive property twice again, gives

$$(x + 3)(x + 7) = x(x + 7) + 3(x + 7)$$
$$= x \cdot x + x \cdot 7 + 3 \cdot x + 3 \cdot 7$$
$$= x^2 + 7x + 3x + 21$$
$$= x^2 + 10x + 21$$

In the same manner,

$$(x + 2)(3x + 4) = x(3x + 4) + 2(3x + 4)$$
$$= x \cdot 3x + x \cdot 4 + 2 \cdot 3x + 2 \cdot 4$$
$$= 3x^2 + 4x + 6x + 8$$
$$= 3x^2 + 10x + 8$$

Similarly,

$$(2x - 1)(x^2 + x - 5) = 2x(x^2 + x - 5) - 1(x^2 + x - 5)$$
$$= 2x \cdot x^2 + 2x \cdot x + 2x(-5) - 1 \cdot x^2 - 1 \cdot x - 1(-5)$$
$$= 2x^3 + 2x^2 - 10x - x^2 - x + 5$$
$$= 2x^3 + x^2 - 11x + 5$$

One quick way to check if your products are correct is to substitute some convenient number for x into the original two factors and into the product. Choose any nonzero number that you like. The values of both expressions should be the same. For example, let $x = 1$. Then,

$$(x + 2)(3x + 4) = (1 + 2)(3 \cdot 1 + 4) = (3)(7) = 21$$

and

$$3x^2 + 10x + 8 = 3 \cdot 1^2 + 10 \cdot 1 + 8 = 3 + 10 + 8 = 21$$

The product $(x + 2)(3x + 4) = 3x^2 + 10x + 8$ seems to be correct. We could double check by letting $x = 5$.

$$(x + 2)(3x + 4) = (5 + 2)(3 \cdot 5 + 4) = (7)(19) = 133$$

and

$$3x^2 + 10x + 8 = 3 \cdot 5^2 + 10 \cdot 5 + 8 = 75 + 50 + 8 = 133$$

Convinced? This is just a quick check, however, and is not foolproof unless you try this process with more values of x than indicated by the degree of the product.

The product of two polynomials can also be found by writing one polynomial under the other. **The distributive property is applied by multiplying each term of one polynomial by each term of the other.** Consider the product $(2x^2 + 3x - 4)(3x + 7)$. Now, writing one polynomial under the other and applying the distributive property, we obtain

Multiply by $+7$:

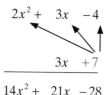

$$14x^2 + 21x - 28$$

Multiply by $3x$:

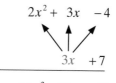

$$14x^2 + 21x - 28$$ Align the like terms so that

$$6x^3 + 9x^2 - 12x$$ they can be easily combined.

Finally, combine like terms:

$$
\begin{array}{r}
2x^2 + 3x - 4 \\
3x + 7 \\
\hline
14x^2 + 21x - 28 \\
6x^3 + 9x^2 - 12x \\
\hline
6x^3 + 23x^2 + 9x - 28
\end{array}
$$ Combine like terms.

> **Example 1: Multiply** •

Find each product.

a. $-4x(x^2 - 3x + 12)$

Solution: $-4x(x^2 - 3x + 12) = -4x \cdot x^2 - 4x(-3x) - 4x \cdot 12$

$$= -4x^3 + 12x^2 - 48x$$

b. $(3a + 4)(2a^2 + a + 5)$

Solution: $(3a+4)(2a^2+a+5) = (3a+4)2a^2 + (3a+4)a + (3a+4)5$

$$= 6a^3 + 8a^2 + 3a^2 + 4a + 15a + 20$$

$$= 6a^3 + 11a^2 + 19a + 20$$

Continued on next page ...

c. $7y^2 - 3y + 2$

$\underline{\qquad 2y + 3}$

Solution:

$$7y^2 - 3y + 2$$
$$\underline{\qquad\qquad 2y + 3}$$
$$21y^2 - 9y + 6 \qquad\qquad \text{Multiply by 3.}$$
$$\underline{14y^3 - 6y^2 + 4y \qquad\qquad\quad \text{Multiply by } 2y.}$$
$$14y^3 + 15y^2 - 5y + 6 \qquad\qquad \text{Combine like terms.}$$

d. $(x - 5)(x + 2)(x - 1)$

Solution: First multiply $(x - 5)(x + 2)$, then multiply this result by $(x - 1)$.

$$(x - 5)(x + 2) = x(x + 2) - 5(x + 2)$$
$$= x^2 + 2x - 5x - 10$$
$$= x^2 - 3x - 10$$

$$(x^2 - 3x - 10)(x - 1) = (x^2 - 3x - 10)x + (x^2 - 3x - 10)(-1)$$
$$= x^3 - 3x^2 - 10x - x^2 + 3x + 10$$
$$= x^3 - 4x^2 - 7x + 10$$

● ●

Practice Problems

Find each product.

1. $2x(3x^2 + x - 1)$

2. $(x + 3)(x - 7)$

3. $(x - 1)(x^2 + x - 4)$

4. $(x + 1)(x - 2)(x + 2)$

Answers to Practice Problems: 1. $6x^3 + 2x^2 - 2x$ **2.** $x^2 - 4x - 21$ **3.** $x^3 - 5x + 4$ **4.** $x^3 + x^2 - 4x - 4$

6.5 Exercises

Multiply as indicated in Exercises 1 – 40 and simplify if possible.

1. $-3x^2(2x^3 + 5x)$ **2.** $4x^5(x^2 - 3x + 1)$ **3.** $5x^2(-4x^2 + 6)$

4. $9x^3(2x^3 - x^2 + 5x)$ **5.** $-1(y^5 - 8y + 2)$ **6.** $-7(2y^4 + 3y^2 + 1)$

7. $-4x^3(x^5 - 2x^4 + 3x)$ **8.** $a^2(a^5 + 2a^4 - 5a + 1)$ **9.** $7t^3(-5t^2 + 2t + 1)$

10. $5x^3(5x^2 - x + 2)$ **11.** $-x(x^3 + 5x - 4)$ **12.** $-2x^4(x^3 - x^2 + 2x)$

13. $3x(2x + 1) - 2(2x + 1)$ **14.** $x(3x + 4) + 7(3x + 4)$ **15.** $3a(3a - 5) + 5(3a - 5)$

16. $6x(x - 1) + 5(x - 1)$ **17.** $5x(-2x + 7) - 2(-2x + 7)$

18. $y(y^2 + 1) - 1(y^2 + 1)$ **19.** $x(x^2 + 3x + 2) + 2(x^2 + 3x + 2)$

20. $4x(x^2 - x + 1) + 3(x^2 - x + 1)$

21. $(x + 4)(x - 3)$ **22.** $(x + 7)(x - 5)$ **23.** $(a + 6)(a - 8)$

24. $(x + 2)(x - 4)$ **25.** $(x - 2)(x - 1)$ **26.** $(x - 7)(x - 8)$

27. $3(t + 4)(t - 5)$ **28.** $-4(x + 6)(x - 7)$ **29.** $x(x + 3)(x + 8)$

30. $t(t - 4)(t - 7)$ **31.** $(2x + 1)(x - 4)$ **32.** $(3x - 1)(x + 4)$

33. $(6x - 1)(x + 3)$ **34.** $(3t + 5)(3t - 5)$ **35.** $(2x + 3)(2x - 3)$

36. $(8x + 15)(x + 1)$ **37.** $(4x + 1)(4x + 1)$ **38.** $(5x - 2)(5x - 2)$

39. $(y + 3)(y^2 - y + 4)$ **40.** $(2x + 1)(x^2 - 7x + 2)$

In Exercises 41 – 45, multiply as indicated and simplify.

41. $3x + 7$
$\underline{\quad x - 5\quad}$

42. $x^2 + 3x + 1$
$\underline{\quad\quad 5x - 9\quad}$

43. $8x^2 + 3x - 2$
$\underline{\quad\quad -2x + 7\quad}$

44. $2x^2 + 3x + 5$
$\underline{\quad x^2 + 2x - 3\quad}$

45. $6x^2 - x + 8$
$\underline{\quad 2x^2 + 5x + 6\quad}$

Find the product in Exercises 46 – 70 and simplify if possible. Check by letting the variable equal 2.

46. $(3x - 4)(x + 2)$ **47.** $(t + 6)(4t - 7)$ **48.** $(2x + 5)(x - 1)$

49. $(5a - 3)(a + 4)$ **50.** $(2x + 1)(3x - 8)$ **51.** $(x - 2)(3x + 8)$

52. $(7x + 1)(x - 2)$ **53.** $(3x + 7)(2x - 5)$ **54.** $(2x + 3)(2x + 3)$

55. $(5y + 2)(5y + 2)$ **56.** $(x + 3)(x^2 - 4)$ **57.** $(y^2 + 2)(y - 4)$

58. $(2x + 7)(2x - 7)$ **59.** $(3x - 4)(3x + 4)$ **60.** $(x + 1)(x^2 - x + 1)$

61. $(x - 2)(x^2 + 2x + 4)$ **62.** $(7a - 2)(7a - 2)$ **63.** $(5a - 6)(5a - 6)$

64. $(2x + 3)(x^2 - x - 1)$ **65.** $(3x + 1)(x^2 - x + 9)$ **66.** $(x + 1)(x + 2)(x + 3)$

67. $(t - 1)(t - 2)\, t - 3)$ **68.** $(a^2 + a - 1)(a^2 - a + 1)$ **69.** $(y^2 + y + 2)(y^2 + y - 2)$

70. $(t^2 + 3t + 2)^2$

Simplify the expressions in Exercises 71 – 75.

71. $(x-3)(x+5) - (x+3)(x+2)$

72. $(y-2)(y-4) + (y-1)(y+1)$

73. $(2a+1)(a-5) + (a-4)(a-4)$

74. $(2t+3)(2t+3) - (t-2)(t-2)$

75. $(y+6)(y-6) + (y+5)(y-5)$

Hawkes Learning Systems: Introductory Algebra

Multiplying a Polynomial by a Monomial
Multiplying Two Polynomials

Special Products of Binomials

After completing this section, you will be able to:

1. Multiply binomials using the FOIL method.

2. Multiply binomials, finding products that are the difference of squares.

3. Square binomials, finding products that are perfect square trinomials.

4. Identify the difference of two squares and perfect square trinomials.

In Section 6.5, we emphasized the use of the distributive property in finding products of polynomials. This approach is correct and serves to provide a solid understanding of multiplication. However, there is a more efficient way to multiply binomials which is used by most students. After some practice, you can probably find these products mentally.

The FOIL Method

Consider finding the product $(2x + 3)(5x + 2)$ using the distributive property, then investigating the positions of the terms in the product.

$$(2x + 3)(5x + 2) = 2x(5x + 2) + 3(5x + 2)$$
$$= 2x \cdot 5x + 2x \cdot 2 + 3 \cdot 5x + 3 \cdot 2$$

First terms	Outside terms	Inside terms	Last terms
F	O	I	L

This technique, illustrated by the following diagrammed equations, is called the **FOIL method** of mentally multiplying two binomials.

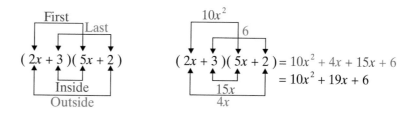

The same layout is shown again to help you get a mental image of the process.

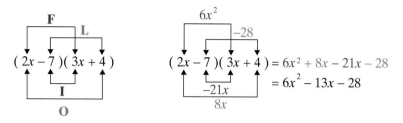

Example 1: FOIL Method

Use the FOIL method to find each product of binomials.

a. $(2x + 1)(4x + 3)$

Solution:

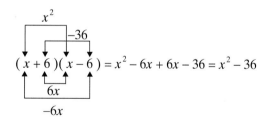

$$(2x + 1)(4x + 3) = 8x^2 + 10x + 3$$

Combining $6x + 4x$.

b. $(x + 6)(x - 6)$

Solution:

$$(x + 6)(x - 6) = x^2 - 6x + 6x - 36 = x^2 - 36$$

The Difference of Two Squares:
$(x + a)(x - a) = x^2 - a^2$

In Example 1b, the middle terms, $-6x$ and $+6x$, are opposites of each other and their sum is 0. Therefore, the resulting product has only two terms.

$$(x + 6)(x - 6) = x^2 - 36$$

The simplified product is in the form of a difference and both terms are squares. Thus, we have the following special case called the **difference of two squares**.

Difference of Two Squares

$$(x + a)(x - a) = x^2 - a^2$$

When the two binomials are in the form of the sum and difference of the same two terms, the product will always be the difference of the squares of the terms. In such a case, we can write the answer directly with no calculations. You should memorize the following squares of the positive integers from 1 to 20. The squares of integers are called **perfect squares**.

Perfect Squares

1, 4, 9, 16, 25, 36, 49, 64, 81, 100, 121, 144, 169, 196, 225, 256, 289, 324, 361, 400

Example 2: Difference of Two Squares ● ● ● ● ● ● ● ● ● ● ● ● ●

Find each product.
a. $(x + 5)(x - 5)$

Solution: The two binomials represent the sum and difference of x and 5. So, the product is the difference of their squares.
$$(x + 5)(x - 5) = x^2 - 5^2 = x^2 - 25$$

b. $(8x + 3)(8x - 3)$

Solution: $(8x + 3)(8x - 3) = (8x)^2 - (3)^2 = 64x^2 - 9$

c. $(x^3 + 2)(x^3 - 2)$

Solution: $(x^3 + 2)(x^3 - 2) = (x^3)^2 - 2^2 = x^6 - 4$

● ●

Perfect Square Trinomials: $(x + a)^2 = x^2 + 2ax + a^2$
$(x - a)^2 = x^2 - 2ax + a^2$

We now want to consider the case where the two binomials being multiplied are the same. That is, we want to consider the **square of a binomial**. As the following discussion shows, there is a pattern that, after some practice, allows us to go directly to the product,

$$
\begin{aligned}
(x + 5)^2 = (x + 5)(x + 5) &= x^2 + 5x + 5x + 25 \\
&= x^2 + 2 \cdot 5x + 25 \\
&= x^2 + 10x + 25
\end{aligned}
$$

$$
\begin{aligned}
(x + 7)^2 = (x + 7)(x + 7) &= x^2 + 2 \cdot 7x + 49 \\
&= x^2 + 14x + 49
\end{aligned}
$$

$$
(x + 10)^2 = (x + 10)(x + 10) = x^2 + 20x + 100
$$

and the basic pattern is,

$$
(x + a)^2 = (x + a)(x + a) = x^2 + 2ax + a^2
$$

The simplified product $x^2 + 2ax + a^2$ is called a **perfect square trinomial** because it is the result of squaring a binomial.

One interesting device for remembering the result of squaring a binomial is the square shown in Figure 6.2 where the total area is the sum of the shaded areas.

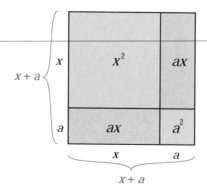

For the area of the square:

$$
\begin{aligned}
(x + a)^2 &= x^2 + ax + ax + a^2 \\
&= x^2 + 2ax + a^2
\end{aligned}
$$

Figure 6.2

Another perfect square trinomial results if an expression of the form $(x - a)$ is squared. The sign between x and a is – instead of +.

$$(x - 5)^2 = (x - 5)(x - 5) = x^2 - 5x - 5x + 25$$
$$= x^2 - 2 \cdot 5x + 25$$
$$= x^2 - 10x + 25$$

$$(x - 7)^2 = (x - 7)(x - 7) = x^2 - 2 \cdot 7x + 49 = x^2 - 14x + 49$$

$$(x - 10)^2 = (x - 10)(x - 10) = x^2 - 20x + 100$$

In general,

$$(x - a)^2 = (x - a)(x - a) = x^2 - 2ax + a^2.$$

Example 3: Perfect Square Trinomials ● ● ● ● ● ● ● ● ● ● ● ●

Find the following products.
a. $(2x + 3)^2$

> **Solution:** The pattern for squaring a binomial gives
> $$(2x + 3)^2 = (2x)^2 + 2 \cdot 3 \cdot 2x + (3)^2$$
> $$= 4x^2 + 12x + 9$$

b. $(5x - 1)^2$

> **Solution:** $(5x - 1)^2 = (5x)^2 - 2(1)(5x) + (1)^2$
> $$= 25x^2 - 10x + 1$$

c. $(9 - x)^2$

> **Solution:** $(9 - x)^2 = (9)^2 - 2(9)(x) + x^2$
> $$= 81 - 18x + x^2$$

d. $(y^3 + 1)^2$

> **Solution:** $(y^3 + 1)^2 = (y^3)^2 + 2(1)(y^3) + 1^2$
> $$= y^6 + 2y^3 + 1$$

● ●

NOTES

Many beginning algebra students make the following error:

$$(x + a)^2 = x^2 + a^2 \qquad \text{WRONG}$$
$$(x + 6)^2 = x^2 + 36 \qquad \text{WRONG}$$

Avoid this error by remembering that **the square of a binomial is a trinomial.**

$$(x + 6)^2 = x^2 + 2 \cdot 6x + 36$$
$$= x^2 + 12x + 36 \qquad \text{RIGHT}$$

Practice Problems

Find the indicated products.

1. $(x + 10)(x - 10)$

2. $(x + 3)^2$

3. $(2x - 1)(x + 3)$

4. $(2x - 5)^2$

5. $(x^2 + 4)(x^2 - 3)$

6.6 Exercises

Find the product in Exercises 1 – 30, and identify those that are the difference of two squares or perfect square trinomials.

1. $(x + 3)(x - 3)$ **2.** $(x - 7)^2$ **3.** $(x - 5)^2$

4. $(x + 4)(x + 4)$ **5.** $(x - 6)(x + 6)$ **6.** $(x + 9)(x - 9)$

7. $(x + 8)(x + 8)$ **8.** $(x + 12)(x - 12)$ **9.** $(2x + 3)(x - 1)$

10. $(3x + 1)(2x + 5)$ **11.** $(3x - 4)^2$ **12.** $(5x + 2)(5x - 2)$

13. $(2x + 1)(2x - 1)$ **14.** $(3x + 1)^2$ **15.** $(3x - 2)(3x - 2)$

16. $(4x + 5)(4x - 5)$ **17.** $(3 + x)^2$ **18.** $(8 - x)(8 - x)$

19. $(5 - x)(5 - x)$ **20.** $(11 - x)(11 + x)$ **21.** $(5x - 9)(5x + 9)$

22. $(4 - x)^2$ **23.** $(2x + 7)(2x + 7)$ **24.** $(3x + 2)^2$

25. $(9x + 2)(9x - 2)$ **26.** $(6x + 5)(6x - 5)$ **27.** $(5x^2 + 2)(2x^2 - 3)$

28. $(4x^2 + 7)(2x^2 + 1)$ **29.** $(1 + 7x)^2$ **30.** $(2 - 5x)^2$

Answers to Practice Problems: **1.** $x^2 - 100$ **2.** $x^2 + 6x + 9$ **3.** $2x^2 + 5x - 3$ **4.** $4x^2 - 20x + 25$
5. $x^4 + x^2 - 12$

Write the indicated products in Exercises 31 – 58.

31. $(x + 2)(5x + 1)$ **32.** $(7x - 2)(x - 3)$ **33.** $(4x - 3)(x + 4)$

34. $(x + 11)(x - 8)$ **35.** $(3x - 7)(x - 6)$ **36.** $(x + 7)(2x + 9)$

37. $(5 + x)(5 + x)$ **38.** $(3 - x)(6 - x)$ **39.** $(x^2 + 1)(x^2 - 1)$

40. $(x^2 + 5)(x^2 - 5)$ **41.** $(x^2 + 3)(x^2 + 3)$ **42.** $(x^2 - 4)^2$

43. $(x^3 - 2)^2$ **44.** $(x^3 + 8)(x^3 - 8)$ **45.** $(x^2 - 6)(x^2 + 9)$

46. $(x^2 + 3)(x^2 - 5)$ **47.** $\left(x + \dfrac{2}{3}\right)\left(x - \dfrac{2}{3}\right)$ **48.** $\left(x - \dfrac{1}{2}\right)\left(x + \dfrac{1}{2}\right)$

49. $\left(x + \dfrac{3}{4}\right)\left(x - \dfrac{3}{4}\right)$ **50.** $\left(x + \dfrac{3}{8}\right)\left(x - \dfrac{3}{8}\right)$ **51.** $\left(x + \dfrac{3}{5}\right)\left(x + \dfrac{3}{5}\right)$

52. $\left(x + \dfrac{4}{3}\right)\left(x + \dfrac{4}{3}\right)$ **53.** $\left(x - \dfrac{5}{6}\right)^2$ **54.** $\left(x - \dfrac{2}{7}\right)^2$

55. $\left(x + \dfrac{1}{4}\right)\left(x - \dfrac{1}{2}\right)$ **56.** $\left(x - \dfrac{1}{5}\right)\left(x + \dfrac{2}{3}\right)$ **57.** $\left(x + \dfrac{1}{3}\right)\left(x + \dfrac{1}{2}\right)$

58. $\left(x - \dfrac{4}{5}\right)\left(x - \dfrac{3}{10}\right)$

Calculator Problems

Write the indicated products in Exercises 59 – 71.

59. $(x + 1.4)(x - 1.4)$ **60.** $(x - 2.1)(x + 2.1)$

61. $(x - 2.5)^2$ **62.** $(x + 1.7)^2$

63. $(x + 2.15)(x - 2.15)$ **64.** $(x + 1.36)(x - 1.36)$

65. $(x + 1.24)^2$ **66.** $(x - 1.45)^2$

67. $(1.42x + 9.6)^2$ **68.** $(0.46x - 0.71)^2$

69. $(11.4x + 3.5)(11.4x - 3.5)$ **70.** $(2.5x + 11.4)(1.3x - 16.9)$

71. $(12.6x - 6.8)(7.4x + 15.3)$

72. A square is 20 inches on each side. A square x inches on each side is cut from each corner of the square.

a. Represent the area of the remaining portion of the square in the form of a polynomial function $A(x)$.

b. Represent the perimeter of the remaining portion of the square in the form of a polynomial function $P(x)$.

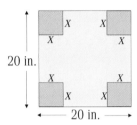

20 in.

20 in.

73. In the case of binomial probabilities, if x is the probability of success in one trial of an event, then the expression $f(x) = 15x^4(1-x)^2$ is the probability of 4 successes in 6 trials where $0 \le x \le 1$.

a. Represent the expression $f(x)$ as a single polynomial.

b. If a fair coin is tossed, the probability of heads occurring is $\frac{1}{2}$. That is, $x = \frac{1}{2}$. Find the probability of 4 heads occurring in 6 tosses.

74. A rectangle has sides $(x+3)$ ft. and $(x+5)$ ft. If a square x feet on a side is cut from the rectangle, represent the remaining area (light shaded area in the figure shown) in the form of a polynomial function $A(x)$.

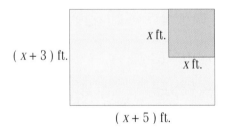

$(x+3)$ ft.

x ft.

x ft.

$(x+5)$ ft.

75. A pool, 20 meters by 50 meters, is surrounded by a concrete deck that is x meters wide.

   a. Represent the area covered by the deck and the pool in the form of a polynomial function.

   b. Represent the area covered by the deck only in the form of a polynomial function.

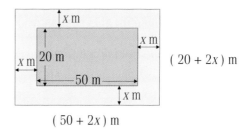

($50 + 2x$) m

76. A rectangular piece of cardboard that is 10 inches by 15 inches has squares of length x inches on a side cut from each corner. (Assume that $0 < x < 5$.)

   a. Represent the remaining area in the form of a polynomial function $A(x)$.

   b. Represent the perimeter of the remaining figure in the form of a polynomial function $P(x)$.

   c. If the flaps of the figure are folded up, an open box is formed. Represent the volume of this box in the form of a polynomial function $V(x)$.

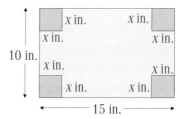

Hawkes Learning Systems: Introductory Algebra

The FOIL Method
Special Products

6.7   Dividing Polynomials

After completing this section, you will be able to:

1. Divide polynomials by monomials.

2. Divide polynomials by other polynomials using long division.

Fractions, such as $\dfrac{135}{8}$ and $\dfrac{7}{8}$, in which the numerator and denominator are integers are called rational numbers. Fractions in which the numerator and denominator are polynomials are called **rational expressions**. (No denominator can be 0).

$$\text{Rational numbers: } \frac{2}{3},\ \frac{-5}{16},\ \frac{22}{7},\ \frac{0}{6},\ \frac{3}{4},\ -\frac{3}{4}.$$

$$\text{Rational expressions: } \frac{x}{x^2+1},\ \frac{x^2+5x+6}{x^2+7x+12},\ \frac{x^2+2x+1}{x}\ \text{and}\ \frac{1}{5x}.$$

In this section, we want to treat a rational expression as an indicated division problem. From this basis, there are two situations to consider:

1. the denominator (divisor) is a monomial, or
2. the denominator (divisor) is not a monomial.

Division by a Monomial

We know from arithmetic that the sum of fractions with the same denominator can be written as a single fraction by adding the numerators and using the common denominator. For example,

$$\frac{3}{a}+\frac{2b}{a}+\frac{5c}{a}=\frac{3+2b+5c}{a}$$

If, instead of adding the fractions, we want to divide the numerator by the denominator (with a monomial in the denominator), we divide each term in the numerator by the monomial denominator and simplify each fraction.

$$\frac{3x^3+6x^2+9x}{3x}=\frac{3x^3}{3x}+\frac{6x^2}{3x}+\frac{9x}{3x}=x^2+2x+3$$

Similarly,

$$\frac{x^2 + 2x + 1}{x} = \frac{x^2}{x} + \frac{2x}{x} + \frac{1}{x} = x + 2 + \frac{1}{x}$$

and

$$\frac{3xy + 6xy^2 - 18}{2y} = \frac{3xy}{2y} + \frac{6xy^2}{2y} - \frac{18}{2y} = \frac{3x}{2} + 3xy - \frac{9}{y}$$

Example 1: Divide by Monomial • • • • • • • • • • • • • • • •

Divide each polynomial by the monomial denominator by writing a sum of fractions. Reduce each fraction, if possible.

a. $\dfrac{8x^2 - 14x + 1}{2}$

Solution: $\dfrac{8x^2 - 14x + 1}{2} = \dfrac{8x^2}{2} - \dfrac{14x}{2} + \dfrac{1}{2} = 4x^2 - 7x + \dfrac{1}{2}$

b. $\dfrac{12x^3 + 3x^2 - 9x}{3x^2}$

Solution: $\dfrac{12x^3 + 3x^2 - 9x}{3x^2} = \dfrac{12x^3}{3x^2} + \dfrac{3x^2}{3x^2} - \dfrac{9x}{3x^2}$

$$= 4x + 1 - \frac{3}{x}$$

c. $\dfrac{10x^2y + 25xy + 3y^2}{5xy^2}$

Solution: $\dfrac{10x^2y + 25xy + 3y^2}{5xy^2} = \dfrac{10x^2y}{5xy^2} + \dfrac{25xy}{5xy^2} + \dfrac{3y^2}{5xy^2}$

$$= \frac{2x}{y} + \frac{5}{y} + \frac{3}{5x}$$

• •

The Division Algorithm

In arithmetic, the process (or series of steps) that we follow in dividing two numbers is called the **division algorithm** (or **long division**). By this division algorithm, we can find $135 \div 8$ as follows:

$$
\begin{array}{r}
16 \\
8\overline{)135} \\
\underline{8} \\
55 \\
\underline{48} \\
7
\end{array}
$$

Divisor $\longrightarrow$ $8\overline{)135}$

16 $\longleftarrow$ Quotient
Dividend
8 $\longleftarrow$ Subtract
55
48 $\longleftarrow$ Subtract
7 $\longleftarrow$ Remainder (The remainder is always smaller than the divisor.)

Check: $8 \cdot 16 + 7 = 128 + 7 = 135$ (Multiply the divisor times the quotient and add the remainder. The result should be the original dividend.)

We can also write the division in fraction form and the remainder over the divisor, giving a mixed number.

$$135 \div 8 = \frac{135}{8} = 16\frac{7}{8}$$

In algebra, the division algorithm with polynomials is quite similar. If we divide one polynomial by another, the quotient will be another polynomial with a remainder. Symbolically, if we have $P \div D$, then

$$P = Q \cdot D + R$$

or

$$\frac{P}{D} = Q + \frac{R}{D}$$

where P is the dividend, D is the divisor, Q is the quotient, and R is the remainder. **The remainder must be of smaller degree than the divisor. If the remainder is 0, then the divisor and quotient are factors of the dividend**.

In the following two examples, the **division algorithm** is illustrated in a step-by-step form with first-degree divisors. Each step is explained in detail and should be studied carefully.

Example 2: The Division Algorithm • • • • • • • • • • • • •

a. $\dfrac{3x^2 - 5x + 2}{x - 4}$ or $\left(3x^2 - 5x + 2\right) \div \left(x - 4\right)$

Solution:

	Steps	**Explanation**

Step 1: $x - 4 \overline{\smash{)}3x^2 - 5x + 2}$

Write both polynomials in order of descending powers. If any powers are missing, fill in with 0's.

Step 2:
$$\begin{array}{r} 3x \\ x-4\overline{\smash{)}\;3x^2 -\; 5x+2} \end{array}$$

Mentally divide $3x^2$ by x:
$\dfrac{3x^2}{x} = 3x$. Write $3x$ above $3x^2$.

Step 3:
$$\begin{array}{r} 3x \\ x-4\overline{\smash{)}\;3x^2 -\; 5x+2} \\ -(3x^2 - 12x) \end{array}$$

Multiply $3x$ times $(x - 4)$ and write the terms under the like terms in the dividend. Use a '−' sign to indicate that the product is to be subtracted.

Step 4:
$$\begin{array}{r} 3x \\ x-4\overline{\smash{)}\;3x^2 -\; 5x+2} \\ -3x^2 + 12x \\ \hline + 7x \end{array}$$

Subtract $3x^2 - 12x$ by changing signs and adding.

Step 5:
$$\begin{array}{r} 3x \\ x-4\overline{\smash{)}\;3x^2 -\; 5x + 2} \\ -3x^2 + 12x \\ \hline + 7x + 2 \end{array}$$

Bring down the +2.

Step 6:
$$\begin{array}{r} 3x + 7 \\ x-4\overline{\smash{)}\;3x^2 -\; 5x+2} \\ -3x^2 + 12x \\ \hline + 7x + 2 \end{array}$$

Mentally divide $+7x$ by x:
$\dfrac{+7x}{x} = +7$.

Write +7 in the quotient.

Continued on next page ...

411

Step 7:

$$\begin{array}{r} 3x + 7 \\ x-4{\overline{\smash{\big)}\,3x^2 - 5x+2}} \\ \underline{-3x^2 +12x} \\ + 7x+ 2 \\ \underline{-(+7x-28)} \end{array}$$

Multiply +7 times ($x - 4$) and write the terms under the like terms in the expression $+7x + 2$. Use a '−' sign to indicate that the product is to be subtracted.

Step 8:

$$\begin{array}{r} 3x + 7 \\ x-4{\overline{\smash{\big)}\,3x^2 - 5x+2}} \\ \underline{-3x^2 +12x} \\ + 7x+ 2 \\ \underline{-7x+28} \\ 30 \end{array}$$

Subtract $7x - 28$ by changing signs and adding.

Step 9:   **Check:** Multiply the divisor and quotient, and add the remainder. The result should be the dividend.

$$Q \cdot D + R = (3x + 7)(x - 4) + 30$$
$$= 3x^2 - 12x + 7x - 28 + 30$$
$$= 3x^2 - 5x + 2$$

Thus, the quotient is $3x + 7$ and the remainder is 30.

The answer can also be written in the form $Q+\dfrac{R}{D}$ as $3x+7+\dfrac{30}{x-4}$.

b. $\dfrac{4x^3 - 7x - 5}{2x+1}$ or $\left(4x^3 - 7x - 5 \right) \div \left(2x+1 \right)$

This example will be done in fewer steps. Note that $0x^2$ is inserted so that like terms will be aligned.

Solution:

Steps	Explanation

Step 1: $2x+1{\overline{\smash{\big)}\,4x^3 +0x^2 -7x-5}}$ x^2 is a missing power, so $0x^2$ is supplied.

Continued on next page ...

Step 2:

$$\begin{array}{r} 2x^2 \\ 2x+1\overline{)\ 4x^3 +0x^2 -7x-5} \\ \underline{-(4x^3 +2x^2)} \end{array}$$

$\dfrac{4x^3}{2x} = 2x^2$ and $2x^2\,(\,2x\,+\,1\,) =$ $4x^3 + 2x^2.$

Step 3:

$$\begin{array}{r} 2x^2 \\ 2x+1\overline{)4x^3 +0x^2 -7x-5} \\ \underline{-4x^3 -2x^2} \\ -2x^2 -7x \end{array}$$

Subtract $4x^3 + 2x^2$ and bring down $-7x.$

Step 4:

$$\begin{array}{r} 2x^2 -x \\ 2x+1\overline{)4x^3 +0x^2 -7x-5} \\ \underline{-4x^3 -2x^2} \\ -2x^2 -7x \\ \underline{-(-2x^2 -\ x)} \end{array}$$

$-\dfrac{2x^2}{2x} = -x.$ Multiply $-x$ times $(\,2x+1\,).$

Step 5:

$$\begin{array}{r} 2x^2\ -x\ \ -3 \\ 2x+1\overline{)4x^3 +0x^2 -7x-5} \\ \underline{-4x^3 -2x^2} \\ -2x^2\ -7x \\ \underline{+2x^2\ +\ x} \\ -6x\ -5 \\ \underline{+6x\ +3} \\ -2 \end{array}$$

Continue dividing, using the same procedure. The remainder is $-2.$

Step 6: **Check:** $(2x+1)\left(2x^2 -x-3\right)-2 = 4x^3 -2x^2 -6x+2x^2 -x-3-2$

$$= 4x^3 -7x-5$$

The answer can also be written in the form

$$2x^2 -x-3+\frac{-2}{2x+1}\ \text{ or }\ 2x^2 -x-3-\frac{2}{2x+1}$$

● ●

The division algorithm can be used to divide two polynomials whenever the degree of the dividend, P (the numerator), is greater than or equal to the degree of the divisor, D (the denominator). The remainder, R, must be of smaller degree than the divisor, D.

Example 3: Long Division

Use the division algorithm (or long division) to divide:

$$\left(x^3 - 3x^2 - 5x - 8\right) \div \left(x^2 + 2x + 3\right)$$

Solution:

$$
\begin{array}{r}
x - 5 \\
x^2 + 2x + 3 \overline{\smash{)}\, x^3 - 3x^2 - 5x - 8} \quad \text{Dividend} \\
\underline{-\left(x^3 + 2x^2 + 3x\right)} \\
-5x^2 - 8x - 8 \\
\underline{-\left(-5x^2 - 10x - 15\right)} \\
2x + 7 \quad \text{Remainder is first-degree, which is}
\end{array}
$$

smaller degree than the divisor.

$$Q + \frac{R}{D} = x - 5 + \frac{2x + 7}{x^2 + 2x + 3}$$

Check: $\left(x^2 + 2x + 3\right)\left(x - 5\right) + \left(2x + 7\right) = x^3 + 2x^2 + 3x - 5x^2 - 10x - 15 + 2x + 7$

$$= x^3 - 3x^2 - 5x - 8$$

Practice Problems

1. *Express the quotient as a sum of fractions:* $\dfrac{4x^2 + 6x + 1}{2x}$.

Use the division algorithm to divide.

2. $(3x^2 + 8x - 4) \div (x + 2)$

3. $(x^3 + 4x^2 - 5) \div (x^2 + x - 1)$

Answers to Practice Problems: **1.** $2x + 3 + \dfrac{1}{2x}$ **2.** $3x + 2 - \dfrac{8}{x + 2}$ **3.** $x + 3 + \dfrac{-2x - 2}{x^2 + x - 1}$

6.7 Exercises

Express each quotient in Exercises 1 – 20 as a sum of fractions and simplify if possible.

1. $\dfrac{4x^2 + 8x + 3}{4}$ **2.** $\dfrac{6x^2 - 10x + 1}{2}$ **3.** $\dfrac{10x^2 - 15x - 3}{5}$ **4.** $\dfrac{9x^2 - 12x + 5}{3}$

5. $\dfrac{2x^2 + 5x}{x}$ **6.** $\dfrac{8x^2 - 7x}{x}$ **7.** $\dfrac{x^2 + 6x - 3}{x}$ **8.** $\dfrac{-2x^2 - 3x + 8}{x}$

9. $\dfrac{4x^2 + 6x - 3}{2x}$ **10.** $\dfrac{3x^3 - 2x^2 + x}{x^2}$ **11.** $\dfrac{6x^3 - 9x^2 - 3x}{3x^2}$

12. $\dfrac{5x^2y - 10xy^2 - 3y}{5xy}$ **13.** $\dfrac{7x^2y^2 + 21xy^3 - 11y^3}{7xy^2}$ **14.** $\dfrac{12x^3y + 6x^2y^2 - 3xy}{6x^2y}$

15. $\dfrac{3x^3y - 8xy^2 - 4y^3}{4xy}$ **16.** $\dfrac{2x^3y^2 - 6x^2y^3 + 15xy}{3xy^2}$ **17.** $\dfrac{5x^2y^2 - 8xy^2 + 16xy}{8xy^2}$

18. $\dfrac{3x^3y - 14x^2y - 7xy}{7x^2y}$ **19.** $\dfrac{8x^3y^2 - 9x^2y^3 + 5xy}{9xy^2}$ **20.** $\dfrac{24x^2y^2 + 12xy - 6xy}{3xy}$

Divide in Exercises 21 – 55 by using the long division procedure. Write your answers in the form $Q + \dfrac{R}{D}$. Check each answer by showing that $P = Q \cdot D + R$.

21. $278 \div 23$ **22.** $326 \div 64$

23. $437 \div 59$ **24.** $(x^2 + 3x + 2) \div (x + 2)$

25. $(x^2 + x - 6) \div (x + 3)$ **26.** $(y^2 + 8y + 15) \div (y + 4)$

27. $(a^2 - 2a - 15) \div (a - 2)$ **28.** $(x^2 - 7x - 18) \div (x + 5)$

29. $(y^2 - y - 42) \div (y + 4)$ **30.** $(4a^2 - 21a + 2) \div (a - 6)$

31. $(5y^2 + 14y - 7) \div (y + 5)$ **32.** $(8x^2 + 10x - 4) \div (2x + 3)$

33. $(8c^2 + 2c - 14) \div (2c + 3)$ **34.** $(6x^2 + x - 4) \div (2x - 1)$

35. $(10m^2 - m - 6) \div (5m - 3)$ **36.** $(x^2 - 6) \div (x + 2)$

37. $(x^2 + 3x) \div (x + 5)$ **38.** $(2x^3 + 4x^2 - x + 1) \div (x - 3)$

39. $(y^3 - 9y^2 + 26y - 24) \div (y - 2)$ **40.** $(3t^3 + 10t^2 + 6t + 3) \div (3t + 1)$

41. $(12a^3 - 3a^2 + 4a + 1) \div (4a - 1)$ **42.** $(2x^3 + x^2 - 6) \div (x + 4)$

43. $(x^3 + 2x^2 - 5) \div (x - 5)$ **44.** $(x^3 - 8) \div (x - 2)$

45. $(x^3 + 27) \div (x + 3)$ **46.** $(x^3 + 7x^2 + x - 2) \div (x^2 - x + 1)$

47. $(2x^3 - x + 3) \div (x^2 - 2)$ **48.** $(x^3 + 3x^2 + 1) \div (x^2 + 2x + 3)$

49. $\dfrac{8x^3 - 27}{2x - 3}$ **50.** $\dfrac{27a^3 - 64}{3a - 4}$ **51.** $\dfrac{4x^2 - x + 5}{x - 1}$ **52.** $\dfrac{5x^3 - 4x^2 + 81}{x + 3}$

53. $\dfrac{3x^3 - 10x^2 - 3x - 20}{x - 4}$ **54.** $\dfrac{x^4 + 2x^2 - 5}{x^2 + 1}$ **55.** $\dfrac{2a^3 + 3a^2 + 6}{a^2 + 2}$

Hawkes Learning Systems: Introductory Algebra

Division by a Monomial
The Division Algorithm

Chapter 6 Index of Key Ideas and Terms

Summary of the Rules for Exponents — page 367, 373
   1. The Exponent 1: $a = a^1$ — page 359
   2. The Exponent 0: $a^0 = 1 \ (a \neq 0)$ — pages 359 - 360
   3. The Product Rule: $a^m \cdot a^n = a^{m+n}$ — pages 356 - 357
   4. The Quotient Rule: $\dfrac{a^m}{a^n} = a^{m-n}$ — pages 360 - 361
   5. Negative Exponents: $a^{-n} = \dfrac{1}{a^n}$ — pages 362
   6. Power Rule: $(a^m)^n = a^{mn}$ — pages 367 - 368
   7. Power of a Product: $(ab)^n = a^n b^n$ — page 368 - 369
   8. Power of a Quotient: $\left(\dfrac{a}{b}\right)^n = \dfrac{a^n}{b^n}$ — page 370

Scientific Notation — pages 373 - 374

In **scientific notation**, all decimal numbers are written as the product of a number between 1 and 10 and some power of 10. Most calculators use some form of scientific notation.

Monomial — page 380

A **monomial in x** is a term of the form kx^n where k is a real number and n is a whole number. n is called the **degree** of the term, and k is called the **coefficient**.

Polynomial — page 381

A **polynomial in x** is an expression of the form
$$a_n x^n + a_{n-1}x^{n-1} + \ldots + a_1 x + a_0$$
where $n, n-1, n-2, \ldots, 1, 0$ are whole numbers and $a_n, a_{n-1}, \ldots, a_1, a_0$ are real numbers with $a_n \neq 0$.
n is called the **degree** of the polynomial.
a_n is called the **leading coefficient** of the polynomial.

Classification of Polynomials — page 382
   Monomial: a polynomial with one term
   Binomial: a polynomial with two terms
   Trinomial: a polynomial with three terms

Chapter 6 Review

For a review of the topics and problems from Chapter 6, look at the following lessons from *Hawkes Learning Systems: Introductory Algebra*

Simplifying Integer Exponents I
Simplifying Integer Exponents II
Identifying Polynomials
Evaluating Polynomials
Adding and Subtracting Polynomials
Multiplying a Polynomial By a Monomial
Multiplying Two Polynomials
The FOIL Method
Special Products
Division by a Monomial
The Division Algorithm

Chapter 6 Test

Use the rules for exponents to simplify each expression in Exercises 1 – 6. Each answer should have only positive exponents.

1. $(5a^2b^5)(-2a^3b^{-5})$

2. $(-7x^4y^{-3})^0$

3. $\dfrac{\left(-8x^2y^{-3} \right)^2}{16xy}$

4. $\dfrac{3x^{-2}}{9x^{-3}y^2}$

5. $\left(\dfrac{4xy^2}{x^3} \right)^{-1}$

6. $\left(\dfrac{2x^0y^3}{x^{-1}y} \right)^2$

7. Write each of the following numbers in decimal notation.
   a. 1.35×10^5
   b. 2.7×10^{-6}

8. Perform the indicated operations by writing each number in scientific notation and write the result in scientific notation.

   a. $250 \cdot 500,000$
   b. $\dfrac{65 \cdot 0.012}{1500}$

In Exercises 9 – 11, simplify the polynomials and state the degree and type of the polynomial.

9. $3x + 4x^2 - x^3 + 4x^2 + x^3$

10. $2x^2 + 3x - x^3 + x^2 - 1$

11. $17 - 14 + 4x^5 - 3x + 2x^4 + x^5 - 8x$

12. For the polynomial $p(x) = 2x^3 - 5x^2 + 7x + 10$, find (a) $p(2)$ and (b) $p(-3)$.

Simplify each expression in Exercises 13 and 14.

13. $7x + [2x - 3(4x + 1) + 5]$

14. $12x - 2[5 - (7x + 1) + 3x]$

In Exercises 15 – 26, perform the indicated operations and simplify each expression. Tell which, if any, answers are the difference of two squares and which, if any, are perfect square trinomials.

15. $(5x^3 - 2x + 7) + (-x^2 + 8x - 2)$

16. $(4x^2 + 3x - 1) - (6x^2 + 2x + 5)$

17. $(x^4 + 3x^2 + 9) - (-6x^4 - 11x^2 + 5)$

18. $(3x^3 - 2x^2) + (4x^3 - 7x + 1)$

19. $5x^2(3x^5 - 4x^4 + 3x^3 - 8x^2 - 2)$

20. $(7x + 3)(7x - 3)$

21. $(4x + 1)^2$

22. $(6x - 5)^2$

23. $(2x+5)(6x-3)$

24. $3x(x-7)(2x-9)$

25. $(3x+1)(3x-1)-(2x+3)(x-5)$

26.
$$\begin{array}{r} 2x^3-3x-7 \\ \times\; 5x+2 \\ \hline \end{array}$$

In Exercises 27 and 28, express each quotient as a sum of fractions and simplify.

27. $\dfrac{4x^3+3x^2-6x}{2x^2}$

28. $\dfrac{5a^2b+6a^2b^2+3ab^3}{3a^2b}$

In Exercises 29 and 30, divide by using the division algorithm and write the answers in the form $Q+\dfrac{R}{D}$.

29. $(2x^2-9x-20)\div(2x+3)$

30. $\dfrac{x^3-8x^2+3x+15}{x^2+x-3}$

Cumulative Review: Chapters 1 – 6

Use the distributive property to complete the expression in Exercises 1 and 2.

1. $3x + 45 = 3(\quad)$

2. $6x + 16 = 2(\quad)$

Find the LCM for each set of numbers or terms in Exercises 3 and 4.

3. $12, 15, 54$

4. $6a^2, 24ab^3, 30ab, 40a^2b^2$

For each pair of numbers in Exercises 5 – 8, find two factors of the first number whose sum is the second number.

5. $32, 18$

6. $36, -13$

7. $-56, 10$

8. $-48, -8$

In Exercises 9 – 12, simplify each polynomial and state its degree and type. Write the values for each of the coefficients $a_n, a_{n-1}, \ldots, a_1, a_0$.

9. $6x^2 + 10x - (x^2 - 6x)$

10. $4(2x - 3) - 3(x + 2)$

11. $4x - [(6x + 7) - (3x + 4)] - 6$

12. $3x^2 - 8(x - 5) + x^4 - 2x^3 + x^2 - 2x$

Solve the equations in Exercises 13 – 16.

13. $7(4 - x) = 3(x + 4)$

14. $-2(5x + 1) + 2x = 4(x + 1)$

15. $\dfrac{2}{3}x + \dfrac{1}{2} = \dfrac{3}{4}$

16. $1.5x - 3.7 = 3.6x + 2.6$

Solve each formula in Exercises 17 and 18 for the indicated variable.

17. $C = \pi d$; solve for d.

18. $3x + 5y = 10$; solve for y.

In Exercises 19 and 20, solve the inequality and graph the solution set on a real number line.

19. $5x + 3 \geq 2x - 15$

20. $-16 < 3x + 5 < 17$

21. Write an equation for the line that has slope $m = \dfrac{2}{3}$ and contains the point $(-1, 2)$. Graph the line.

22. Write an equation for the line passing through the two points $(-5, 1)$ and $(3, -4)$. Graph the line.

23. Write an equation for the line that is parallel to the line $2x + 5y = 3$ and passes through the point $(0, 7)$. Graph both lines.

24. For the function $f(x) = 3x^2 - 4x + 2$, find (a) $f(-2)$ and (b) $f(5)$.

Use a graphing calculator to graph and solve the systems of equations in Exercises 25 and 26. You may need to estimate the answers.

25. $\begin{cases} 2x + 3y = 18 \\ 3x - 2y = -12 \end{cases}$

26. $\begin{cases} y = \dfrac{1}{3}x + 5 \\ y = -2x + 1 \end{cases}$

Solve each system of equations in Exercises 27 and 28 by using an algebraic method.

27. $\begin{cases} 3x - 4y = -25 \\ 2x + y = 9 \end{cases}$

28. $\begin{cases} x + 8y = -22 \\ 3x - y = -9 \end{cases}$

Simplify each expression in Exercises 29 – 36 so that it has no exponents or only positive exponents.

29. $\dfrac{4x^3}{2x^{-2}x^4}$

30. $(4x^2y)^3$

31. $(7x^5y^{-2})^2$

32. $\left(\dfrac{6x^2}{y^5}\right)^2$

33. $(a^{-3}b^2)^{-2}$

34. $\left(\dfrac{3xy^4}{x^2}\right)^{-1}$

35. $\left(\dfrac{8x^{-3}y^2}{xy^{-1}}\right)^0$

36. $\left(\dfrac{3^{-1}x^3y^{-1}}{x^{-1}y^2}\right)^2$

37. Write each number in decimal notation.

    a. 2.8×10^{-7}

    b. 3.51×10^4

Perform the indicated operations in Exercises 38 – 40 by writing each number in scientific notation and write the answer in scientific notation.

38. $0.0015 \cdot 4200$

39. $\dfrac{840}{0.00021}$

40. $\dfrac{0.005 \cdot 77}{0.011 \cdot 3500}$

In Exercises 41 – 44, (a) simplify each polynomial, (b) state its degree and type, (c) find $p(3)$ and (d) find $p(-5)$.

41. $p(x) = -x^2 + 5x + 2x^2 - x$

42. $p(x) = 9x - x^3 + 3x^2 - x + x^3$

43. $p(x) = -x^4 + 4x^3 - 3x + x^2 - x^3 + x$

44. $p(x) = 8x - 7x^2 + x^2 - x^3 - 6x - 4$

Perform the indicated operations in Exercises 45 – 65 and simplify each expression.

45. $(-2x^2 - 11x + 1) + (x^3 + 3x - 7) + (2x - 1)$
46. $(4x^2 + 2x - 7) - (5x^2 + x - 2)$
47. $(x^3 + 4x^2 - x) - (-2x^3 + 6x + 3)$
48. $(6x^2 + x - 10) - (x^3 - x^2 + x - 4)$
49. $(x^2 + 2x + 6) + (5x^2 - x - 2) - (8x + 3)$
50. $(2x^2 - 5x - 7) - (3x^2 - 4x + 1) + (x^2 - 9)$

51. $-3x(x^2 - 4x + 1)$ **52.** $5x^3(x^2 + 2x)$ **53.** $(x + 6)(x - 6)$
54. $(x + 4)(x - 3)$ **55.** $(3x + 7)(3x + 7)$ **56.** $(2x - 1)(2x + 1)$
57. $(x^2 + 5)(x^2 - 5)$ **58.** $(x^2 - 2)(x^2 - 2)$ **59.** $-1(x^2 - 5x + 2)$
60. $(x - 4)^2$ **61.** $(2x - 9)(x + 4)$ **62.** $(x - 6)(5x + 3)$
63. $(3x - 4)(2x + 3)$ **64.** $(3x + 8)(3x - 8)$ **65.** $(4x + 1)(x^2 - x)$

In Exercises 66 – 68, express each quotient as a sum of fractions and simplify if possible.

66. $\dfrac{8x^2 - 14x + 6}{2x}$

67. $\dfrac{13x^3 y + 10x^2 y^2 + 5xy^3}{5x^2 y}$

68. $\dfrac{x^2 y^2 - 21x^2 y^3 + 7xy - 28x^2 y^4}{7x^2 y^2}$

In Exercises 69 – 71, divide by using the division algorithm and express each answer in the form $Q + \dfrac{R}{D}$.

69. $\dfrac{x^2 + 7x - 18}{x + 9}$

70. $\dfrac{4x^2 + 5x - 2}{4x - 3}$

71. $\dfrac{2x^3 + 5x^2 + 7}{x + 3}$

72. Find three consecutive even integers such that the sum of the second and third is equal to three times the first decreased by 14.

73. The length of a rectangle is 9 centimeters more than its width. The perimeter is 82 centimeters. Find the dimensions of the rectangle.

74. Twice the difference between a number and 16 is equal to four times the number increased by 6. What is the number?

75. An isosceles triangle has two equal sides. The third side is 4 feet more than the length of the two equal sides. If each side is increased by 3 feet in length, the perimeter of the triangle will be 127 feet. What were the lengths of each side of the original triangle?

76. Determine the values of a and b such that the straight line $ax + by = 9$ passes through the two points $(-1, -3)$ and $(1, -1.5)$.

77. Sylvester has two investments that total $100,000. One investment earns interest at 6% and the other investment earns interest at 8%. If the total amount of interest from the two investments is $6700 in one year, how much money does he have invested at each rate?

78. A chemist needs 8 ounces of a mixture that is 15% iodine. He has a 10% mixture and a 30% mixture in stock. How many ounces of each mixture must he use to get the mixture he wants?

Factoring Polynomials and Solving Quadratic Equations

Did You Know?

You have noticed by now that almost every algebraic skill somehow relates to equation solving and applied problems. The emphasis on equation solving has always been a part of classical algebra.

In Italy, during the Renaissance, it was the custom for one mathematician to challenge another mathematician to an equation-solving contest. A large amount of money, often in gold, was supplied by patrons or sponsoring cities as the prize. At that time, it was important not to publish equation-solving methods, since mathematicians could earn large amounts of money if they could solve problems that their competitors could not. Equation-solving techniques were passed down from a mathematician to an apprentice, but they were never shared.

A Venetian mathematician, Niccolo Fontana (1500? – 1557) known as Tartaglia, "the stammerer," discovered how to solve third-degree or cubic equations. At that time, everyone could solve first- and second-degree equations and special kinds of equations of higher degree. Tartaglia easily won equation-solving contests simply by giving his opponents third-degree equations to solve.

Tartaglia planned to keep his method secret, but after receiving a pledge of secrecy, he gave his method to Girolamo Cardano (1501 – 1576). Cardano broke his promise by publishing one of the first successful Latin algebra texts, *Ars Magna*, "The Great Art." In it, he included not only Tartaglia's solution to the third-degree equations but also a pupil's (Ferrari) discovery of the general solution to fourth-degree equations. Until recently, Cardano received credit for discovering both methods.

It was not until 300 years later that it was shown that there are no general algebraic methods for solving fifth- or higher-degree equations. As you can see, a great deal of time and energy has gone into developing the methods of equation solving that you are learning.

"Algebra is the intellectual instrument which has been created for rendering clear the quantitative aspect of the world."

Alfred North Whitehead (1861-1947)

Factoring is the reverse of multiplication. That is, to factor polynomials, you need to remember how you multiplied them. In this way, the concept of factoring is built on your previous knowledge and skills with multiplication. For example, if you are given a product, such as $x^2 - a^2$ (the difference of two squares), you must recall the factors of $(x + a)$ and $(x - a)$ from your work in multiplying polynomials in Chapter 6.

Studying mathematics is a building process with each topic dependent on previous topics with a few new ideas added each time. The equations and applications in Chapter 7 involve many of the concepts studied earlier, yet you will find them a step higher and more interesting and more challenging.

7.1 Greatest Common Factor and Factoring by Grouping

Objectives

After completing this section, you will be able to:

1. Find the greatest common factor of a set of terms.

2. Factor polynomials by finding the greatest common monomial factor.

3. Factor polynomials by grouping.

In Chapter 6, we used the distributive property and the FOIL method to multiply polynomials and other algebraic expressions. The result of multiplication is called the **product**; and the numbers or expressions being multiplied are called **factors** of the product. In this chapter we will study the reverse of this process, which is called **factoring**. That is, given a product, we want to find the factors. For example,

Multiplying Polynomials

$$2x\ (x + 3) = 2x^2 + 6x$$

factor   factor   product

Factoring Polynomials

$$5x^3 + 30x^2 = 5x^2\ (x + 6)$$

product   factor   factor

Greatest Common Factor

The **greatest common factor (GCF)** of two or more integers is the largest integer that is a factor (or divisor) of all of the integers. For example, the GCF of 30 and 40 is 10. Note that 5 is also a common factor of 30 and 40, but 5 is not the greatest common factor. The number 10 is the largest number that will divide into both 30 and 40.

One way of finding the GCF is to use the prime factorization of each number. For example, to find the GCF for 36 and 60, we can write

$$36 = 4 \cdot 9 = \mathbf{2} \cdot \mathbf{2} \cdot \mathbf{3} \cdot 3$$
$$60 = 4 \cdot 15 = \mathbf{2} \cdot \mathbf{2} \cdot \mathbf{3} \cdot 5$$

The common factors are 2, 2, and 3 and their product is the GCF:

$$GCF = 2 \cdot 2 \cdot 3 = 12$$

Writing the prime factorizations using exponents gives

$$36 = 2^2 \cdot 3^2 \text{ and } 60 = 2^2 \cdot 3 \cdot 5$$

We can see that the GCF is the product of the greatest power of each factor that is common to both numbers. That is,

$$GCF = 2^2 \cdot 3 = 12$$

This procedure can be used to find the GCF for any set of integers or algebraic terms with integer exponents.

Procedure for Finding the GCF of a Set of Terms

1. *Find the prime factorization of all integers and integer coefficients.*
2. *List all factors common to all terms, including variables.*
3. *Choose the greatest power of each factor common to all terms.*
4. *Multiply these powers to find the GCF.*

*(**Note:** If there is no common prime factor or variable, then the GCF is 1.)*

> ### Example 1: GCF ●

Find the GCF for each of the following sets of algebraic terms.

a. $30, 45, 75$

> **Solution:** Find the prime factorization of each number:
> $$30 = 2 \cdot 3 \cdot 5, \quad 45 = 3^2 \cdot 5, \quad \text{and} \quad 75 = 3 \cdot 5^2$$
> The common factors are 3 and 5 and the greatest power of each common to all numbers is 1.
>
> Thus, $\text{GCF} = 3^1 \cdot 5^1 = 15$

b. $20x^4y, 15x^3y, 10x^5y^2$

> **Solution:** Writing each integer coefficient in prime factored form gives:
> $$20x^4y = 2^2 \cdot 5 \cdot x^4 \cdot y$$
> $$15x^3y = 3 \cdot 5 \cdot x^3 \cdot y$$
> $$10x^5y^2 = 2 \cdot 5 \cdot x^5 \cdot y^2$$
> The factors common to each term are $5, x,$ and y. The greatest power of each common to all terms is $5^1, x^3,$ and y^1. Thus, $\text{GCF} = 5^1 \cdot x^3 \cdot y^1 = 5x^3y$.

● ●

Common Monomial Factors

Now consider the polynomial $3n + 15$. We want to write this polynomial as a product of two factors. Since
$$3n + 15 = \mathbf{3} \cdot n + \mathbf{3} \cdot 5$$

we see that 3 is a common factor of the two terms in the polynomial. By using the distributive property, we can write

$$3n + 15 = \mathbf{3} \cdot n + \mathbf{3} \cdot 5 = 3(n + 5)$$

In this way, the polynomial $3n + 15$ has been **factored** into the product of 3 and $(n + 5)$.

For a more general approach to finding factors of a polynomial, we can use the Quotient Property of Exponents:

$$\frac{a^m}{a^n} = a^{m-n}$$

This property is used when dividing terms. For example,

$$\frac{35x^8}{5x^2} = 7x^6 \text{ and } \frac{16a^5}{-8a} = -2a^4$$

To divide a polynomial by a monomial, a procedure that we discussed in Chapter 6, each term in the polynomial is divided by the monomial. For example,

$$\frac{8x^3 - 14x^2 + 10x}{2x} = \frac{8x^3}{2x} - \frac{14x^2}{2x} + \frac{10x}{2x} = 4x^2 - 7x + 5$$

Note: With practice, this division can be done mentally.

This concept of dividing each term by a monomial is part of finding a monomial factor of a polynomial. Finding the greatest common monomial factor (GCF) of a polynomial means to **find the GCF of the terms of the polynomial**. This monomial will be one factor, and the sum of the various quotients found by dividing each term by the GCF will be the other factor. Thus, using the quotients just found and the fact that $2x$ is the GCF of the terms,

$$8x^3 - 14x^2 + 10x = 2x(4x^2 - 7x + 5)$$

We say that $2x$ is **factored out** and $2x$ and $(4x^2 - 7x + 5)$ are the factors of $8x^3 - 14x^2 + 10x$.

Now factor $24x^6 - 12x^4 - 18x^3$. The GCF is $6x^3$ and $6x^3$ is factored out as follows:

$$24x^6 - 12x^4 - 18x^3 = 6x^3 \cdot 4x^3 + 6x^3(-2x) + 6x^3(-3)$$
$$= 6x^3(4x^3 - 2x - 3)$$

If all the terms are negative or if the leading coefficient is negative, we will choose the greatest common factor to be negative. This will leave a positive coefficient for the first term in parentheses. Factoring out a monomial is essentially using the distributive property in a sort of reverse sense:

$ab + ac = a(b + c)$ where a is the greatest common factor of the two terms ab and ac.

Example 2: GCF of a Polynomial ● ● ● ● ● ● ● ● ● ● ● ● ● ●

Factor each polynomial by factoring out the greatest common monomial factor.

a. $6n + 30$

Solution: $6n + 30 = 6 \cdot n + 6 \cdot 5 = 6(n + 5)$

b. $x^3 + x$

Solution: $x^3 + x = x \cdot x^2 + x(+1) = x(x^2 + 1)$ **Note:** $+1$ is the coefficient of $+x$.

c. $5x^3 - 15x^2$

Solution: $5x^3 - 15x^2 = 5x^2 \cdot x + 5x^2(-3) = 5x^2(x - 3)$

d. $2x^4 - 3x^2 + 2$

Solution: $2x^4 - 3x^2 + 2$ No common monomial factor other than 1.

e. $-4a^5 + 2a^3 - 6a^2$

Solution: Since the leading term is negative, we factor out $-2a^2$,

$$-4a^5 + 2a^3 - 6a^2 = -2a^2(2a^3 - a + 3)$$

Note: we could have factored out $2a^2$, instead of $-2a^2$, and been correct. However, when the leading coefficient is negative, factoring out a negative term is preferred because the leading term of the factor in the parentheses is then positive.

● ●

A polynomial may be in more than one variable. For example, $5x^2y + 10xy^2$ is in the two variables x and y. Thus, the GCF may have more than one variable.

$$5x^2y + 10xy^2 = 5xy \cdot x + 5xy \cdot 2y$$
$$= 5xy(x + 2y)$$

Similarly,

$$4xy^3 - 2x^2y^2 + 8xy^2 = 2xy^2 \cdot 2y + 2xy^2(-x) + 2xy^2 \cdot 4$$
$$= 2xy^2(2y - x + 4)$$

Example 3: GCF of a Polynomial ● ● ● ● ● ● ● ● ● ● ● ● ● ●

Factor each polynomial by finding the GCF.

a. $4ax^3 + 4ax$

> **Solution:** $4ax^3 + 4ax = 4ax(x^2 + 1)$ Note that $4ax = 1 \cdot 4ax$

b. $3x^2y^2 - 6xy^2$

> **Solution:** $3x^2y^2 - 6xy^2 = 3xy^2(x - 2)$

c. $14by^3 + 7b^2y - 21by^2$

> **Solution:** $14by^3 + 7b^2y - 21by^2 = 7by(2y^2 + b - 3y)$

d. $13a^4b^5 - 3a^2b^9 - 4a^6b^3$

> **Solution:** $13a^4b^5 - 3a^2b^9 - 4a^6b^3 = a^2b^3(13a^2b^2 - 3b^6 - 4a^4)$

● ●

Factoring by Grouping

Consider the expression

$$y(x + 4) + 2(x + 4)$$

as the sum of two "terms", $y(x + 4)$ and $2(x + 4)$. Each of these "terms" has the common **binomial** factor $(x + 4)$. Factoring out this common binomial factor by using the distributive property gives,

$$y(x + 4) + 2(x + 4) = (x + 4)(y + 2).$$

Similarly,

$$3(x - 2) - a(x - 2) = (x - 2)(3 - a).$$

Now, consider the product

$$(x+3)(y+5) = (x+3)y + (x+3)5$$
$$= xy + 3y + 5x + 15$$

which has four terms and no like terms. Yet the product has two factors, namely $(x + 3)$ and $(y + 5)$. Factoring polynomials with four or more terms can sometimes be accomplished by grouping the terms and using the distributive property, as in the above discussion and the following examples. Keep in mind that the common factor can be a binomial or other polynomial.

> ### Example 4: Factoring Out a Common Binomial Factor   ● ● ● ● ●

Factor each polynomial.

a. $3x^2(5x + 1) - 2(5x + 1)$

Solution: $3x^2(5x + 1) - 2(5x + 1) = (5x + 1)(3x^2 - 2)$

b. $7a(2x - 3) + (2x - 3)$

Solution: $7a(2x-3)+(2x-3)=7a(2x-3)+1\cdot(2x-3)$ **Note**: 1 is the under-

$$= (2x-3)(7a+1)$$

stood coefficient of $(2x - 3)$

● ●

In the examples just discussed, the common binomial factor was in parentheses. However, many times the expression to be factored is in a form with four or more terms. For example, by multiplying in Example 4a, we get the following expression:

$$3x^2(5x + 1) - 2(5x + 1) = 15x^3 + 3x^2 - 10x - 2$$

The expression $(15x^3 + 3x^2 - 10x - 2)$ has four terms with no common monomial factor; yet, we know that it has the two binomial factors $(5x + 1)$ and $(3x^2 - 2)$. We can find these binomial factors by **grouping** and looking for common monomial factors in each "group" and then looking for common binomial factors. The process is illustrated in Example 5.

> ### Example 5: Factoring by Grouping   ● ● ● ● ● ● ● ● ● ● ● ● ● ● ●

Factor each polynomial by grouping.

a. $xy + 5x + 3y + 15$

Solution:

$$xy+5x+3y+15=(xy+5x)+(3y+15)$$ Group terms that have a common monomial factor.

$$= x(y+5)+3(y+5)$$ Using the distributive property.

$$= (y+5)(x+3)$$ **Note:** $y+5$ is a common binomial factor.

b. $ax + ay + bx + by$

Solution:

$$ax+ay+bx+by=(ax+ay)+(bx+by)$$ Group.

$$= a(x+y)+b(x+y)$$ Use the distributive property.

$$= (x+y)(a+b)$$ Use the distributive property again.

Continued on next page ...

Some other grouping may yield the same results. For example,

$$ax + ay + bx + by = (ax + bx) + (ay + by)$$
$$= x(a+b) + y(a+b)$$
$$= (a+b)(x+y)$$

c. $x^2 - xy - 5x + 5y$

Solution: $x^2 - xy - 5x + 5y = (x^2 - xy) + (-5x + 5y)$
$$= x(x-y) + 5(-x+y)$$

This does not work because $(x - y) \neq (-x + y)$. This problem is solved by factoring -5 from the last two terms.

$$x^2 - xy - 5x + 5y = (x^2 - xy) + (-5x + 5y)$$
$$= x(x-y) - 5(x-y)$$
$$= (x-y)(x-5) \qquad \text{Success!}$$

d. $x^2 + ax + 3x + 3y$

Solution: $x^2 + ax + 3x + 3y = x(x + a) + 3(x + y)$
But $x + a \neq x + y$ and there is no common factor.
So, $x^2 + ax + 3x + 3y$ is **not factorable**.

e. $xy + 5x + y + 5$

Solution: $xy + 5x + y + 5 = x(y+5) + (y+5)$ **Note**: 1 is the understood coefficient of $(y+5)$.

$$= (y+5)(x+1)$$

● ●

Practice Problems

Factor each expression.

1. $2x - 16$

2. $-5x^2 - 5x$

3. $7ax^2 - 7ax$

4. $a^4b^5 + 2a^3b^2 - a^3b^3$

5. $9x^2y^2 + 12x^2y - 6x^3$

6. $6a^3(x + 3) - (x + 3)$

7. $5x + 35 - xy - 7y$

Answers to Practice Problems: 1. $2(x-8)$ **2.** $-5x(x+1)$ **3.** $7ax(x-1)$ **4.** $a^3b^2(ab^3 + 2 - b)$
5. $3x^2(3y^2 + 4y - 2x)$ **6.** $(6a^3 - 1)(x+3)$ **7.** $(5-y)(x+7)$

7.1 Exercises

Find the GCF for each set of terms in Exercises 1 – 14.

1. $10, 15, 20$

2. $25, 30, 75$

3. $16, 40, 56$

4. $30, 42, 54$

5. $9, 14, 22$

6. $44, 66, 88$

7. $30x^3, 40x^5$

8. $15y^4, 25y$

9. $26ab^2, 39a^2b, 52a^2b^2$

10. $-8a^3, -16a^4, -20a^2$

11. $-28c^2d^3, -14c^3d^2, -42cd^2$

12. $45x^2y^2z^2, 75xy^2z^3$

13. $36xy, 48xy, 60xy$

14. $21a^5b^4c^3, 28a^3b^4c^3, 35a^3b^4c^2$

Simplify the expressions in Exercises 15 – 22. Assume that no denominator is equal to 0.

15. $\dfrac{x^7}{x^3}$

16. $\dfrac{x^8}{x^3}$

17. $\dfrac{-8y^3}{2y^2}$

18. $\dfrac{12x^2}{2x}$

19. $\dfrac{9x^5}{3x^2}$

20. $\dfrac{-10x^5}{2x}$

21. $\dfrac{4x^3y^2}{2xy}$

22. $\dfrac{21x^4y^3}{-3xy^2}$

Complete the factoring of the polynomial as indicated in Exercises 23 – 30.

23. $3m + 27 = 3(\quad)$

24. $2x + 18 = 2(\quad)$

25. $5x^2 - 30x = 5x(\quad)$

26. $6y^3 - 24y^2 = 6y^2(\quad)$

27. $13ab^2 + 13ab = 13ab(\quad)$

28. $8x^2y - 4xy = 4xy(\quad)$

29. $-15xy^2 - 20x^2y - 5xy = -5xy(\quad)$

30. $-9m^3 - 3m^2 - 6m = -3m(\quad)$

Factor each of the polynomials in Exercises 31 – 50 by finding the greatest common monomial factor.

31. $11x - 121$

32. $14x + 21$

33. $16y^3 + 12y$

34. $4ax - 8ay$

35. $-8a - 16b$

36. $-3x^2 + 6x$

37. $-6ax + 9ay$

38. $10x^2y - 25xy$

39. $16x^4y - 14x^2y$

40. $18y^2z^2 - 2yz$

41. $-14x^2y^3 - 14x^2y$

42. $8y^2 - 32y + 8$

43. $5x^2 - 15x - 5$

44. $ad^2 + 10ad + 25a$

45. $8m^2x^3 - 12m^2y + 4m^2z$

46. $36t^2x^4 - 45t^2x^3 + 24t^2x^2$

47. $34x^5y^6 - 51x^3y^5 + 17x^5y^6$

48. $-56x^4z^3 - 98x^3z^4 - 35x^2z^5$

49. $15x^4y^2 + 24x^6y^6 - 32x^7y^3$

50. $-3x^2y^4 - 6x^3y^4 - 9x^2y^3$

In Exercises 51 – 60, factor each expression by factoring out the common binomial factor.

51. $7y^2(y + 3) + 2(y + 3)$ **52.** $6a(a - 7) - 5(a - 7)$ **53.** $3x(x - 4) + (x - 4)$

54. $2x^2(x + 5) + (x + 5)$ **55.** $4x^3(x - 2) - (x - 2)$ **56.** $9a(x + 1) - (x + 1)$

57. $10y(2y + 3) - 7(2y + 3)$ **58.** $a(x + 5) + b(x + 5)$

59. $a(x - 2) - b(x - 2)$ **60.** $3a(x - 10) + 5b(x - 10)$

Factor each of the polynomials in Exercises 61 – 80 by grouping. If a polynomial cannot be factored, write *not factorable*.

61. $bx + b + cx + c$ **62.** $3x + 3y + ax + ay$ **63.** $x^3 + 3x^2 + 6x + 18$

64. $2z^3 - 14z^2 + 3z - 21$ **65.** $x^2 - 4x + 6xy - 24y$ **66.** $3x + 3y - bx - by$

67. $5xy + yz - 20x - 4z$ **68.** $x - 3xy + 2z - 6zy$ **69.** $24y + 2xz - 3yz - 16x$

70. $10xy - 2y^2 + 7yz - 35xz$ **71.** $ax + 5ay + 3x + 15y$

72. $6ax + 12x + a + 2$ **73.** $4xy + 3x - 4y - 3$

74. $xy + x + y + 1$ **75.** $xy + x - y - 1$

76. $z^2 + 3 + az^2 + 3a$ **77.** $x^2 - 5 + x^2y + 5y$

78. $ab^2 + 6ab + b^2 + 12$ **79.** $7xy - 3y + 2x^2 - 3x$

80. $10a^2 - 5az + 2a + z$

Hawkes Learning Systems: Introductory Algebra

Greatest Common Factor of Two or More Terms
Greatest Common Factor of a Polynomial
Factoring Expressions by Grouping

7.2   Factoring Trinomials: Leading Coefficient 1

Objectives

After completing this section, you will be able to:

1. Factor trinomials with a leading coefficient of 1 (of the form $x^2 + bx + c$).

2. Factor out a common monomial factor and then factor such trinomials.

In Chapter 6, we learned to use the FOIL method to multiply two binomials. In many cases, the simplified form of the product was a trinomial. In the next two sections, we will learn to factor trinomials by reversing the FOIL method or by using the grouping method. In this section, we consider only methods for factoring trinomials in one variable with leading coefficient 1.

Factoring Trinomials with Leading Coefficient 1

Using the FOIL method to multiply $(x + 5)$ and $(x + 3)$, we find

$$(x + 5)(x + 3) = x^2 + 3x + 5x + 15 = x^2 + 8x + 15$$

Note that the constant term 15 is the product of 5 and 3 and the coefficient 8 is the sum of 5 and 3. That is,

$$15 = 5 \cdot 3 \text{ and } 8 = 5 + 3$$
$$x^2 + 8x + 15$$
$$\uparrow \quad \uparrow$$
$$5 + 3 \quad 5 \cdot 3$$

This relationship is true in general. Consider the following general product:

$$(x + a)(x + b) = x^2 + ax + bx + ab$$
$$= x^2 + (a + b)x + ab$$
$$\uparrow \qquad \uparrow$$

sum of   product of
constants constants
a and b a and b

Now, given a trinomial with leading coefficient 1, we want to find the binomial factors, if any. Reversing the relationship between a and b, as shown previously, we can proceed as follows:

To factor a trinomial with leading coefficient 1, find two factors of the constant term whose sum is the coefficient of the middle term.

For example,

(a) To factor $x^2 + 9x + 14$, find two factors of $+14$ whose sum is $+9$.

Since, $14 = 2 \cdot 7$ and $9 = 2 + 7$, we have $a = 2$ and $b = 7$ for

$$x^2 + 9x + 14 = (x + 2)(x + 7)$$

(b) To factor $x^2 - 6x + 5$, find two factors of $+5$ whose sum is -6.

Since, $5 = (-1)(-5)$ and $-6 = (-1) + (-5)$, we have $a = -1$ and $b = -5$ for

$$x^2 - 6x + 5 = (x - 1)(x - 5)$$

Note: Because of the commutative property of multiplication, the order of the factors does not matter. We could just as easily write $(x - 5)(x - 1)$.

To check that the factors are correct, multiply the factors to see that outer and inner products add to the middle term.

$$(x + 2)(x + 7) = x^2 + 9x + 14 \quad \text{and} \quad (x - 1)(x - 5) = x^2 - 6x + 5$$

Example 1: Factoring Trinomials with Leading Coefficient 1

Factor the following trinomials.

a. $x^2 + 8x + 12$

Solution: 12 has three pairs of positive integer factors: $1 \cdot 12, 2 \cdot 6$, and $3 \cdot 4$. Of these three pairs, only $2 + 6 = 8$. Thus,

$$x^2 + 8x + 12 = (x + 2)(x + 6)$$

Note: If the middle term had been $-8x$, then we would have wanted pairs of negative integer factors to find a sum of -8.

Continued on next page...

b. $y^2 - 8y - 20$

    Solution: We want a pair of integer factors of –20 whose sum is –8.

$$-20 = -10 \cdot 2 \quad \text{and} \quad -10 + 2 = -8$$

(Two other pairs are –5 and +4, –1 and +20. But, their sums are not –8. This stage is **a trial and error stage**. That is, you try different pairs of factors until you find a pair whose sum is the coefficient of the middle term.)

Thus,

$$y^2 - 8y - 20 = (y - 10)(y + 2)$$

Note that by the commutative property of multiplication, the order of the factors does not matter. That is, we can write

$$(y - 10)(y + 2)$$
or
$$(y + 2)(y - 10)$$

as the factored form of $y^2 - 8y - 20$.

● ●

To Factor Trinomials of the Form $x^2 + bx + c$

To factor $x^2 + bx + c$, if possible, find an integer pair of factors of c whose sum is b.

1. If c is positive, then both factors must have the same sign.

    a. Both will be positive if b is positive.

    b. Both will be negative if b is negative.

2. If c is negative, then one factor must be positive and the other negative.

Finding a Common Monomial Factor First

If a trinomial does not have a leading coefficient of 1, then we look for a common monomial factor. If there is a common monomial factor, factor out this common monomial factor and factor the remaining trinomial factor, if possible. (We will discuss factoring trinomials with leading coefficients other than 1 in the next section.) A polynomial is **completely factored** if none of its factors can be factored.

Example 2: Finding a Common Monomial Factor First • • • • •

Completely factor the following trinomials by first factoring out the GCF in the form of a common monomial factor.

a. $5x^3 - 15x^2 + 10x$

Solution: First factor out the GCF, $5x$.

$$5x^3 - 15x^2 + 10x = 5x(x^2 - 3x + 2) \qquad \text{Factored, but not completely factored.}$$

Now factor the trinomial $x^2 - 3x + 2$. Look for factors of $+2$ that add up to -3. Since $(-1)(-2) = +2$ and $(-1) + (-2) = -3$, we have

$$5x^3 - 15x^2 + 10x = 5x(x^2 - 3x + 2)$$
$$= 5x(x - 1)(x - 2). \quad \text{Completely factored.}$$

b. $10y^5 - 20y^4 - 80y^3$

Solution: First factor out the GCF, $10y^3$.

$$10y^5 - 20y^4 - 80y^3 = 10y^3(y^2 - 2y - 8)$$

Now factor the trinomial $y^2 - 2y - 8$. Look for factors of -8 that add up to -2. Since $(-4)(+2) = -8$ and $(-4) + (+2) = -2$, we have

$$10y^5 - 20y^4 - 80y^3 = 10y^3(y^2 - 2y - 8) = 10y^3(y - 4)(y + 2).$$

• •

NOTES

When factoring polynomials, always look for a common monomial factor first. Then, if there is one, remember to include this common monomial factor as part of the answer. Not all polynomials are factorable. For example, no matter what combinations are tried, $x^2 + 3x + 4$ does not have two binomial factors with integer coefficients. (There are no factors of $+4$ that will add to $+3$.) We say that the polynomial is **irreducible** (or **not factorable** or **prime**). **An irreducible polynomial is one that cannot be factored as the product of polynomials with integer coefficients.**

7.2 Exercises

In Exercises 1 – 10, list all pairs of integer factors for the given integer. Remember to include negative integers as well as positive integers.

1. 15	**2.** 12	**3.** 20	**4.** 30	**5.** −6
6. −7	**7.** 16	**8.** −18	**9.** −10	**10.** 25

In Exercises 11 – 20, find the pair of integers whose product is the first integer and whose sum is the second integer.

11. 12, 7	**12.** 25, 26	**13.** −14, −5	**14.** −30, −1	**15.** −8, 7
16. −40, −6	**17.** 36, −12	**18.** 16, −10	**19.** 20, −9	**20.** 4, −5

In Exercises 21 – 26, complete the factorization.

21. $x^2 + 6x + 5 = (x + 5)($ $)$

22. $y^2 - 7y + 6 = (y - 1)($ $)$

23. $p^2 - 9p - 10 = (p + 1)($ $)$

24. $m^2 + 4m - 45 = (m - 5)($ $)$

25. $a^2 + 12a + 36 = (a + 6)($ $)$

26. $n^2 - 2n - 3 = (n - 3)($ $)$

Factor the trinomials in Exercises 27 – 40. If the trinomial cannot be factored, write not factorable.

27. $x^2 - x - 12$

28. $x^2 + 7x + 12$

29. $y^2 + y - 30$

30. $y^2 - 3y + 2$

31. $a^2 + a + 2$

32. $m^2 + 3m - 1$

33. $x^2 + 3x + 5$

34. $x^2 - 8x + 16$

35. $x^2 + 3x - 18$

36. $y^2 + 12y + 35$

37. $y^2 - 14y + 24$

38. $a^2 + 10a + 25$

39. $x^2 - 6x - 27$

40. $x^2 + 5x - 36$

Completely factor the polynomials in Exercises 41 – 55.

41. $x^3 + 10x^2 + 21x$

42. $x^3 + 8x^2 + 15x$

43. $5x^2 - 5x - 60$

44. $6x^2 + 24x + 18$

45. $10y^2 - 10y - 60$

46. $7y^3 - 70y^2 + 168y$

47. $4p^4 + 36p^3 + 32p^2$

48. $15m^5 - 30m^4 + 15m^3$

49. $2x^4 - 14x^3 - 36x^2$

50. $3y^6 + 33y^5 + 90y^4$

51. $2x^2 - 2x - 72$

52. $3x^2 - 18x + 30$

53. $a^2 - 30a - 216$

54. $a^4 + 30a^3 + 81a^2$

55. $3y^5 - 21y^4 - 24y^3$

In Exercises 56 – 60, the polynomials are in more than one variable. Completely factor each polynomial.

56. $x^2 - 10xy + 16y^2$ **57.** $x^2 - 2xy - 3y^2$ **58.** $5a^2 + 10ab - 30b^2$

59. $20a^2 + 40ab + 20b^2$ **60.** $6x^2 - 12xy + 6y^2$

61. The area (in square inches) of the rectangle shown is given by the polynomial function $A(x) = 4x^2 + 20x$. If the width of the rectangle is $4x$ inches, what is the length?

$$A(x) = 4x^2 + 20x \qquad 4x$$

62. The area (in square feet) of the rectangle shown is given by the polynomial function $A(x) = x^2 + 11x + 24$. If the length of the rectangle is $(x + 8)$ feet, what is the width?

$$A(x) = x^2 + 11x + 24$$

$$x + 8$$

63. The volume of an open box is found by cutting equal squares (x units on a side) from a sheet of cardboard that is 10 inches by 40 inches. The function representing this volume is $V(x) = 4x^3 - 100x^2 + 400x$, where $0 < x < 5$. Factor this function and use the factors to explain, in your own words, how the function represents the volume.

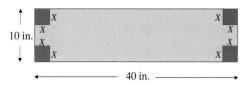

64. The area of a triangle is $\dfrac{1}{2}$ the product of its base and its height. If the area (in square meters) of the triangle shown is given by the function $A(x) = \dfrac{1}{2}x^2 + 24x$ find representations for the lengths of its base and its height.

$$A(x) = \dfrac{1}{2}x^2 + 24x$$

Writing and Thinking About Mathematics

65. It is true that $2x^2 + 10x + 12 = (2x + 6)(x + 2) = (2x + 4)(x + 3)$. Explain how the trinomial can be factored in two ways. Is there some kind of error?

66. It is true that $5x^2 - 5x - 30 = (5x - 15)(x + 2)$. Explain why this is not the completely factored form of the trinomial.

Hawkes Learning Systems: Introductory Algebra

 Factoring Trinomials: Leading Coefficient 1

<table>
<tr><td>**7.3**</td><td></td></tr>
</table>

More on Factoring Trinomials

After completing this section, you will be able to:

1. Factor trinomials by using the ac-method.

2. Factor trinomials by using reverse FOIL (or trial-and-error method).

In this section, we will investigate two methods for factoring trinomials in the general form

$$ax^2 + bx + c \quad \text{where the coefficients } a, b, \text{ and } c \text{ are integers.}$$

In this general form, the coefficient a may be an integer other than 1. The two methods are related to two methods of factoring that we have already studied: factoring by grouping and the FOIL method. The **ac-method** involves factoring by grouping and the **trial-and-error method** is a reverse FOIL method.

The *ac*-Method (Grouping)

The *ac*-method is in reference to the coefficients a and c in the general form stated above:

$$ax^2 + bx + c \quad \text{where the coefficients } a, b, \text{ and } c \text{ are integers.}$$

The method is best explained by analyzing an example and explaining each step.

Consider the problem of factoring the trinomial

$$3x^2 + 17x + 10 \qquad \text{where } a = 3, b = 17, \text{ and } c = 10.$$

Analysis of Factoring by the *ac*-Method

$$3x^2 + 17x + 10 \qquad\qquad ax^2 + bx + c$$

Step 1: Multiply $3 \cdot 10 = 30$.   Multiply $a \cdot c$.

Step 2: Find two integers whose product is 30 and whose sum is 17. (In this case, $2 \cdot 15 = 30$ and $2 + 15 = 17$.)

Find two integers whose product is ac and whose sum is b. If this is not possible, then the trinomial is **not factorable**.

Step 3: Rewrite the middle term $(+17x)$ using $+2$ and $+15$ as coefficients.
$$3x^2 + 17x + 10 = 3x^2 + 2x + 15x + 10$$

Rewrite the middle term (bx) using the two numbers found in Step 2 as coefficients.

Step 4: Factor by grouping the first two terms and the last two terms.
$$3x^2 + 2x + 15x + 10 = x(3x + 2) + 5(3x + 2)$$

Factor by grouping the first terms and the last two terms.

Step 5: Factor out the common binomial factor $(3x + 2)$. Thus,
$$3x^2 + 17x + 10 = 3x^2 + 2x + 15x + 10$$
$$= x(3x + 2) + 5(3x + 2)$$
$$= (3x + 2)(x + 5)$$

Factor out the common binomial factor to find two binomial factors of the trinomial $ax^2 + bx + c$.

Example 1: *ac*-Method ● ● ● ● ● ● ● ● ● ● ● ● ● ● ● ● ● ●

a. Factor $4x^2 + 33x + 35$ using the *ac*-method.

Solution: $a = 4, b = 33, c = 35$

Step 1: Find the product ac: $4 \cdot 35 = 140$.

Step 2: Find two integers whose product is 140 and whose sum is 33:
$$(+5)(+28) = 140 \quad \text{and} \quad (+5) + (+28) = +33$$

Continued on next page...

Note: this step may take some experimenting with factors. You might try prime factoring. For example

$$140 = 10 \cdot 14 = 2 \cdot 5 \cdot 2 \cdot 7.$$

With combinations of these prime factors we can write

$$140 = 1 \cdot 140 = 2 \cdot 70 = 5 \cdot 28 = 7 \cdot 20 = 4 \cdot 35 = 10 \cdot 14.$$

Since $+5 + 28 = +33$, $+5$ and $+28$ are the desired constants.

Step 3: Rewrite $+33x$ as $+5x + 28x$, giving

$$4x^2 + 33x + 35 = 4x^2 + 5x + 28x + 35$$

Step 4: Factor by grouping:

$$4x^2 + 33x + 35 = 4x^2 + 5x + 28x + 35$$
$$= 4x^2 + 28x + 5x + 35 \qquad \text{Rearrange the terms for easy grouping.}$$
$$= 4x(x+7) + 5(x+7)$$

Step 5: Factor out the common binomial factor $(x + 7)$.

$$4x^2 + 33x + 35 = 4x(x + 7) + 5(x + 7) = (x + 7)(4x + 5)$$

b. Factor $12x^3 - 26x^2 + 12x$ using the *ac*-method.

Solution: First factor out the greatest common factor $2x$:

$$12x^3 - 26x^2 + 12x = 2x(6x^2 - 13x + 6)$$

Now factor the trinomial $6x^2 - 13x + 6$ with $a = 6$, $b = -13$, and $c = 6$.

Step 1: Find the product *ac*: $6(6) = 36$.

Step 2: Find two integers whose product is 36 and whose sum is -13: (**Note**: This may take some time and experimentation. We do know that both numbers must be negative since the product is positive and the sum is negative.)

$$(-9)(-4) = +36 \text{ and } -9 + (-4) = -13$$

Continued on next page...

**Steps 3
and 4:** Factor by grouping.

$$6x^2 - 13x + 6 = 6x^2 - 9x - 4x + 6$$
$$= 3x(2x - 3) - 2(2x - 3)$$

[**Note:** −2 is factored from the last two terms so that there will be a common binomial factor $(2x - 3)$.]

Step 5: Factor out the common binomial factor $(2x - 3)$:

$$6x^2 - 13x + 6 = 6x^2 - 9x - 4x + 6$$
$$= 3x(2x - 3) - 2(2x - 3)$$
$$= (2x - 3)(3x - 2)$$

Thus, for the original expression,

$$12x^3 - 26x^2 + 12x = 2x(6x^2 - 13x + 6)$$
$$= 2x(2x - 3)(3x - 2)$$

Do not forget to write the common monomial factor, $2x$, in the answer.

● ●

NOTES   As illustrated in Example 1b, the first step in factoring should always be to factor out any common monomial factor. Also, if the leading coefficient is negative, factor out a negative monomial even if it is just −1. For example,

$$-12x^3 + 26x^2 - 12x = -2x(6x^2 - 13x + 6) = -2x(2x - 3)(3x - 2)$$

Remember that the product of the factors must always equal the original polynomial. This is a good way to check your work.

The Trial-and-Error Method

The key to the trial-and-error method of factoring trinomials is the FOIL method of multiplication of two binomials that we studied in Section 6.6. The FOIL method is used in a reverse sense, since the product is given and the object is to find the factors. Reviewing the FOIL method of multiplication, we can find the product

$$(2x + 5)(3x + 1) = 6x^2 + 17x + 5$$

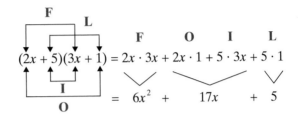

F   The product of the **first** two terms is $6x^2$.

O
I   The sum of **inner** and **outer** products is $17x$.

L   The product of the **last** two terms is 5.

We have already discussed, in Section 7.2, the situation of factoring a trinomial with leading coefficient 1. For example, to factor $x^2 + 7x + 12$, we find factors of 12 whose sum is 7. This is a simpler form of the reverse FOIL method with $\mathbf{F} = x^2 = x \cdot x$ and $\mathbf{L} = 12$.

$$x^2 + 7x + 12 = (x \qquad)(x \qquad)$$

So, the only problem is to find the factors of 12 whose sum is 7. The factors are 4 and 3 since $4 \cdot 3 = 12$ and $4 + 3 = 7$.

$$x^2 + 7x + 12 = (x + 3)(x + 4)$$

$$\mathbf{I} = 3x$$
$$\mathbf{O} = 4x$$

$3x + 4x = 7x$ Add to get the middle term.

Now consider the problem of factoring

$$6x^2 + 31x + 5$$

as a product of two binomials.

$$\mathbf{F} = 6x^2 \qquad \mathbf{L} = +5$$

$$6x^2 + 31x + 5 = (\qquad)(\qquad)$$

For **F** $= 6x^2$, we know that $6x^2 = 3x \cdot 2x$ and $6x^2 = 6x \cdot x$.

For **L** $= +5$, we know that $+5 = (+1)(+5)$ and $+5 = (-1)(-5)$.

Now we use various combinations for **F** and **L** in the **trial-and-error method** as follows:

1. List all the possible combinations of factors of $6x^2$ and $+5$ in their respective **F** and **L** positions.

2. Check the sum of the products in the **O** and **I** positions until you find the sum to be $31x$.

3. If none of these sums is $31x$, the trinomial is not factorable.

 NOTES   No matter which method you use (the *ac*-method or the trial-and-error method), factoring trinomials takes time. With practice, you will become more efficient with either method. Make sure to be patient and observant.

---

a. $(3x + 1)(2x + 5)$

b. $(3x + 5)(2x + 1)$

c. $(6x + 1)(x + 5)$

d. $(6x + 5)(x + 1)$

e. $(3x - 1)(2x - 5)$

f. $(3x - 5)(2x - 1)$

g. $(6x - 1)(x - 5)$

h. $(6x - 5)(x - 1)$

We really don't need to check these last four because the **O** and **I** sum would have to be negative, and we are looking for $+31x$. In this manner the trial-and-error method is more efficient than it first appears to be.

Now, investigating only the possibilities in the list with positive constants, we need to check the sums of the outer (**O**) and inner (**I**) products to find $+31x$.

    a. $(3x + 1)(2x + 5)$; $15x + 2x = 17x$

$$\underset{\underset{15x}{\underline{}}}{\overset{2x}{\boxed{}}}$$

    b. $(3x + 5)(2x + 1)$; $3x + 10x = 13x$

$$\underset{\underset{3x}{\underline{}}}{\overset{10x}{\boxed{}}}$$

    c. $(6x + 1)(x + 5)$; $30x + x = 31x$

$$\underset{\underset{30x}{\underline{}}}{\overset{1x}{\boxed{}}}$$

The correct factors, $(6x + 1)$ and $(x + 5)$, have been found so we need not take the time to try the next product in the list $(6x + 5)(x + 1)$. Thus, even though the list of possibilities of factors may be long, the actual time may be quite short if the correct factors are found early in the trial-and-error method. So, we have

$$6x^2 + 31x + 5 = (6x + 1)(x + 5).$$

Look at the constant term to determine what signs to use for the constants in the factors. The following guidelines will help limit the trial-and-error search.

Guidelines for the Trial-and-Error Method

1.   If the sign of the constant term is positive (+), the signs in both factors will be the same, either both positive or both negative.

2.   If the sign of the constant term is negative (–), the signs in the factors will be different, one positive and one negative.

Example 2: Trial-and-Error Method ● ● ● ● ● ● ● ● ● ● ● ● ● ●

Factor by using the trial-and-error method.

a.   Factor $6x^2 - 31x + 5$.

Solution: Since the middle term is $-31x$ and the constant is $+5$, we know that the two factors of 5 must both be negative, -5 and -1. Also, from the previous discussion, $\mathbf{F} = 6x^2 = 6x \cdot x$.

$$6x^2 - 31x + 5 = (6x - 1)(x - 5) \text{ thus, } -30x - x = -31x$$

$$\underset{\substack{-x \\ -30x}}{}$$

b. $4x^2 - 4x - 15$

Solution: For $\mathbf{F}$: $4x^2 = 4x \cdot x$ and $4x^2 = 2x \cdot 2x$.
For $\mathbf{L}$: $-15 = -15 \cdot 1 = -1 \cdot 15$ and $-15 = -3 \cdot 5 = -5 \cdot 3$.

Trials: $(2x - 15)(2x + 1)$ $-30x + 2x = -28x$ is the wrong middle term.

$$\underset{\substack{-30x \\ +2x}}{}$$

$(2x - 3)(2x + 5)$ ~~$-6x + 10x = +4x$ is the wrong middle term only~~ because the sign is wrong. So, just switch the signs and the factors will be right.

$$\underset{\substack{-6x \\ 10x}}{}$$

$(2x + 3)(2x - 5)$ $+6x - 10x = -4x$ is the right middle term.

$$\underset{\substack{6x \\ -10x}}{}$$

Thus, $4x^2 - 4x - 15 = (2x + 3)(2x - 5)$.

There is no need to try all the possibilities with $(4x \quad)(x \quad)$.

Continued on next page...

c. Factor $2x^2 + 12x + 10$ completely.

Solution: First factor the common monomial factor, 2.

$$2x^2 + 12x + 10 = 2\left(x^2 + 6x + 5\right)$$
$$= 2(x+5)(x+1); \ x + 5x = 6x$$

$$5x$$
$$x$$

● ●

NOTES

Reminder: To factor completely means to find factors of the polynomial, none of which are themselves factorable. Thus,

$$2x^2 + 12x + 10 = (2x + 10)(x + 1)$$

is not factored completely since $2x + 10 = 2(x + 5)$.

We could write

$$2x^2 + 12x + 10 = (2x + 10)(x + 1)$$
$$= 2(x + 5)(x + 1).$$

The sum of two squares, $a^2 + b^2$, is an important polynomial that is not factorable. For example, $x^2 + 4$ is not factorable. There are no factors with integer coefficients whose product is $x^2 + 4$. To understand this situation, write $x^2 + 4 = x^2 + 0x + 4$ and note that there are no factors of $+4$ that will add to 0.

Example 3: Factor Completely ● ● ● ● ● ● ● ● ● ● ● ● ● ●

Factor completely. Be sure to look first for the greatest common monomial factor.

a. $4x^2 + 100$

Solution: $4x^2 + 100 = 4\left(x^2 + 25\right)$ $x^2 + 25$ is the sum of two squares and is not factorable.

b. $6x^3 - 8x^2 + 2x$

Solution: $6x^3 - 8x^2 + 2x = 2x\left(3x^2 - 4x + 1\right) = 2x(3x - 1)(x - 1)$

c. $-2x^2 - x + 6$

Solution: $-2x^2 - x + 6 = -1(2x^2 + x - 6) = -1(2x - 3)(x + 2)$

d. $x^2 + x + 1$

   Solution: $x^2 + x + 1 = x^2 + x + 1$ This trinomial is not factorable.

● ●

Practice Problems

Factor completely.

1. $3x^2 + 7x - 6$ **2.** $x^2 + 49$

3. $3x^2 + 15x + 18$ **4.** $10x^2 - 41x - 18$

5. $x^2 + 11x + 28$ **6.** $-3x^3 + 6x^2 - 3x$

7.3 Exercises

In Exercises 1 – 70, factor completely, if possible. If a polynomial cannot be factored, say "not factorable."

1. $x^2 + 5x + 6$ **2.** $x^2 - 6x + 8$ **3.** $2x^2 - 3x - 5$

4. $3x^2 - 4x - 7$ **5.** $6x^2 + 11x + 5$ **6.** $4x^2 - 11x + 6$

7. $-x^2 + 3x - 2$ **8.** $-x^2 - 5x - 6$ **9.** $x^2 - 3x - 10$

10. $x^2 - 11x + 10$ **11.** $-x^2 + 13x + 14$ **12.** $-x^2 + 12x - 36$

13. $x^2 + 8x + 64$ **14.** $x^2 + 2x + 3$ **15.** $-2x^3 + x^2 + x$

16. $-2y^3 - 3y^2 - y$ **17.** $4t^2 - 3t - 1$ **18.** $2x^2 - 3x - 2$

19. $5a^2 - a - 6$ **20.** $3a^2 + 4a + 1$ **21.** $7x^2 + 5x - 2$

22. $8x^2 - 10x - 3$ **23.** $4x^2 + 23x + 15$ **24.** $6x^2 + 23x + 21$

25. $x^2 + 6x - 16$ **26.** $9x^2 - 3x - 20$ **27.** $12x^2 - 38x + 20$

28. $12b^2 - 12b + 3$ **29.** $3x^2 - 7x + 2$ **30.** $7x^2 - 11x - 6$

31. $9x^2 - 6x + 1$ **32.** $4x^2 + 40x + 25$ **33.** $4x^2 + 25$

34. $x^2 + 81$ **35.** $12y^2 - 7y - 12$ **36.** $x^2 - 46x + 45$

37. $5x^2 + 45$ **38.** $4x^2 + 64$ **39.** $8a^2b - 22ab + 12b$

40. $12m^3n - 146m^2n + 24mn$ **41.** $x^2 + x + 1$ **42.** $x^2 + 2x + 2$

43. $16x^2 - 8x + 1$ **44.** $3x^2 - 11x - 4$ **45.** $64x^2 - 48x + 9$

46. $9x^2 - 12x + 4$ **47.** $6x^2 + 2x - 20$ **48.** $12y^2 - 15y + 3$

49. $10x^2 + 35x + 30$ **50.** $24y^2 + 4y - 4$ **51.** $-18x^2 + 72x - 8$

52. $-45y^2 + 30y + 120$ **53.** $7x^4 - 5x^3 + 3x^2$ **54.** $12x^2 - 60x + 75$

Answers to Practice Problems: 1. $(3x - 2)(x + 3)$ **2.** Not factorable **3.** $3(x + 2)(x + 3)$
4. $(5x + 2)(2x - 9)$ **5.** $(x + 7)(x + 4)$ **6.** $-3x(x - 1)(x - 1)$

55. $-12m^2 + 22m + 4$ **56.** $32y^2 + 50$ **57.** $6x^3 + 9x^2 - 6x$

58. $-5y^2 + 40y - 60$ **59.** $9x^3y^3 + 9x^2y^3 + 9xy^3$ **60.** $30a^3 + 51a^2 + 9a$

61. $12x^3 - 108x^2 + 243x$ **62.** $48x^2y - 354xy + 126y$ **63.** $48xy^3 - 100xy^2 + 48xy$

64. $24a^2x^2 + 72a^2x + 243x$ **65.** $21y^4 - 98y^3 + 56y^2$ **66.** $72a^3 - 306a^2 + 189a$

67. $ax + ay + 3x + 3y$ **68.** $2x^2 + 2y^2 + ax^2 + ay^2$ **69.** $5x^2 + 5y^2 + bx^2 + by^2$

70. $ax + bx - 4a - 4b$

Writing and Thinking About Mathematics

71. A ball is dropped from the top of a building that is 784 feet high. The height of the ball above ground level is given by the polynomial function $h(t) = -16t^2 + 784$ where t is measured in seconds.

  a. How high is the ball after 3 seconds? 5 seconds?

  b. How far has the ball traveled in 3 seconds? 5 seconds?

  c. When will the ball hit the ground? Explain your reasoning in terms of factors.

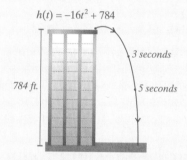

72. Explain, in your own words, why the sum of two squares is not factorable. Use two specific examples and a general expression to make your point.

Hawkes Learning Systems: Introductory Algebra

Factoring Trinomials by Trial and Error
Factoring Trinomials by Grouping

7.4   Factoring Special Products

After completing this section, you will be able to:

1. *Factor the differences of two squares.*

2. *Factor perfect square trinomials.*

3. *Complete the square of trinomials by determining the missing terms that make incomplete trinomials perfect squares.*

In Chapter 6, we discussed the following three special products.

I. $(x + a)(x - a) = x^2 - a^2$ Difference of two squares

II. $(x + a)^2 = x^2 + 2ax + a^2$ Perfect square trinomial

III. $(x - a)^2 = x^2 - 2ax + a^2$ Perfect square trinomial

The objective in this section is to learn to factor products of these types without referring to the *ac*-method or the trial-and-error method. That is, with practice, you will learn to be able to recognize the "form" of the special product and go directly to the factors.

Factoring Special Products

Memorize the products and the names of the products listed above. These products are like formulas. If you are familiar enough with them, you will be able see how some polynomials fit the forms; and you will be able to write the factors without using the techniques learned in Section 7.3. Obviously, many polynomials do not fit any of these forms and you will need to use the *ac*-method or the trial-and-error method to factor these polynomials. However, the difference of two squares and perfect square trinomials occur frequently enough to deserve special treatment.

Difference of Two Squares:

Consider the polynomial $x^2 - 9$. By recognizing this expression as **the difference of two squares** we can go directly to the factors:

$$x^2 - 9 = (x + 3)(x - 3)$$

Similarly,

$$x^2 - y^2 = (x + y)(x - y)$$

$$25 - y^2 = (5 + y)(5 - y)$$

$$36x^2 - 49 = (6x + 7)(6x - 7)$$

Perfect Square Trinomials:

In a **perfect square trinomial**, both the first and last terms must be perfect squares. If we think of the first term in the form of x^2 and the last term in the form of a^2, then the middle term must be of the form $2ax$ or $-2ax$. For example,

$$x^2 + 6x + 9 = (x+3)^2, \text{ here } 9 = 3^2 = a^2 \text{ and } 6 = 2 \cdot 3 = 2a$$

$$x^2 - 14x + 49 = (x-7)^2, \text{ here } 49 = 7^2 = a^2 \text{ and } -14 = -2 \cdot 7 = -2a$$

Recognizing the form of the polynomial is the key to factoring the three special products discussed here. Sometimes the form may be disguised by a common monomial factor or by a rearrangement of the terms. **Always look for a common monomial factor first.** For example:

$$5x^2 y - 20y = 5y\left(x^2 - 4\right) \qquad \text{Factoring out } 5y.$$

$$= 5y(x+2)(x-2) \qquad \text{Difference of two squares.}$$

Example 1: Factoring Special Products • • • • • • • • • • • • •

Factor completely each of the following polynomials.

a. $y^2 - 10y + 25$

   Solution: $y^2 - 10y + 25 = (y-5)^2$ Perfect square trinomial.

b. $6a^2 b - 6b$

   Solution: $6a^2 b - 6b = 6b\left(a^2 - 1\right)$ Common monomial factor.

        $= 6b(a+1)(a-1)$ Difference of two squares.

c. $3x^2 + 12 + 12x$

   Solution: $3x^2 + 12 + 12x = 3x^2 + 12x + 12$ Rearrange terms in order of descending powers.

        $= 3\left(x^2 + 4x + 4\right)$ Factor out common monomial factor.

        $= 3(x+2)^2$ Perfect square trinomial.

d. $a^6 - 100$ Even powers, such as a^6, can always be treated as squares: $a^6 = \left(a^3\right)^2$

   Solution: $a^6 - 100 = \left(a^3 + 10\right)\left(a^3 - 10\right)$ Difference of two squares.

e. $x^2 + 36$

   Solution: $x^2 + 36$ is the **sum** of two squares and is **not** factorable.

• •

Summary of Procedures to Follow in Factoring

1. Look for a common factor.
2. Check the number of terms:
   a. Two terms:
      (1) difference of two squares? - factorable
      (2) sum of two squares? - not factorable
   b. Three terms:
      (1) perfect square trinomial?
      (2) use trial-and-error method?
      (3) use the ac-method?
   c. Four terms:
      (1) group terms with a common factor
3. Check the possibility of factoring any of the factors.

Completing the Square

Closely related to factoring special products is the procedure of **completing the square**. This procedure helps in understanding the nature of perfect square trinomials, and we will use it again in Chapter 10 in solving second-degree equations (called quadratic equations). The procedure involves adding a square term to a binomial so that the resulting trinomial is a perfect square trinomial, thus "completing the square." For example, we want

$$x^2 + 10x + \underline{\quad} = (\quad)^2 \qquad \text{to be of the form} \qquad x^2 + 2ax + a^2 = (x+a)^2$$

The middle coefficient, 10, is twice the number that is to be squared. That is, $10 = 2a$. So, taking half of this coefficient gives

$$\frac{1}{2}(10) = 5 = a$$

and squaring the result gives

$$5^2 = 25 = a^2$$

Thus, $x^2 + 10x + \underline{\quad} = (\quad)^2$ becomes $x^2 + 10x + \underline{25} = (x + 5)^2$

For $x^2 + 18x$, we have $\frac{1}{2}(18) = 9$ and $9^2 = 81$. So,

$$x^2 + 18x + \underline{\quad} = (\quad)^2 \quad \text{becomes} \quad x^2 + 18x + \underline{81} = (x + 9)^2$$

Example 2: Completing the Square

Complete the square as indicated. In Examples 2b and 2c, fractions are introduced into the process of completing the square to help in understanding that algebraic concepts need not be restricted to integers.

a. $y^2 + 20y +$ ____ $= ($ $)^2$

Solution: Since $\frac{1}{2}(20) = 10$ and $10^2 = 100$, then

$$y^2 + 20y + \underline{100} = (y+10)^2$$

b. $x^2 + 3x +$ ____ $= ($ $)^2$

Solution: In this example, $2a = 3$; and since 3 is an odd number, a will be a fraction, not an integer.

$$\text{Thus, } \frac{1}{2}(3) = \frac{3}{2} \text{ and } \left(\frac{3}{2}\right)^2 = \frac{9}{4}$$

$$\text{This gives } x^2 + 3x + \frac{9}{\underline{4}} = \left(x + \frac{3}{2}\right)^2$$

c. $a^2 - 5a +$ ____ $= ($ $)^2$

Solution: We have $2a = -5$. Thus, $\frac{1}{2}(-5) = -\frac{5}{2}$ and $\left(-\frac{5}{2}\right)^2 = \frac{25}{4}$.

$$\text{So, } a^2 - 5a + \frac{25}{\underline{4}} = \left(a - \frac{5}{2}\right)^2$$

To help in understanding the relationships between the terms in a perfect square trinomial, we look at completing the square from a slightly different perspective. Suppose that the middle x-term is missing and we want to find this term so that the resulting trinomial is the square of a binomial. For example,

$$x^2 + \underline{\quad} + 36 = ($$ $)^2$

From the formulas,

$$(x+a)^2 = x^2 + \underline{2ax} + a^2 \text{ and } (x-a)^2 = x^2 - \underline{2ax} + a^2$$

we know that the middle (missing) term is to have coefficient $2a$ (or $-2a$) and, in this case,

$$a^2 = 36 = 6^2 \text{ and } 2a = 2 \cdot 6 = 12.$$

So,

$$x^2 + \underline{12x} + 36 = (x+6)^2.$$

Similarly, for

$$x^2 - \underline{\quad} + 16 = (\quad\quad)^2$$

$a^2 = 16$ and $a = 4$. Thus, $-2a = -2(4) = -8$, and we get $x^2 - \underline{8x} + 16 = (x-4)^2$.

Example 3: Finding the Missing Term

Find the missing term in the trinomial and complete the square as indicated.

$$x^2 - \underline{\quad} + \frac{9}{16} = (\quad\quad)^2$$

Solution: Since $\dfrac{9}{16} = \left(\dfrac{3}{4}\right)^2$ and $-2 \cdot \dfrac{3}{4} = -\dfrac{3}{2}$, we have $x^2 - \underline{\dfrac{3}{2}}x + \dfrac{9}{16} = \left(x - \dfrac{3}{4}\right)^2$

Practice Problems

Factor completely.

1. $x^2 - 16$
2. $25 - x^2$
3. $3y^2 + 6y + 3$
4. $y^{10} - 81$

Complete the square.

5. $x^2 + 6x + \underline{\quad} = (\quad\quad)^2$
6. $x^2 + \underline{\quad} + 49 = (\quad\quad)^2$

Answers to Practice Problems: 1. $(x+4)(x-4)$ **2.** $(5+x)(5-x)$ **3.** $3(y+1)^2$ **4.** $(y^5+9)(y^5-9)$
5. $x^2 + 6x + \underline{9} = (x+3)^2$ **6.** $x^2 + \underline{14x} + 49 = (x+7)^2$

7.4 Exercises

Factor completely each of the polynomials in Exercises 1 – 40.

1. $x^2 - 1$ **2.** $x^2 - 4$ **3.** $x^2 - 49$

4. $x^2 - 144$ **5.** $x^2 + 4x + 4$ **6.** $x^2 + 10x + 25$

7. $x^2 - 12x + 36$ **8.** $x^2 - 20x + 100$ **9.** $16x^2 - 9$

10. $4x^2 - 49$ **11.** $9x^2 - 1$ **12.** $36x^2 - 1$

13. $25 - 4x^2$ **14.** $16 - 9x^2$ **15.** $x^2 - 14x + 49$

16. $x^2 - 2x + 1$ **17.** $x^2 + 6x + 9$ **18.** $x^2 + 8x + 16$

19. $3x^2 - 27y^2$ **20.** $27x^2 - 48y^2$ **21.** $x^3 - xy^2$

22. $12ax^2 - 75ay^2$ **23.** $x^2 - \dfrac{1}{4}$ **24.** $x^2 - \dfrac{4}{9}$

25. $x^2 - \dfrac{9}{16}$ **26.** $x^2 - \dfrac{25}{49}$ **27.** $x^4 - 1$

28. $x^4 - 16y^4$ **29.** $2x^2 - 32x + 128$ **30.** $3x^2 - 30x + 75$

31. $ay^2 + 2ay + a$ **32.** $ax^2 + 18ax + 81a$ **33.** $4x^2y^2 - 24xy^2 + 36y^2$

34. $2ax^2 - 44ax + 242a$ **35.** $16x^2 + 81$ **36.** $4y^2 + 49$

37. $x^2 + x + 1$ **38.** $y^2 + 4y + 5$ **39.** $y^2 + 2y + 2$

40. $x^2 + 4x + 1$

In Exercises 41 – 56, complete the square by adding the correct missing term on the left, then factor as indicated.

41. $x^2 - 6x + \underline{\quad} = (\quad\quad)^2$ **42.** $x^2 - 12x + \underline{\quad} = (\quad\quad)^2$

43. $x^2 - 4x + \underline{\quad} = (\quad\quad)^2$ **44.** $x^2 + 20x + \underline{\quad} = (\quad\quad)^2$

45. $x^2 + \underline{\quad} + 16 = (\quad\quad)^2$ **46.** $x^2 - \underline{\quad} + 49 = (\quad\quad)^2$

47. $x^2 - \underline{\quad} + 81 = (\quad\quad)^2$ **48.** $x^2 + \underline{\quad} + 121 = (\quad\quad)^2$

49. $x^2 + x + \underline{\quad} = (\quad\quad)^2$ **50.** $x^2 - 7x + \underline{\quad} = (\quad\quad)^2$

51. $x^2 - 9x + \underline{\quad} = (\quad\quad)^2$ **52.** $x^2 + 3x + \underline{\quad} = (\quad\quad)^2$

53. $x^2 + \underline{\quad} + \dfrac{25}{4} = (\quad\quad)^2$ **54.** $x^2 - \underline{\quad} + \dfrac{121}{4} = (\quad\quad)^2$

55. $x^2 - \underline{\quad} + \dfrac{9}{4} = (\quad\quad)^2$ **56.** $x^2 + \underline{\quad} + \dfrac{81}{4} = (\quad\quad)^2$

Hawkes Learning Systems: Introductory Algebra

 Special Factorizations - Squares

7.5

Solving Quadratic Equations by Factoring

After completing this section, you will be able to:

1. Solve quadratic equations by factoring.

In Chapter 7, the emphasis has been on second-degree polynomials and functions. Such polynomials are particularly evident in physics with the path of thrown objects (or projectiles) affected by gravity, in mathematics involving area (area of a circle, area of a rectangle, square, and triangle), and in many situations in higher level mathematics. The methods of factoring we have studied (*ac*-method, trial-and-error method, difference of two squares, and perfect square trinomials) have been related, in general, to second-degree polynomials. Second-degree polynomials are called **quadratic polynomials** (or **quadratics**) and, as just discussed, they play a major role in many applications of mathematics. In fact, quadratic polynomials and techniques for solving quadratic equations are central topics in the first two courses in algebra.

Solving Quadratic Equations by Factoring

Equations of the form

$$ax^2 + bx + c = 0 \quad \text{where } a, b, \text{ and } c \text{ are constants and } a \neq 0$$

are called **quadratic equations**. The form $ax^2 + bx + c = 0$ is called the **standard form** (or **general form**) of a quadratic equation. In the standard form, a quadratic polynomial is on one side of the equation and 0 is on the other side. Of the equations

$$x^2 - 8x + 12 = 0, \quad 3x^2 - 2x = 27, \quad \text{and} \quad x^2 = 25x$$

only the first is in standard form. The other two can be manipulated algebraically so that one side is 0. Having the constant 0 on one side of a quadratic equation is necessary to be able to solve the equation by factoring because of the following zero-factor property.

Zero-Factor Property

If a product is 0, then at least one of the factors must be 0. That is, if a and b are real numbers, then

$$\textbf{if } a \cdot b = 0, \textbf{ then } a = 0 \textbf{ or } b = 0.$$

Example 1: Solving Factored Quadratic Equations ● ● ● ● ● ● ●

Solve the following quadratic equation.

a. $(x - 5)(2x - 7) = 0$

Solution: Since the quadratic is already factored and the other side of the equation is 0, we use the zero-factor property and set each factor equal to 0. This process yields two linear equations, which can, in turn, be solved.

$$
\begin{array}{llll}
x - 5 = 0 & \text{or} & 2x - 7 = 0 & \\
x = 5 & \text{or} & 2x = 7 & \text{Add 7 to both sides.} \\
& & x = \dfrac{7}{2} & \text{Divide both sides by 2.}
\end{array}
$$

Thus, the two solutions to the original equation are $x = 5$ and $x = \dfrac{7}{2}$. The solutions can be **checked** by substituting them, one at a time, for x in the equation. That is, there will be two "checks."

Substituting $x = 5$ gives

$$(x - 5)(2x - 7) = (5 - 5)(2 \cdot 5 - 7) = (0)(3) = 0$$

Substituting $x = \dfrac{7}{2}$ gives

$$(x - 5)(2x - 7) = \left(\frac{7}{2} - 5\right)\left(2 \cdot \frac{7}{2} - 7\right) = \left(-\frac{3}{2}\right)(7 - 7) = \left(-\frac{3}{2}\right)(0) = 0$$

Therefore, both 5 and $\dfrac{7}{2}$ are solutions to the original equation.

● ●

In Example 1, the polynomial was factored and the other side of the equation was 0. The equation was solved by setting each factor, in turn, equal to 0 and solving the resulting linear equations. These solutions tell us that the original equation has two solutions. In general, a quadratic equation has two solutions. In the special cases, where the two factors are the same, there is only one solution and it is called a **double solution** (or **double root**). The following examples show how to solve quadratic equations by factoring. Remember that the equation must be in standard form with one side of the equation equal to 0. Study these examples carefully.

Example 2: Solving Quadratic Equations ● ● ● ● ● ● ● ● ● ● ●

Solve the following equations by writing the equation in standard form with one side 0 and factoring the polynomial. Then set each factor equal to 0 and solve. Checking is left as an exercise for the student.

a. $3x^2 = 6x$

Solution:

$$3x^2 = 6x$$

$$3x^2 - 6x = 0$$ Write the equation in standard form with 0 on one side.

$$3x(x-2) = 0$$ Factor.

$$3x = 0 \quad \text{or} \quad x - 2 = 0$$ Set each factor equal to 0.

$$x = 0 \quad \text{or} \quad x = 2$$ Solve each linear equation.

The solutions are 0 and 2.

b. $x^2 - 8x + 16 = 0$

Solution: $x^2 - 8x + 16 = 0$ The trinomial is a perfect square.

$$(x-4)^2 = 0$$ Both factors are the same. So there is

$$x - 4 = 0 \quad \text{or} \quad x - 4 = 0$$ only one distinct solution.

$$x = 4 \qquad\qquad x = 4$$

The only solution is 4, and it is called a double solution (or double root).

Continued on next page ...

c. $4x^2 - 4x = 24$

Solution:

$$4x^2 - 4x = 24$$

$$4x^2 - 4x - 24 = 0 \qquad \text{One side must be 0.}$$

$$4\left(x^2 - x - 6\right) = 0 \qquad \text{4 is a common monomial factor.}$$

$$4(x - 3)(x + 2) = 0 \qquad \text{The constant factor 4 can never be 0 and does not affect the solution.}$$

$$x - 3 = 0 \quad \text{or} \quad x + 2 = 0$$

$$x = 3 \qquad\qquad x = -2$$

The solutions are 3 and –2.

d. $\dfrac{2x^2}{15} - \dfrac{x}{3} = -\dfrac{1}{5}$

Solution:

$$\dfrac{2x^2}{15} - \dfrac{x}{3} = -\dfrac{1}{5}$$

$$15 \cdot \dfrac{2x^2}{15} - \dfrac{x}{3} \cdot 15 = -\dfrac{1}{5} \cdot 15 \qquad \text{Multiply each term by 15, the LCM of the denominators, to get integer coefficients.}$$

$$2x^2 - 5x = -3 \qquad \text{Simplify.}$$

$$2x^2 - 5x + 3 = 0 \qquad \text{One side must be 0.}$$

$$(2x - 3)(x - 1) = 0 \qquad \text{Factor by trial-and-error method or the } ac \text{ - method.}$$

$$2x - 3 = 0 \quad \text{or} \quad x - 1 = 0$$

$$2x = 3 \qquad\qquad x = 1$$

$$x = \dfrac{3}{2}$$

The solutions are $\dfrac{3}{2}$ and 1.

e. $(x + 5)^2 = 36$

Solution:

$$(x + 5)^2 = 36$$

$$x^2 + 10x + 25 = 36 \qquad (x + 5)^2 \text{ gives a perfect square trinomial.}$$

$$x^2 + 10x - 11 = 0 \qquad \text{Add –36 to both sides so that one side is 0.}$$

$$(x + 11)(x - 1) = 0 \qquad \text{Factor.}$$

$$x + 11 = 0 \qquad \text{or} \quad x - 1 = 0$$

$$x = -11 \qquad\qquad x = 1$$

The solutions are –11 and 1.

Continued on next page ...

f. $3x(x-1) = 2(5-x)$

Solution:

$$3x(x-1) = 2(5-x)$$

$$3x^2 - 3x = 10 - 2x \qquad \text{Use the distributive property.}$$

$$3x^2 - 3x + 2x - 10 = 0 \qquad \text{Arrange terms so that 0 is on one side.}$$

$$3x^2 - x - 10 = 0 \qquad \text{Simplify.}$$

$$(3x + 5)(x - 2) = 0 \qquad \text{Factor.}$$

$$3x + 5 = 0 \quad \text{or} \quad x - 2 = 0$$

$$x = -\frac{5}{3} \qquad x = 2$$

The solutions are $-\dfrac{5}{3}$ and 2.

In the next example, we show that factoring can be used to solve equations of a degree higher than second-degree. There will be more than two factors. But, just as with quadratics, if the product is 0, then the solutions are found by setting each factor equal to 0.

Example 3: Solving Higher Degree Equations

Solve the following equation: $2x^3 - 4x^2 - 6x = 0$.

Solution:

$$2x^3 - 4x^2 - 6x = 0$$

$$2x(x^2 - 2x - 3) = 0 \qquad \text{Factor out the common monomial } 2x.$$

$$2x(x - 3)(x + 1) = 0 \qquad \text{Factor the trinomial.}$$

$$2x = 0 \quad \text{or} \quad x - 3 = 0 \quad \text{or} \quad x + 1 = 0 \qquad \text{Set each factor equal to 0 and solve.}$$

$$x = 0 \qquad x = 3 \qquad x = -1$$

The solutions are 0, 3, and –1.

The procedures to solve a quadratic equation can be summarized as follows.

To Solve a Quadratic Equation by Factoring

1. *Add or subtract terms as necessary so that 0 is on one side of the equation and the equation is in the standard form $ax^2 + bx + c = 0$ where a, b, and c are real constants and $a \neq 0$.*

2. *Factor completely. (If there are any fractional coefficients, multiply each term by the least common denominator so that all coefficients will be integers.)*

3. *Set each nonconstant factor equal to 0 and solve each linear equation for the unknown.*

4. *Check each solution, one at a time, in the original equation.*

Special Comment: All of the quadratic equations in this section can be solved by factoring. That is, all of the quadratic polynomials are factorable. However, as we have seen in some of the previous sections, not all polynomials are factorable. In Chapter 10, we will develop techniques (other than factoring) for solving quadratic equations whether the quadratic polynomial is factorable or not.

NOTES

COMMON ERROR

A **common error** is to divide both sides of an equation by the variable x. This error can be illustrated by using the equation in Example 2.

$$3x^2 = 6x$$

$$\frac{3x^2}{x} = \frac{6x}{x} \qquad \textbf{WRONG} \text{ DO NOT divide by } x$$

because you lose the solution $x = 0$.

$$3x = 6$$

$$x = 2$$

Factoring is the method to use. By factoring, you will find all solutions as shown in the previous examples.

Practice Problems

Solve each equation by factoring.

1. $x^2 - 6x = 0$

2. $6x^2 - x - 1 = 0$

3. $(x - 2)^2 - 25 = 0$

4. $x^3 - 8x^2 + 16x = 0$

Answers to Practice Problems: 1. $x = 0, x = 6$ **2.** $x = \dfrac{1}{2}, x = -\dfrac{1}{3}$ **3.** $x = 7, x = -3$ **4.** $x = 0, x = 4$

7.5 Exercises

Solve the equations in Exercises 1 – 10 by setting each factor equal to 0 and solving the resulting linear equations.

1. $(x - 3)(x - 2) = 0$

2. $(x + 5)(x - 2) = 0$

3. $(2x - 9)(x + 2) = 0$

4. $(x + 7)(3x - 4) = 0$

5. $0 = (x + 3)(x + 3)$

6. $0 = (x + 10)(x - 10)$

7. $(x + 5)(x + 5) = 0$

8. $(x + 5)(x - 5) = 0$

9. $2x(x - 2) = 0$

10. $3x(x + 3) = 0$

In Exercises 11 – 72, solve the equations by factoring.

11. $x^2 - 3x = 4$

12. $x^2 + 7x = -12$

13. $0 = 5x^2 + 15x$

14. $x^2 - 11x = -18$

15. $2x^2 - 24 = 2x$

16. $0 = 4x^2 - 12x$

17. $x^2 + 8 = 6x$

18. $x^2 = x + 30$

19. $2x^2 + 2x - 24 = 0$

20. $9x^2 + 63x + 90 = 0$

21. $0 = 2x^2 - 5x - 3$

22. $0 = 2x^2 - x - 3$

23. $3x^2 - 4x - 4 = 0$

24. $3x^2 - 8x + 5 = 0$

25. $2x^2 - 7x - 4 = 0$

26. $4x^2 + 8x + 3 = 0$

27. $0 = 3x^2 + 2x - 8$

28. $0 = 6x^2 + 7x + 2$

29. $4x^2 - 12x + 9 = 0$

30. $25x^2 - 60x + 36 = 0$

31. $(x - 1)^2 = 4$

32. $(x - 3)^2 = 1$

33. $(x + 5)^2 = 9$

34. $(x + 4)^2 = 16$

35. $(x + 4)(x - 1) = 6$

36. $(x - 5)(x + 3) = 9$

37. $27 = (x + 2)(x - 4)$

38. $-1 = (x + 2)(x + 4)$

39. $x(x + 7) = 3(x + 4)$

40. $x(x + 9) = 6(x + 3)$

41. $3x(x + 1) = 2(x + 1)$

42. $2x(x - 1) = 3(x - 1)$

43. $x(2x + 1) = 6(x + 2)$

44. $3x(x + 3) = 2(2x - 1)$

45. $8x = 5x^2$

46. $15x = 3x^2$

47. $9x^2 = 36$

48. $4x^2 - 16 = 0$

49. $5x^2 + 10x + 5 = 0$

50. $2x^2 = 4x + 6$

51. $8x^2 + 32 = 32x$

52. $-6x^2 + 18x + 24 = 0$

53. $\dfrac{x^2}{3} - 2x + 3 = 0$

54. $\dfrac{x^2}{9} = 1$

55. $\dfrac{x^2}{5} - x - 10 = 0$

56. $\dfrac{2}{3}x^2 + 2x - \dfrac{20}{3} = 0$

57. $\dfrac{x^2}{8} + x + \dfrac{3}{2} = 0$

58. $\dfrac{x^2}{6} - \dfrac{1}{2}x - 3 = 0$

59. $x^2 - x + \dfrac{1}{4} = 0$

60. $x^2 - \dfrac{7}{6}x + \dfrac{1}{3} = 0$

61. $x^3 + 8x = 6x^2$

62. $x^3 = x^2 + 30x$

63. $6x^3 + 7x^2 = -2x$

64. $3x^3 = 8x - 2x^2$

65. $0 = x^2 - 100$

66. $0 = x^2 - 121$

67. $3x^2 - 75 = 0$

68. $5x^2 - 45 = 0$

69. $x^2 + 8x + 16 = 0$

70. $x^2 + 14x + 49 = 0$

71. $3x^2 = 18x - 27$

72. $5x^2 = 10x - 5$

Hawkes Learning Systems: Introductory Algebra

   Solving Quadratic Equations By Factoring

7.6

Applications of Quadratic Equations

After completing this section, you will be able to:

1. Solve word problems by writing quadratic equations that can be factored and solved.

Whether or not word problems cause you difficulty depends a great deal on your personal experiences and general reasoning abilities. These abilities are developed over a long period of time. A problem that is easy for you, possibly because you have had experience in a particular situation, might be quite difficult for a friend, and vice versa.

Most problems do not say specifically to add, subtract, multiply, or divide. You are to know from the nature of the problem what to do. You are to ask yourself, "What information is given? What am I trying to find? What tools, skills, and abilities do I need to use?"

Word problems should be approached in an orderly manner. Have an "attack plan."

Attack Plan for Word Problems

1. *Read the problem carefully at least twice.*
2. *Decide what is asked for and assign a variable or variable expression to the unknown quantities.*
3. *Organize a chart, or a table, or a diagram relating all the information provided.*
4. *Form an equation. (A formula of some type may be necessary.)*
5. *Solve the equation.*
6. *Check your solution with the wording of the problem to be sure it makes sense.*

Several types of problems lead to quadratic equations. The problems in this section are set up so that the equations can be solved by factoring. More general problems and approaches to solving quadratic equations are discussed in Chapter 10.

Example 1: Applications • • • • • • • • • • • • • • • • • •

a. One number is four more than another and the sum of their squares is 296. What are the numbers?

Solution: Let x = smaller number. Then, $x + 4$ = larger number

$$x^2 + (x + 4)^2 = 296 \quad \text{Add the squares.}$$

$$x^2 + x^2 + 8x + 16 = 296$$

$$2x^2 + 8x - 280 = 0 \quad \text{Write the equation in standard form.}$$

$$2(x^2 + 4x - 140) = 0 \quad \text{Factor out the GCF.}$$

$$(x + 14)(x - 10) = 0 \quad \text{The constant factor 2 does not affect the solution.}$$

$$x + 14 = 0 \quad \text{or} \quad x - 10 = 0$$

$$x = -14 \qquad\qquad x = 10$$

$$x + 4 = -10 \qquad\qquad x + 4 = 14$$

There are two sets of answers to the problem: 10 and 14 or −14 and −10.

Check:
$$10^2 + 14^2 = 100 + 196 = 296$$
$$\text{and } (-14)^2 + (-10)^2 = 196 + 100 = 296$$

b. In an orange grove, there are 10 more trees in each row than there are rows. How many rows are there if there are 96 trees in the grove?

Solution: Let r = number of rows. Then, $r + 10$ = number of trees per row.

Set up the equation and solve.

$$r(r + 10) = 96$$

$$r^2 + 10r = 96$$

$$r^2 + 10r - 96 = 0$$

$$(r - 6)(r + 16) = 0$$

$$r - 6 = 0 \quad \text{or} \quad r + 16 = 0$$

$$r = 6 \qquad\qquad r = -16$$

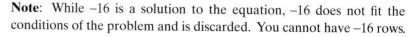

There are 6 rows in the grove ($6 \cdot 16 = 96$ trees).

Note: While −16 is a solution to the equation, −16 does not fit the conditions of the problem and is discarded. You cannot have −16 rows.

Continued on next page...

c. A rectangle has an area of 135 square meters and perimeter of 48 meters. What are the dimensions of the rectangle?

Solution: The area of a rectangle is the product of its length and width ($A = lw$).
The perimeter of a rectangle is given by $P = 2l + 2w$.
Since the perimeter is 48 meters, then the length plus the width must be 24 meters (one-half of the perimeter).
Let w = width. Then, $24 - w$ = length.
Set up the equation and solve.

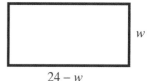

$24 - w$

$$w(24 - w) = 135 \qquad \text{Area} = lw = \text{length times width}$$

$$24w - w^2 = 135$$

$$0 = w^2 - 24w + 135 \qquad \textbf{Note: } 0 \text{ can be on either side of}$$

$$0 = (w - 9)(w - 15) \qquad \text{the equation.}$$

$$w - 9 = 0 \quad or \quad w - 15 = 0$$

$$w = 9 \qquad\qquad w = 15$$

$$24 - 9 = 15 \qquad 24 - 15 = 9$$

The width is 9 meters and the length is 15 meters ($9 \cdot 15 = 135$).

d. A man wants to build a block wall shaped like a rectangle along three sides of his property. If 180 feet of fencing is needed and the area of the lot is 4000 square feet, what are the dimensions of the lot?

Solution: Let x = one of two equal sides. Then, $180 - 2x$ = third side.
Set up the equation and solve.

$$x(180 - 2x) = 4000$$

$$180x - 2x^2 = 4000 \qquad\qquad\qquad 180 - 2x$$

$$0 = 2x^2 - 180x + 4000$$

$$0 = 2(x^2 - 90x + 2000) \qquad x \qquad\qquad\qquad x$$

$$0 = 2(x - 50)(x - 40)$$

$$x - 50 = 0 \quad or \quad x - 40 = 0$$

$$x = 50 \qquad\qquad x = 40$$

$$180 - 2(50) = 80 \qquad 180 - 2(40) = 100$$

From this information, there are two possible answers: the lot is 50 ft. by 80 ft., or the lot is 40 ft. by 100 ft.   *Continued on next page...*

e. The sum of the squares of two positive consecutive odd integers is 202. Find the integers.

Solution: Let n = first odd integer. Then, $n + 2$ = next consecutive odd integer.
Set up the equation and solve.

$$n^2 + (n + 2)^2 = 202$$

$$n^2 + n^2 + 4n + 4 = 202$$

$$2n^2 + 4n - 198 = 0$$

$$2(n^2 + 2n - 99) = 0$$

$$2(n - 9)(n + 11) = 0$$

$$n = 9 \quad \text{or} \quad n = -11 \qquad \text{Since the problem asked for positive}$$

$$n + 2 = 11 \qquad\qquad\qquad\qquad \text{integers, } -11 \text{ is discarded.}$$

The first integer is 9 and the next consecutive odd integer is 11.

Checking: $9^2 + 11^2 = 81 + 121 = 202$

7.6 Exercises

Determine a quadratic equation for each of the following problems in Exercises 1 – 44. Then solve the equation.

1. The square of an integer is equal to seven times the integer. Find the integer.

2. The square of an integer is equal to twice the integer. Find the integer.

3. The square of a positive integer is equal to the sum of the integer and 12. Find the integer.

4. If the square of a positive integer is added to three times the integer, the result is 28. Find the integer.

5. One number is seven more than another. Their product is 78. Find the numbers.

6. One positive number is three more than twice another. If the product is 27, find the numbers.

7. If the square of a positive integer is added to three times the number, the result is 54. Find the number.

8. One number is six more than another. The difference between their squares is 132. What are the numbers?

9. The difference between two positive integers is 8. If the smaller is added to the square of the larger, the sum is 124. Find the numbers.

10. One number is three less than twice another. The sum of their squares is 74. Find the numbers.

11. One number is five less than another. The sum of their squares is 97. Find the numbers.

12. Find a positive integer such that the product of the integer with a number three less than the integer is equal to the integer increased by 32.

13. The product of two consecutive positive integers is 72. Find the integers.

14. Find two consecutive integers whose product is 110.

15. Find two consecutive positive integers such that the sum of their squares is 85.

16. Find two consecutive positive integers such that the square of the second integer added to four times the first is equal to 41.

17. Find two consecutive positive integers such that the difference between their squares is 17.

18. The product of two consecutive odd integers is 63. Find the integers.

19. The product of two consecutive even integers is 120. Find the integers.

20. The product of two consecutive even integers is 168. Find the integers.

21. The length of a rectangle is twice the width. The area is 72 square inches. Find the length and width of the rectangle.

22. The length of a rectangle is three times the width. If the area is 147 square centimeters, find the length and width of the rectangle.

23. The length of a rectangular yard is 12 meters greater than the width. If the area of the yard is 85 square meters, find the length and width of the yard.

24. The length of a rectangle is three centimeters greater than the width. The area is 108 square centimeters. Find the length and width of the rectangle.

25. The width of a rectangle is 4 feet less than the length. The area is 117 square feet. Find the length and width of the rectangle.

26. The height of a triangle is 4 feet less than the base. The area of the triangle is 16 square feet. Find the length of the base and height of the triangle.

27. The base of a triangle exceeds the height by 5 meters. If the area is 42 square meters, find the length of the base and height of the triangle.

28. The base of a triangle is 15 inches greater than the height. If the area is 63 square inches, find the length of the base.

29. The base of a triangle is 6 feet less than the height. The area is 56 square feet. Find the length of the height.

30. The perimeter of a rectangle is 32 inches. The area of the rectangle is 48 square inches. Find the dimensions of the rectangle.

31. The area of a rectangle is 24 square centimeters. If the perimeter is 20 centimeters, find the length and width of the rectangle.

32. The perimeter of a rectangle is 40 meters and the area is 96 square meters. Find the dimensions of the rectangle.

33. An orchard has 140 orange trees. The number of rows exceeds the number of trees per row by 13. How many trees are there in each row?

34. One formation for a drill team is rectangular. The number of members in each row exceeds the number of rows by 3. If there is a total of 108 members in the formation, how many rows are there?

35. A theater can seat 144 people. The number of rows is 7 less than the number of seats in each row. How many rows of seats are there?

36. The length of a rectangle is 7 centimeters greater than the width. If 4 centimeters are added to both the length and width, the new area would be 98 square centimeters. Find the dimensions of the original rectangle.

37. The width of a rectangle is 5 meters less than the length. If 6 meters are added to both the length and width, the new area will be 300 square meters. Find the dimensions of the original rectangle.

38. Susan is going to fence a rectangular flower garden in her back yard. She has 50 feet of fencing and she plans to use the house as the fence on one side of the garden. If the area is 300 square feet, what are the dimensions of the flower garden?

39. A rancher is going to build a corral with 52 yards of fencing. He is planning to use the barn as one side of the corral. If the area is 320 square yards, what are the dimensions?

40. The area of a square is 81 square centimeters. How long is each side of the square?

41. The length of a rectangle is three times the width. If the area is 48 square feet, find the length and width of the rectangle.

42. The length of a rectangle is five times the width. If the area is 180 square inches, find the length and width of the rectangle.

43. One number is eight more than another. Their product is −16. What are the numbers?

44. One number is 10 more than another. If their product is −25, find the numbers.

Hawkes Learning Systems: Introductory Algebra

 Applications of Quadratic Equations

7.7

Additional Applications of Quadratic Equations

After completing this section, you will be able to:

1. Solve the applied formulas.

2. Solve the resulting quadratic equations.

The following applications provide practice with equations found in real-life settings. The formulas are provided without any discussion of how or why they fit the described situation. Further studies in other fields will show how some of these formulas are generated.

Note that this section consists of brief descriptions and then exercises.

Volume of a Cylinder

The volume of a cylinder is given by the formula

$$V = \pi r^2 h$$

where V is the volume;
  $\pi = 3.14$ (3.14 is an approximation for π);
  r is the radius;
  h is the height.

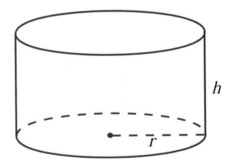

1. Find the volume of a cylinder with a radius of 6 in. and a height of 20 in.
2. Find the height of a cylinder if the volume is 282.6 cu. in. and the radius in 3 in.
3. A cylinder has a height of 14 in. and a volume of 1099 cu. in. Find the radius.
4. Find the radius of a cylinder whose volume is 2512 cu. cm and whose height is 8 cm.

Load on a Wooden Beam

The safe load, *L,* of a horizontal wooden beam supported at both ends is expressed by the formula

$$L = \frac{kbd^2}{l}$$

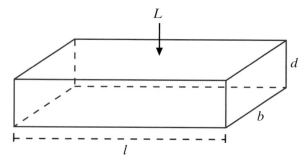

where *L* is expressed in pounds;

b is the breadth (or width) in inches;

d is the depth in inches;

l is the length in inches;

k, a constant, depends on the grade of the beam and is expressed in pounds per square inch.

5. What is the safe maximum load of a white pine beam 180 in. long, 3 in. wide, and 4 in. deep if $k = 3000$ lb. per sq. in.?

6. Find the constant *k* if a beam 144 in. long, 2 in. wide, and 6 in. deep supports a maximum load of 1100 lb.

7. A solid oak beam is required to support a load of 12,000 lbs. It can be no more than 8 in. deep and is 192 in. long. For this grade of oak, $k = 6000$ lb. per sq. in. How wide should the beam be?

8. The safe maximum load of a white pine beam 4 in. wide and 150 in. long is 2880 lb. For white pine, $k = 3000$ lb. per sq. in. Find the depth of the beam.

9. A Douglas fir beam is required to support a load of 20,000 lb. For Douglas fir, $k = 4800$ lb. per sq. in. If the beam is 6 in. wide and 144 in. long, what is the minimum depth for the beam to support the required load?

Height of a Projectile

The equation

$$h = -16t^2 + v_0 t$$

gives the height h, in feet, that a body will be above the earth at time t, in seconds, if it is projected upward with an initial velocity v_0, in feet per second. (The initial velocity v_0, is the velocity when $t = 0$.)

10. Find the height of an object 3 seconds after it has been projected upward at a rate of 56 feet per second.

11. Find the height of an object 5 seconds after it has been projected upward at a rate of 120 feet per second.

12. A ball is thrown upward with a velocity of 144 feet per second. When will it strike the ground?

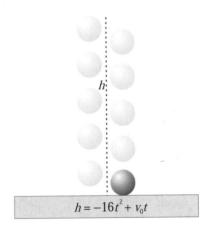

$$h = -16t^2 + v_0 t$$

13. An object is projected upward at a rate of 160 feet per second. Find the time when it is 384 feet above the ground.

14. An object is projected upward at a rate of 96 feet per second. Find the time when it is 144 feet above the ground.

Electrical Power

The power output of a generator is given by the equation

$$P_0 = E_g I - r_g I^2$$

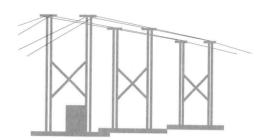

where P_0 is measured in kilowatts;
        E_g is measured in volts;
        r_g is measured in ohms;
        I is measured in amperes.

15. Find I if $P_0 = 120$ kilowatts, $E_g = 16$ volts, and $r_g = \dfrac{1}{2}$ ohm.

16. Find I if $P_0 = 180$ kilowatts, $E_g = 22$ volts, and $r_g = \dfrac{2}{3}$ ohm.

Consumer Demand

The demand for a product is the number of units of the product that consumers are willing to buy when the market price is p dollars. The consumers' total expenditure for the product is found by multiplying the price times the demand.

17. When fishing reels are priced at p dollars, local consumers will buy $36 - p$ fishing reels. What is the price if total sales were $320?

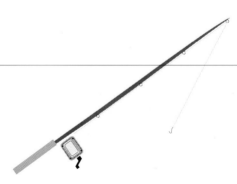

18. A manufacturer can sell $100 - 2p$ lamps at p dollars each. If the receipts from the lamps total $1200, what is the price of the lamps?

Consumer Demand

The demand for a certain commodity is given by

$$D = -20p^2 + ap + 1200 \text{ units per month}$$

where p is the selling price and a is a constant.

19. Find the selling price if 1120 units are sold and $a = 60$.

20. Find the selling price if 1860 units are sold and $a = 232$.

Hawkes Learning Systems: Introductory Algebra

You may wish to review:
Applications of Quadratic Equations

7.8   Additional Factoring Review

As we have seen throughout Chapter 7, factoring is a basic algebraic skill. Factoring is useful in solving equations and, as we will see in Chapter 8, necessary for simplifying algebraic fractions. This section is provided as extra practice in factoring. The following guidelines are provided for easy reference.

General Guidelines for Factoring Polynomials

1. **Always look for a common monomial factor first.** *If the leading coefficient is negative, factor out a negative monomial even if it is just −1.*

2. **Check the number of terms.**

   a. Two terms:

       (1) difference of two squares? – factorable

       (2) sum of two squares? – not factorable

   b. Three terms:

       (1) perfect square trinomial?

       (2) use ac-method?

       Guidelines for the ac-method

         (a) Multiply $a \cdot c$.

         (b) Find two integers whose product is ac and whose sum is b. If this is not possible, then the trinomial is not factorable.

         (c) Rewrite the middle terms (bx) using the two numbers found in Step (b) as coefficients.

         (d) Factor by grouping.

       (3) use trial-and-error method?

       Guidelines for the Trial-and-Error Method

         (a) If the sign of the constant term is positive (+), the signs in both factors will be the same, either both positive or both negative.

         (b) If the sign of the constant term is negative (−), the signs in the factors will be different, one positive and one negative.

Continued on next page...

c. *Four terms:*

   (1) group terms with a common factor and factor out any common binomial
   factor.

3. Check the possibility of factoring any of the factors.

7.8 Exercises

Factor completely, if possible.

1. $m^2 + 7m + 6$ **2.** $a^2 - 4a + 3$ **3.** $x^2 + 11x + 18$

4. $y^2 + 8y + 15$ **5.** $x^2 - 100$ **6.** $n^2 - 8n + 12$

7. $m^2 - m - 6$ **8.** $y^2 - 49$ **9.** $a^2 + 2a + 24$

10. $-x^2 - 12x - 35$ **11.** $64x^2 - 1$ **12.** $x^2 - 4x - 21$

13. $x^2 + 10x + 25$ **14.** $x^2 + 3x - 10$ **15.** $x^2 + 9x - 36$

16. $x^2 + 17x + 72$ **17.** $x^2 + 13x + 36$ **18.** $-2y^2 + 24y - 70$

19. $-5x^2 + 70x - 240$ **20.** $7x^2 + 14x - 168$ **21.** $200 + 20x - 4x^2$

22. $64 + 49x^2$ **23.** $3x^2 - 147$ **24.** $x^3 - 4x^2 - 12x$

25. $3x^3 + 15x^2 + 18x$ **26.** $112x - 2x^2 - 2x^3$ **27.** $16x^3 - 100x$

28. $2x^2 - 3x + 1$ **29.** $-3x^2 + 17x - 10$ **30.** $2x^2 + 7x + 3$

31. $4x^2 - 14x + 6$ **32.** $6x^2 - 11x + 4$ **33.** $12x^2 - 32x + 5$

34. $12x^2 + x - 6$ **35.** $6x^2 + x - 35$ **36.** $-4x^2 + 18x - 20$

37. $8x^2 + 6x - 35$ **38.** $12x^2 + 5x - 3$ **39.** $20x^2 - 21x - 54$

40. $-150x^2 + 96$ **41.** $12x^2 - 60x - 75$ **42.** $14 + 11x - 15x^2$

43. $24 + x - 3x^2$ **44.** $21x^2 - x - 10$ **45.** $-8x^2 + 22x - 15$

46. $63x^2 - 40x - 12$ **47.** $20x^2 + 9x - 20$ **48.** $35x^2 - x - 6$

49. $18x^2 - 15x + 2$ **50.** $12x^2 - 47x + 11$ **51.** $252x - 175x^3$

52. $-12x^3 - 2x^2 - 70x$ **53.** $21x^3 - 13x^2 - 2x$ **54.** $13x^3 + 120x^2 + 100x$

55. $36x^3 + 21x^2 - 30x$ **56.** $63x - 3x^2 - 30x^3$ **57.** $16x^3 - 52x^2 + 22x$

58. $24x^3 - 4x^2 - 160x$ **59.** $75 + 10x + 120x^2$ **60.** $144x^3 - 10x^2 - 50x$

61. $xy + 3y - 4x - 12$ **62.** $2xz + 10x + z + 5$ **63.** $x^2 + 2xy - 6x - 12y$

64. $2y^2 + 6yz + 5y + 15z$ **65.** $-x^3 + 8x^2 + 5x - 40$ **66.** $2x^3 - 14x^2 - 3x + 21$

Hawkes Learning Systems: Introductory Algebra

You may wish to review:
Greatest Common Factor of Two or More Terms
Greatest Common Factor of a Polynomial
Factoring Expressions by Grouping
Factoring Trinomials: Leading Coefficient 1
Factoring Trinomials by Trial and Error
Factoring Trinomials by Grouping
Special Factorizations - Squares
Solving Quadratic Equations By Factoring

Chapter 7 Index of Key Ideas and Terms

2. If the sign of the constant term is negative (–), the signs in the factors will be different, one positive and one negative.

Add a square term to a binomial so that the resulting trinomial is a perfect square trinomial.

Standard form: $ax^2 + bx + c = 0 \ (a \neq 0)$

If a product is 0, then at least one of the factors must be 0. For real numbers a and b, if $a \cdot b = 0$, then $a = 0$ or $b = 0$.
1. Add or subtract terms as necessary so that 0 is on one side of the equation and the equation is in the standard form $ax^2 + bx + c = 0$.
2. Factor completely. (If there are any fractional coefficients, multiply each term by the least common denominator so that all coefficients will be integers.)
3. Set each nonconstant factor equal to 0 and solve each linear equation for the unknown.
4. Check each solution, one at a time, in the original equation.

1. Read the problem carefully at least twice.
2. Decide what is asked for and assign a variable or variable expression to the unknown quantities.

3. Organize a chart, or a table, or a diagram relating all the information provided.
4. Form an equation.
(A formula of some type may be necessary.)
5. Solve the equation.
6. Check your solution with the wording of the problem to be sure it makes sense.

General Guidelines for Factoring Polynomials   pages 482 - 483

1. Always look for a common monomial factor first. If the leading coefficient is negative, factor out a negative monomial even if it is just −1.
2. Check the number of terms.
    a. Two terms:
        (1) difference of two squares? - factorable
        (2) sum of two squares? - not factorable
    b. Three terms:
        (1) perfect square trinomial?
        (2) use *ac*-method?
        (3) use trial-and-error method?
    c. Four terms:
        (1) group terms with a common factor and factor out any common binomial factor.
3. Check the possibility of factoring any of the factors.

Chapter 7 Review

For a review of the topics and problems from Chapter 7, look at the following lessons from *Hawkes Learning Systems: Introductory Algebra*

Greatest Common Factor of Two or More Terms
Greatest Common Factor of a Polynomial
Factoring Expressions By Grouping
Factoring Trinomials: Leading Coefficient 1
Factoring Trinomials By Trial and Error
Factoring Trinomials By Grouping
Special Factorizations - Squares
Solving Quadratic Equations By Factoring
Applications of Quadratic Equations

Chapter 7 Test

Find the GCF for each of the sets of terms in Exercises 1 and 2.

1. $30, 75, 90$

2. $40x^2y,\ 48x^2y^3,\ 56x^2y^2$

In Exercises 3 and 4, simplify each expression. Assume that no denominator is 0.

3. $\dfrac{16x^4}{8x}$

4. $\dfrac{42x^3y^3}{-6x^3y}$

Factor completely, if possible, each of the polynomials in Exercises 5 – 16.

5. $20x^3y^2 + 30x^3y + 10x^3$

6. $x^2 - 9x + 20$

7. $-x^2 - 14x - 49$

8. $6x^2 - 6$

9. $12x^2 + 2x - 10$

10. $3x^2 + x - 24$

11. $16x^2 - 25y^2$

12. $2x^3 - x^2 - 3x$

13. $6x^2 - 13x + 6$

14. $2xy - 3y + 14x - 21$

15. $4x^2 + 25$

16. $-3x^3 + 6x^2 - 6x$

In Exercises 17 and 18, complete the square by adding the correct missing term on the left, then factor as indicated.

17. $x^2 - 10x + \underline{\hspace{1em}} = (\hspace{2em})^2$

18. $x^2 + 16x + \underline{\hspace{1em}} = (\hspace{2em})^2$

Solve the equations in Exercises 19 – 25.

19. $(x + 2)(3x - 5) = 0$

20. $x^2 - 7x - 8 = 0$

21. $-3x^2 = 18x$

22. $\dfrac{2x^2}{5} - 6 = \dfrac{4x}{5}$

23. $0 = 4x^2 - 17x - 15$

24. $0 = 8x^2 - 2x - 15$

25. $(2x - 7)(x + 1) = 6x - 19$

26. One number is 10 less than five times another number. Their product is 120. Find the numbers.

27. The length of a rectangle is 7 centimeters less than twice the width. If the area of the rectangle is 165 square centimeters, find the length and width.

28. The product of two consecutive positive integers is 342. Find the two integers.

29. The difference between two positive numbers is 9. If the smaller is added to the square of the larger, the result is 147. Find the numbers.

30. A sheet of metal is in the shape of a rectangle with width 12 inches and length 20 inches. A slot (see the figure) of width x inches and length $x + 3$ inches is cut from the top of the rectangle.

    a. Write a polynomial function, $A(x)$, that represents the area of the remaining figure.

    b. Write a polynomial function, $P(x)$, that represents the perimeter of the remaining figure.

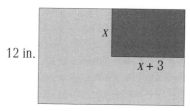

12 in.

x

$x + 3$

20 in.

Cumulative Review: Chapter 1 – 7

Find the LCM for each set of terms in Exercises 1 and 2.

1. $20, 12, 24$

2. $8x^2,\ 14x^2y,\ 21xy$

Perform the indicated operation in Exercises 3 – 6. Reduce all answers to lowest terms.

3. $\dfrac{7}{12} + \dfrac{9}{16}$

4. $\dfrac{11}{15a} - \dfrac{5}{12a}$

5. $\dfrac{6x}{25} \cdot \dfrac{5}{4x}$

6. $\dfrac{40}{92} \div \dfrac{2}{15x}$

Simplify each of the expressions in Exercises 7 – 10 so that it has only positive exponents.

7. $\dfrac{-24x^4y^2}{3x^2y}$

8. $\dfrac{36x^3y^5}{9xy^4}$

9. $\dfrac{-4x^2y}{12xy^{-3}}$

10. $\dfrac{21xy^2}{3x^{-1}y}$

11. Given the relation $r = \{(2, -3), (3, -2), (5, 0), (7.1, 3.2)\}$.
  a. What is the domain of the relation?
  b. What is the range of the relation?
  c. Is the relation a function? Explain.

12. For the function $f(x) = x^2 - 3x + 4$, find
  a. $f(-6)$
  b. $f(0)$
  c. $f\left(\dfrac{1}{2}\right)$

13. Find the equation of the line determined by the two points $(-5, 3)$ and $(2, -4)$. Graph the line.

14. Find the equation of the line parallel to the line $2x - y = 7$ and passing through the point $(-1, 6)$. Graph both lines.

In Exercises 15 and 16, use the vertical line test to determine whether each of the graphs does or does not represent a function.

15.

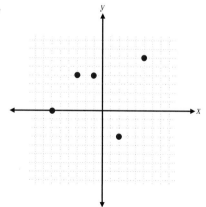

16.

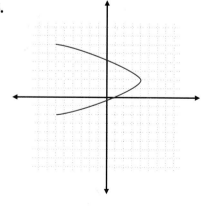

In Exercises 17 and 18, use a graphing calculator to estimate the solutions to each system of equations.

17. $\begin{cases} 5x - y = 1.2 \\ 4x + 3y = -2.1 \end{cases}$

18. $\begin{cases} -3x + y = 5 \\ x + 2y = 4 \end{cases}$

Solve the systems of equations in Exercises 19 – 21. State whether the system is (a) consistent, (b) inconsistent, or (c) dependent.

19. $\begin{cases} 3x - y = 12 \\ x + 2y = -3 \end{cases}$

20. $\begin{cases} y = 4x - 1 \\ 5x + 2y = -2 \end{cases}$

21. $\begin{cases} y = 2x - 7 \\ 4x - 2y = 3 \end{cases}$

22. Use a graphing calculator to graph the two functions $f(x) = \dfrac{1}{3}x$ and $g(x) = x^2 - 5$. Estimate the points of intersection.

Perform the indicated operations in Exercises 23 – 32 and simplify by combining like terms.

23. $2(4x + 3) + 5(x - 1)$

24. $2x(x + 5) - (x + 1)(x - 3)$

25. $\left(x^2 + 3x - 1 \right) + \left(2x^2 - 8x + 4 \right)$

26. $\left(2x^2 + 6x - 7 \right) + \left(2x^2 - x - 1 \right)$

27. $\left(2x^2 - 5x + 3 \right) - \left(x^2 - 2x + 8 \right)$

28. $(x + 1) - \left(4x^2 + 3x - 2 \right)$

29. $(2x - 7)(x + 4)$

30. $-(x + 6)(3x - 1)$

31. $(x + 6)^2$

32. $(2x - 7)^2$

In Exercises 33 and 34, divide by using long division and write the answers in the form $Q + \dfrac{R}{D}$.

33. $\left(3x^2 - 5x + 20 \right) \div (x - 2)$

34. $\dfrac{x^3 - 4x^2 + 6x + 10}{x + 4}$

Factor each expression as completely as possible in Exercises 35 – 54.

35. $8x - 20$

36. $6x - 96$

37. $2x^2 - 15x + 18$

38. $6x^2 - x - 12$

39. $5x - 10$

40. $-12x^2 - 16x$

41. $16x^2y - 24xy$

42. $10x^4 - 25x^3 + 5x^2$

43. $4x^2 - 1$

44. $y^2 - 20y + 100$

45. $3x^2 - 48y^2$

46. $x^2 - 7x - 18$

47. $5x^2 + 40x + 80$

48. $x^2 + x + 3$

49. $25x^2 + 20x + 4$

50. $3x^2 + 5x + 2$

51. $2x^3 - 20x^2 + 50x$

52. $4x^3 + 100x$

53. $xy + 3x + 2y + 6$

54. $ax - 2a - 2b + bx$

Complete the square by adding the correct term in Exercises 55 – 59 so that the trinomials in each will factor as indicated.

55. $x^2 - 4x + \underline{\quad} = (\quad)^2$

56. $x^2 + 18x + \underline{\quad} = (\quad)^2$

57. $x^2 - 8x + \underline{\quad} = (\quad)^2$

58. $x^2 - \underline{\quad} + 25 = (\quad)^2$

59. $x^2 - \underline{\quad} + \dfrac{25}{4} = (\quad)^2$

Solve the equations in Exercises 60 – 74.

60. $(x - 7)(x + 1) = 0$

61. $x(3x + 5) = 0$

62. $4x^2 + 9x - 9 = 0$

63. $21x - 3x^2 = 0$

64. $0 = x^2 + 8x + 12$

65. $x^2 = 3x + 28$

66. $x^3 + 5x^2 - 6x = 0$

67. $\dfrac{1}{4}x^2 + x - 15 = 0$

68. $0 = 15 - 12x - 3x^2$

69. $x^3 + 14x^2 + 49x = 0$

70. $8x = 12x + 2x^2$

71. $6x^2 = 24x$

72. $2x(x - 2) = x + 25$

73. $(4x - 3)(x + 1) = 2$

74. $2x(x + 5)(x - 2) = 0$

75. A boat can travel 21 miles downstream in 3 hours. The return trip takes 7 hours. Find the speed of the boat in still water and the speed of the current.

76. Karl invested a total of $10,000 in two separate accounts. One account paid 6% interest and the other paid 8% interest. If the annual income from both accounts was $650, how much did he have invested in each account?

77. The perimeter of a rectangle is 60 inches. The area of the rectangle is 221 square inches. Find the length and width of the rectangle. (**Hint**: After substituting and simplifying, you will have a quadratic equation.)

78. The difference between two positive numbers is 9. If the smaller is added to the square of the larger, the result is 147. Find the numbers.

79. Find two consecutive integers such that the sum of their squares is equal to 145.

80. A ball is thrown upward with an initial velocity of 88 feet per second $(v_0 = 88)$. When will the ball be 120 feet above the ground? $\left[h(t) = -16t^2 + v_0 t. \right]$

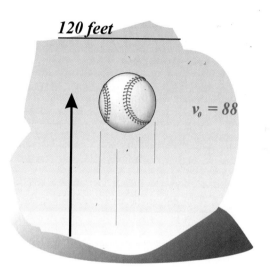

120 feet

$v_0 = 88$

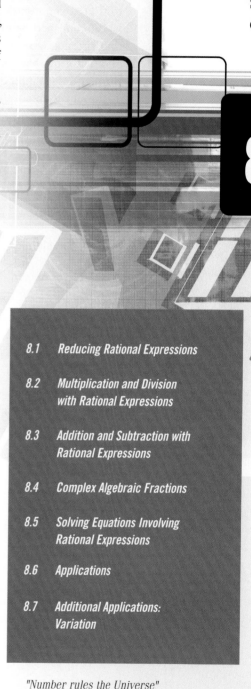
Rational Expressions

Did You Know?

Pythagoras (c. 550 BC) was a Greek mathematician who founded a secret brotherhood whose objective was to investigate music, astronomy, and mathematics, especially geometry. The Pythagoreans believed that all physical phenomena were explainable in terms of "arithmos," the basic properties of whole numbers and their ratios.

In Chapter 8, you will be studying rational numbers (fractions), numbers written as the ratio of two integers. The Pythagorean Society attached mystical significance to these rational numbers. Their investigations of musical harmony revealed that a vibrating string must be divided exactly into halves, thirds, fourths, fifths, and so on, to produce tones in harmony with the string vibrating as a whole. The recognition that the chords which are pleasing to the ear correspond to exact divisions of the vibrating string by whole numbers stimulated Pythagoras to propose that all physical properties could be described using rational numbers. The Society attempted to compute the orbits of the planets by relating them to the musical intervals. The Pythagoreans thought that as the planets moved through space they produced music, the music of the spheres.

Unfortunately, a scandalous idea soon developed within the Pythagorean Society. It became apparent that there were some numbers that could not be represented as the ratio of two whole numbers. This shocking idea came about when the Pythagoreans investigated their master's favorite theorem, the Pythagorean Theorem:

Given a right triangle, the length of the hypotenuse squared is equal to the sum of the lengths squared of the other two sides (as illustrated in the diagrams).

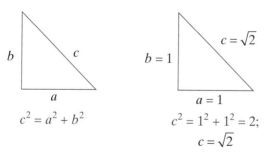

$$c^2 = a^2 + b^2$$

$$c^2 = 1^2 + 1^2 = 2;$$
$$c = \sqrt{2}$$

The hypotenuse of the second triangle is some number that, when squared, gives two. It can be proven (and it was by the Pythagoreans) that no such rational number exists. Imagine the amazement within the Brotherhood. Legend has it that the Pythagoreans attempted to keep the "irrational" numbers a secret. Hippaes, the Brother who first told an outsider about the new numbers, supposedly was expelled from the Society and punished by death for his unfaithfulness.

"Number rules the Universe"

The Pythagoreans

R ational expressions are fractions. So, if you know how to work with fractions in arithmetic, you will find the same techniques used in Chapter 8. You will need to find a common denominator to add (or subtract) rational expressions, and division is accomplished by multiplying by the reciprocal of the divisor. The basic difference is that the numerators and denominators are polynomials and the skills of multiplying and factoring that you learned in Chapters 6 and 7 will be needed. If you keep in mind throughout Chapter 8 that you are following the same rules you learned for fractions in arithmetic, some of the difficult-looking expressions may seem considerably easier.

8.1   Reducing Rational Expressions

Objectives

After completing this section, you will be able to:

1. Find the numerical value of a rational expression.

2. Reduce rational expressions to lowest terms.

3. Determine what values of the variable, if any, will make a rational expression undefined.

Introduction to Rational Expressions

The term rational number is the technical name for a fraction in which both the numerator and denominator are integers. Similarly, the term **rational expression** is the technical name for a fraction in which both the numerator and denominator are polynomials.

Rational Expression

A **rational expression** *is an algebraic expression that can be written in the form*

$$\frac{P}{Q} \text{ where P and Q are polynomials and } Q \neq 0.$$

Examples of rational expressions are

$$\frac{5x^2}{7}, \quad \frac{m^2 - 9}{m^2 + 9}, \quad \text{and} \quad \frac{x^2 + x - 6}{x^3 - 2x^2}$$

As with rational numbers, the denominators of rational expressions cannot be 0. If a numerical value is substituted for the variable in a rational expression and the denominator assumes a value of 0, we say that the expression is **undefined** for that value of the variable.

To determine which values, if any, of the variable will make the expression undefined,

1. set the denominator equal to 0 and
2. solve the resulting equation. Then,
3. the solutions to this equation will make the expression undefined.

Example 1: Undefined • • • • • • • • • • • • • • • • • • •

Determine what values of the variable, if any, will make the rational expression undefined.

a. $\dfrac{3}{2x-1}$

Solution: Set the denominator equal to 0: $\quad 2x-1=0$

Solve the equation: $\qquad 2x=1$

$$x=\frac{1}{2}$$

Thus, the expression $\dfrac{3}{2x-1}$ is undefined for $x=\dfrac{1}{2}$. Any other real number may be substituted for x in the expression. We write $x\neq\dfrac{1}{2}$.

b. $\dfrac{m^2-4}{m^2-2m-3}$

Solution: Set the denominator equal to 0: $\qquad m^2-2m-3=0$

Solve the equation: $\qquad (m-3)(m+1)=0$

$$m-3=0 \quad\text{or}\quad m+1=0$$
$$m=3 \qquad\qquad m=-1$$

Thus, the expression $\dfrac{m^2-4}{m^2-2m-3}$ is undefined for $m=3$ and $m=-1$.

We write $m\neq 3,-1$.

c. $\dfrac{x-5}{x^2+1}$

Solution: Set the denominator equal to 0: $\quad x^2+1=0$

Solve the equation: $\quad x^2+1=0$

$$x^2=-1$$

But there are no real numbers whose square is -1. Therefore, this equation has no real solutions and the denominator will not be 0. There are **no restrictions** on the variable in the real number system.

• •

NOTES

Special Comment About the Numerator Being 0:

If the numerator of a fraction has a value of 0 and the denominator is not 0, then the fraction is defined and has a value of 0. For example, if, in Example 1c, $x = 5$, then $\dfrac{x-5}{x^2+1} = \dfrac{5-5}{5^2+1} = \dfrac{0}{26} = 0$. If both numerator and denominator are 0, then the fraction is undefined just as in the case where only the denominator is 0.

While we cannot substitute values for the variable that will make a denominator 0, other values may be substituted. The following examples illustrate how to find the numerical value of a rational expression by substituting a value for the variable.

Example 2: Determine the Value

Find the value of each rational expression for the given value of the variable.

a. $\dfrac{2x}{x^2+1}$; $x = 2$

Solution: $\dfrac{2x}{x^2+1} = \dfrac{2 \cdot 2}{2^2+1} = \dfrac{4}{4+1} = \dfrac{4}{5}$

Note that any real number may be substituted for x since the denominator x^2+1 is never 0 for any real number.

b. $\dfrac{a-3}{a^2-5}$; $a = 3$

Solution: $\dfrac{a-3}{a^2-5} = \dfrac{3-3}{3^2-5} = \dfrac{0}{9-5} = \dfrac{0}{4} = 0$ Note that a numerator may be 0.

Example 2b illustrates an important fact about fractions. **If the numerator is 0 and the denominator is not 0, then the value of the fraction is 0.**

Reducing Rational Expressions

The rules for operating with rational expressions are essentially the same as those for operating with fractions in arithmetic. That is, adding, subtracting, multiplying and dividing with rational expressions are operations involving factoring and common denominators just as with fractions in arithmetic. The following rules for arithmetic with fractions were discussed in Chapter 2.

Summary of Arithmetic Rules for Rational Numbers (or Fractions)

A *fraction* (or *rational number*) is a number that can be written in the form $\frac{a}{b}$ where a and b are integers and $b \neq 0$. No denominator can be 0.

The Fundamental Principle: $\frac{a}{b} = \frac{a \cdot k}{b \cdot k}$ $b, k \neq 0$.

The **reciprocal** of $\frac{a}{b}$ is $\frac{b}{a}$ and $\frac{a}{b} \cdot \frac{b}{a} = 1$ where $a, b \neq 0$.

Multiplication: $\frac{a}{b} \cdot \frac{c}{d} = \frac{a \cdot c}{b \cdot d}$ where $b, d \neq 0$.

Division: $\frac{a}{b} \div \frac{c}{d} = \frac{a}{b} \cdot \frac{d}{c}$ where $b, c, d \neq 0$.

Addition: $\frac{a}{b} + \frac{c}{b} = \frac{a + c}{b}$ where $b \neq 0$.

Subtraction: $\frac{a}{b} - \frac{c}{b} = \frac{a - c}{b}$ where $b \neq 0$.

These same rules are valid for rational expressions. Each can be restated by replacing a and b with P and Q where P and Q represent polynomials. In particular, the Fundamental Principle can be restated as follows:

The Fundamental Principle of Rational Expressions

If $\frac{P}{Q}$ is a rational expression and $Q \neq 0$ and K is a polynomial and $K \neq 0$, then

$$\frac{P}{Q} = \frac{P \cdot K}{Q \cdot K}$$

The Fundamental Principle can be used to build a rational expression to higher terms or to reduce a rational expression. We will want to build to higher terms when adding and subtracting rational expressions in Section 8.3. At this time, we will discuss reducing rational expressions by factoring the numerators and denominators, just as we did with reducing rational numbers.

To reduce a rational expression, factor both the numerator and denominator and use the Fundamental Principle to "divide out" any common factors. A rational expression is reduced to lowest terms if the greatest common factor (GCF) of the numerator and denominator is 1.

For example, the following expressions are reduced to lowest terms.

$$\text{Fraction: } \frac{15}{35} = \frac{\cancel{5} \cdot 3}{\cancel{5} \cdot 7} = 1 \cdot \frac{3}{7} = \frac{3}{7}$$

$$\text{Rational expression: } \frac{x^2 + 5x + 6}{x^2 + 7x + 12} = \frac{(x+3)(x+2)}{(x+3)(x+4)} = 1 \cdot \frac{x+2}{x+4} = \frac{x+2}{x+4}$$

Remember that **no denominator can be 0**. Thus, in the illustration here, $x + 3 \neq 0$ and $x + 4 \neq 0$ (or $x \neq -3$ and $x \neq -4$). Therefore, in the expression $\dfrac{x^2 + 5x + 6}{x^2 + 7x + 12}$, any real number, except -3 and -4, may be substituted for x. We will see that this idea is particularly important when solving equations.

NOTES

COMMON ERROR

The key word when reducing is **factor**. Many students incorrectly "divide out" terms or numbers that are not factors.

Consider the following incorrect statement that $2 = 4$:

$$2 = \frac{6}{3} = \frac{\cancel{3} + 3}{\cancel{3}} = \frac{1+3}{1} = \frac{4}{1} = 4 \qquad \text{WRONG}$$

Obviously, incorrect reducing can lead to ridiculous results.

"Divide out" only common factors.

WRONG

$$\frac{5x}{x+5} = \frac{\cancel{5}x}{x+\cancel{5}} = \frac{x}{x+1}$$

5 is not a factor of the denominator.

WRONG

$$\frac{5x}{x+5} = \frac{5\cancel{x}}{\cancel{x}+5} = \frac{5}{1+5}$$

x is not a factor of the denominator.

RIGHT

$$\frac{5x+15}{5x+20} = \frac{\cancel{5}(x+3)}{\cancel{5}(x+4)} = \frac{x+3}{x+4}$$

5 is a common **factor**.

Example 3: Reduce ● ● ● ● ● ● ● ● ● ● ● ● ● ● ● ● ●

Reduce each rational expression to lowest terms. State any restrictions on the variable **even before reducing**. **Remember, no denominator can be 0**.

a. $\dfrac{2x+4}{3x+6}$

Solution: $\dfrac{2x+4}{3x+6} = \dfrac{2\,\overset{1}{\cancel{(x+2)}}}{3\,\underset{1}{\cancel{(x+2)}}} = \dfrac{2}{3}$ ($x \ne -2$)   Since $x + 2 \ne 0$, $x \ne -2$.

b. $\dfrac{x^2-16}{x-4}$

Solution: $\dfrac{x^2-16}{x-4} = \dfrac{(x+4)\,\overset{1}{\cancel{(x-4)}}}{\underset{1}{\cancel{x-4}}} = x+4$ ($x \ne 4$) Since $x - 4 \ne 0$, $x \ne 4$.

c. $\dfrac{a}{a^2-5a}$

Solution: $\dfrac{a}{a^2-5a} = \dfrac{\overset{1}{\cancel{a}}}{\underset{1}{\cancel{a}}(a-5)} = \dfrac{1}{a-5}$ ($a \ne 0, 5$)   Note that these restrictions are determined before reducing.

This example illustrates the importance of writing 1 in the numerator if all the factors divide out. Remember that 1 is an understood factor.

d. $\dfrac{3-y}{y-3}$

Solution: $\dfrac{3-y}{y-3} = \dfrac{-y+3}{y-3}$ First write both numerator and denominator in descending order. Now factor out -1 in the numerator.

$= \dfrac{-1(y-3)}{(y-3)}$

$= \dfrac{-1\,\overset{1}{\cancel{(y-3)}}}{\underset{1}{\cancel{(y-3)}}} = -1$ ($y \ne 3$)

● ●

In Example 3d, the result was -1. The expression $3 - y$ is the opposite of $y - 3$ for any value of y, except 3. That is,

$$3 - y = -y + 3 = -1(y-3).$$

When nonzero opposites are divided, the quotient is always -1.

$$\dfrac{-8}{+8} = -1, \quad \dfrac{14}{-14} = -1, \quad \dfrac{x-5}{5-x} = \dfrac{\overset{1}{\cancel{(x-5)}}}{-1\,\underset{1}{\cancel{(x-5)}}} = \dfrac{1}{-1} = -1$$

501

Opposites

In general, for $b \neq a$, $(a - b)$ and $(b - a)$ are opposites and

$$\frac{a-b}{b-a} = \frac{a-b}{-1(a-b)} = -1$$

Now consider the problem of placement of the negative sign in a fraction. For example,

$$-\frac{6}{2} = -3, \quad \frac{-6}{2} = -3, \quad \text{and} \quad \frac{6}{-2} = -3$$

Thus,

$$-\frac{6}{2} = \frac{-6}{2} = \frac{6}{-2}$$

The following fact about negative fractions should be clear. Without changing the value of the fraction, the negative sign can be placed,

1. in front of the fraction, or
2. in the numerator, or
3. in the denominator.

The following general statements about the placement of negative signs can be made.

$$\text{For integers } a \text{ and } b, b \neq 0, \quad -\frac{a}{b} = \frac{-a}{b} = \frac{a}{-b}$$

$$\text{For polynomials } P \text{ and } Q, Q \neq 0, \quad -\frac{P}{Q} = \frac{-P}{Q} = \frac{P}{-Q}$$

At this stage of the discussion, we will assume that no denominator is 0. You should keep in mind that for any rational expression, there are certain restrictions on the variable; but these restrictions are generally not stated, just understood.

Example 4: Simplify

Simplify the following expressions.

a. $-\dfrac{-2x+6}{x^2-3x}$

Solution: $-\dfrac{-2x+6}{x^2-3x} = \dfrac{-(-2x+6)}{x^2-3x} = \dfrac{2x-6}{x^2-3x} = \dfrac{2\overset{1}{\cancel{(x-3)}}}{x\underset{1}{\cancel{(x-3)}}} = \dfrac{2}{x}$

Continued on next page...

b. $-\dfrac{5-10x}{6x-3}$

Solution: $-\dfrac{5-10x}{6x-3} = -\dfrac{5\,(1-2x)}{3\,(2x-1)} = -\dfrac{-5}{3} = \dfrac{5}{3}$

or

$-\dfrac{5-10x}{6x-3} = \dfrac{-(5-10x)}{6x-3} = \dfrac{-(5)(1-2x)}{3(2x-1)} = \dfrac{5}{3}$

or

$-\dfrac{5-10x}{6x-3} = \dfrac{(5-10x)}{-(6x-3)} = \dfrac{(5)(1-2x)}{-1(3)(2x-1)} = \dfrac{5}{3}$

• •

8.1 Exercises

In Exercises 1 – 10, determine what values of the variable, if any, will make the rational expression undefined. (Tell what the variable cannot equal.)

1. $\dfrac{-3}{5x}$

2. $\dfrac{17}{4y}$

3. $\dfrac{5y^2}{3y-4}$

4. $\dfrac{20a}{a^2+2}$

5. $\dfrac{x+1}{x^2-4x}$

6. $\dfrac{2x+3}{x^2-7x-8}$

7. $\dfrac{y^2-2y+3}{y^2-3y-10}$

8. $\dfrac{x-1}{2x-5}$

9. $\dfrac{x^2+2x}{x^2+4}$

10. $\dfrac{5m^2+2m}{3m^2-m-2}$

In Exercises 11 – 20, evaluate each rational expression for the given value of the variable.

11. $\dfrac{x-3}{3x^2}; x=5$

12. $\dfrac{2x+1}{3x-2}; x=1$

13. $\dfrac{5x^2}{x^2-4}; x=-3$

14. $\dfrac{3y-4}{y^2+25}; y=3$

15. $\dfrac{2x^2+5x}{x^2-1}; x=0$

16. $\dfrac{n^3}{n^2-5n+6}; n=-1$

17. $\dfrac{2m-7}{m^2+8m+12}; m=2$

18. $\dfrac{-x+3}{x-3}; x=-10$

19. $-\dfrac{15-x}{x-15}; x=1000$

20. $-\dfrac{16+x}{x^2-16}; x=20$

In Exercises 21 – 46, reduce each rational expression to lowest terms and state any restrictions on the variable.

21. $\dfrac{4x-12}{4x}$

22. $\dfrac{3y+15}{3y}$

23. $\dfrac{x^2+2x}{2x}$

24. $\dfrac{x^3+5x^2}{5x^2}$

25. $\dfrac{2y+3}{4y+6}$

26. $\dfrac{x-2}{x^2-6x+8}$

27. $\dfrac{3x-6}{6x+3}$

28. $\dfrac{5x+20}{6x+24}$

29. $\dfrac{x}{x^2-4x}$

30. $\dfrac{4-x}{x-4}$

31. $\dfrac{7x-14}{2-x}$

32. $\dfrac{3x^2-3x}{2x-2}$

33. $\dfrac{5+3x}{3x+5}$

34. $\dfrac{4xy+y^2}{3y^2+2y}$

35. $\dfrac{4-4x^2}{4x^2-4}$

36. $\dfrac{4x-8}{(x-2)^2}$

37. $\dfrac{x^2+2x}{x^2+4x+4}$

38. $\dfrac{x^2-4}{2x+4}$

39. $\dfrac{x^2+7x+10}{x^2-25}$

40. $\dfrac{x^2-3x-10}{x^2-7x+10}$

41. $\dfrac{x^2-3x-18}{x^2+6x+9}$

42. $\dfrac{8x^2+6x-9}{16x^2-9}$

43. $\dfrac{x^2-5x+6}{8x-2x^3}$

44. $\dfrac{6x^2-11x+3}{4x^2-12x+9}$

45. $\dfrac{16x^2+40x+25}{4x^2+13x+10}$

46. $\dfrac{6x^2-11x+4}{3x^2-7x+4}$

Writing and Thinking About Mathematics

47. The following "proof" that $2 = 0$ contains an error. Discuss the error in your own words.

$$a = b$$
$$2a = 2b$$
$$2a - 2b = 0$$
$$2(a - b) = 0$$
$$\frac{2(a - b)}{a - b} = \frac{0}{a - b}$$
$$2 = 0$$

 Defining Rational Expressions

8.2 Multiplication and Division with Rational Expressions

Objectives

After completing this section, you will be able to:

1. Multiply rational expressions.

2. Divide rational expressions.

Multiplying Rational Expressions

To multiply any two rational expressions, multiply the numerators and multiply the denominators, **keeping the expressions in factored form**. Then "divide out" any common factors.

Multiplying Rational Expressions

If P, Q, R, and S are polynomials with Q, S ≠ 0, then

$$\frac{P}{Q} \cdot \frac{R}{S} = \frac{P \cdot R}{Q \cdot S}$$

Example 1: Multiply

Multiply and reduce, if possible.

a. $\dfrac{x+3}{x} \cdot \dfrac{x-3}{x+5}$

Solution: $\dfrac{x+3}{x} \cdot \dfrac{x-3}{x+5} = \dfrac{(x+3)(x-3)}{x(x+5)} = \dfrac{x^2-9}{x(x+5)}$ There are no common factors in the numerator and denominator.

b. $\dfrac{x+5}{7x} \cdot \dfrac{49x^2}{x^2-25}$

Solution: $\dfrac{x+5}{7x} \cdot \dfrac{49x^2}{x^2-25} = \dfrac{\overset{7x}{\cancel{49x^2}}\,\cancel{(x+5)}}{\cancel{7x}\,\cancel{(x+5)}(x-5)} = \dfrac{7x}{x-5}$

Continued on next page...

c. $\dfrac{x^2-7x+12}{2x+6} \cdot \dfrac{x^2-4}{x^2-2x-8}$

Solution: $\dfrac{x^2-7x+12}{2x+6} \cdot \dfrac{x^2-4}{x^2-2x-8} = \dfrac{(x-4)(x-3)(x+2)(x-2)}{2(x+3)(x-4)(x+2)}$

$$= \dfrac{(x-3)(x-2)}{2(x+3)} \quad \text{or} \quad \dfrac{x^2-5x+6}{2(x+3)}$$

● ●

As shown in Examples 1a and 1c, we will multiply the factors in the numerator yet leave the denominator in factored form. This form is not necessary and other forms of the answers are correct. However, as we will see in Section 8.3, this form is useful for adding and subtracting.

Dividing Rational Expressions

To **divide** any two rational expressions, multiply by the **reciprocal** of the divisor.

Dividing Rational Expressions

If P, Q, R, and S are polynomials with Q, R, S ≠ 0, then

$$\frac{P}{Q} \div \frac{R}{S} = \frac{P}{Q} \cdot \frac{S}{R}$$

Example 2: Divide ●

Divide and reduce, if possible.

a. $\dfrac{a^2-49}{12a^2} \div \dfrac{a^2+8a+7}{18a}$

Solution: $\dfrac{a^2-49}{12a^2} \div \dfrac{a^2+8a+7}{18a} = \dfrac{a^2-49}{12a^2} \cdot \dfrac{18a}{a^2+8a+7}$

$$= \dfrac{(a+7)(a-7) \cdot 6 \cdot 3 \cdot a}{6 \cdot 2 \cdot \underset{a}{a^2} \, (a+7)(a+1)}$$

$$= \dfrac{3(a-7)}{2a(a+1)}$$

$$= \dfrac{3a-21}{2a(a+1)}$$

Continued on next page...

b. $\dfrac{2x^2+3x-2}{x^2+3x+2} \div \dfrac{1-2x}{x-2}$

Solution: $\dfrac{2x^2+3x-2}{x^2+3x+2} \div \dfrac{1-2x}{x-2} = \dfrac{2x^2+3x-2}{x^2+3x+2} \cdot \dfrac{x-2}{1-2x}$

$$= \dfrac{\overset{-1}{\cancel{(2x-1)}}\,\cancel{(x+2)}(x-2)}{(x+1)\cancel{(x+2)}\,\cancel{(1-2x)}}$$

$$= \dfrac{-1(x-2)}{x+1}$$

$$= \dfrac{-x+2}{x+1} \ \text{ or } \ \dfrac{2-x}{x+1}$$

Practice Problems

Perform the indicated operations and simplify. Assume that no denominators are 0.

1. $\dfrac{2y^2-16y}{6y^2+7y-3} \cdot \dfrac{2y^2+11y+12}{y^2-9y+8}$

2. $\dfrac{a-b}{b-a} \div \dfrac{a^2+2ab+b^2}{a^2+ab}$

3. $\dfrac{x^2+3x-4}{x^2-1} \div \dfrac{x^2+6x+8}{x+1}$

Answers to Practice Problems: **1.** $\dfrac{2y^2+8y}{(3y-1)(y-1)}$ **2.** $\dfrac{-a}{a+b}$ **3.** $\dfrac{1}{x+2}$

8.2 Exercises

Perform the indicated operations in Exercises 1 – 40. Assume that no denominator is 0.

1. $\dfrac{2x-4}{3x+6} \cdot \dfrac{3x}{x-2}$

2. $\dfrac{5x+20}{2x} \cdot \dfrac{4x}{2x+4}$

3. $\dfrac{4x^2}{x^2+3x} \cdot \dfrac{x^2-9}{2x-2}$

4. $\dfrac{x^2-x}{x-1} \cdot \dfrac{x+1}{x}$

5. $\dfrac{x^2-4}{x} \cdot \dfrac{3x}{x+2}$

6. $\dfrac{x-3}{15x} \div \dfrac{3x-9}{30x^2}$

7. $\dfrac{x^2-1}{5} \div \dfrac{x^2+2x+1}{10}$

8. $\dfrac{x-5}{3} \div \dfrac{x^2-25}{6x+30}$

9. $\dfrac{10x^2-5x}{6x^2+12x} \div \dfrac{2x-1}{x^2+2x}$

10. $\dfrac{x^2+2x-8}{4x} \div \dfrac{2x^2+5x+2}{3x^2}$

11. $\dfrac{x^2+x}{x^2+2x+1} \cdot \dfrac{x^2-x-2}{x^2-1}$

12. $\dfrac{x^2-x-6}{x^2-4} \cdot \dfrac{x^2-25}{x^2+2x-15}$

13. $\dfrac{x+3}{x^2+3x-4} \cdot \dfrac{x^2+x-2}{x+2}$

14. $\dfrac{x^2+5x+6}{2x+4} \cdot \dfrac{5x}{x+3}$

15. $\dfrac{6x^2-7x-3}{x^2-1} \cdot \dfrac{x-1}{2x-3}$

16. $\dfrac{x^2-9}{2x^2+7x+3} \cdot \dfrac{2x^2+11x+5}{x^2-3x}$

17. $\dfrac{x^2-8x+15}{x^2-9x+14} \cdot \dfrac{7-x}{x^2+4x-21}$

18. $\dfrac{x^2-16x+39}{6+x-x^2} \cdot \dfrac{4x+8}{x+1}$

19. $\dfrac{3x^2-7x+2}{1-9x^2} \cdot \dfrac{3x+1}{x-2}$

20. $\dfrac{16-x^2}{x^2+2x-8} \cdot \dfrac{4-x^2}{x^2-2x-8}$

21. $\dfrac{4x^2-1}{x^2-16} \div \dfrac{2x+1}{x^2-4x}$

22. $\dfrac{4x^2-13x+3}{16x^2-4x} \div \dfrac{x^2-6x+9}{8x^2}$

23. $\dfrac{x^2+x-6}{2x^2+6x} \div \dfrac{x^2-5x+6}{8x^2}$

24. $\dfrac{x^2-4}{x^2-5x+6} \div \dfrac{x^2+3x+2}{x^2-2x-3}$

25. $\dfrac{2x^2-5x-12}{x^2-10x+24} \div \dfrac{4x^2-9}{x^2-9x+18}$

26. $\dfrac{2x^2-7x+3}{x^2-3x+2} \div \dfrac{x^2-6x+9}{x^2-4x+3}$

27. $\dfrac{6x^2-x-2}{12x^2+5x-2} \div \dfrac{4x^2-1}{8x^2-6x+1}$

28. $\dfrac{8x^2+6x-9}{8x^2-26x+15} \div \dfrac{4x^2+12x+9}{16x^2+18x-9}$

29. $\dfrac{3x^2+11x+6}{4x^2+16x+7} \div \dfrac{9x^2+12x+4}{2x^2-x-28}$

30. $\dfrac{2x^2+5x-3}{5x^2+17x-12} \div \dfrac{2x^2+3x-2}{5x^2+7x-6}$

31. $\dfrac{x^2-16}{x^3-8x^2+16x} \cdot \dfrac{x^2}{x^2+4x} \cdot \dfrac{2x^2-2x}{x^2-2x+1}$

32. $\dfrac{10x^2+3x-1}{6x^2+x-2} \cdot \dfrac{2x^2-x}{2x^2-x-1} \cdot \dfrac{3x+2}{5x^2-x}$

33. $\dfrac{x^2-3x}{x^2+2x+1} \cdot \dfrac{x-3}{x^2-9} \cdot \dfrac{x^2-3x-18}{x-6}$

34. $\dfrac{x^2+2x-3}{2x^2+3x-2} \cdot \dfrac{3x+5}{2x-3} \div \dfrac{3x^2+2x-5}{2x^2+x-6}$

35. $\dfrac{2x^2+5x-3}{x^2+2x-3} \cdot \dfrac{x^2+3x-4}{4x^2-1} \div \dfrac{x^2+4x}{x^3+5x^2}$

36. $\dfrac{6x^2+7x-3}{2x^2+5x+3} \div \dfrac{x^2+5x+6}{4x^2+3x-1} \cdot \dfrac{x^2+2x-3}{4x^2+7x-2}$

37. $\dfrac{x^2+4x+3}{x^2+8x+7} \div \dfrac{x^2-7x-8}{35+12x+x^2} \cdot \dfrac{x^2+8x+15}{x^2-9x+8}$

38. $\dfrac{12x^2}{x^2-1} \cdot \dfrac{4x^2+4x-3}{2x^2-5x+3} \div \dfrac{6x^2-9x}{1-4x^2}$

39. $\dfrac{2x^2-7x+3}{36x^2-1} \div \dfrac{6x^2+5x+1}{2x+1} \div \dfrac{2x^2-5x-3}{18x^2+3x-1}$

40. $\dfrac{12x^2-8x-15}{4x^2-8x+3} \cdot \dfrac{6x^2-13x+6}{4x^2+5x+1} \div \dfrac{18x^2+3x-10}{8x^2-2x-1}$

41. The area of a rectangle (in square feet) is represented by the polynomial function $A(x)=4x^2-4x-15$. If the length of the rectangle is $(2x+3)$ feet, find a representation for the width.

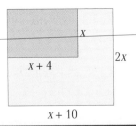

$A(x)=4x^2-4x-15$

$2x+3$

42. A rectangle with dimensions x by $x+4$ centimeters is cut from a rectangle with dimensions $2x$ by $x+10$ centimeters. (a) Represent the remaining area as a polynomial function. (b) Find this area if $x=8$ cm.

x

$2x$

$x+4$

$x+10$

Hawkes Learning Systems: Introductory Algebra

Multiplication and Division of Rational Expressions

8.3

Addition and Subtraction with Rational Expressions

After completing this section, you will be able to:

1. Add rational expressions.

2. Subtract rational expressions.

Adding Rational Expressions

To add rational expressions with a common denominator, proceed just as with fractions: add the numerators and keep the common denominator. For example,

$$\frac{5}{x+1} + \frac{6}{x+1} = \frac{5+6}{x+1} = \frac{11}{x+1}$$

Sometimes the sum can be reduced:

$$\frac{x^2}{x+1} + \frac{2x+1}{x+1} = \frac{x^2+2x+1}{x+1} = \frac{(x+1)^2}{x+1} = x+1$$

Adding Rational Expressions

For polynomials P, Q, and R, with Q ≠ 0,

$$\frac{P}{Q} + \frac{R}{Q} = \frac{P+R}{Q}$$

Example 1: Addition •

Find the indicated sums and reduce if possible.

a. $\dfrac{x^2+5}{x+5} + \dfrac{6x}{x+5}$

Continued on next page...

Solution: $\dfrac{x^2+5}{x+5}+\dfrac{6x}{x+5}=\dfrac{x^2+5+6x}{x+5}=\dfrac{x^2+6x+5}{x+5}$

$$=\dfrac{(x+5)(x+1)}{x+5}=x+1$$

b. $\dfrac{3}{x^2+3x+2}+\dfrac{2x+1}{x^2+3x+2}$

Solution: $\dfrac{3}{x^2+3x+2}+\dfrac{2x+1}{x^2+3x+2}=\dfrac{3+2x+1}{x^2+3x+2}=\dfrac{2x+4}{(x+2)(x+1)}$

$$=\dfrac{2(x+2)}{(x+2)(x+1)}=\dfrac{2}{x+1}$$

● ●

The rational expressions added in Examples 1a and 1b had common denominators. To add expressions with different denominators, we need to find the least common multiple (LCM) of the denominators. The LCM was discussed in Section 2.2. We list the procedure here for polynomials and use this in finding the common denominator for rational expressions.

To Find the LCM for a Set of Polynomials

1. Completely factor each polynomial (including prime factors for numerical factors).

2. Form the product of all factors that appear, using each factor the most number of times it appears in any one polynomial.

Example 2: Find LCM and Add ● ● ● ● ● ● ● ● ● ● ● ● ● ● ● ●

Find the indicated sums by finding the LCM of the denominators and changing each expression to an equivalent expression with that LCM as the denominator.

a. $\dfrac{5}{x^2}+\dfrac{3}{x^2-4x}$

Solution: First find the LCM. Factoring, we have

$$\left.\begin{array}{l} x^2=x^2 \\[4pt] x^2-4x=x(x-4) \end{array}\right\}\text{LCM}=x^2(x-4)$$

Continued on next page...

Thus, $x^2(x-4)$ is the common denominator. Multiply the numerator and denominator of each fraction so that it has denominator $x^2(x-4)$

$$\frac{5}{x^2}+\frac{3}{x^2-4x}=\frac{5(x-4)}{x^2(x-4)}+\frac{3\cdot x}{x(x-4)\cdot x}$$

$$=\frac{5x-20}{x^2(x-4)}+\frac{3x}{x^2(x-4)}$$

$$=\frac{5x-20+3x}{x^2(x-4)}=\frac{8x-20}{x^2(x-4)}$$

b. $\dfrac{5}{x^2-1}+\dfrac{4}{(x-1)^2}$

Solution: $\left.\begin{array}{l}x^2-1=(x+1)(x-1)\\[6pt](x-1)^2=(x-1)^2\end{array}\right\}$ $LCM=(x+1)(x-1)^2$

$$\frac{5}{x^2-1}+\frac{4}{(x-1)^2}=\frac{5(x-1)}{(x+1)(x-1)(x-1)}+\frac{4(x+1)}{(x-1)^2(x+1)}$$

$$=\frac{5x-5+4x+4}{(x-1)^2(x+1)}=\frac{9x-1}{(x-1)^2(x+1)}$$

● ●

Subtracting Rational Expressions

To subtract one rational expression from another with the same denominator, simply subtract the numerators and use the common denominator. Since the numerators will be polynomials, **a good idea is to put both numerators in parentheses so that all changes in signs will be done correctly**.

Subtracting Rational Expressions

For polynomials P, Q, and R, with Q ≠ 0

$$\frac{P}{Q}-\frac{R}{Q}=\frac{P-R}{Q}$$

Example 3: Subtraction ● ● ● ● ● ● ● ● ● ● ● ● ● ● ● ●

Find the indicated difference and reduce, if possible.

a. $\dfrac{3}{2x-4} - \dfrac{5}{2x-4}$

Solution: $\dfrac{3}{2x-4} - \dfrac{5}{2x-4} = \dfrac{3-5}{2x-4} = \dfrac{-2}{2x-4} = \dfrac{\overset{-1}{\cancel{-2}}}{\cancel{2}(x-2)} = \dfrac{-1}{x-2}$

b. $\dfrac{x+4}{x+3} - \dfrac{5x+1}{x+3}$

Solution: $\dfrac{x+4}{x+3} - \dfrac{5x+1}{x+3} = \dfrac{(x+4)-(5x+1)}{x+3} = \dfrac{x+4-5x-1}{x+3} = \dfrac{-4x+3}{x+3}$

● ●

NOTES

COMMON ERROR

Many beginning students make a **mistake** in subtracting fractions by not subtracting the entire numerator. They make a mistake similar to the following.

WRONG $\dfrac{7}{x+5} - \dfrac{2-x}{x+5} = \dfrac{7-2-x}{x+5}$

By using parentheses, you can avoid such mistakes.

RIGHT $\dfrac{7}{x+5} - \dfrac{2-x}{x+5} = \dfrac{7-(2-x)}{x+5} = \dfrac{7-2+x}{x+5} = \dfrac{5+x}{x+5} = \dfrac{x+5}{x+5} = 1$

As with addition, if the rational expressions do not have the same denominator, find the LCM of the denominators and change each fraction to an equivalent fraction with the LCM as denominator.

Example 4: Find LCM and Subtract ● ● ● ● ● ● ● ● ● ● ● ●

Subtract the following expression.

$$\frac{2x}{x^2-9} - \frac{1}{x^2+7x+12}$$

Solution:

$$\left.\begin{array}{l} x^2-9=(x+3)(x-3) \\ x^2+7x+12=(x+4)(x+3) \end{array}\right\} \text{LCM}=(x+3)(x-3)(x+4)$$

$$\frac{2x}{x^2-9} - \frac{1}{x^2+7x+12} = \frac{2x(x+4)}{(x+3)(x-3)(x+4)} + \frac{-1(x-3)}{(x+4)(x+3)(x-3)}$$

$$= \frac{2x^2+8x-x+3}{(x+3)(x-3)(x+4)} = \frac{2x^2+7x+3}{(x+3)(x-3)(x+4)}$$

$$= \frac{(2x+1)\cancel{(x+3)}}{\cancel{(x+3)}(x-3)(x+4)} = \frac{2x+1}{(x-3)(x+4)}$$

● ●

Whenever the denominators are opposites of each other or contain factors that are opposites of each other, it is important to use -1 as a factor or to multiply both the numerator and denominator by -1. This will simplify the work considerably.

Example 5: -1 as a Factor ● ● ● ● ● ● ● ● ● ● ● ● ● ● ● ●

Simplify the following expression.

$$\frac{10}{x-2} + \frac{8}{2-x}$$

Solution: Note that $x-2$ and $2-x$ are opposites and $x-2=-1(2-x)$. So, if we multiply,

$$\frac{8}{2-x} \cdot \frac{(-1)}{(-1)} = \frac{-8}{x-2}$$

the fractions will have the same denominator and we do not have to find an LCM.

Continued on next page...

$$\frac{10}{x-2}+\frac{8}{2-x}=\frac{10}{x-2}+\frac{8}{2-x}\cdot\frac{(-1)}{(-1)}=\frac{10}{x-2}+\frac{-8}{x-2}=\frac{10-8}{x-2}=\frac{2}{x-2}$$

Do not try to reduce this last fraction, since 2 is **not a factor** of the denominator.

● ●

Practice Problems

Perform the indicated operations and reduce, if possible.

1. $\dfrac{5}{x-1}-\dfrac{4+x}{x-1}$

2. $\dfrac{5}{x+1}+\dfrac{10x}{x^2+4x+3}$

3. $\dfrac{1}{x^2+x}+\dfrac{4}{x^2}-\dfrac{2}{x^2-x}$

4. $\dfrac{x}{2x-1}-\dfrac{2}{1-2x}$

8.3 Exercises

Perform the indicated operations and reduce if possible in Exercises 1 – 60.

1. $\dfrac{4x}{x+2}+\dfrac{8}{x+2}$

2. $\dfrac{x-1}{x+4}+\dfrac{x+9}{x+4}$

3. $\dfrac{3x-1}{2x-6}+\dfrac{x-11}{2x-6}$

4. $\dfrac{2x+5}{4(x+1)}+\dfrac{-x+3}{4x+4}$

5. $\dfrac{x^2}{x^2+2x+1}+\dfrac{-2x-3}{x^2+2x+1}$

6. $\dfrac{2x-1}{x^2-x-6}+\dfrac{1-x}{x^2-x-6}$

7. $\dfrac{-(3x+2)}{x^2-7x+6}+\dfrac{4x-4}{x^2-7x+6}$

8. $\dfrac{x+1}{x^2-2x-3}+\dfrac{-(3x-5)}{x^2-2x-3}$

9. $\dfrac{x^2+2}{x^2-4}+\dfrac{-(4x-2)}{x^2-4}$

10. $\dfrac{2x+5}{2x^2-x-1}+\dfrac{-(4x+2)}{2x^2-x-1}$

11. $\dfrac{3x+2}{4x-2}-\dfrac{x+2}{4x-2}$

12. $\dfrac{2x+1}{x^2-x-6}-\dfrac{x-1}{x^2-x-6}$

13. $\dfrac{2x^2-3x+1}{x^2-3x-4}-\dfrac{x^2+x+6}{x^2-3x-4}$

Answers to Practice Problems: 1. -1 **2.** $\dfrac{15}{x+3}$ **3.** $\dfrac{3x^2-3x-4}{x^2(x+1)(x-1)}$ **4.** $\dfrac{x+2}{2x-1}$

14. $\dfrac{x^2-x-2}{x^2-4}-\dfrac{x^2+x-2}{x^2-4}$ **15.** $\dfrac{3}{x-3}-\dfrac{2}{2x-6}$ **16.** $\dfrac{2x}{3x+6}+\dfrac{5}{2x+4}$

17. $\dfrac{8}{5x-10}-\dfrac{6}{3x-6}$ **18.** $\dfrac{7}{4x-20}-\dfrac{1}{3x-15}$ **19.** $\dfrac{2}{x}+\dfrac{1}{x+4}$

20. $\dfrac{8x+3}{x+1}+\dfrac{x-1}{x}$ **21.** $\dfrac{x}{x+4}+\dfrac{2}{x-4}$ **22.** $\dfrac{5}{x-2}+\dfrac{x}{x+3}$

23. $\dfrac{3}{2-x}+\dfrac{6}{x-2}$ **24.** $\dfrac{5}{2x-3}+\dfrac{2}{3-2x}$ **25.** $\dfrac{x}{5-x}+\dfrac{2x+3}{x-5}$

26. $\dfrac{x-1}{3x-1}+\dfrac{4}{x+2}$ **27.** $\dfrac{x}{x-1}-\dfrac{1}{x+2}$ **28.** $\dfrac{x+2}{x+3}+\dfrac{4}{3-x}$

29. $\dfrac{x+1}{x+4}+\dfrac{2x}{4-x}$ **30.** $\dfrac{8}{x}-\dfrac{x+1}{x-6}$ **31.** $\dfrac{8}{x-2}-\dfrac{4}{x+2}$

32. $\dfrac{8}{2x+2}+\dfrac{3}{3x+3}$ **33.** $\dfrac{x}{4x-8}-\dfrac{3x+1}{3x-6}$ **34.** $\dfrac{7x-1}{x^2-25}+\dfrac{4}{x+5}$

35. $\dfrac{2x+3}{x^2+4x-5}-\dfrac{4}{x-1}$ **36.** $\dfrac{2x-3}{x+1}-\dfrac{x+1}{2x-2}$ **37.** $\dfrac{3x}{6+x}-\dfrac{2x}{x^2-36}$

38. $\dfrac{x}{7+x}-\dfrac{7-13x}{x^2-49}$ **39.** $\dfrac{2x+1}{x-7}+\dfrac{3x}{x^2-8x+7}$ **40.** $\dfrac{x-4}{x-2}-\dfrac{x-7}{2x-10}$

41. $\dfrac{x+1}{x+2}-\dfrac{x+2}{2x+6}$ **42.** $\dfrac{3x-4}{x^2-x-20}-\dfrac{2}{x-5}$ **43.** $\dfrac{x+2}{x^2-16}-\dfrac{x+1}{2x-8}$

44. $\dfrac{7}{x-9}-\dfrac{x-1}{3x+6}$ **45.** $\dfrac{1}{x^2+x-2}-\dfrac{1}{x^2-1}$ **46.** $\dfrac{4}{x^2+x-6}+\dfrac{4}{x^2+5x+6}$

47. $\dfrac{2x}{x^2+x-12}+\dfrac{3x}{x^2-9}$ **48.** $\dfrac{x}{x^2+4x-21}+\dfrac{1-x}{x^2+8x+7}$

49. $\dfrac{x-3}{x^2+4x+4}-\dfrac{3x}{x^2+3x+2}$ **50.** $\dfrac{x-1}{2x^2+3x-2}+\dfrac{x}{2x^2-3x+1}$

51. $\dfrac{x}{x^2-16}-\dfrac{3x}{x^2+5x+4}$ **52.** $\dfrac{2x}{x^2-2x-15}-\dfrac{5}{x^2-6x+5}$

53. $\dfrac{x-2}{2x^2+5x-3}+\dfrac{2x-5}{2x^2-9x+4}$

54. $\dfrac{6}{x^2-4x-12}+\dfrac{x+1}{x^2-3x-18}$

55. $\dfrac{x}{x^2+3x-10}+\dfrac{3x}{x^2-4}$

56. $\dfrac{2x+3}{x^2-6x-7}+\dfrac{x-1}{x^2-5x-14}$

57. $\dfrac{3x}{x-4}+\dfrac{7x}{x+4}-\dfrac{x+3}{x^2-16}$

58. $\dfrac{x}{x+3}-\dfrac{x+1}{x-3}+\dfrac{x^2+4}{x^2-9}$

59. $\dfrac{1}{x-2}-\dfrac{1}{x-1}+\dfrac{x}{x^2-3x+2}$

60. $\dfrac{2}{x^2-9}-\dfrac{3}{x^2-4x+3}+\dfrac{x-1}{x^2+2x-3}$

Hawkes Learning Systems: Introductory Algebra

Addition and Subtraction of Rational Expressions

8.4

Complex Algebraic Fractions

After completing this section, you will be able to:

1. Simplify complex fractions and complex algebraic expressions.

Simplifying Complex Algebraic Fractions (First Method)

A fraction that contains various combinations of addition, subtraction, multiplication, and division with rational expressions is called a **complex algebraic fraction**. Examples of complex algebraic fractions are

$$\dfrac{\dfrac{1}{x}+\dfrac{1}{y}}{x+y} \quad \text{and} \quad \dfrac{\dfrac{1}{x+2}-\dfrac{1}{x}}{1+\dfrac{2}{x}}$$

An expression such as

$$\dfrac{4-x}{x+3}+\dfrac{x}{x+3}\div\dfrac{x}{x-3}$$

that involves rational expressions with more than one operation and is not written in fraction form is called a **complex algebraic expression**. The objective here is to develop techniques for simplifying complex algebraic fractions and expressions so that they are written in the form of a **single reduced rational expression**.

In a complex fraction such as $\dfrac{\dfrac{1}{x}+\dfrac{1}{y}}{x+y}$, the large fraction bar is a symbol of inclusion. The expression could be written as follows:

$$\dfrac{\dfrac{1}{x}+\dfrac{1}{y}}{x+y}=\left(\dfrac{1}{x}+\dfrac{1}{y}\right)\div\left(x+y\right)$$

Similarly,

$$\frac{\dfrac{1}{x+2}-\dfrac{1}{x}}{1+\dfrac{2}{x}}=\left(\frac{1}{x+2}-\frac{1}{x}\right)\div\left(1+\frac{2}{x}\right)$$

Thus, a complex fraction indicates that the numerator is to be divided by the denominator. So, a complex fraction can be simplified as follows.

Simplify Complex Fractions

1. *Simplify the numerator so that it is a single rational expression.*
2. *Simplify the denominator so that it is a single rational expression.*
3. *Divide the numerator by the denominator and reduce to lowest terms.*

This method is used to simplify the following two complex fractions. Study the examples closely so that you understand what happens at each step.

Example 1: First Method ● ● ● ● ● ● ● ● ● ● ● ● ● ● ● ● ●

Simplify the following expression.

a. $\dfrac{\dfrac{1}{x}+\dfrac{1}{y}}{x+y}$

Solution: $\dfrac{\dfrac{1}{x}+\dfrac{1}{y}}{x+y}=\dfrac{\dfrac{1\cdot y}{x\cdot y}+\dfrac{1\cdot x}{y\cdot x}}{x+y}=\dfrac{\dfrac{y+x}{xy}}{\dfrac{x+y}{1}}=\dfrac{\cancel{y+x}}{xy}\cdot\dfrac{1}{\cancel{x+y}}=\dfrac{1}{xy}$

b. $\dfrac{\dfrac{1}{x+2}-\dfrac{1}{x}}{1+\dfrac{2}{x}}$

Continued on next page...

Solution:
$$\frac{\dfrac{1}{x+2}-\dfrac{1}{x}}{1+\dfrac{2}{x}}=\frac{\dfrac{1\cdot x}{(x+2)x}-\dfrac{1(x+2)}{x(x+2)}}{1\cdot\dfrac{x}{x}+\dfrac{2}{x}}$$

$$=\frac{\dfrac{x-(x+2)}{x(x+2)}}{\dfrac{x+2}{x}}=\frac{\dfrac{x-x-2}{x(x+2)}}{\dfrac{x+2}{x}}$$

$$=\frac{-2}{\cancel{x}(x+2)}\cdot\frac{\cancel{x}}{x+2}=\frac{-2}{(x+2)^2}$$

● ●

Simplifying Complex Algebraic Fractions (Second Method)

A second method is to find the LCM of the denominators in the fractions in both the original numerator and the original denominator, and then multiply both the numerator and the denominator by this LCM.

Example 2: Second Method ● ● ● ● ● ● ● ● ● ● ● ● ● ● ● ● ●

Simplify the following expression.

a.
$$\frac{\dfrac{1}{x}+\dfrac{1}{y}}{x+y}$$

Solution:
$$\left.\begin{array}{c}x\\y\\1\end{array}\right\}\ \text{LCM}=xy$$

$$\frac{\dfrac{1}{x}+\dfrac{1}{y}}{x+y}=\frac{\left(\dfrac{1}{x}+\dfrac{1}{y}\right)xy}{\left(\dfrac{x+y}{1}\right)xy}=\frac{\dfrac{1}{x}\cdot xy+\dfrac{1}{y}\cdot xy}{(x+y)xy}=\frac{\cancel{y+x}}{\cancel{(x+y)}xy}=\frac{1}{xy}$$

b.
$$\frac{\dfrac{1}{x+2}-\dfrac{1}{x}}{1+\dfrac{2}{x}}$$

Continued on next page...

Solution: $\left.\begin{array}{c} x \\ x+2 \end{array}\right\}$ LCM $= x(x+2)$

$$\frac{\dfrac{1}{x+2}-\dfrac{1}{x}}{1+\dfrac{2}{x}} = \frac{\left(\dfrac{1}{x+2}-\dfrac{1}{x}\right)\cdot x(x+2)}{\left(1+\dfrac{2}{x}\right)\cdot x(x+2)} = \frac{\dfrac{1}{x+2}\cdot x(x+2) - \dfrac{1}{x}\cdot x(x+2)}{1\cdot x(x+2)+\dfrac{2}{x}\cdot x(x+2)}$$

$$= \frac{x-(x+2)}{x(x+2)+2(x+2)} = \frac{x-x-2}{(x+2)(x+2)} = \frac{-2}{(x+2)^2}$$

• •

Each of the techniques, just described, is valid. Sometimes one is easier to use than the other, but the choice is up to you.

Example 3: Simplify •

Simplify the following complex algebraic fraction: $\dfrac{\dfrac{1}{x+y}-\dfrac{1}{x-y}}{\dfrac{2y}{x^2-y^2}}$

Solution:

$$\frac{\dfrac{1}{x+y}-\dfrac{1}{x-y}}{\dfrac{2y}{x^2-y^2}} = \frac{\dfrac{1(x-y)}{(x+y)(x-y)}-\dfrac{1(x+y)}{(x-y)(x+y)}}{\dfrac{2y}{x^2-y^2}} = \frac{\dfrac{(x-y)-(x+y)}{(x-y)(x+y)}}{\dfrac{2y}{x^2-y^2}}$$

$$= \frac{x-y-x-y}{(x-y)(x+y)}\cdot\frac{x^2-y^2}{2y} = \frac{\overset{-1}{\cancel{-2y}}}{\cancel{(x-y)}\,\cancel{(x+y)}}\cdot\frac{\cancel{(x-y)}\,\cancel{(x+y)}}{\cancel{2y}}$$

$$= -1$$

Or use the technique of multiplying the numerator and the denominator by the LCM of the denominators of the various fractions.

$$\left.\begin{array}{l} x+y \\ x-y \\ x^2-y^2 = (x+y)(x-y) \end{array}\right\}\ \text{LCM} = (x+y)(x-y)$$

Continued on next page...

$$\frac{\dfrac{1}{x+y}-\dfrac{1}{x-y}}{\dfrac{2y}{x^2-y^2}}=\frac{\left(\dfrac{1}{x+y}-\dfrac{1}{x-y}\right)(x+y)(x-y)}{\left(\dfrac{2y}{x^2-y^2}\right)(x+y)(x-y)}$$

$$=\frac{\dfrac{1}{\cancel{x+y}}\,\cancel{(x+y)}(x-y)-\dfrac{1}{\cancel{x-y}}(x+y)\cancel{(x-y)}}{\dfrac{2y}{\cancel{(x+y)}\,\cancel{(x-y)}}\,\cancel{(x+y)}\,\cancel{(x-y)}}$$

$$=\frac{(x-y)-(x+y)}{2y}=\frac{x-y-x-y}{2y}=\frac{-2y}{2y}=-1$$

● ●

Simplifying Complex Algebraic Expressions

A complex algebraic expression (as stated earlier) is an expression that involves rational expressions and more than one operation. In simplifying such expressions, the rules for order of operations apply. The objective is to simplify the expression so that it is written in the form of **a single reduced rational expression**.

> **Example 4: Simplifying Complex Algebraic Expressions** ● ● ● ●

Simplify the following expression.

$$\frac{4-x}{x+3}+\frac{x}{x+3}\div\frac{x}{x-3}$$

Solution: In a complex algebraic expression such as

$$\frac{4-x}{x+3}+\frac{x}{x+3}\div\frac{x}{x-3}$$

the Rules for Order of Operations indicate that the division is to be done first.

$$\frac{4-x}{x+3}+\frac{x}{x+3}\div\frac{x}{x-3}=\frac{4-x}{x+3}+\frac{\cancel{x}}{x+3}\cdot\frac{x-3}{\cancel{x}}=\frac{4-x}{x+3}+\frac{x-3}{x+3}$$

$$=\frac{4-x+x-3}{x+3}=\frac{1}{x+3}$$

● ●

Practice Problems

Simplify the following expressions.

1. $\dfrac{\dfrac{1}{x}}{1+\dfrac{1}{x}}$

2. $\dfrac{1+\dfrac{3}{x-3}}{x-\dfrac{x^2}{x-3}}$

8.4 Exercises

Simplify the complex algebraic fractions in Exercises 1 – 32 so that each is a rational expression in reduced form.

1. $\dfrac{\dfrac{4}{5}}{\dfrac{7}{10}}$

2. $\dfrac{\dfrac{5}{6}}{\dfrac{2}{3}}$

3. $\dfrac{\dfrac{1}{2}+\dfrac{1}{3}}{\dfrac{5}{6}-\dfrac{1}{4}}$

4. $\dfrac{\dfrac{3}{4}-\dfrac{7}{8}}{\dfrac{1}{3}-\dfrac{1}{6}}$

5. $\dfrac{1+\dfrac{1}{3}}{2+\dfrac{2}{3}}$

6. $\dfrac{2+\dfrac{5}{7}}{3-\dfrac{2}{7}}$

7. $\dfrac{\dfrac{3x}{5}}{\dfrac{x}{5}}$

8. $\dfrac{\dfrac{x}{2y}}{\dfrac{3x}{y}}$

9. $\dfrac{\dfrac{4}{xy^2}}{\dfrac{6}{x^2y}}$

10. $\dfrac{\dfrac{x}{2y^2}}{\dfrac{5x^2}{6y}}$

11. $\dfrac{\dfrac{8x^2}{5y}}{\dfrac{x}{10y^2}}$

12. $\dfrac{\dfrac{15y^2}{2x^3}}{\dfrac{8y}{}}$

13. $\dfrac{\dfrac{12x^3}{7y^4}}{\dfrac{3x^5}{2y}}$

14. $\dfrac{\dfrac{9x^2}{5y^3}}{\dfrac{3xy}{10}}$

15. $\dfrac{\dfrac{3x}{x+1}}{\dfrac{x-2}{x+1}}$

16. $\dfrac{\dfrac{x-3}{x+4}}{\dfrac{x-2}{x+4}}$

17. $\dfrac{\dfrac{x+3}{2x}}{\dfrac{2x-1}{4x^2}}$

18. $\dfrac{\dfrac{x-2}{6x}}{\dfrac{x+3}{3x^2}}$

19. $\dfrac{\dfrac{x^2-9}{x}}{x-3}$

20. $\dfrac{\dfrac{x^2-3x+2}{x-4}}{x-2}$

Answers to Practice Problems: 1. $\dfrac{1}{x+1}$ **2.** $-\dfrac{1}{3}$

21. $\dfrac{\dfrac{x+2}{x-2}}{\dfrac{4x^2+8x}{x^2-4}}$

22. $\dfrac{\dfrac{x-1}{x+3}}{\dfrac{x+2}{x^2+2x-3}}$

23. $\dfrac{\dfrac{1}{x}-\dfrac{1}{3x}}{\dfrac{x+6}{x^2}}$

24. $\dfrac{\dfrac{1}{3}+\dfrac{1}{x}}{\dfrac{1}{2}-\dfrac{1}{x}}$

25. $\dfrac{1+\dfrac{1}{x}}{1-\dfrac{1}{x^2}}$

26. $\dfrac{\dfrac{3}{x}-\dfrac{6}{x^2}}{\dfrac{1}{x}-\dfrac{2}{x^2}}$

27. $\dfrac{\dfrac{1}{x}-\dfrac{1}{y}}{\dfrac{y}{x^2}-\dfrac{1}{y}}$

28. $\dfrac{\dfrac{x}{y}-\dfrac{1}{3}}{\dfrac{6}{y}-\dfrac{2}{x}}$

29. $\dfrac{2-\dfrac{4}{x}}{\dfrac{x^2-4}{x^2+x}}$

30. $\dfrac{\dfrac{1}{x}}{1-\dfrac{1}{x-2}}$

31. $\dfrac{x-\dfrac{2}{x+1}}{x+\dfrac{x-3}{x+1}}$

32. $\dfrac{x-\dfrac{2x-3}{x-2}}{2x-\dfrac{x+3}{x-2}}$

Write each of the expressions as a single fraction reduced to lowest terms in Exercises 33 – 40.

33. $\dfrac{1}{x+1}-\dfrac{3}{2x}\cdot\dfrac{4x}{x+1}$

34. $\dfrac{4}{x}-\dfrac{2}{x^2-2x}\cdot\dfrac{x-2}{5}$

35. $\left(\dfrac{8}{x}-\dfrac{3}{4x}\right)\div\dfrac{4x+5}{x}$

36. $\left(\dfrac{2}{x}+\dfrac{5}{x-3}\right)\cdot\dfrac{2x-6}{x}$

37. $\dfrac{x}{x-1}-\dfrac{3}{x-1}\cdot\dfrac{x+2}{x}$

38. $\dfrac{x+3}{x+2}+\dfrac{x}{x+2}\div\dfrac{x^2}{x-3}$

39. $\dfrac{x-1}{x+4}+\dfrac{x-6}{x^2+3x-4}\div\dfrac{x-4}{x-1}$

40. $\dfrac{x}{x+3}-\dfrac{3}{x-5}\cdot\dfrac{x^2-3x-10}{x-2}$

Writing and Thinking About Mathematics

41. Some complex fractions involve the sum (or difference) of complex fractions. Beginning with the "farthest" denominator, simplify each of the following expressions.

a. $1+\dfrac{1}{1+\dfrac{1}{1+\dfrac{1}{1+1}}}$

b. $2-\dfrac{1}{2-\dfrac{1}{2-\dfrac{1}{2-1}}}$

c. $x+\dfrac{1}{x+\dfrac{1}{x+\dfrac{1}{x+1}}}$

Hawkes Learning Systems: Introductory Algebra

Complex Algebraic Fractions

8.5 Solving Equations Involving Rational Expressions

After completing this section, you will be able to:

1. Solve equations involving rational expressions.

2. Solve proportions.

3. Solve word problems by using proportions and rational expressions.

Solving Equations Involving Rational Expressions

In Section 3.2, we solved equations containing fractions. The approach there was to perform the arithmetic indicated by the fractions. Another approach is to "clear" the fractions by multiplying all terms on both sides of the equation by the LCM of the denominators. The result will be an equation with integer coefficients, which then may be solved. For example, to solve

$$\frac{x+1}{3} + \frac{x}{2} = 1$$

multiply both sides of the equation by 6, the LCM of the denominators. The procedure looks like this:

$$\frac{x+1}{3} + \frac{x}{2} = 1$$

$$6\left(\frac{x+1}{3} + \frac{x}{2}\right) = 1 \cdot 6 \quad \text{Multiply both sides by 6.}$$

$$\overset{2}{\cancel{6}}\left(\frac{x+1}{\cancel{3}}\right) + \overset{3}{\cancel{6}}\left(\frac{x}{\cancel{2}}\right) = 1 \cdot 6 \quad \text{Multiply all terms on both sides by 6.}$$

$$2x + 2 + 3x = 6 \quad \begin{array}{l}\text{This equation has only integer coefficients}\\\text{and constants.}\end{array}$$

$$5x = 4$$

$$x = \frac{4}{5}$$

If the denominators have variables, we proceed in a similar manner as follows.

To Solve an Equation Containing Rational Expressions

1. *Find the LCM of the denominators.*

2. *Multiply both sides of the equation by this LCM and simplify.*

3. *Solve the resulting equation. (This equation will have only polynomials on both sides.)*

4. *Check to see that no solution makes a denominator 0. Note: Such solutions, if there are any, are called **extraneous solutions**.*

Checking is particularly important when equations have rational expressions, because a solution to the new equation found by multiplying by the LCM may not be a solution of the original equation. In effect, multiplying by the LCM might be multiplication by 0. In such a case, a solution of the equation containing only polynomials may not be a solution of the original equation containing rational expressions. Such solutions are called **extraneous solutions** and occur because multiplication by a variable expression increases the degree of the equation.

Example 1: Solve ●

First state any restrictions on the variable, then solve each of the following equations.

a. $\dfrac{3}{x}+\dfrac{1}{2}=\dfrac{7}{x}$

Solution:

$$\frac{3}{x}+\frac{1}{2}=\frac{7}{x} \qquad x \neq 0 \text{ and } LCM = 2x$$

$$2x\left(\frac{3}{x}+\frac{1}{2}\right)=2x\left(\frac{7}{x}\right) \qquad \text{Multiply both sides by } 2x.$$

$$\overset{2}{2x}\left(\frac{3}{x}\right)+\overset{x}{2x}\left(\frac{1}{2}\right)=\overset{2}{2x}\left(\frac{7}{x}\right) \qquad \text{Multiply each fraction by } 2x.$$

$$6+x=14 \qquad \text{Simplify. This equation has only polynomials}$$

$$x=8 \qquad \text{on both sides.}$$

Since no denominator can be 0, $x \neq 0$. Since $8 \neq 0$, 8 is a possible solution. Checking, we have

$$\frac{3}{8}+\frac{1}{2}=\frac{7}{8}$$

$$\frac{3}{8}+\frac{4}{8}=\frac{7}{8}$$

$$\frac{7}{8}=\frac{7}{8}$$

So, $x = 8$ is the solution.

Continued on next page...

b. $\dfrac{3}{x-2} + 2 = \dfrac{18}{x+1}$

Solution:

$$\dfrac{3}{x-2} + 2 = \dfrac{18}{x+1} \qquad x \neq 2,\ -1 \text{ and}$$

$$\text{LCM} = (x-2)(x+1).$$

$$(x-2)(x+1)\left[\dfrac{3}{x-2} + 2\right] = \dfrac{18}{x+1}(x-2)(x+1) \qquad \text{Multiply both sides by}$$

$$(x-2)(x-1).$$

$$\cancel{(x-2)}(x+1)\left(\dfrac{3}{\cancel{x-2}}\right) + 2(x-2)(x+1) = \left(\dfrac{18}{\cancel{x+1}}\right)(x-2)\cancel{(x+1)} \quad \text{Multiply each term,}$$

$$\text{including } 2, \text{ by } (x-2)(x+1).$$

$$3(x+1) + 2(x^2 - x - 2) = 18(x-2) \qquad \text{Both sides of the new}$$

$$\text{equation are polynomials.}$$

$$3x + 3 + 2x^2 - 2x - 4 = 18x - 36$$

$$2x^2 + x - 1 = 18x - 36$$

$$2x^2 - 17x + 35 = 0$$

$$(2x - 7)(x - 5) = 0$$

$$2x - 7 = 0 \quad \text{or} \quad x - 5 = 0$$

$$x = \dfrac{7}{2} \qquad\qquad x = 5$$

There are two solutions: $\dfrac{7}{2}$ and 5. Both will check. (Note that neither of these is 2 or −1, the restricted values.)

c. $\dfrac{x}{x-2} + \dfrac{x-6}{x(x-2)} = \dfrac{5x}{x-2} - \dfrac{10}{x-2}$

Solution: $\dfrac{x}{x-2} + \dfrac{x-6}{x(x-2)} = \dfrac{5x}{x-2} - \dfrac{10}{x-2} \qquad \text{LCM} = x(x-2).$ Also, $x \neq 0,\ 2.$

$$\dfrac{x}{\cancel{x-2}} \cdot x\cancel{(x-2)} + \dfrac{x-6}{\cancel{x(x-2)}} \cdot \cancel{x(x-2)} = \dfrac{5x}{\cancel{x-2}} \cdot x\cancel{(x-2)} - \dfrac{10}{\cancel{x-2}} \cdot x\cancel{(x-2)}$$

Multiply each term by $x(x-2)$.

Continued on next page...

$$x^2 + x - 6 = 5x^2 - 10x$$

$$0 = 4x^2 - 11x + 6$$

$$0 = (4x - 3)(x - 2)$$

$$4x - 3 = 0 \quad \text{or} \quad x - 2 = 0$$

$$4x = 3 \qquad\qquad x = 2$$

$$x = \frac{3}{4}$$

The only solution is $x = \frac{3}{4}$. Since $x \neq 2$, 2 is **not** a solution. No denominator can be 0.

d. $\dfrac{4}{x-4} - \dfrac{x}{x-4} = 1$

Solution: $\dfrac{4}{x-4} - \dfrac{x}{x-4} = 1$ $x \neq 4$ and LCM $= x - 4$.

$$(x-4)\left[\frac{4}{x-4} - \frac{x}{x-4}\right] = 1(x-4) \quad \text{Multiply both sides by } x - 4.$$

$$(x-4)\left(\frac{4}{x-4}\right) - (x-4)\left(\frac{x}{x-4}\right) = 1(x-4)$$

$$4 - x = x - 4$$

$$8 = 2x$$

$$4 = x$$

But, $x \neq 4$ was the restriction, because this value will give denominators of 0. Therefore, **this equation has no solution**.

● ●

Proportions

A **ratio** is a comparison of two numbers by division. Ratios are written in the form

$$a : b \quad \text{or} \quad \frac{a}{b} \quad \text{or} \quad a \text{ to } b.$$

For example, suppose the ratio of female to male faculty at the local high school is 3 to 2. We can also write this ratio in the form $3 : 2$ or in fraction form $\dfrac{3}{2}$. This ratio does not mean that there are only 5 teachers in the high school. There are many ratios (fractions) that reduce to $\dfrac{3}{2}$. If there are 35 teachers in the high school, then there are 21 women and 14 men teachers because $21 + 14 = 35$ and $\dfrac{21}{14} = \dfrac{3}{2}$.

A **proportion** is an equation stating that two ratios are equal. Such equations can involve rational expressions with variables. The solutions to such proportions can be found by multiplying both sides of the equation by the LCM of the denominators.

> **Example 2: Proportions** ● ● ● ● ● ● ● ● ● ● ● ● ● ● ● ● ●

Solving the following proportion.

a. $\dfrac{4}{x-5} = \dfrac{2}{x+3}$

$$\dfrac{4}{x-5} = \dfrac{2}{x+3} \qquad \text{Note: } x \neq 5, -3.$$

$$(x-5)(x+3)\dfrac{4}{x-5} = \dfrac{2}{x+3}(x-5)(x+3) \quad \text{Multiply both sides by the LCM } (x-5)(x+3).$$

$$4(x+3) = 2(x-5)$$

$$4x+12 = 2x-10$$

$$4x+12-2x-12 = 2x-10-2x-12$$

$$2x = -22$$

$$x = -11$$

● ●

Proportions can be used to solve many everyday types of word problems. We must be very careful with the units in the setup of a proportion. One of the following conditions must be true.

1. **The numerators agree in type and the denominators agree in type**.
2. **The numerators correspond and the denominators correspond**.

Both arrangements are illustrated in the following example. Consider the situation where tires are on sale at 2 for \$75, but you need 5 new tires. What would you pay for the 5 tires? Setting up a proportion gives one of the following types of equations:

Solving, we have

$$\dfrac{2 \text{ tires}}{\$75} = \dfrac{5 \text{ tires}}{\$x} \qquad\qquad \dfrac{\$75}{\$x} = \dfrac{2 \text{ tires}}{5 \text{ tires}}$$

$$75x \cdot \dfrac{2}{75} = 75x \cdot \dfrac{5}{x} \quad \text{or} \quad 5x \cdot \dfrac{75}{x} = 5x \cdot \dfrac{2}{5}$$

$$2x = 375 \qquad\qquad 375 = 2x$$

$$x = 187.50 \qquad\qquad 187.50 = x$$

TIRE SALE
2 for \$75

531

The cost of 5 tires would be $187.50.

Be careful to follow one of the patterns illustrated. Other arrangements will give the wrong answer. For example,

$$\frac{75 \ \text{dollars}}{x \ \text{dollars}} = \frac{5 \ \textit{tires}}{2 \ \textit{tires}} \quad \text{is \textbf{WRONG}}$$

Looking at the numerators we see that $75 does not correspond to 5 tires.

Example 3: Proportion Word Problems

a. On an architect's scale drawing of a building, $\frac{1}{2}$ inch represents 12 feet. What does 3 inches represent?

Solution: Set up a proportion representing the information.

$$\frac{\frac{1}{2} \ inch}{12 \ \text{feet}} = \frac{3 \ inches}{x \ \text{feet}}$$

$$12x \cdot \frac{\frac{1}{2}}{12} = 12x \cdot \frac{3}{x} \qquad \text{Multiply both sides by the LCM } 12x.$$

$$\frac{1}{2}x = 36$$

$$2 \cdot \frac{1}{2}x = 2 \cdot 36$$

$$x = 72$$

Three inches represents 72 feet.

b. Making a statistical analysis, Mike finds 3 defective computer disks in a sample of 20 disks. If this ratio is consistent, how many bad disks does he expect to find in an order of 2400?

Solution: $\dfrac{3 \ \textit{defective disks}}{20 \ \text{disks}} = \dfrac{x \ \textit{defective disks}}{2400 \ \text{disks}}$ Set up a proportion representing the given information.

$$2400 \cdot \frac{3}{20} = 2400 \cdot \frac{x}{2400}$$

$$360 = x$$

He expects to find 360 defective disks. (**Note:** He probably will return the order to the manufacturer.)

Practice Problems

Solve the following equations.

1. $\dfrac{10}{x} = \dfrac{15}{48}$

2. $\dfrac{x-4}{7} = \dfrac{3}{5}$

3. $\dfrac{3}{5x+2} = \dfrac{2}{x-1}$

4. $\dfrac{3}{x} + \dfrac{2}{3} = \dfrac{23}{3x}$

5. $\dfrac{5}{8}x + 1 = x - \dfrac{1}{6}$

6. $\dfrac{x-5}{x} + \dfrac{7}{x} = x$

8.5 Exercises

First state any restriction on the variable, then solve each equation in Exercises 1 – 44.

1. $\dfrac{4x}{3} - \dfrac{3}{4} = \dfrac{5x}{6}$

2. $\dfrac{5x}{8} - \dfrac{3}{4} = \dfrac{5}{16}$

3. $\dfrac{x-2}{3} - \dfrac{x-3}{5} = \dfrac{13}{15}$

4. $\dfrac{2+x}{4} - \dfrac{5x-2}{12} = \dfrac{8-2x}{5}$

5. $\dfrac{4}{5x} = \dfrac{2}{25}$

6. $\dfrac{2}{15} = \dfrac{8}{3x}$

7. $\dfrac{21}{x+4} = \dfrac{7}{8}$

8. $\dfrac{5}{x+3} = \dfrac{6}{11}$

9. $\dfrac{9}{3x} = \dfrac{1}{4} - \dfrac{1}{6x}$

10. $\dfrac{3}{8x} - \dfrac{7}{10} = \dfrac{1}{5x}$

11. $\dfrac{1}{x} - \dfrac{8}{21} = \dfrac{3}{7x}$

12. $\dfrac{3}{4x} = \dfrac{7}{8x} + \dfrac{2}{3}$

13. $\dfrac{2x}{x-1} = \dfrac{3}{2}$

14. $\dfrac{x+7}{x-4} = \dfrac{5}{6}$

15. $\dfrac{10}{x} = \dfrac{5}{x-2}$

16. $\dfrac{8}{x-3} = \dfrac{12}{2x-3}$

17. $\dfrac{6}{x-4} = \dfrac{5}{x+7}$

18. $\dfrac{4}{x-6} = \dfrac{11}{5x+2}$

19. $\dfrac{5}{x-6} - 5 = \dfrac{2x}{x-6}$

20. $\dfrac{2x}{x-2} + 3 = \dfrac{9}{x-2}$

21. $3 + \dfrac{2x}{x-3} = \dfrac{4}{x-3}$

Answers to Practice Problems: 1. $x = 32$ **2.** $x = \dfrac{41}{5}$ **3.** $x = -1$ **4.** $x = 7$ **5.** $x = \dfrac{28}{9}$ **6.** $x = 2$ or $x = -1$

22. $\dfrac{x}{x+3}+\dfrac{1}{x+2}=1$

23. $\dfrac{1}{3}+\dfrac{1}{x+4}=\dfrac{1}{x}$

24. $\dfrac{5}{x+3}-\dfrac{2}{x+1}=\dfrac{1}{4}$

25. $\dfrac{4}{x}-\dfrac{2}{x+1}=\dfrac{4}{3}$

26. $\dfrac{2}{x+2}+\dfrac{5}{x-2}=-3$

27. $\dfrac{6}{x+4}+\dfrac{2}{x-3}=-1$

28. $\dfrac{12}{x-4}=6+\dfrac{3x}{x-4}$

29. $\dfrac{6}{x-3}+1=\dfrac{2x}{x-3}$

30. $\dfrac{4}{x+2}+\dfrac{5}{x(x+2)}=\dfrac{3x+5}{x(x+2)}$

31. $\dfrac{x}{x+3}+\dfrac{1}{x+2}=1$

32. $\dfrac{2x}{x-4}+\dfrac{2}{x+1}=2$

33. $\dfrac{x}{x-4}-\dfrac{4}{2x-1}=1$

34. $\dfrac{x}{x-2}-\dfrac{x+1}{x+4}=1$

35. $\dfrac{2}{4x-1}+\dfrac{1}{x+1}=\dfrac{3}{x+1}$

36. $\dfrac{x-2}{x+4}-\dfrac{3}{2x+1}=\dfrac{x-7}{x+4}$

37. $\dfrac{x}{x-4}-\dfrac{12x}{x^2+x-20}=\dfrac{x-1}{x+5}$

38. $\dfrac{x-2}{x-3}+\dfrac{x-3}{x-2}=\dfrac{2x^2}{x^2-5x+6}$

39. $\dfrac{12}{x+1}+1=\dfrac{5}{x-1}$

40. $\dfrac{8}{x-4}-3=\dfrac{1}{x-2}$

41. $\dfrac{x}{x-3}+\dfrac{8}{x^2-2x-3}=\dfrac{-2}{x+1}$

42. $\dfrac{x}{x+4}+\dfrac{8x}{x^2-16}=\dfrac{1}{x-4}$

43. $\dfrac{x}{x+5}-\dfrac{5}{x^2+9x+20}=\dfrac{2}{x+4}$

44. $\dfrac{3}{x-1}-\dfrac{5}{x+4}=\dfrac{11}{14}$

45. The local sales tax is 6 cents for each dollar. What is the price of an item if the sales tax is 51 cents?

46. The property tax rate is \$10.87 on every \$100 of assessed valuation. Find the property tax on a house assessed at \$38,000.

47. At the Bright-As-Day light bulb plant, 3 out of each 100 bulbs produced are defective. If the daily production is 4800 bulbs, how many are defective?

48. On a map, each inch represents $7\dfrac{1}{2}$ miles. What is the distance represented by 6 inches?

49. An elementary school has a ratio of 1 teacher for every 24 children. If the school presently has 21 teachers, how many students are enrolled?

50. A floor plan is drawn to scale in which 1 inch represents 4 feet. What size will the drawing be for a room that is 26 feet by 32 feet?

51. Pete is averaging 17 hits for every 50 times at bat. If he maintains this average, how many "at bats" will he need in order to get 204 hits?

52. Every 50 pounds of a certain alloy contains 6 pounds of copper. To have 27 pounds of copper, how many pounds of alloy will you need?

53. Instructions for Never-Ice Antifreeze state that 4 quarts of antifreeze are needed for every 10 quarts of radiator capacity. If Sal's car has a 22-quart radiator, how many quarts of antifreeze will it need?

54. The negative of a rectangular-shaped picture is $1\frac{1}{2}$ in. by 2 in. If the largest paper available is 11 in. long, what would be the maximum width of the enlargement?

55. A flagpole casts a 30-foot shadow at the same time as a 6-foot man casts a $4\frac{1}{2}$ foot shadow. How tall is the flagpole?

56. In a group of 30 people, there are 3 left-handed people. How many left-handed people would you except to find in a group of 180 people?

Writing and Thinking About Mathematics

In simplifying rational expressions, the result is a rational expression. However, in solving equations that contain rational expressions, the goal is to find a value (or values) for the variable that will make the equation a true statement. Many students confuse these two ideas. To avoid confusing the techniques for adding and subtracting rational expressions with the techniques for solving equations, simplify the expression in part (a) and solve the equation in part (b). Explain, in your own words, the differences in your procedures.

57. a. $\dfrac{5}{x}+\dfrac{2}{x-1}+\dfrac{2x}{x-1}$ **b.** $\dfrac{5}{x}+\dfrac{2}{x-1}=\dfrac{2x}{x-1}$

58. a. $\dfrac{9}{x^2-9}+\dfrac{x}{3x+9}-\dfrac{1}{3}$ **b.** $\dfrac{9}{x^2-9}+\dfrac{x}{3x+9}=\dfrac{1}{3}$

Hawkes Learning Systems: Introductory Algebra

 Solving Equations Involving Rational Expressions

8.6   Applications

After completing this section, you will be able to:

1. Solve word problems using proportions.

2. Solve word problems related to distance, rate, and time.

3. Solve word problems related to the time it takes to complete a job.

In Section 8.5, we solved equations involving rational expressions (fractions) and some word problems using proportions. In this section, we continue to deal with rational expressions in solving a variety of applications involving numbers and proportions, time (using the relationship $d = rt$ in the form $t = \dfrac{d}{r}$), and work.

More on Proportions

Continuing our discussion of proportions from Section 8.5, we are interested in situations where a total amount is known and this total is to be separated into two parts according to some ratio. The key idea here is to recognize the relationship between the total and the two parts. In general,

> If T = total amount
> and x = one part
> then $T - x$ = second part.

For example, suppose that you have $20 and you give $5 to your brother and the remainder to your sister. What amount will you give to your sister? Answer: $20 – $5 = $15.

Example 1: More on Proportions ● ● ● ● ● ● ● ● ● ● ● ● ● ●

a.   The sum of two integers is 104 and they are in the ratio of 8 to 5. Find the two numbers.

Solution: Here we know the total sum of the two numbers is 104. So,

Continued on next page...

let n = one number

and $104 - n$ = second number.

Then set up a proportion using the ratio 8 to 5.

$$\frac{n}{104-n} = \frac{8}{5}$$

$$5n = 8(104-n)$$

$$5n = 832 - 8n$$

$$\frac{13n}{13} = \frac{832}{13}$$

$$n = 64$$

$$104 - n = 40$$

The two numbers are 64 and 40.

(**Checking**, we get the sum $64 + 40 = 104$ and the ratio $\dfrac{64}{40} = \dfrac{8\cdot 8}{8\cdot 5} = \dfrac{8}{5}$.)

b. Suppose that an artist expects that for every 9 special brushes he orders, 7 will be good and 2 will be defective. If he orders 45 brushes, how many will he expect to be defective?

Solution: Let n = number of defective brushes

and $45 - n$ = number of good brushes (Total $- n$).

Then, using the ratio of good to defective,

$$\frac{7 \ good}{2 \ defective} = \frac{45 - n \ good}{n \ defective}$$

$$7n = 2(45 - n)$$

$$7n = 90 - 2n$$

$$9n = 90$$

$$n = 10$$

He expects 10 defective brushes.

• •

Time $\left(\boldsymbol{d = rt} \ \textbf{or} \ \boldsymbol{t = \dfrac{d}{r}} \right)$

Another application of rational expressions and proportions involves the formula $d = rt$ (distance equals rate times time). The same formula can be written in three forms:

$$d = rt, \quad r = \frac{d}{t}, \quad \text{and} \quad t = \frac{d}{r}.$$

The last formula, $t = \dfrac{d}{r}$, indicates that time can be represented as a ratio of distance divided by rate.

Example 2: Time ●

a. Janice can run 32 kilometers (about 20 miles) in the same amount of time that her twin sister, Jill, can run 24 kilometers. If Janice runs 2 kilometers per hour faster than Jill, how fast does each girl run?

Solution: Let r = Jill's rate
and $r + 2$ = Janice's rate (She runs faster than Jill.)

	Rate	Time = $\dfrac{\text{distance}}{\text{rate}}$	Distance
Jill	r	$\dfrac{24}{r}$	24
Janice	$r + 2$	$\dfrac{32}{r+2}$	32

Using the formula $t = \dfrac{d}{r}$ and the fact that **their times are equal**, we can write the equation:

$$\text{Jill's time} \Big\} \rightarrow \frac{24}{r} = \frac{32}{r+2} \leftarrow \Big\{ \text{Janice's time}$$

$$24(r+2) = 32r$$

$$24r + 48 = 32r$$

$$48 = 8r$$

$$6 = r$$

So, Jill's rate = r = 6 km/hr.
and Janice's rate = $r + 2$ = 8 km/hr.

Continued on next page...

b. A small plane can fly at 120 mph in still air. If it can fly 490 miles with a tailwind in the same time that it can fly 350 miles against a headwind, what is the speed of the wind? (**Note:** A tailwind is a wind that blows in the same direction that the plane is flying and, therefore increases the speed of the plane. A headwind is a wind that blows from the opposite direction the plane is flying and therefore decreases the speed of the plane.)

Solution: Let w = speed of the wind

	Rate	Time $= \dfrac{\text{distance}}{\text{rate}}$	Distance
Tailwind	$120 + w$	$\dfrac{490}{120 + w}$	490
Headwind	$120 - w$	$\dfrac{350}{120 - w}$	350

Using the formula $t = \dfrac{d}{r}$ and the fact that the times are equal, the equation is

$$\text{time with the wind} \Big\} \rightarrow \frac{490}{120 + w} = \frac{350}{120 - w} \leftarrow \Big\{ \text{time against the wind}$$

$$490(120 - w) = 350(120 + w)$$

$$58{,}800 - 490w = 42{,}000 + 350w$$

$$16{,}800 = 840w$$

$$20 = w$$

The speed of the wind is 20 mph.

● ●

Work

Problems involving "work" can be very sophisticated and require calculus and physics. The problems we will be concerned with are related to the time involved to complete the work on a particular job. These problems involve only the idea of what fraction of the job is done in one unit of time (hours, minutes, days, weeks, and so on). For example, if a man can dig a ditch in 4 hours, what part (of the ditch-digging job) can he do in one hour?

The answer is $\dfrac{1}{4}$. If the job took 5 hours, he would do $\dfrac{1}{5}$ in one hour. If the job took x hours, he would do $\dfrac{1}{x}$ in one hour.

Example 3: Work ●

a. Mike can clean his family's pool in 2 hours. His younger sister, Stacey, can do it in 3 hours. If they work together, how long will it take them to clean the pool?

Solution: Let x = number of hours working together. Then

	Hours to Clean Pool	Part Cleaned in 1 hour
Mike	2	$\dfrac{1}{2}$
Stacey	3	$\dfrac{1}{3}$
Together	x	$\dfrac{1}{x}$

$$\underbrace{\frac{\text{Part done in}}{\text{1 hr. by Mike}}}_{} + \underbrace{\frac{\text{Part done in}}{\text{1hr. by Stacey}}}_{} = \underbrace{\frac{\text{Part done in}}{\text{1hr. together}}}_{}$$

$$\frac{1}{2} \quad + \quad \frac{1}{3} \quad = \quad \frac{1}{x}$$

$\dfrac{1}{2}(6x)+\dfrac{1}{3}(6x)=\dfrac{1}{x}(6x)$ Multiply each term on both sides of the equation by $6x$, the LCM of the denominators.

$$3x+2x=6$$

$$5x=6$$

$$x=\frac{6}{5}$$

Together they can clean the pool in $\dfrac{6}{5}$ hours, or 1 hour, 12 minutes.

(Note that this answer seems reasonable since it is less time than either person would take working alone.)

Continued on next page...

b. A man was told that his new Jacuzzi® pool would fill through an inlet valve in 3 hours. He knew something was wrong when the pool took 8 hours to fill. He found he had left the drain valve open. How long will it take to drain the pool once it is filled and only the drain valve is opened?

Solution: Let t = time to drain pool. (**Note**: We use the information gained when the pool was filled with both valves open. In that situation, the inlet and outlet valves worked against each other.)

	Hours to Fill or Drain	Part Filled or Drained in 1 Hour
Inlet	3	$\dfrac{1}{3}$
Outlet	t	$\dfrac{1}{t}$
Together	8	$\dfrac{1}{8}$

$$\underbrace{\text{Part filled} \atop \text{by inlet}} - \underbrace{\text{Part emptied} \atop \text{by outlet}} = \underbrace{\text{Part filled} \atop \text{together}}$$

$$\frac{1}{3} \quad - \quad \frac{1}{t} \quad = \quad \frac{1}{8}$$

$$\frac{1}{3}(24t) - \frac{1}{t}(24t) = \frac{1}{8}(24t)$$

$$8t - 24 = 3t$$

$$5t = 24$$

$$t = \frac{24}{5}$$

The pool will drain in $\dfrac{24}{5}$ hours, or 4 hours, 48 minutes. (Note that this is more time than the inlet valve would take to fill the pool. If the outlet valve worked faster than the inlet valve, then the pool would never have filled in the first place.)

8.6 Exercises

1. The ratio of two numbers is 7 to 9. If the sum of the numbers is 48, find the numbers.

2. The ratio of two numbers is 3 to 7. If the sum of the numbers is 60, find the numbers.

3. In a mathematics class, the ratio of women to men is 6 to 5. If there are 44 students in the class, how many men are there?

4. In a certain class, the ratio of success to failure is 7 to 2. How many can be expected to successfully complete the class if 36 people are enrolled?

5. The denominator of a fraction is two more than the numerator. If both numerator and denominator are increased by three, the result is $\frac{5}{6}$. Find the original fraction.

6. In an iodine-alcohol mixture, there are 3 parts of iodine for each 8 parts of alcohol. If you want a total of 220 milliliters of mixture, how many milliliters of iodine and alcohol are needed?

7. An office has two copiers. One makes 8 copies per minute and the other makes 10 copies per minute. If both are used at the same time to make a total of 675 copies, how many copies are made by each machine? (**Hint**: Both machines are on for the same amount of time.)

8. The denominator of a fraction is one more than twice the numerator. If both numerator and denominator are increased by seven, the result is $\frac{3}{4}$. Find the original fraction.

9. Julie travels 15 miles per hour faster than Derek. She can travel 180 miles in the same amount of time that he can travel 135 miles. Find the two rates.

	Rate	Time	Distance
Derek	x	?	135
Julie	$x + 15$	?	180

10. An airliner travels 1820 miles in the same amount of time it takes a private plane to travel 520 miles. If the airliner's average speed is 52 mph more than three times the average speed of the private plane, find both rates.

	Rate	Time	Distance
Private plane	x	?	520
Airliner	$3x + 54$	?	1820

11. An airplane can travel 320 mph in still air. If it travels 690 miles with the wind in the same length of time it travels 510 miles against the wind, what is the speed of the wind?   48mph

12. A boat can travel 5 miles downstream in the same length of time it travels 3 miles upstream. If the rate of the current is 3 mph, find the speed of the boat in still water.

13. After traveling 750 miles, the pilot of an airliner increased the speed by 20 mph and continued on for another 780 miles. If the same amount of time was spent at each rate, find the rates.

14. Mr. Snyder had a sales meeting scheduled 80 miles from his home. After traveling the first 30 miles at a rather slow rate due to traffic, he found that in order to make the meeting on time, he had to increase his speed by 25 mph. If he traveled the same length of time at each rate, find the rates.

15. An airliner leaves Phoenix traveling at an average rate of 420 mph. Thirty minutes later, a second airliner leaves Phoenix traveling along a parallel route. If the second plane overtakes the first at a point 1680 miles from Phoenix, find its rate. (**Hint:** The difference in the times is $\frac{1}{2}$ hr.)

	Rate	Time	Distance
First airliner	420	?	1680
Second airliner	x	?	1680

16. Mrs. Green leaves El Paso on a business trip to Los Angeles, traveling at 56 mph. Fifteen minutes later, Mr. Green discovers that she had forgotten her briefcase and, jumping into their second car, he begins to pursue her. What was his average speed if he caught her 84 miles outside of El Paso? (**Hint:** The difference in times is $\frac{1}{4}$ hr.)

	Rate	Time	Distance
Mrs. Green	56	?	84
Mr. Green	x	?	84

17. A boat traveled 9 miles downstream and returned. The total trip took $2\frac{1}{4}$ hours. If the speed of the current was 3 mph, find the speed of the boat in still water.

18. A boat travels 15 miles upstream and returns. If the boat travels 14 mph in still water, find the speed of the current if the total trip takes 2 hours, 20 minutes.

19. Miles rides the ski lift to the top of Snowy Peak, a distance of $1\frac{3}{4}$ miles. He then skis directly down the slope. If he skis five times as fast as the lift travels and the total trip takes 45 minutes, find the rate at which he skis.

20. Bonnie lives 45 miles from town. Recently she went to town and returned home. She traveled 6 mph faster returning home than she did going to town. If her total travel time was 1 hour, 35 minutes, find the two rates.

21. Tom can fix a car in 6 hours. Ellen can do it in 4 hours. How long would it take them working together?

22. One company can clear a parcel of land in 25 hours. Another company requires 30 hours to do the job. If both are hired, how long will it take them to clear the parcel working together?

23. One pipe can fill a tank in 15 minutes. The tank can be drained by another pipe in 45 minutes. If both pipes are opened, how long will it take to fill the tank?

24. One pipe can fill a tank three times as fast as another. When both pipes are used, it takes $1\frac{1}{2}$ hours to fill the tank. How long does it take each pipe alone?

25. Working together, Rick and Rod can clean the snow from the driveway in 20 minutes. It would have taken Rick, working alone, 36 minutes. How long would it have taken Rod alone?

26. A carpenter and his partner can put up a patio cover in $3\frac{3}{7}$ hours. If the partner needs 8 hours to complete the patio alone, how long will it take the carpenter working alone?

27. It takes Maria twice as long to complete a certain type of research project as it takes Judy to do the same project. They both worked on a similar type of project and they finished in 1 hour. How long would each of them taken to finish this project working alone?

28. It takes Bob four hours longer to repair a car than it takes Ken. Working together, they can complete the job in $1\frac{1}{2}$ hours. How long would each of them take working alone?

29. A power plant has two coal-burning boilers. If both boilers are in operation, a load of coal is burned in 6 days. If only one boiler is used, a load of coal will last the new "fuel-efficient" boiler 5 days longer than the old boiler. How long will a load of coal last each boiler?

30. Working together, Alice and Sharon can prepare the company's sales report in $2\frac{2}{5}$ hours. Working alone, Sharon would need two hours longer to prepare the report than Alice would. How long would it take each of them working alone?

Hawkes Learning Systems: Introductory Algebra

Applications Involving Rational Expressions

8.7   Additional Applications: Variation

After completing this section, you will be able to:

1. Understand direct and inverse variation.

2. Solve applied problems.

Direct Variation

Suppose that you ride your bicycle at a steady rate of 10 miles per hour. If you ride for 1 hour, the distance you travel would be 10 miles. If you ride for 2 hours, the distance you travel would be 20 miles. This relationship can be written in the form of the formula $d = 10t$ (or $\frac{d}{t} = 10$). We say that distance and time **vary directly** (or are in **direct variation** or are **directly proportional**). The term directly proportional implies that the ratio is constant. In this example, 10 is the constant and is called the **constant of variation**. When two variables vary directly, an increase in the value of one variable indicates an increase in the other and the ratio of the two quantities is constant.

Direct Variation

*A variable quantity y **varies directly** as (or is **directly proportional to**) a variable x if there is a constant k such that*

$$\frac{y}{x} = k \quad or \quad y = kx$$

*The constant k is called the **constant of variation**.*

Example 1: Direct Variation • • • • • • • • • • • • • • • • •

The length, s, that a hanging spring stretches varies directly as the weight placed at the end of the spring. If a weight of 5 mg stretches a certain spring 3 cm, how far will the spring stretch with a weight of 6 mg? (**Note:** We assume that the weight is not so great as to break the spring.)

Continued on next page...

Solution: Since the two variables are directly proportional, they are related as

$s = kw$ where s is the distance the spring stretches and w is the weight placed at the end of the spring.

First substitute the given information to find the value for k. (The value for k will depend on the particular spring. Springs made of different material or which are wound tighter will have different values for k.)

$$s = kw$$
$$3 = k \cdot 5$$
$$\frac{3}{5} = k$$

So, substitute $\frac{3}{5}$ for k in the equation. Thus, for this spring

$$s = \frac{3}{5}w \ \ (\text{ or } s = 0.6w).$$

Now, for $w = 6\text{mg}$,

$$s = 0.6(\,6\,) = 3.6$$

The spring will stretch 3.6 cm if a weight of 6 mg is placed at its end.

Inverse Variation

When two variables vary in such a way that their product is constant, we say that the two variables **vary inversely** (or are **inversely proportional**). For example, if a gas is placed in a container (as in an automobile engine) and pressure is increased on the gas, then the product of the pressure and the volume of the gas will be constant. That is, pressure and volume are related by the formula $V \cdot P = k$ $\left(\text{or } V = \dfrac{k}{P} \right)$. Note that if a product of two variables is to remain constant, then an increase in the value of one variable must be accompanied by a decrease in the other. Or, in the case of a fraction with a constant numerator, if the denominator increases in value, then the fraction decreases in value. For the gas in an engine, an increase in pressure indicates a decrease in the volume of gas.

Inverse Variation

*A variable quantity y **varies inversely** as (or is **inversely proportional to**) a variable x if there is a constant k such that*

$$xy = k \quad or \quad y = \frac{k}{x}$$

*The constant k is called the **constant of variation.***

Example 2: Inverse Variation

The volume of gas in a container varies inversely to the pressure on the gas. If a gas has a volume of 200 in.^{3} under a pressure of 4 lb. per in.^{2}, what will be its volume if the pressure is increased to 5 lb. per in.^{2}?

Solution:

$$V = \frac{k}{P}$$

$$200 = \frac{k}{4} \qquad \text{Substitute 200 for } V \text{ and 4 for } P \text{ and solve for } k.$$

$$k = 800$$

Now, substitute 800 for k in the equation.

$$V = \frac{800}{P}$$

$$V = \frac{800}{5} = 160$$

The volume will be 160 in.^{3}.

PROPANE

8.7 Exercises

Write an equation or formula that represents the general relationship indicated in Exercises 1 – 15. Then use the given information to find the unknown value.

1. If y varies directly as x and $y = 2$ when $x = 10$, find y if $x = 7$.

2. If y is directly proportional to x and $y = 3$ when $x = 18$, find y if $x = 10$.

3. If y is directly proportional to x^2 and $y = 9$ when $x = 2$, what is y when $x = 4$?

4. If y varies inversely as x^2 and $y = -8$ when $x = 2$, find y when $x = 3$.

5. If y is inversely proportional to x and $y = 5$ when $x = 4$, find the value of y when $x = 2$.

6. If y varies inversely as x^3 and $y = 40$ when $x = \frac{1}{2}$, what is the value of y when $x = \frac{1}{3}$?

7. The total price, P, of gasoline purchased varies directly with the number of gallons purchased. If 20 gallons are purchased for $25, what will be the price of 30 gallons?

8. The total price, P, of pumpkin pies purchased is directly proportional to the number of pies purchased. If 3 pumpkin pies are bought for $19.50, what will 5 pumpkin pies cost?

9. The gravitational force, F, between an object and the earth is inversely proportional to the square of the distance from the object to the center of the earth. If an astronaut weighs 200 pounds on the surface of the earth, what will the astronaut weigh 100 miles above the earth? (Assume that the radius of the earth is 4000 miles and round to the nearest tenth.)

10. The distance a free falling object falls is directly proportional to the square of the time it falls (before it hits the ground). If an object fell 64 ft. in 2 seconds, how far will it have fallen by the end of 3 seconds?

11. The volume of gas in a container is 300 cm^3 when the pressure on the gas is 15 gm per cm^2. What will be the volume of the gas if the pressure is increased to 20 gm per cm^2?

12. A hanging spring will stretch 4 in. if a weight of 8 lbs. is placed at its end. How far will the spring stretch if the weight is increased to 10 lbs.?

13. The circumference of a circle varies directly as the diameter. A circular clock with a diameter of 1 foot has a circumference of 3.14 feet. What will be the circumference of a clock with a diameter of 2 feet?

14. The area of a circle varies directly as the square of its radius. A pebble is dropped into a pool of water and the concentric circles created by the waves are observed. If the area of the circle with radius 2 feet is 12.56 ft.^{2}, what is the area of the circle with a radius of 4 feet?

15. Several triangles are to have the same area. In this set of triangles, the height and base are inversely proportional. In one such triangle, the height is 6 in. when the base is 10 in. Find the height of the triangle in this set with a base of 15 in.

Lifting Force

The lifting force (or lift), P, in pounds exerted by the atmosphere on the wings of an airplane is related to the area, A, of the wings in square feet and the speed (or velocity), v, of the plane in miles per hour by the formula

$P = kAv^2$ where k is the constant of variation.

16. If the lift is 9600 lbs. for a wing area of 120 ft.^{2} and a speed of 80 mph, find the lift of the same airplane at a speed of 100 mph.

17. If the lift is 12,000 lbs. for a wing area of 110 ft.^{2} and a speed of 90 mph, what would be the necessary wing area to attain the same lift at 80 mph? (Round to the nearest thousandth.)

18. The lift for a wing of area 280 ft.^{2} is 34,300 lbs. when the plane is going 210 mph. What is the lift if the speed is decreased to 180 mph?

Pressure

Boyle's Law states that if the temperature of a gas sample remains the same, the pressure of the gas is related to the volume by the formula

$P = \dfrac{k}{V}$ where k is the constant of proportionality.

19. The temperature of a gas remains the same. The pressure of the gas is 16 lbs. per ft.^{2} when the volume is 300 cu. ft. What will be the pressure if the gas is compressed into 25 cu. ft.?

20. A pressure of 1600 lbs. per ft.^{2} is exerted by 2 cu. ft. of air in a cylinder. If a piston is pushed into the cylinder until the pressure is 1800 lbs. per ft.^{2}, what will be the volume of the air? (Round to the nearest tenth.)

21. If 1000 cu. ft. of gas exerting a pressure of 140 lb. per ft.^{2} must be placed in a container which has a capacity of 200 cu. ft., what is the new pressure exerted by the gas?

Electricity

The resistance, R (in ohms), in a wire is given by the formula

$$R = \frac{kL}{d^2}$$

where k is the constant of variaition, L is the length of
the wire and d is the diameter.

22. The resistance of a wire 500 ft. long with a diameter of 0.01 in. is 20 ohms. What is
the resistance of a wire 1500 ft. long with a diameter of 0.02 in.?

23. The resistance of a wire 100 ft. long and 0.01 in. in diameter is 8 ohms. What is the
resistance of a piece of the same type of wire with a length of 150 ft. and a diameter
of 0.015 in.?

24. The resistance is 2.6 ohms when the diameter of a wire is 0.02 in. and the wire is 10
ft. long. Find the resistance of the same type of wire with a diameter of 0.01 in. and
a length of 5 ft.

Gears

If a gear with T_1 teeth is meshed with a gear having T_2 teeth, the numbers of revolutions
of the two gears are related by the proportion

$$\frac{T_1}{T_2} = \frac{R_2}{R_1} \quad \text{or} \quad T_1R_1 = T_2R_2$$

where R_1 and R_2 are the numbers of revolutions.

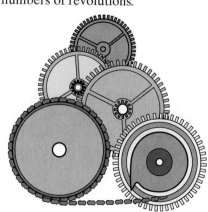

25. A gear having 96 teeth is meshed with a gear having 27 teeth. Find the speed (in revolutions per minute) of the small gear if the large one is traveling at 108 revolutions per minute.

26. What size gear (how many teeth) is needed to turn a pulley on a feed grinder at 24 revolutions per minute if it is driven by a 15 tooth gear turning at 56 revolutions per minute?

27. What size gear (how many teeth) is needed to turn the reel on a harvester at 20 revolutions per minute if it is driven by a 12-tooth gear turning at 60 revolutions per minute?

Lever

If a lever is balanced with weight on opposite sides of the balance points, then the following proportion exists:

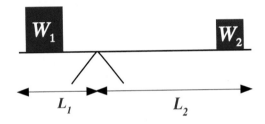

$$\frac{W_1}{W_2} = \frac{L_2}{L_1} \text{ or } W_1L_1 = W_2L_2$$

28. How much weight can be raised at one end of a bar 8 ft. long by the downward force of 60 lbs. when the balance point is $\frac{1}{2}$ ft. from the unknown weight?

29. A force of 40 lbs. at one end of a bar 5 ft. long is to balance 160 lbs. at the other end of the bar. Ignoring the weight of the bar, how far from the 40 lb. weight should a hole be drilled in the bar for a bolt to serve as a balance point?

30. Where should the balance point of a bar 12 ft. long be located if a 120 lb. force is to raise a load weighing 960 lbs.?

Hawkes Learning Systems: Introductory Algebra

Additional Applications: Variation

Chapter 8 Index of Key Ideas and Terms

Rational Expression page 496

A **rational expression** is an algebraic expression that can be written in the form $\dfrac{P}{Q}$ where P and Q are polynomials and $Q \neq 0$.

Fundamental Principle of Fractions page 499

If $\dfrac{P}{Q}$ is a rational expression and $Q \neq 0$ and K is a polynomial and $K \neq 0$, then $\dfrac{P}{Q} = \dfrac{P \cdot K}{Q \cdot K}$.

Opposites page 502

In general, for $b \neq a$, $(a - b)$ and $(b - a)$ are **opposites** and $\dfrac{a-b}{b-a} = \dfrac{a-b}{-1(a-b)} = -1$.

Multiplying Rational Expressions page 506

If $P, Q, R,$ and S are rational expressions with $Q, S \neq 0$, then

$$\frac{P}{Q} \cdot \frac{R}{S} = \frac{P \cdot R}{Q \cdot S}$$

Dividing Rational Expressions page 507

If $P, Q, R,$ and S are rational expressions with $Q, R, S \neq 0$, then

$$\frac{P}{Q} \div \frac{R}{S} = \frac{P}{Q} \cdot \frac{S}{R}$$

Adding Rational Expressions page 511

For polynomials $P, Q,$ and R with $Q \neq 0$, $\dfrac{P}{Q} + \dfrac{R}{Q} = \dfrac{P+R}{Q}$.

LCM for a Set of Polynomials page 512

To find the LCM for a set of polynomials,

1. Completely factor each polynomial (including prime factors for numerical factors).
2. Form the product of all factors that appear, using each factor the most number of times it appears in any one polynomial.

  A variable quantity y **varies directly** as (or is **directly
  proportional to**) a variable x if there is a constant k such that

$$\frac{y}{x} = k \text{ or } y = kx$$

  The constant k is called the **constant of variation**.

Inverse Variation   page 549

A variable quantity y **varies inversely** as (or is **inversely proportional to**) a variable x if there is a constant k such that

$$xy = k \quad \text{or} \quad y = \frac{k}{x}$$

The constant k is called the constant of variation.

Chapter 8 Review

For a review of the topics and problems from Chapter 8, look at the following lessons from *Hawkes Learning Systems: Introductory Algebra*

Defining Rational Expressions
Multiplication and Division of Rational Expressions
Addition and Subtraction of Rational Expressions
Complex Algebraic Expressions
Solving Equations Involving Rational Expressions
Applications Involving Rational Expressions
Additional Applications: Variation

Chapter 8 Test

In Exercises 1 – 3, reduce each of the rational expressions to lowest terms, and state any restrictions on the variable.

1. $\dfrac{2x^2 + 5x - 3}{2x^2 - x}$

2. $\dfrac{8x^2 - 2x - 3}{4x^2 + x - 3}$

3. $\dfrac{15x^4 - 18x^3}{36x - 30x^2}$

4. Evaluate the rational expression for $x = 2$: $\dfrac{x - 6}{6 - x}$

5. Evaluate the rational expression for $y = -3$: $\dfrac{2y^3}{y^2 - 25}$

In Exercises 6 – 11, perform the indicated operations and reduce all answers. Also, state any restrictions on the variables.

6. $\dfrac{16x^2 + 24x + 9}{3x^2 + 14x + 8} \cdot \dfrac{21x + 14}{4x^2 + 3x}$

7. $\dfrac{3x - 3}{3x^2 - x - 2} \div \dfrac{24x^2 - 6}{6x^2 + x - 2}$

8. $\dfrac{3}{x^2 - 1} + \dfrac{7}{x^2 - x - 2}$

9. $\dfrac{8x}{x^2 - 25} - \dfrac{3}{x^2 + 3x - 10}$

10. $\dfrac{1}{x - 1} + \dfrac{2x}{x^2 - x - 12} - \dfrac{5}{x^2 - 5x + 4}$

11. $\dfrac{x^2 + 8x + 15}{x^2 + 6x + 5} \cdot \dfrac{x - 3}{2x} \div \dfrac{x + 3}{x^2}$

Simplify the complex fractions in Exercises 12 and 13.

12. $\dfrac{\dfrac{1}{x} - \dfrac{1}{x^2}}{1 - \dfrac{1}{x^2}}$

13. $\dfrac{\dfrac{6}{x - 5}}{\dfrac{x + 18}{x^2 - 2x - 15}}$

Solve the equations in Exercises 14 – 16.

14. $\dfrac{9}{3 - x} = \dfrac{8}{2x + 1}$

15. $\dfrac{3}{4} = \dfrac{2x + 5}{5 - x}$

16. $\dfrac{2x}{x - 1} - \dfrac{3}{x^2 - 1} = \dfrac{2x + 5}{x + 1}$

17. Simplify the expression in part (a) and solve the equation in part (b).

a. $\dfrac{3}{x} - \dfrac{2}{x + 1} + \dfrac{5}{2x}$

b. $\dfrac{3}{x} - \dfrac{2}{x + 1} = \dfrac{5}{2x}$

18. Two numbers are in a ratio of 5 to 7. Their sum is 156. Find the numbers.

19. Lisa can travel 228 miles in the same time that Kim travels 168 miles. If Lisa's speed is 15 mph faster than Kim's, find their rates.

20. A plumber can complete a job in 2 hours. If an apprentice helps, it takes only $1\frac{1}{3}$ hours. How long would it take the apprentice working alone?

21. The pressure on one side of a metal plate submerged horizontally in water varies directly as the depth of the water. If a plate 10 ft. below the surface has a total pressure of 625 lbs. on one face, how much pressure will then be on that same plate 25 ft. below the surface?

22. The lifting force on the wings of an airplane is given by the formula $P = kAv^2$. If the lift on the wing of a certain plane is 16,000 lbs. for a wing area of 180 ft.^{2} at a speed of 120 mph, find the lift of the same plane at a speed of 150 mph.

23. The resistance, R (in ohms), in a wire is given by the formula $R = \dfrac{kL}{d^2}$. The resistance of a wire with a diameter of 0.01 in. and 250 ft. long is 10 ohms. What is the resistance of a piece of the same type of wire with a length of 300 ft. and a diameter of 0.02 in.?

24. A swimming pool has two inlet pipes. One inlet pipe alone can fill the pool in 3 hours and the two pipes open together can fill the pool in 2 hours. How long would it take for the second inlet pipe alone to fill the pool?

Cumulative Review: Chapters 1 – 8

1. Perform the indicated operations and reduce.

a. $\dfrac{5}{6} + \dfrac{3}{4}$ **b.** $\dfrac{1}{2} - \dfrac{5}{7}$ **c.** $\dfrac{1}{2} \cdot \dfrac{1}{2}$ **d.** $\dfrac{7}{32} \div \dfrac{21}{8}$

2. Find 12.5% of 40.

In Exercises 3 – 6, perform the indicated operations and simplify the expressions.

3. $(3x + 7)(x - 4)$ **4.** $(2x - 5)^2$

5. $-4(x + 2) + 5(2x - 3)$ **6.** $x(x + 7) - (x + 1)(x - 4)$

7. In the formula $A = P + Prt$, solve for t.

8. In the formula $C = 2\pi r$, solve for r.

Factor each expression in Exercises 9 and 10.

9. $4x^2 + 16x + 15$ **10.** $2x^2 + 6x - 20$

11. Given the relation $r = \{(-1, 5), (0, 2), (-1, 0), (5, 0), (6, 0)\}$.

  a. What is the domain of the relation?
  b. What is the range of the relation?
  c. Is the relation a function? Explain.

12. For the function $f(x) = x^3 - 2x^2 - 1$, find

  a. $f(3)$ **b.** $f(0)$ **c.** $f\left(-\dfrac{2}{3} \right)$

13. Find the equation of the line that has slope $\dfrac{3}{4}$ and passes through the point $(-2, 4)$. Graph the line.

14. Find the equation of the line perpendicular to the line $x - 2y = 3$ and passing through the point $(-4, 0)$. Graph both lines.

In Exercises 15 and 16, use the vertical line test to determine whether each of the graphs does or does not represent a function.

15.

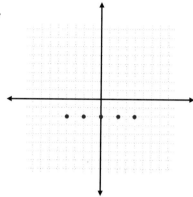

16.

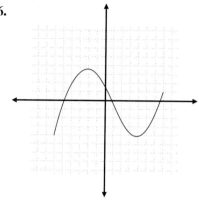

In Exercises 17 and 18, use a graphing calculator to estimate the solutions to each system of equations.

17. $\begin{cases} x+y=5.2 \\ 2x-y=-1.5 \end{cases}$

18. $\begin{cases} -x+2y=6 \\ 3x+2y=-4 \end{cases}$

Solve the systems of equations in Exercises 19 – 21. State whether the system is (a) consistent, (b) inconsistent, or (c) dependent.

19. $\begin{cases} 2x-y=10 \\ x+2y=5 \end{cases}$

20. $\begin{cases} y=3x+2 \\ -6x+2y=4 \end{cases}$

21. $\begin{cases} y=\dfrac{1}{2}x+3 \\ 6x-2y=3 \end{cases}$

22. Use a graphing calculator to graph the two functions $f(x)=\dfrac{2}{5}x+1$ and $g(x)=-2x^2+6$. Estimate the points of intersection.

Simplify each expression in Exercises 23 – 24.

23. $\dfrac{32x^3y^2}{4xy^2}$

24. $\dfrac{15xy^{-1}}{3x^{-2}y^{-3}}$

In Exercises 25 and 26, express each quotient as a sum of fractions and simplify if possible.

25. $\dfrac{3x^3y + 10x^2y^2 + 5xy^3}{5x^2y}$

26. $\dfrac{7x^2y^2 - 21x^2y^3 + 8x^2y^4}{7x^2y^2}$

Divide in Exercises 27 and 28 by using the long division algorithm.

27. $(x^2 - 14x - 16) \div (x + 1)$

28. $\dfrac{2x^3 + 5x^2 + 7}{x + 3}$

In Exercises 29 – 36, reduce the rational expressions to lowest terms and indicate any restrictions on the variable.

29. $\dfrac{8x^3 + 4x^2}{2x^2 - 5x - 3}$

30. $\dfrac{4x^2 + 5x + 9}{x - 2}$

31. $\dfrac{x^2 - 9}{x + 3}$

32. $\dfrac{x}{x^2 + x}$

33. $\dfrac{4 - x}{3x - 12}$

34. $\dfrac{2x + 6}{3x + 9}$

35. $\dfrac{x^2 + 3x}{x^2 + 7x + 12}$

36. $\dfrac{x^2 + 2x - 15}{2x^2 - 12x + 18}$

Perform the indicated operations and simplify in Exercises 37 – 46. Assume that no denominator is 0.

37. $\dfrac{x^2}{x + y} - \dfrac{y^2}{x + y}$

38. $\dfrac{7}{x} + \dfrac{9}{x - 4}$

39. $\dfrac{4}{y + 2} - \dfrac{4}{y + 3}$

40. $\dfrac{4x}{3x + 3} - \dfrac{x}{x + 1}$

41. $\dfrac{4x}{x - 4} \div \dfrac{12x^2}{x^2 - 16}$

42. $\dfrac{x^2 + 3x + 2}{x + 3} \div \dfrac{x + 1}{3x^2 + 6x}$

43. $\dfrac{2x + 1}{x^2 + 5x - 6} \cdot \dfrac{x^2 + 6x}{x}$

44. $\dfrac{8}{x^2 + x - 6} + \dfrac{2x}{x^2 - 3x + 2}$

45. $\dfrac{x}{x^2 + 3x - 4} + \dfrac{x + 1}{x^2 - 1}$

46. $\dfrac{x + 1}{x^2 + 4x + 4} \div \dfrac{x^2 - x - 2}{x^2 - 2x - 8}$

Simplify the complex algebraic fractions in Exercises 47 – 49.

47. $\dfrac{\dfrac{3}{x} + \dfrac{1}{6x}}{\dfrac{7}{3x}}$

48. $\dfrac{\dfrac{1}{x} - \dfrac{1}{x^2}}{\dfrac{1}{x} + \dfrac{1}{x^2}}$

49. $\dfrac{\dfrac{4}{3x} + \dfrac{1}{6x}}{\dfrac{1}{x^2} - \dfrac{1}{2x}}$

Solve each of the equations in Exercises 50 – 54.

50. $4(x+2)-7=-2(3x+1)-3$

51. $(x+4)(x-5)=10$

52. $0=x^2-7x+10$

53. $x^2+8x-4=16$

54. $\dfrac{7}{2x-1}=\dfrac{3}{x+6}$

55. a. Simplify the following expression: $\dfrac{3}{x}-\dfrac{5}{x+3}$

b. Solve the following equation: $\dfrac{3}{x}=\dfrac{5}{x+3}$

56. a. Simplify the following expression: $\dfrac{3x}{x+2}+\dfrac{2}{x-4}+3$

b. Solve the following equation: $\dfrac{3x}{x+2}+\dfrac{2}{x-4}=3$

57. An airplane can travel 1035 miles in the same time that a train travels 270 miles. The speed of the plane is 50 mph more than three times the speed of the train. Find the speed of each.

58. A man can wax his car three times as fast as his daughter can. Together they can complete the job in 2 hours. How long would it take each of them working alone?

59. The illumination (in foot-candles, fc) of a light source varies inversely as the square of the distance from the source. If a certain light source provides an illumination of 10 fc at a distance of 20 ft., what is the illumination at a distance of 40 ft.?

60. Two numbers are in a ratio of 2 to 5. If their sum is 84, find the numbers.

61. A family travels 18 miles downriver and returns. It takes 8 hours to make the round trip. Their rate in still water is twice the rate of the current. Find the rate of the current.

62. If a hanging spring stretches 5 cm when a weight of 13 gm is placed at its end, how far will the spring stretch if a weight of 20 gm is placed at its end?

Real Numbers and Radicals

Did You Know?

An important method of reasoning related to mathematical proofs is proof by contradiction. See if you can follow the reasoning in the following "proof" that $\sqrt{2}$ is an irrational number. This proof uses the fact that the square of an integer is even if and only if the integer is even.

Proof:

$\sqrt{2}$ is either an irrational number or a rational number. Suppose that $\sqrt{2}$ is a rational number and

$\dfrac{a}{b} = \sqrt{2}$ Where a and b are integers and $\dfrac{a}{b}$ is reduced. (±1 are the only factors common to a and b.)

$\dfrac{a^2}{b^2} = 2$ Square both sides.

$a^2 = 2b^2$ This means a^2 is an even integer.

So,

$a = 2n$ Since a^2 is even, a must be even.

$a^2 = 4n^2$ Square both sides.

$a^2 = 4n^2 = 2b^2$ Substituting.

$2n^2 = b^2$ This means b^2 is an even integer.

Therefore, b is an even integer.

But if a and b are both even, 2 is a common factor. This contradicts the statement that $\dfrac{a}{b}$ is reduced. Thus, our original supposition that $\sqrt{2}$ is rational is false, and $\sqrt{2}$ is an irrational number.

"The number of grains of sand on the beach at Coney Island is much less than a googol: 10,000, 000,000,000,000,000,000,000,000,000,000, 000,000,000,000,000,000,000,000,000,000, 000,000,000,000,000,000,000,000,000."

Edward Kasner

Real numbers and real number lines have been discussed and used throughout the text. In Chapter 9, the related concepts of real numbers, rational numbers, and irrational numbers are going to be presented in some depth. These are extremely important concepts since the real number system forms the foundation for most of the mathematical ideas we use today.

Calculators have helped a great deal in making operations with both rational numbers and irrational numbers seem easier. However, you should understand that in many cases calculators give only approximate values. To fully grasp this fact, as with much of mathematics, you need to understand some of the more abstract theoretical concepts, such as those that will be introduced in Chapter 9.

9.1   Real Numbers and Simplifying Radicals

Objectives

After completing this section, you will be able to:

1. Determine if a real number is rational or irrational.

2. Understand the meaning of radical expressions.

3. Use a calculator to evaluate radicals.

Rational Numbers

Real numbers were discussed in Chapter 1 and again in Chapter 3 when we graphed solutions to inequalities, such as $x < 3$, on real number lines. In this section, we are going to expand on our development of the real number system, and we begin by repeating the definitions for **integers** and **rational numbers**.

Integers

Integers are whole numbers and their opposites.

$$..., -4, -3, -2, -1, 0, 1, 2, 3, 4, ...$$

> *A **rational number** is any number that can be written in the form* $\dfrac{a}{b}$ *where a and b are integers and* $b \neq 0$.

In decimal form, all rational numbers can be written as **terminating decimals** or **infinite repeating decimals**. That is, if an integer is divided by a nonzero integer, then the decimal form of the quotient will be either terminating or infinite repeating. For example,

$$\frac{1}{4} = 0.250000... = 0.25\overline{0} \qquad \text{Repeating 0's}$$

or $\quad \dfrac{1}{4} = 0.25 \qquad\qquad\qquad$ Terminating decimal

$$\frac{1}{3} = 0.3333... = 0.\overline{3}$$

$$\frac{1}{7} = 0.142857142857142857... = 0.\overline{142857}$$

The bar over the digits indicates that the pattern of digits is to be repeated without end. The pattern can be found by long division. As examples,

$$\frac{1}{3} = 1 \div 3 \qquad\qquad\qquad \frac{1}{7} = 1 \div 7$$

$$
\begin{array}{r}
.333 \\
3\overline{)1.0000} \\
\underline{9} \\
10 \\
\underline{9} \\
10 \\
\underline{9} \\
1
\end{array}
\qquad\qquad
\begin{array}{r}
.142857 \\
7\overline{)1.000000} \\
\underline{7} \\
30 \\
\underline{28} \\
20 \\
\underline{14} \\
60 \\
\underline{56} \\
40 \\
\underline{35} \\
50 \\
\underline{49} \\
1
\end{array}
$$

Perfect Squares and Square Roots

A number is squared when it is multiplied by itself. For example,

$$5^2 = 5 \cdot 5 = 25$$

$$(-30)^2 = (-30) \cdot (-30) = 900$$

$$\text{and } (2.1)^2 = (2.1) \cdot (2.1) = 4.41$$

If an integer is squared, the result is called a **perfect square**. The squares for the integers from 1 to 20 are shown here for easy reference.

Table 9.1   Squares of Integers from 1 to 20 (Perfect Squares)

$1^2 = 1$	$2^2 = 4$	$3^2 = 9$	$4^2 = 16$	$5^2 = 25$
$6^2 = 36$	$7^2 = 49$	$8^2 = 64$	$9^2 = 81$	$10^2 = 100$
$11^2 = 121$	$12^2 = 144$	$13^2 = 169$	$14^2 = 196$	$15^2 = 225$
$16^2 = 256$	$17^2 = 289$	$18^2 = 324$	$19^2 = 361$	$20^2 = 400$

Now, we want to reverse the process of squaring. That is, given a number, we want to find a number that when squared will result in the given number. This is called **finding the square root** of the given number. For example,

since $7^2 = 49$, then 7 is the **square root** of 49

and, since $10^2 = 100$, then 10 is the **square root** of 100.

We write

$$\sqrt{49} = 7 \quad \text{and} \quad \sqrt{100} = 10$$

Terminology

The symbol $\sqrt{}$ is called a **radical sign**.

The number under the radical sign is called the **radicand**.

The complete expression, such as $\sqrt{49}$, is called a **radical** or **radical expression**.

Every positive real number has two square roots, one positive and one negative. The positive square root is called the **principal square root**. For example, since $(-9)^2 = 81$ and $(+9)^2 = 81$, we see that 81 has the two square roots –9 and +9. We write

$$\sqrt{81} = 9 \quad \longleftarrow \quad \textbf{the principal square root}$$

$$\text{and} \ -\sqrt{81} = -9 \quad \longleftarrow \quad \textbf{the negative square root}$$

The number 0 has only one square root, namely 0.

Square Root

If b is a nonnegative real number and a is a real number such that

$$a^2 = b, \text{ then a is called a } \textbf{square root } \textit{of b.}$$

If a is nonnegative, then we write $\sqrt{b} = a$ *and* $-\sqrt{b} = -a.$

The radicand may or may not be a perfect square. Example 1 shows radicals that involve fractions.

Example 1: Square Roots

a. Since $6^2 = 36$, $\sqrt{36} = 6$.

b. Since $\left(\dfrac{2}{3}\right)^2 = \dfrac{4}{9}$, $\sqrt{\dfrac{4}{9}} = \dfrac{2}{3}$.

c. Since $8^2 = 64$, $\sqrt{64} = 8$.

d. 49 has two square roots, one positive and one negative. The $\sqrt{}$ sign is understood to represent the positive square root or the principal square root. Therefore,

$$\sqrt{49} = 7 \qquad \text{The negative square root must have a} - \text{sign}$$
$$\text{and} \quad -\sqrt{49} = -7 \qquad \text{in front of the radical sign.}$$

e. $-\sqrt{\dfrac{25}{121}} = -\dfrac{5}{11}$

f. $\sqrt{100} = 10$ and $-\sqrt{100} = -10$

g. $\sqrt{4.41} = 2.1$

> **NOTES** The square roots of negative real numbers are **not real numbers**. (There is no real number whose square is negative.) Thus, $\sqrt{-4}$ is not a real number and $\sqrt{-25}$ is not a real number. Such numbers are called **imaginary numbers** or **nonreal complex numbers**. These numbers and their properties will be studied in later courses in mathematics.

Cube Roots

There are roots other than square roots. For example, $5^3 = 125$, so we say that 5 is the **cube root** of 125. We write

$$\sqrt[3]{125} = 5$$

Cube Root

If a and b are real numbers such that

$$a^3 = b, \quad \text{then a is called the **cube root** of b.}$$

We write $\sqrt[3]{b} = a$.

Example 2: Cube Roots •

a. Since $2^3 = 8$, $\sqrt[3]{8} = 2$.

b. Since $4^3 = 64$, $\sqrt[3]{64} = 4$.

c. Since $(-6)^3 = -216$, $\sqrt[3]{-216} = -6$.

d. Since $\left(\dfrac{1}{3}\right)^3 = \dfrac{1}{27}$, $\sqrt[3]{\dfrac{1}{27}} = \dfrac{1}{3}$.

• •

Irrational Numbers

In Examples 1 and 2, all the roots were integers or other rational numbers. However, not all roots are integers or other rational numbers. In fact, most roots, written in decimal form, are infinite nonrepeating decimals called **irrational numbers**.

Irrational Numbers

Irrational numbers are real numbers that are not rational numbers. Written in decimal form, irrational numbers are **infinite, nonrepeating decimals**.

Examples of irrational numbers are:

$$\sqrt{5}, \sqrt{7}, \sqrt{39}, \sqrt[3]{2}, -\sqrt{10}, \text{ and } \pi.$$

Example 3 shows the decimal form of some irrational numbers.

Example 3: Irrational Numbers ● ● ● ● ● ● ● ● ● ● ● ● ● ● ● ●

The following numbers are all irrational numbers. Your calculator will show them rounded off to nine or more decimal places; however, the digits continue without end and there is no pattern to the digits.

a. $\sqrt{2} = 1.4142136...$ See "Did You Know?" at the beginning of this chapter for a formal proof of the fact that $\sqrt{2}$ is irrational.

b. $\sqrt[3]{4} = 1.5874011...$

c. $\pi = 3.14159265358979...$ π is the ratio of the circumference to the diameter of any circle.

d. $e = 2.7182818284590...$ e is the base of natural logarithms and will be studied in detail in later courses.

● ●

Irrational numbers are just as important as rational numbers and just as useful in solving equations. As we discussed in Chapter 1, real number lines have points corresponding to irrational numbers as well as rational numbers. For example, in decimal form,

$$-\frac{1}{3} = -0.33333... = -0.\overline{3}$$ Repeating and never-ending decimal; a rational number

and $\sqrt{2} = 1.4142135...$ Nonrepeating and never-ending decimal; an irrational number

However, each number corresponds to just one point on a line (Figure 9.1).

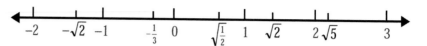

Figure 9.1

The following statement forms the basis for understanding number lines:

Every point on a number line is represented by one real number, and every real number is represented by one point on a number line.

To further emphasize the idea that irrational numbers correspond to points on a number line, consider a circle with a diameter of 1 unit rolling on a line. If the circle touches the line at the point 0, at what point on the line will the same point on the circle again touch the line?

The point will be at π on the number line because π is the circumference of the circle (Figure 9.2).

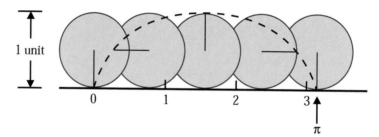

Figure 9.2

Real numbers are numbers that are either rational or irrational. That is, every rational number and every irrational number is also a real number, just as every integer is also a rational number. (Of course, every integer is a real number, too.)

The relationships between the various types of numbers we have discussed are shown in the following diagram.

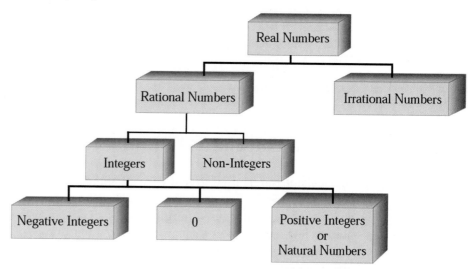

Now, consider the case where $a = \sqrt{3}$. By the definition of square root, $a^2 = 3$. That is,

$$a^2 = \left(\sqrt{3}\right)^2 = 3$$

Thus, even though $\sqrt{3}$ is an irrational number and it is an infinite nonrepeating decimal, its square is the integer 3. The square of the square root of a positive real number is that number.

Example 4: Find the Square ● ● ● ● ● ● ● ● ● ● ● ● ● ● ● ● ● ●

Use the definition of square root to find the square of each radical expression.

a. $\sqrt{15}$

 Solution: $\left(\sqrt{15}\right)^2 = 15$

b. $\sqrt{7}$

 Solution: $\left(\sqrt{7}\right)^2 = 7$

c. $-\sqrt{23}$

 Solution: $\left(-\sqrt{23}\right)^2 = 23$ The square of a negative number is positive.

● ●

Summary of Square Roots

In general, if a is a positive real number, then

1. $\sqrt{a}$ *is rational if a is a perfect square,*

2. $\sqrt{a}$ *is irrational if a is not a perfect square,*

3. $\sqrt{-a}$ *is nonreal.*

Using a Calculator to Evaluate Radicals

A calculator can be used to find decimal approximations for radicals and expressions containing radicals. For a scientific calculator, the $\sqrt{x}$ key can be used for square roots and the y^x key can be used for other roots. Example 5 shows the steps that might be used to find the values of a square root and a fourth root of a number. (**Note:** Not all scientific calculators operate exactly as discussed here. You may need to make some adjustments for your particular calculator.)

Example 5: Using a Calculator ● ● ● ● ● ● ● ● ● ● ● ● ● ● ● ●

Find an approximation for each of the following radicals using a scientific calculator.

a. $\sqrt{3}$

Solution: **Step 1:** Enter 3.
Step 2: Press the $\sqrt{x}$ key.

The result should read 1.732050808. (**Note:** On some calculators, you may have to press (2nd) then (x^2).)

Thus, $\sqrt{3} \approx 1.732050808$.

b. $\sqrt[4]{100}$

Solution: (We will find in Section 9.6 that radicals can be written with fractional exponents. Thus, $\sqrt[4]{100} = 100^{1/4} = 100^{0.25}$.)
Step 1: Enter 100.
Step 2: Press the y^x key.
Step 3: Enter 0.25.
Step 4: Press the = key.

The result should read 3.16227766.

Thus, $\sqrt[4]{100} \approx 3.16227766$.

Continued on next page ...

Note: To verify this result, with 3.16227766 still on the display, press the y^x key, then enter 4, then press the = key. This gives $(3.16227766)^4$, which should be 100 or very close to 100.

c. $\sqrt[3]{5}$

Solution: (Later, we will learn that $\sqrt[3]{5} = 5^{\frac{1}{3}}$. Here we use $\frac{1}{3} \approx 0.333333333$.)

> **Step 1:** Enter 5.
> **Step 2:** Press the y^x key.
> **Step 3:** Enter 0.333333333.
> **Step 4:** Press the = key.

The result should read 1.709975946.

Thus, $\sqrt[3]{5} \approx 1.709975946$.

Note: To check, press the y^x key, then enter 3, then press the = key. This gives $(1.709975946)^3$, which should be 5 or very close to 5.

• •

In Example 6, we show how a TI-83 Plus graphing calculator can be used to evaluate radicals and more complicated expressions that contain radicals. Note the advantage of seeing expressions being evaluated as it appears on the display.

Example 6: Using a Calculator • • • • • • • • • • • • • • •

Use a TI-83 Plus graphing calculator to evaluate each of the following radical expressions.

a. $3\sqrt{10}$

Solution: Press the keys in the following sequence:

Note: When you press the square root key on the TI-83 Plus, the square root symbol will appear with a left-hand parenthesis. You should press the right-hand parenthesis to close the square root operation. In this example, the calculator will operate correctly even if you do not. However, in more complicated situations, not putting the right-hand parenthesis might lead to an error.

Continued on next page ...

The display should appear as follows:

b. $1 + 2\sqrt{3}$

Solution: Press the keys in the following sequence:

Note: The times key is not necessary with the TI-83 Plus. The multiplication is understood. In this example, the times sign is included to emphasize the fact that a number written next to a radical indicates multiplication.

The display should appear as follows:

Or, without the times sign, the display should appear as follows:

Continued on next page ...

c. $\dfrac{2 + \sqrt{5}}{4}$

Solution: The fraction bar indicates that the numerator must be treated as a quantity. To ensure the correct answer, we must put the numerator in parentheses. Press the keys in the following sequence:

The display should appear as follows:

Now, you should have some idea of about how big (in decimal form) a radical number (rational or irrational) is and be able to estimate its place on a number line. Of course, you can always check your estimate with your calculator. For example, since 22 is between 16 and 25, $\sqrt{22}$ should be between 4 and 5. Symbolically,

$$16 < 22 < 25$$

$$\text{and} \quad \sqrt{16} < \sqrt{22} < \sqrt{25}$$

$$\text{so} \quad 4 < \sqrt{22} < 5$$

You might guess $\sqrt{22}$ to be about 4.5 or 4.6. With a calculator, $\sqrt{22} = 4.69041576$ (accurate to 8 decimal places).

9.1 Exercises

Find the terminating or infinite repeating decimal form for each of the fractions by performing the long division or using your calculator. Write your answers in the form with a bar over the repeating part of the decimal.

1. $\dfrac{3}{8}$ **2.** $\dfrac{5}{16}$ **3.** $\dfrac{3}{11}$ **4.** $-\dfrac{7}{18}$ **5.** $-\dfrac{6}{7}$ **6.** $\dfrac{12}{13}$

7. First use your knowledge of radicals to estimate the value of each number. Use your estimates to place all of the numbers on one number line and then use your calculator to check that your estimate makes sense.

    a. $\sqrt{32}$ **b.** $-\sqrt{18}$ **c.** $-\sqrt{\pi}$ **d.** $\sqrt[3]{20}$ **e.** $\sqrt[4]{60}$

8. First use your knowledge of radicals to estimate the value of each number. Use your estimates to place all of the numbers on one number line and then use your calculator to check that your estimate makes sense.

    a. $-\sqrt{27}$ **b.** $\sqrt{29}$ **c.** $\sqrt[3]{-20}$ **d.** $\sqrt{40}$ **e.** $\sqrt[4]{85}$

In Exercises 9 – 25, determine whether each number is rational, irrational, or nonreal.

9. $\sqrt{4}$ **10.** $\sqrt{17}$ **11.** $\sqrt{169}$ **12.** $\sqrt{\dfrac{4}{9}}$ **13.** $\sqrt{8}$ **14.** $-\sqrt{\dfrac{1}{4}}$

15. $\sqrt{-36}$ **16.** $\sqrt[3]{-27}$ **17.** $\sqrt{1.69}$ **18.** $\sqrt{5.29}$ **19.** $-\sqrt[3]{125}$ **20.** $\sqrt{-10}$

21. $0.1010010001\ldots$ (nonrepeating)   **22.** $4.\overline{23}$ (repeating)

23. $1.\overline{93471}$ (repeating)   **24.** $-6.\overline{051}$ (repeating)

25. $3.1415926535\ldots$ (nonrepeating)

Use your calculator to find the value (accurate to four decimal places) of each of the square root expressions in Exercises 26 – 35. Test your understanding by first writing down your own estimate of the value of each expression then comparing your estimate with your calculator results.

26. $1+2\sqrt{2}$ **27.** $3+2\sqrt{6}$ **28.** $-1-5\sqrt{3}$ **29.** $4-3\sqrt{5}$

30. $-2+2\sqrt{8}$ **31.** $\dfrac{1+\sqrt{10}}{2}$ **32.** $\dfrac{-2-\sqrt{12}}{3}$ **33.** $\dfrac{-10+\sqrt[3]{16}}{3}$

34. $\dfrac{4-\sqrt[3]{30}}{10}$ **35.** $\dfrac{5+\sqrt{90}}{6}$

In Exercises 36 – 40, find the value of each indicated square.

36. $\left(\sqrt{5}\right)^2$ **37.** $\left(\sqrt{25}\right)^2$ **38.** $\left(\sqrt{36}\right)^2$ **39.** $\left(-\sqrt{6}\right)^2$ **40.** $\left(-\sqrt{2}\right)^2$

In Exercises 41 – 60, find the value (accurate to four decimal places) of each radical expression. Use your calculator when necessary. Test your understanding by first writing down your own estimate of the value of each expression then comparing your estimate with your calculator results. (**Note**: Use the following definition in these exercises: If $a^n = b$, then a is called the nth root of b. We write $\sqrt[n]{b} = a$ and $\sqrt[n]{b} = b^{1/n}$. (See more about n^{th} roots in Section 9.6.)

41. $\sqrt[3]{27}$ **42.** $\sqrt[3]{125}$ **43.** $\sqrt[3]{-8}$ **44.** $\sqrt[3]{-64}$

45. $\sqrt[4]{16}$ **46.** $\sqrt[4]{81}$ **47.** $\sqrt[3]{-216}$ **48.** $\sqrt[4]{625}$

49. $\sqrt[3]{\dfrac{27}{8}}$ **50.** $\sqrt[5]{\dfrac{32}{243}}$ **51.** $\sqrt{65}$ **52.** $\sqrt{127}$

53. $\sqrt{37.9}$ **54.** $\sqrt{82.5}$ **55.** $\sqrt[3]{39.4}$ **56.** $\sqrt[3]{155.3}$

57. $\sqrt[4]{75}$ **58.** $\sqrt[5]{100}$ **59.** $\sqrt[5]{10,000}$ **60.** $\sqrt[3]{1000}$

Writing and Thinking About Mathematics

61. Discuss, in your own words, why the square root of a negative number is nonreal. Is the same reasoning true about the fourth root of a negative number?

Hawkes Learning Systems: Introductory Algebra

Evaluating Radicals

9.2 # Simplifying Radicals

After completing this section, you will be able to:

1. Simplify radicals, including square roots and cube roots.

Simplifying Square Roots

Various roots, rational and irrational, can be related to solutions of equations, and we want such numbers to be in a **simplified form** for easier calculations and algebraic manipulations. We need the two properties of radicals stated here for square roots. (Similar properties are true for other radicals.)

Properties of Square Roots

If *a* and *b* are **positive** real numbers, then

1. $\quad \sqrt{ab} = \sqrt{a}\sqrt{b}$

2. $\quad \sqrt{\dfrac{a}{b}} = \dfrac{\sqrt{a}}{\sqrt{b}}$

For example, we know that 144 is a perfect square and $\sqrt{144} = 12$. But using property 1 of square roots, we can also write

$$\sqrt{144} = \sqrt{36} \cdot \sqrt{4} = 6 \cdot 2 = 12$$

Similarly,

$$\sqrt{\frac{9}{25}} = \frac{\sqrt{9}}{\sqrt{25}} = \frac{3}{5}$$

The number 450 is not a perfect square, and to simplify $\sqrt{450}$ we can use property 1 and any of the following three approaches.

Approach I: Factor 450 as $25 \cdot 18$ because 25 is a perfect square. Then

$$\sqrt{450} = \sqrt{25 \cdot 18} = \sqrt{25} \cdot \sqrt{18} = 5\sqrt{18}$$

However, $5\sqrt{18}$ is not in simplest form because 18 has a perfect square factor, 9. Thus, to complete the process, we have

$$\sqrt{450} = 5\sqrt{18} = 5\sqrt{9 \cdot 2} = 5\sqrt{9} \cdot \sqrt{2} = 5 \cdot 3\sqrt{2} = 15\sqrt{2}$$

Approach II: Note that 225 is a perfect square factor of 450 and $450 = 225 \cdot 2$. Then

$$\sqrt{450} = \sqrt{225 \cdot 2} = \sqrt{225} \cdot \sqrt{2} = 15\sqrt{2}$$

Approach III: Use prime factors.

$$\sqrt{450} = \sqrt{45 \cdot 10} = \sqrt{3 \cdot 3 \cdot 5 \cdot 2 \cdot 5} = \sqrt{3 \cdot 3 \cdot 5 \cdot 5} \cdot \sqrt{2} = 3 \cdot 5 \cdot \sqrt{2} = 15\sqrt{2}$$

Of these three approaches, the second appears to be the easiest because it has the fewest steps. However, "seeing" the largest square factor may be difficult. If you do not immediately see a square factor, then proceed to find other factors or prime factors as illustrated.

Simplest Form

A square root is considered to be in simplest form when the radicand has no perfect square as a factor.

Example 1: Simplify Square Roots

Simplify the following square roots.

a. $\sqrt{24}$

Solution: Factor 24 so that one factor is a square number.

$$\sqrt{24} = \sqrt{4 \cdot 6} = \sqrt{4} \cdot \sqrt{6} = 2\sqrt{6}$$

Continued on next page ...

b. $\sqrt{72}$

Solution: Find the largest square factor you can before simplifying.

$$\sqrt{72} = \sqrt{36 \cdot 2} = \sqrt{36} \cdot \sqrt{2} = 6\sqrt{2}$$

Or, if you did not notice 36 as a factor, you could write

$$\sqrt{72} = \sqrt{9 \cdot 8} = \sqrt{9 \cdot 4 \cdot 2} = \sqrt{9} \cdot \sqrt{4} \cdot \sqrt{2} = 3 \cdot 2 \cdot \sqrt{2} = 6\sqrt{2}$$

c. $\sqrt{\dfrac{75}{4}}$

Solution: $\sqrt{\dfrac{75}{4}} = \dfrac{\sqrt{75}}{\sqrt{4}} = \dfrac{\sqrt{25 \cdot 3}}{2} = \dfrac{\sqrt{25} \cdot \sqrt{3}}{2} = \dfrac{5\sqrt{3}}{2}$

d. $\dfrac{3 + \sqrt{18}}{3}$

Solution: $\dfrac{3 + \sqrt{18}}{3} = \dfrac{3 + \sqrt{9 \cdot 2}}{3} = \dfrac{3 + \sqrt{9} \cdot \sqrt{2}}{3} = \dfrac{3 + 3\sqrt{2}}{3}$

$$= \dfrac{3}{3} + \dfrac{3\sqrt{2}}{3} = 1 + \sqrt{2}$$

$$\left[\text{or factoring, } \dfrac{3 + 3\sqrt{2}}{3} = \dfrac{\cancel{3}\left(1 + \sqrt{2}\right)}{\cancel{3}} = 1 + \sqrt{2} \right]$$

● ●

Simplifying Square Roots with Variables

Consider simplifying the radical expression $\sqrt{x^2}$. We do not know whether x represents a positive number or a negative number. For example,

$$\text{If } x = 3, \text{ then } \sqrt{x^2} = \sqrt{3^2} = \sqrt{9} = 3 = x.$$

$$\text{But, if } x = -3, \text{ then } \sqrt{x^2} = \sqrt{(-3)^2} = \sqrt{9} = 3 \neq x.$$

$$\text{In fact, if } x = -3, \text{ then } \sqrt{x^2} = \sqrt{(-3)^2} = \sqrt{9} = 3 = |x|.$$

Thus, simplifying radical expressions with variables involves more detailed analysis than simplifying radical expressions with only constants. The following definition indicates the correct way to simplify $\sqrt{x^2}$.

Square Root of x^2

If x is a real number, then $\sqrt{x^2} = |x|$.

Note: If $x \geq 0$ is given, then we can write $\sqrt{x^2} = x$.

Example 2: Simplify Square Roots

Simplify each of the following radical expressions.

a. $\sqrt{36x^2}$ **b.** $\sqrt{75a^2}$ **c.** $\sqrt{8x^2y^2}$

Solution:

a. $\sqrt{36x^2} = \sqrt{36} \cdot \sqrt{x^2} = 6|x|$

b. $\sqrt{75a^2} = \sqrt{25 \cdot 3} \cdot \sqrt{a^2} = \sqrt{25} \cdot \sqrt{3} \cdot \sqrt{a^2} = 5\sqrt{3}|a|$

c. $\sqrt{8x^2y^2} = \sqrt{4 \cdot 2} \cdot \sqrt{x^2} \cdot \sqrt{y^2} = \sqrt{4} \cdot \sqrt{2} \cdot \sqrt{x^2} \cdot \sqrt{y^2} = 2\sqrt{2}|x||y| \ \left(\text{or } 2\sqrt{2}|xy|\right)$

Simplifying Square Roots with Even and Odd Powers of Variables

When expressions with the same base are multiplied, the exponents are added. Now, if an expression is multiplied by itself, the exponents will be doubled and, therefore, even. This means that, to find the square root of an expression with even exponents, the exponents can be divided by 2. For example,

$$x^2 \cdot x^2 = x^4 \qquad a^3 \cdot a^3 = a^6 \qquad y^4 \cdot y^4 = y^8$$

and $\sqrt{x^4} = x^2 \qquad \sqrt{a^6} = |a^3| \qquad \sqrt{y^8} = y^4$

To find the square root of an expression with odd exponents, factor the expression into two terms, one with exponent 1 and the other with an even exponent. For example,

$$x^3 = x^2 \cdot x \ \text{ and } \ x^5 = x^4 \cdot x$$

which means that

$$\sqrt{x^3} = \sqrt{x^2 \cdot x} = \sqrt{x^2} \cdot \sqrt{x} = |x|\sqrt{x} \quad \text{and} \quad \sqrt{x^5} = \sqrt{x^4 \cdot x} = x^2\sqrt{x}$$

Square Roots of Expressions with Even and Odd Exponents

For any real number x and positive integer m,

$$\sqrt{x^{2m}} = \left| x^m \right| \quad \text{and} \quad \sqrt{x^{2m+1}} = \left| x^m \right| \sqrt{x}$$

Note that the absolute value sign is necessary only if m is odd.
Also, note that for $\sqrt{x^{2m+1}}$ to be defined, x cannot be negative.

Comment: *If m is any integer, then 2m is even and 2m + 1 is odd.*

Example 3: Simplify Square Roots

Simplify each of the following radical expressions.

a. $\sqrt{16x^4}$ **b.** $\sqrt{49x^3y^2}$, assume that $x, y \geq 0$ **c.** $\sqrt{18a^6b^4}$ **d.** $\sqrt{\dfrac{9a^{15}}{64b^8}}$

Solution:

a. $\sqrt{16x^4} = 4x^2$

The exponent 4 is divided by 2 and x can be positive or negative.

b. $\sqrt{49x^3y^2} = \sqrt{49 \cdot x^2 \cdot x \cdot y^2} = 7xy\sqrt{x}$

Assuming that $x, y \geq 0$.

c. $\sqrt{18a^6b^4} = \sqrt{9 \cdot 2 \cdot a^6 \cdot b^4} = 3\left| a^3 \right| b^2\sqrt{2}$ Each exponent is divided by 2.

d. $\sqrt{\dfrac{9a^{15}}{64b^8}} = \dfrac{\sqrt{9 \cdot a^{14} \cdot a}}{\sqrt{64b^8}} = \dfrac{3a^7\sqrt{a}}{8b^4}$

Since 15 is odd, we assume that $a \geq 0$.

Simplifying Cube Roots

Cube roots can be simplified if they have a perfect cube factor. Perfect cubes are:

$$0, \ 1, \ 8, \ 27, \ 64, \ 125, \ 216, \text{ and so on.}$$

$$0^3 \ \ 1^3 \ \ 2^3 \ \ 3^3 \ \ 4^3 \ \ 5^3 \ \ 6^3$$

Simplest Form

A cube root is considered to be in simplest form when the radicand has no perfect cube as a factor.

Example 4: Simplify Cube Roots

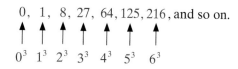

Simplify each of the following cube roots. (Note, since we are taking an odd root, the variables can be positive or negative.)

a. $\sqrt[3]{54x^3}$ **b.** $\sqrt[3]{40x^6y^4}$ **c.** $\sqrt[3]{125x^8y^9}$

Solution:

a. $\sqrt[3]{54x^3} = \sqrt[3]{27 \cdot 2 \cdot x^3} = \sqrt[3]{27x^3} \cdot \sqrt[3]{2} = 3x\sqrt[3]{2}$ **Note:** The exponent is divided by 3 because we are taking a cube root.

b. $\sqrt[3]{40x^6y^4} = \sqrt[3]{8 \cdot 5 \cdot x^6 \cdot y^3 \cdot y} = \sqrt[3]{8x^6y^3} \cdot \sqrt[3]{5y} = 2x^2y\sqrt[3]{5y}$ Find a factor of 3 for each exponent.

c. $\sqrt[3]{125x^8y^9} = \sqrt{125 \cdot x^6 \cdot x^2 \cdot y^9} = \sqrt[3]{125 \cdot x^6 \cdot y^9} \cdot \sqrt[3]{x^2} = 5x^2y^3\sqrt[3]{x^2}$

Practice Problems

Simplify the following radical expressions.

1. $\sqrt{49}$ **2.** $\sqrt[3]{27}$ **3.** $-\sqrt{25}$ **4.** $\sqrt{\dfrac{3}{4}}$

5. $\sqrt{32x^2}$ **6.** $\sqrt[3]{32x^5}$ **7.** $\sqrt{\dfrac{54y^5}{25x^2}}$ **8.** $\dfrac{4-\sqrt{12}}{2}$

9.2 Exercises

Simplify each of the radicals in Exercises 1 – 55.

1. $\sqrt{36x^2}$ **2.** $\sqrt{49y^2}$ **3.** $\sqrt{\dfrac{1}{4}}$ **4.** $-\sqrt{121}$

5. $\sqrt[3]{-1}$ **6.** $\sqrt[3]{216}$ **7.** $\sqrt[3]{1}$ **8.** $\sqrt[3]{-125}$

9. $\sqrt{8x^3}$ **10.** $\sqrt{18a^5b^2}$ **11.** $-\sqrt{56}$ **12.** $\sqrt{162}$

13. $\sqrt{\dfrac{5x^4}{9}}$ **14.** $-\sqrt{\dfrac{7y^6}{16x^4}}$ **15.** $\sqrt{12}$ **16.** $-\sqrt{45}$

17. $\sqrt{288}$ **18.** $-\sqrt{63}$ **19.** $-\sqrt{72}$ **20.** $\sqrt{98}$

21. $-\sqrt{125}$ **22.** $\sqrt[3]{56}$ **23.** $\sqrt[3]{-128}$ **24.** $\sqrt[3]{-250}$

25. $\sqrt[3]{216x^6y^5}$ **26.** $\sqrt[3]{64x^9y^2}$ **27.** $\sqrt[3]{-72}$ **28.** $\sqrt{\dfrac{32}{49}}$

29. $-\sqrt{\dfrac{11}{64}}$ **30.** $-\sqrt{\dfrac{125}{100}}$ **31.** $\sqrt{\dfrac{28}{25}}$ **32.** $\sqrt{\dfrac{147}{100}}$

33. $\sqrt{\dfrac{32a^5}{81b^{16}}}$ **34.** $\sqrt{\dfrac{75x^8}{121y^{12}}}$ **35.** $\dfrac{\sqrt[3]{81}}{6}$ **36.** $\dfrac{\sqrt[3]{192}}{10}$

37. $\sqrt[3]{\dfrac{375}{8}}$ **38.** $\sqrt[3]{-\dfrac{48}{125}}$ **39.** $\dfrac{2-2\sqrt{3}}{4}$ **40.** $\dfrac{4+2\sqrt{6}}{2}$

Answers to Practice Problems: 1. 7 **2.** 3 **3.** -5 **4.** $\dfrac{\sqrt{3}}{2}$ **5.** $4|x|\sqrt{2}$ **6.** $2x\sqrt[3]{4x^2}$ **7.** $\dfrac{3y^2\sqrt{6y}}{5|x|}$ **8.** $2-\sqrt{3}$

4, 9, 16, 25, 36
49

41. $\dfrac{3 + 3\sqrt{6}}{6}$ **42.** $\dfrac{6 - 2\sqrt{3}}{8}$ **43.** $\dfrac{12 + \sqrt{45}}{15}$ **44.** $\dfrac{7 - \sqrt{98}}{14}$

45. $\dfrac{10 - \sqrt{108}}{4}$ **46.** $\dfrac{4 + \sqrt{288}}{20}$ **47.** $\dfrac{16 - \sqrt{60}}{12}$ **48.** $\dfrac{12 - \sqrt{18}}{21}$

49. $\dfrac{12 - \sqrt{192}}{28}$ **50.** $\dfrac{14 + \sqrt{147}}{35}$ **51.** $\sqrt[3]{125x^4}$ **52.** $\sqrt[3]{72a^6b^4}$

53. $\sqrt[3]{\dfrac{125y^{12}}{27x^6}}$ **54.** $\sqrt{\dfrac{200x^8}{289}}$ **55.** $\sqrt{\dfrac{32x^{15}y^{10}}{169}}$

Hawkes Learning Systems: Introductory Algebra

Simplifying Radicals

9.3   Addition, Subtraction, and Multiplication with Radicals

After completing this section, you will be able to:

1. Add and subtract radical expressions.

2. Multiply radical expressions.

Sometimes solving equations and simplifying algebraic expressions can involve operations with radicals (square roots, cube roots, fourth roots, and so on). The emphasis here will be on radicals that are square roots with some work on cube roots. **Like radicals** are terms that either have the same radicals or can be simplified so that the radicals are the same.

Like radicals are similar to like terms. For example,

$7\sqrt{2}$ and $3\sqrt{2}$ are like radicals. The radical $\sqrt{2}$ is the same in both terms.

$-4\sqrt{10}$ and $6\sqrt{10}$ are like radicals. The radical $\sqrt{10}$ is the same in both terms.

$2\sqrt{5}$ and $2\sqrt{3}$ are not like radicals. The radicals $\sqrt{5}$ and $\sqrt{3}$ are different.

Adding Like Radicals

To add (or subtract) like radicals, proceed just as in combining like terms. Use the distributive property, then add the coefficients.

$$7x + 3x = (7 + 3)x = 10x$$

and

$$7\sqrt{2} + 3\sqrt{2} = (7 + 3)\sqrt{2} = 10\sqrt{2}$$

Simplifying radicals will show whether or not they are like radicals. As illustrated here, it may be possible to combine radicals after they have been simplified.

$$\sqrt{75} - \sqrt{27} = \sqrt{25 \cdot 3} - \sqrt{9 \cdot 3} \quad \text{25 and 9 are perfect squares.}$$
$$= 5\sqrt{3} - 3\sqrt{3}$$
$$= 2\sqrt{3}$$

and $\quad\quad\quad \sqrt{16x} + \sqrt{4x} = \sqrt{16 \cdot x} + \sqrt{4 \cdot x} \quad\quad$ Assume $x \geq 0$.

$$= 4\sqrt{x} + 2\sqrt{x}$$

$$= 6\sqrt{x}$$

Combining cube roots is done in the same way.

$$\sqrt[3]{16x} + \sqrt[3]{54x} = \sqrt[3]{8 \cdot 2x} + \sqrt[3]{27 \cdot 2x} \quad\quad \text{8 and 27 are perfect cubes.}$$

$$= \sqrt[3]{8} \cdot \sqrt[3]{2x} + \sqrt[3]{27} \cdot \sqrt[3]{2x}$$

$$= 2\sqrt[3]{2x} + 3\sqrt[3]{2x}$$

$$= 5\sqrt[3]{2x}$$

Example 1: Combine Like Radicals ● ● ● ● ● ● ● ● ● ● ● ● ●

Simplify the following expressions by combining like radicals. (Assume all variables are nonnegative.)

a. $5\sqrt{2} - \sqrt{2}$

Solution: $\quad 5\sqrt{2} - \sqrt{2} = (5-1)\sqrt{2} = 4\sqrt{2}$

b. $3\sqrt{5} + 8\sqrt{5} - 14\sqrt{5}$

Solution: $\quad 3\sqrt{5} + 8\sqrt{5} - 14\sqrt{5} = (3 + 8 - 14)\sqrt{5} = -3\sqrt{5}$

c. $\sqrt{18x^2} + \sqrt{8x^2}$

Solution: $\quad \sqrt{18x^2} + \sqrt{8x^2} = \sqrt{9x^2 \cdot 2} + \sqrt{4x^2 \cdot 2} = 3x\sqrt{2} + 2x\sqrt{2} = 5x\sqrt{2}$

d. $\sqrt{5} - \sqrt{3} + \sqrt{12}$

Solution:

$$\sqrt{5} - \sqrt{3} + \sqrt{12} = \sqrt{5} - \sqrt{3} + \sqrt{4}\sqrt{3} = \sqrt{5} - \sqrt{3} + 2\sqrt{3} = \sqrt{5} + \sqrt{3}$$

Continued on next page ...

e.　$x\sqrt{x} + 7x\sqrt{x}$

　　Solution:　$x\sqrt{x} + 7x\sqrt{x} = (x + 7x)\sqrt{x} = 8x\sqrt{x}$

f.　$\sqrt[3]{125a} - \sqrt[3]{64a}$

　　Solution:　$\sqrt[3]{125a} - \sqrt[3]{64a} = 5\sqrt[3]{a} - 4\sqrt[3]{a} = \sqrt[3]{a}$

● ●

Multiplying Radicals

To find the product of radicals, we proceed just as in multiplying polynomials. To illustrate the similarities, we show multiplication of a polynomial followed by multiplication of a similar radical expression:

$$5(x + y) = 5x + 5y$$

$$\sqrt{2}\left(\sqrt{7} + \sqrt{3}\right) = \sqrt{2}\sqrt{7} + \sqrt{2}\sqrt{3} = \sqrt{14} + \sqrt{6}.　\quad \text{Remember: } \sqrt{a}\sqrt{b} = \sqrt{ab}.$$

And, with binomials,

$$(x + 5)(x - 3) = x^2 - 3x + 5x - 15 = x^2 + 2x - 15$$

$$\left(\sqrt{2} + 5\right)\left(\sqrt{2} - 3\right) = \left(\sqrt{2}\right)^2 - 3\sqrt{2} + 5\sqrt{2} - 15$$

$$= 2 + 2\sqrt{2} - 15 = -13 + 2\sqrt{2}$$

$$\left[\textbf{Note: } \sqrt{2}\sqrt{2} = \sqrt{4} = 2 \ \ or \ \ \left(\sqrt{2}\right)^2 = 2.\right]$$

In general, **if a is positive,** $\sqrt{a}\sqrt{a} = a.$ $\left[Or \ \left(\sqrt{a}\right)^2 = a.\right]$

Example 2: Multiply and Simplify ● ● ● ● ● ● ● ● ● ● ● ● ● ●

Find the following products and simplify.

a. $\sqrt{7}\left(\sqrt{7} - \sqrt{14}\right)$

   Solution: $\sqrt{7}\left(\sqrt{7} - \sqrt{14}\right) = \sqrt{7}\sqrt{7} - \sqrt{7}\sqrt{14}$

   $$= 7 - \sqrt{98}$$

   $$= 7 - \sqrt{49 \cdot 2}$$

   $$= 7 - 7\sqrt{2}$$

b. $\left(\sqrt{2} + 4\right)\left(\sqrt{2} - 4\right)$

   Solution: $\left(\sqrt{2} + 4\right)\left(\sqrt{2} - 4\right) = \left(\sqrt{2}\right)^2 - 4^2 = 2 - 16 = -14$

c. $\left(\sqrt{5} + \sqrt{3}\right)\left(\sqrt{5} + \sqrt{3}\right)$

   Solution: $\left(\sqrt{5} + \sqrt{3}\right)\left(\sqrt{5} + \sqrt{3}\right) = \left(\sqrt{5}\right)^2 + 2\sqrt{5}\sqrt{3} + \left(\sqrt{3}\right)^2$

   $$= 5 + 2\sqrt{15} + 3 = 8 + 2\sqrt{15}$$

d. $\left(\sqrt{x} + 1\right)\left(\sqrt{x} - 1\right)$ (Assume $x \geq 0$.)

   Solution: $\left(\sqrt{x} + 1\right)\left(\sqrt{x} - 1\right) = \left(\sqrt{x}\right)^2 - 1^2 = x - 1$

● ●

Practice Problems

Simplify the following radical expressions.

1. $2\sqrt{3} - \sqrt{3}$

2. $-2\left(\sqrt{8} + \sqrt{2}\right)$

3. $\sqrt{75} - \sqrt{27} + \sqrt{20}$

4. $\left(\sqrt{3} + \sqrt{8}\right)\left(\sqrt{2} - \sqrt{3}\right)$

5. $5\sqrt{x} - 4\sqrt{x} + 2\sqrt{x}$

6. $\sqrt[3]{8} + \sqrt[3]{27}$

7. $\sqrt[3]{16} + \sqrt[3]{250}$

9.3 Exercises

Simplify the radical expressions in Exercises 1 – 50. (Assume all variables are nonnegative.)

1. $3\sqrt{2} + 5\sqrt{2}$

2. $7\sqrt{3} - 2\sqrt{3}$

3. $6\sqrt{5} + \sqrt{5}$

4. $4\sqrt{11} - 3\sqrt{11}$

5. $8\sqrt{10} - 11\sqrt{10}$

6. $6\sqrt{17} - 9\sqrt{17}$

7. $4\sqrt[3]{3} + 9\sqrt[3]{3}$

8. $11\sqrt[3]{14} - 6\sqrt[3]{14}$

9. $6\sqrt{11} - 5\sqrt{11} - 2\sqrt{11}$

10. $\sqrt{7} + 6\sqrt{7} - 2\sqrt{7}$

11. $\sqrt{a} + 4\sqrt{a} - 2\sqrt{a}$

12. $2\sqrt{x} - 3\sqrt{x} + 7\sqrt{x}$

13. $5\sqrt{x} + 3\sqrt{x} - \sqrt{x}$

14. $6\sqrt{xy} - 10\sqrt{xy} + \sqrt{xy}$

15. $3\sqrt{2} + 5\sqrt{3} - 2\sqrt{3} + \sqrt{2}$

16. $\sqrt{5} + \sqrt{4} - 2\sqrt{5} + 6$

17. $2\sqrt{a} + 7\sqrt{b} - 6\sqrt{a} + \sqrt{b}$

18. $4\sqrt{x} - 3\sqrt{x} + 2\sqrt{y} + 2\sqrt{x}$

19. $6\sqrt[3]{x} - 4\sqrt[3]{y} + 7\sqrt[3]{x} + 2\sqrt[3]{y}$

20. $5\sqrt[3]{x} + 9\sqrt[3]{y} - 10\sqrt[3]{y} + 4\sqrt[3]{x}$

21. $\sqrt{12} + \sqrt{27}$

22. $\sqrt{32} - \sqrt{18}$

23. $3\sqrt{5} - \sqrt{45}$

24. $2\sqrt{7} + 5\sqrt{28}$

Answers to Practice Problems: **1.** $\sqrt{3}$ **2.** $-6\sqrt{2}$ **3.** $2\sqrt{3} + 2\sqrt{5}$ **4.** $1 - \sqrt{6}$ **5.** $3\sqrt{x}$ **6.** 5 **7.** $7\sqrt[3]{2}$

25. $3\sqrt[3]{54} + 8\sqrt[3]{2}$

26. $2\sqrt[3]{128} + 5\sqrt[3]{-54}$

27. $\sqrt{50} - \sqrt{18} - 3\sqrt{12}$

28. $2\sqrt{48} - \sqrt{54} + \sqrt{27}$

29. $2\sqrt{20} - \sqrt{45} + \sqrt{36}$

30. $\sqrt{18} - 2\sqrt{12} + 5\sqrt{2}$

31. $\sqrt{8} - 2\sqrt{3} + \sqrt{27} - \sqrt{72}$

32. $\sqrt{80} + \sqrt{8} - \sqrt{45} + \sqrt{50}$

33. $5\sqrt[3]{16} - 4\sqrt[3]{24} + \sqrt[3]{-250}$

34. $\sqrt[3]{192} - 2\sqrt[3]{128} + \sqrt[3]{-81}$

35. $6\sqrt{2x} - \sqrt{8x}$

36. $5\sqrt{3x} + 2\sqrt{12x}$

37. $5y\sqrt{2y} - y\sqrt{18y}$

38. $9x\sqrt{xy} - x\sqrt{16xy}$

39. $4x\sqrt{3xy} - x\sqrt{12xy} - 2x\sqrt{27xy}$

40. $x\sqrt{32x} - x\sqrt{50x} + 2x\sqrt{18x}$

41. $\sqrt{36x^3} + \sqrt{81x^3}$

42. $\sqrt{4a^2b} + \sqrt{9a^2b}$

43. $\sqrt{16x^3y^4} - \sqrt{25x^3y^4}$

44. $\sqrt{72x^{12}y^{15}} + \sqrt{18x^{12}y^{15}} + \sqrt{2x^{12}y^{15}}$

45. $\sqrt{12x^{10}y^{20}} + \sqrt{27x^{10}y^{20}} - \sqrt{3x^{10}y^{20}}$

46. $\sqrt[3]{8a^{12}} + \sqrt[3]{1000a^{12}}$

47. $\sqrt[3]{-27x^{24}y^6} + \sqrt[3]{-125x^{24}y^6}$

48. $\sqrt[3]{27a^{15}b} + \sqrt[3]{8a^{15}b} + \sqrt[3]{64a^{15}b}$

49. $\sqrt[3]{-16x^9y^{12}} - \sqrt[3]{16x^{12}y^9} + \sqrt[3]{54x^3y^6}$

50. $\sqrt[3]{54x^{13}y^3} + \sqrt[3]{8x^{23}y^6} + \sqrt[3]{3x^{13}y^3}$

Multiply the expressions in Exercises 51–70 and simplify the result. (Assume all variables are nonnegative.)

51. $\sqrt{2}\left(3 - 4\sqrt{2}\right)$

52. $2\sqrt{7}\left(\sqrt{7} + 3\sqrt{2}\right)$

53. $3\sqrt{18} \cdot \sqrt{2}$

54. $2\sqrt{10} \cdot \sqrt{5}$

55. $-2\sqrt{6} \cdot \sqrt{8}$

56. $2\sqrt{15} \cdot 5\sqrt{6}$

57. $\sqrt{3}\left(\sqrt{2} + 2\sqrt{12}\right)$

58. $\sqrt{2}\left(\sqrt{3} - \sqrt{6}\right)$

59. $\sqrt{y}\left(\sqrt{x} + 2\sqrt{y}\right)$

60. $\sqrt{x}\left(\sqrt{x} - 3\sqrt{y}\right)$

61. $\left(5 + \sqrt{2}\right)\left(3 - \sqrt{2}\right)$

62. $\left(2\sqrt{3} + 1\right)\left(\sqrt{3} - 3\right)$

63. $\left(4\sqrt{3}+\sqrt{2}\right)\left(\sqrt{3}-2\sqrt{2}\right)$ **64.** $\left(\sqrt{5}-\sqrt{3}\right)\left(2\sqrt{5}+3\sqrt{3}\right)$

65. $\left(\sqrt{x}+3\right)\left(\sqrt{x}-3\right)$ **66.** $\left(\sqrt{a}+b\right)\left(\sqrt{a}+b\right)$

67. $\left(\sqrt{2}+\sqrt{7}\right)\left(\sqrt{2}-\sqrt{7}\right)$ **68.** $\left(\sqrt{x}+\sqrt{y}\right)\left(\sqrt{x}-\sqrt{y}\right)$

69. $\left(\sqrt{x}+5\right)\left(\sqrt{x}-3\right)$ **70.** $\left(\sqrt{3}+\sqrt{7}\right)\left(\sqrt{3}+2\sqrt{7}\right)$

Hawkes Learning Systems: Introductory Algebra

Addition and Subtraction of Radicals
Multiplication of Radicals

9.4 Rationalizing Denominators

After completing this section, you will be able to:

1. Rationalize the denominators of rational expressions containing radicals.

Rationalizing Denominators with One Term in the Denominator

Each of the expressions

$$\frac{5}{\sqrt{3}}, \quad \frac{\sqrt{7}}{\sqrt{8}}, \quad \text{and} \quad \frac{2}{3-\sqrt{2}}$$

contain a radical in the denominator that is an irrational number. Such expressions are not considered in simplest form because they are difficult to operate with algebraically. Calculations of sums and differences are much easier if the denominators are rational expressions. So, in simplifying, the objective is to find an equal fraction that has a rational number for a denominator.

That is, we want to simplify the expression by **rationalizing the denominator**. The following examples illustrate the method. Each fraction is multiplied by 1 in the form $\dfrac{k}{k}$, where multiplication by k rationalizes the denominator.

a. $\dfrac{5}{\sqrt{3}}$

$$\frac{5}{\sqrt{3}} = \frac{5 \cdot \sqrt{3}}{\sqrt{3} \cdot \sqrt{3}} = \frac{5\sqrt{3}}{3}$$

Multiply the numerator and denominator by $\sqrt{3}$ because $\sqrt{3} \cdot \sqrt{3} = 3$, a rational number.

b. $\dfrac{4}{\sqrt{x}}$ (Assume $x > 0$)

$$\frac{4}{\sqrt{x}} = \frac{4 \cdot \sqrt{x}}{\sqrt{x} \cdot \sqrt{x}} = \frac{4\sqrt{x}}{x}$$

Multiply the numerator and denominator by $\sqrt{x}$ because $\sqrt{x} \cdot \sqrt{x} = x$. There is no guarantee that x is rational, but the radical sign does not appear in the denominator of the result.

c. $\dfrac{3}{7\sqrt{2}}$

Multiply the numerator and denominator by $\sqrt{2}$ because $\sqrt{2} \cdot \sqrt{2} = 2$, a rational number.

$$\frac{3}{7\sqrt{2}} = \frac{3 \cdot \sqrt{2}}{7\sqrt{2} \cdot \sqrt{2}} = \frac{3\sqrt{2}}{7 \cdot 2} = \frac{3\sqrt{2}}{14}$$

d. $\dfrac{\sqrt{7}}{\sqrt{8}}$

Multiply the numerator and denominator by $\sqrt{2}$ because $\sqrt{8} \cdot \sqrt{2} = \sqrt{16} = 4$, a rational number. (**Note:** $8 \cdot 2 = 16$ and 16 is a perfect square.)

$$\frac{\sqrt{7}}{\sqrt{8}} = \frac{\sqrt{7} \cdot \sqrt{2}}{\sqrt{8} \cdot \sqrt{2}} = \frac{\sqrt{14}}{4}$$

In Example d, if the numerator and denominator are multiplied by $\sqrt{8}$, the results will be the same. But, one more step will be added because the fraction can be reduced.

$$\frac{\sqrt{7}}{\sqrt{8}} = \frac{\sqrt{7} \cdot \sqrt{8}}{\sqrt{8} \cdot \sqrt{8}} = \frac{\sqrt{56}}{8} = \frac{\sqrt{4} \cdot \sqrt{14}}{8} = \frac{2\sqrt{14}}{8} = \frac{\sqrt{14}}{4}$$

NOTES

Historical Note: An expression with a rational denominator such as $\dfrac{\sqrt{2}}{2}$ is much easier to calculate than its equivalent form $\dfrac{1}{\sqrt{2}}$. That is, with pencil and paper, $1.41421356 \div 2$ is a much easier calculation than $1 \div 1.4121356$. Now, with modern calculators and computers available to everyone, rational denominators are more important for higher level mathematics than for arithmetic.

Rationalizing Denominators with a Sum or Difference in the Denominator

Now, if the denominator of a fraction contains a sum or difference involving square roots, rationalizing the denominator involves the algebraic concept of the difference of squares. For example, consider the fraction

$$\frac{2}{3 - \sqrt{2}}$$

in which the denominator is of the form $a - b = 3 - \sqrt{2}$ where square roots are involved.

Two expressions of the form $(a - b)$ and $(a + b)$ are called **conjugates** of each other and their product is always the difference of two squares.

$$(a - b)(a + b) = a^2 - b^2$$

Consider two cases for rationalizing a denominator:

1. If the denominator is of the form $a - b$, multiply both the numerator and denominator by the conjugate $a + b$.
2. If the denominator is of the form $a + b$, multiply both the numerator and denominator by the conjugate $a - b$.

In either case, the denominator becomes, $a^2 - b^2$ (**the difference of two squares**) and the denominator is a rational number (or at least a rational expression). Thus,

$$\frac{2}{3 - \sqrt{2}} = \frac{2\left(3 + \sqrt{2}\right)}{\left(3 - \sqrt{2}\right)\left(3 + \sqrt{2}\right)}$$ If $a - b = 3 - \sqrt{2}$, then $a + b = 3 + \sqrt{2}$.

$$= \frac{2\left(3 + \sqrt{2}\right)}{3^2 - \left(\sqrt{2}\right)^2}$$ The denominator is the difference of two squares.

$$= \frac{2\left(3 + \sqrt{2}\right)}{9 - 2}$$ The denominator is a rational number.

$$= \frac{2\left(3 + \sqrt{2}\right)}{7}$$

Example 1: Rationalize Denominators ● ● ● ● ● ● ● ● ● ● ● ● ●

Rationalize the denominator of each of the following expressions and simplify.

a. $\sqrt{\dfrac{1}{2}}$

   Solution: $\sqrt{\dfrac{1}{2}} = \dfrac{\sqrt{1}}{\sqrt{2}} = \dfrac{1 \cdot \sqrt{2}}{\sqrt{2} \cdot \sqrt{2}} = \dfrac{\sqrt{2}}{2}$

b. $\dfrac{5}{\sqrt{5}}$

   Solution: $\dfrac{5}{\sqrt{5}} = \dfrac{5 \cdot \sqrt{5}}{\sqrt{5} \cdot \sqrt{5}} = \dfrac{5\sqrt{5}}{5} = \sqrt{5}$

c. $\dfrac{6}{\sqrt{27}}$

   Solution: $\dfrac{6}{\sqrt{27}} = \dfrac{6 \cdot \sqrt{3}}{\sqrt{27} \cdot \sqrt{3}} = \dfrac{6 \cdot \sqrt{3}}{\sqrt{81}} = \dfrac{6 \cdot \sqrt{3}}{9} = \dfrac{2\sqrt{3}}{3}$

Continued on next page ...

d. $\dfrac{-3}{\sqrt{xy}}$

Solution: $\dfrac{-3}{\sqrt{xy}} = \dfrac{-3 \cdot \sqrt{xy}}{\sqrt{xy} \cdot \sqrt{xy}} = \dfrac{-3\sqrt{xy}}{xy}$

e. $\dfrac{3}{\sqrt{5} + \sqrt{2}}$

Solution: $\dfrac{3}{\sqrt{5} + \sqrt{2}} = \dfrac{3\left(\sqrt{5} - \sqrt{2}\right)}{\left(\sqrt{5} + \sqrt{2}\right)\left(\sqrt{5} - \sqrt{2}\right)} = \dfrac{3\left(\sqrt{5} - \sqrt{2}\right)}{\left(\sqrt{5}\right)^2 - \left(\sqrt{2}\right)^2}$

$$= \dfrac{3\left(\sqrt{5} - \sqrt{2}\right)}{5 - 2} = \dfrac{3\left(\sqrt{5} - \sqrt{2}\right)}{3} = \sqrt{5} - \sqrt{2}$$

f. $\dfrac{x}{\sqrt{x} - 3}$

Solution: $\dfrac{x}{\sqrt{x} - 3} = \dfrac{x\left(\sqrt{x} + 3\right)}{\left(\sqrt{x} - 3\right)\left(\sqrt{x} + 3\right)}$

$$= \dfrac{x\left(\sqrt{x} + 3\right)}{\left(\sqrt{x}\right)^2 - (3)^2} = \dfrac{x\left(\sqrt{x} + 3\right)}{x - 9}$$

Practice Problems

Rationalize each denominator and simplify.

1. $\sqrt{\dfrac{5}{2}}$ **2.** $\dfrac{\sqrt{7}}{\sqrt{18}}$ **3.** $\dfrac{4}{\sqrt{7} + \sqrt{3}}$ **4.** $\dfrac{5}{\sqrt{x} + 2}$

Answers to Practice Problems: 1. $\dfrac{\sqrt{10}}{2}$ **2.** $\dfrac{\sqrt{14}}{6}$ **3.** $\sqrt{7} - \sqrt{3}$ **4.** $\dfrac{5\left(\sqrt{x} - 2\right)}{x - 4}$

9.4 Exercises

Rationalize each denominator in Exercises 1 – 60 and simplify.

1. $\dfrac{5}{\sqrt{2}}$

2. $\dfrac{7}{\sqrt{5}}$

3. $\dfrac{-3}{\sqrt{7}}$

4. $\dfrac{-10}{\sqrt{2}}$

5. $\dfrac{6}{\sqrt{3}}$

6. $\dfrac{8}{\sqrt{2}}$

7. $\dfrac{\sqrt{18}}{\sqrt{2}}$

8. $\dfrac{\sqrt{25}}{\sqrt{3}}$

9. $\dfrac{\sqrt{27x}}{\sqrt{3x}}$

10. $\dfrac{\sqrt{45y}}{\sqrt{5y}}$

11. $\dfrac{\sqrt{ab}}{\sqrt{9ab}}$

12. $\dfrac{\sqrt{5}}{\sqrt{12}}$

13. $\dfrac{\sqrt{4}}{\sqrt{3}}$

14. $\sqrt{\dfrac{3}{8}}$

15. $\sqrt{\dfrac{9}{2}}$

16. $\sqrt{\dfrac{3}{5}}$

17. $\sqrt{\dfrac{1}{x}}$

18. $\sqrt{\dfrac{x}{y}}$

19. $\sqrt{\dfrac{2x}{y}}$

20. $\sqrt{\dfrac{x}{4y}}$

21. $\dfrac{2}{\sqrt{2y}}$

22. $\dfrac{-10}{3\sqrt{5}}$

23. $\dfrac{21}{5\sqrt{7}}$

24. $\dfrac{x}{5\sqrt{x}}$

25. $\dfrac{-2y}{5\sqrt{2y}}$

26. $\dfrac{3}{1+\sqrt{2}}$

27. $\dfrac{2}{\sqrt{6}-2}$

28. $\dfrac{-11}{\sqrt{3}-4}$

29. $\dfrac{1}{\sqrt{5}-3}$

30. $\dfrac{7}{3-2\sqrt{2}}$

31. $\dfrac{-6}{5-3\sqrt{2}}$

32. $\dfrac{11}{2\sqrt{3}+1}$

33. $\dfrac{-\sqrt{3}}{\sqrt{2}+5}$

34. $\dfrac{\sqrt{2}}{\sqrt{7}+4}$

35. $\dfrac{7}{1-3\sqrt{5}}$

36. $\dfrac{-3\sqrt{3}}{6+\sqrt{3}}$

37. $\dfrac{1}{\sqrt{3}-\sqrt{5}}$

38. $\dfrac{-4}{\sqrt{7}-\sqrt{3}}$

39. $\dfrac{-5}{\sqrt{2}+\sqrt{3}}$

40. $\dfrac{7}{\sqrt{2}+\sqrt{5}}$

41. $\dfrac{4}{\sqrt{x}+1}$

42. $\dfrac{-7}{\sqrt{x}-3}$

43. $\dfrac{5}{6+\sqrt{y}}$

44. $\dfrac{x}{\sqrt{x}+2}$

45. $\dfrac{8}{2\sqrt{x}+3}$

46. $\dfrac{3\sqrt{x}}{\sqrt{2x}-5}$

47. $\dfrac{\sqrt{4y}}{\sqrt{5y}-\sqrt{3}}$

48. $\dfrac{\sqrt{3x}}{\sqrt{2}+\sqrt{3x}}$

49. $\dfrac{3}{\sqrt{x}-\sqrt{y}}$

50. $\dfrac{4}{2\sqrt{x}+\sqrt{y}}$

51. $\dfrac{x}{\sqrt{x}+2\sqrt{y}}$

52. $\dfrac{y}{\sqrt{x}-\sqrt{3y}}$

53. $\dfrac{\sqrt{3}+1}{\sqrt{3}-2}$

54. $\dfrac{\sqrt{2}+4}{5-\sqrt{2}}$

55. $\dfrac{\sqrt{5}-2}{\sqrt{5}+3}$

56. $\dfrac{1+\sqrt{3}}{3-\sqrt{3}}$

57. $\dfrac{\sqrt{x}+1}{\sqrt{x}-1}$

58. $\dfrac{\sqrt{x}-4}{\sqrt{x}+3}$

59. $\dfrac{\sqrt{x}+2}{\sqrt{3x}+y}$

60. $\dfrac{3-\sqrt{x}}{2\sqrt{x}+y}$

Hawkes Learning Systems: Introductory Algebra

Rationalizing Denominators
Division of Radicals

| 9.5 | **Solving Equations with Radicals** |

After completing this section, you will be able to:

1. Solve equations with radical expressions by squaring both sides of the equations.

If an equation involves radicals with radicands that include variables, then the equation is called a **radical equation**. For example,

$$\sqrt{x} = 5, \quad \sqrt{2x + 1} = 3, \quad \text{and} \quad x + 1 = \sqrt{3x - 1}$$

are all radical equations. To solve such equations, we need the following property of real numbers.

Property of Real Numbers

For real numbers a and b, if a = b, then $a^2 = b^2$.

We use this property by **squaring both sides of a radical equation that contains square roots**. However, because squaring positive numbers and squaring negative numbers can sometimes give the same result, we must be sure to check all potential solutions in the **original equation**. For example, consider the equation

$$x = -2 \qquad \text{Given an equation with one solution.}$$
$$(x)^2 = (-2)^2 \qquad \text{Square both sides.}$$
$$x^2 = 4 \qquad \text{Result is an equation with two solutions.}$$
$$x^2 - 4 = 0$$
$$(x + 2)(x - 2) = 0$$
$$x = -2 \quad \text{or} \quad x = 2$$

Thus, squaring both sides created an equation with more solutions than the original equation.

> **Example 1: Solve** •

Solve the radical equation $\sqrt{18 - x} = x - 6$

Solution: $\sqrt{18 - x} = x - 6$

$\left(\sqrt{18 - x}\right)^2 = (x - 6)^2$ Square both sides.

$18 - x = x^2 - 12x + 36$

$0 = x^2 - 11x + 18$ One side must be 0 to solve by factorng.

$0 = (x - 2)(x - 9)$ Factor.

There are two **potential** solutions, 2 and 9.

Check: $x = 2$:

$\sqrt{18 - 2} \overset{?}{=} 2 - 6$

$\sqrt{16} \overset{?}{=} -4$

$4 \neq -4$

So, 2 is not a solution.

$x = 9$:

$\sqrt{18 - 9} \overset{?}{=} 9 - 6$

$\sqrt{9} \overset{?}{=} 3$

$3 = 3$

So, 9 is a solution.

The only solution to the original equation is 9.

• •

The process of squaring both sides will give a new equation. This equation will have all the solutions of the original equation and possibly, as illustrated in Example 1, some solutions that are not solutions of the original equation. These "extra" solutions, if they exist, are called **extraneous solutions**. (See Section 8.5.)

In Example 1, $x = 2$ is an extraneous solution. It is a solution of the quadratic equation,

$$0 = x^2 - 11x + 18$$

but it is not a solution of the original radical equation,

$$\sqrt{18 - x} = x - 6.$$

The checking shows that substituting $x = 2$ in the radical equation gives

$$4 \neq -4.$$

We can see that squaring both sides does give a new equation that is true. Namely, $(4)^2 = (-4)^2$.

Solving radical equations by squaring does not always yield extraneous solutions, but we emphasize that all solutions must be checked in the original equation. The steps for solving an equation with square roots are summarized here.

To Solve an Equation with a Square Root Radical Expression

1. *Isolate the radical expression on one side of the equation.*

2. *Square both sides of the equation.*

3. *Solve the new equation.*

4. *Check each solution of the new equation in the original equation and eliminate any extraneous solutions.*

The following examples illustrate more possible results when solving radical equations.

Example 2: Solve and Check ● ● ● ● ● ● ● ● ● ● ● ● ● ● ● ● ●

Solve the following radical equations.

a. $\sqrt{2x-3} = 5$

Solution:
$$\sqrt{2x-3} = 5$$
$$\left(\sqrt{2x-3}\right)^2 = (5)^2 \qquad \text{Square both sides.}$$
$$2x - 3 = 25$$
$$2x = 28$$
$$x = 14$$

Check:
$$\sqrt{2\cdot 14 - 3} \overset{?}{=} 5$$
$$\sqrt{28-3} \overset{?}{=} 5$$
$$\sqrt{25} \overset{?}{=} 5$$
$$5 \overset{?}{=} 5$$

Thus, 14 checks and the solution is 14.

b. $\sqrt{x+1} = -3$

Solution: We can stop right here. There is **no real solution** to this equation because the radical on the left is nonnegative and cannot possibly equal −3, a negative number. Suppose we did not notice this relationship. Then, proceeding as usual, we will find an answer that does not check.

Continued on next page ...

$$\sqrt{x+1} = -3$$

$$\left(\sqrt{x+1}\right)^2 = (-3)^2 \qquad \text{Square both sides.}$$

$$x+1 = 9$$

$$x = 8$$

Check:
$$\sqrt{8+1} \overset{?}{=} -3$$

$$\sqrt{9} \overset{?}{=} -3$$

$$3 \neq -3$$

So, 8 does **not** check, and there are no solutions.

c. $a - 3 = \sqrt{3a - 9}$

Solution:
$$a - 3 = \sqrt{3a - 9}$$

$$(a-3)^2 = \left(\sqrt{3a-9}\right)^2 \qquad \text{Square both sides.}$$

$$a^2 - 6a + 9 = 3a - 9$$

$$a^2 - 9a + 18 = 0$$

$$(a-6)(a-3) = 0 \qquad \text{Factor.}$$

$$a - 6 = 0 \quad \text{or} \quad a - 3 = 0$$

$$a = 6 \qquad\qquad a = 3$$

Both 6 and 3 check and there are two solutions: 6 and 3.

d. $\sqrt{2x+5} + 3 = 18$

Solution: $\sqrt{2x+5} + 3 = 18$

$$\sqrt{2x+5} = 15 \qquad \text{Isolate the radical on one side.}$$

$$\left(\sqrt{2x+5}\right)^2 = 15^2 \qquad \text{Square both sides.}$$

$$2x + 5 = 225$$

$$2x = 220$$

$$x = 110$$

Check:
$$\sqrt{2 \cdot 110 + 5} + 3 \overset{?}{=} 18$$

$$\sqrt{225} + 3 \overset{?}{=} 18$$

Continued on next page ...

$$15 + 3 \overset{?}{=} 18$$

$$18 = 18$$

The solution is 110.

9.5 Exercises

Solve the equations in Exercises 1 – 45. Check all solutions in the original equation.

1. $\sqrt{x+3} = 6$ **2.** $\sqrt{x-1} = 4$ **3.** $\sqrt{2x-1} = 3$

4. $\sqrt{3x+1} = 5$ **5.** $\sqrt{3x+4} = -5$ **6.** $\sqrt{2x-5} = -1$

7. $\sqrt{5x+4} = 7$ **8.** $\sqrt{3x-2} = 4$ **9.** $\sqrt{6-x} = 3$

10. $\sqrt{11-x} = 5$ **11.** $\sqrt{x-4} + 6 = 2$ **12.** $\sqrt{2x-7} + 5 = 3$

13. $\sqrt{4x+1} - 2 = 3$ **14.** $\sqrt{6x+4} - 3 = 5$ **15.** $\sqrt{2x+11} = 3$

16. $\sqrt{3x+7} = 1$ **17.** $\sqrt{5-2x} = 7$ **18.** $\sqrt{4-3x} = 5$

19. $\sqrt{5x-4} = 14$ **20.** $\sqrt{7x-5} = 4$ **21.** $\sqrt{2x-1} = \sqrt{x+1}$

22. $\sqrt{3x+2} = \sqrt{x+4}$ **23.** $\sqrt{x+2} = \sqrt{2x-5}$ **24.** $\sqrt{2x-5} = \sqrt{3x-9}$

25. $\sqrt{4x-3} = \sqrt{2x+5}$ **26.** $\sqrt{4x-6} = \sqrt{3x-1}$ **27.** $\sqrt{x+6} = x+4$

28. $\sqrt{x+4} = x-8$ **29.** $\sqrt{x-2} = x-2$ **30.** $\sqrt{x+8} = x-4$

31. $x+1 = \sqrt{x+7}$ **32.** $x+2 = \sqrt{x+4}$ **33.** $\sqrt{4x+1} = x-5$

34. $\sqrt{3x+1} = x-3$ **35.** $\sqrt{2-x} = x+4$ **36.** $x+2 = \sqrt{x+14}$

37. $x-3 = \sqrt{27-3x}$ **38.** $x+1 = 2\sqrt{x+4}$ **39.** $2x+3 = \sqrt{3x+7}$

40. $3x+1 = \sqrt{x+5}$ **41.** $x-2 = \sqrt{3x-6}$ **42.** $\sqrt{8-4x} = x+1$

43. $\sqrt{x+3} = x+3$ **44.** $\sqrt{-x-5} = x+5$ **45.** $x+6 = \sqrt{2x+12}$

Hawkes Learning Systems: Introductory Algebra

Solving Radical Equations

9.6 Rational Exponents

After completing this section, you will be able to:

1. Write radical expressions with fractional exponents in radical form.

2. Simplify expressions with fractional exponents.

n^{th} Roots

In Section 9.1, we discussed radicals involving square roots and cube roots. For example,

$$\sqrt{36} = 6 \text{ because } 6^2 = 36$$

and

$$\sqrt[3]{125} = 5 \text{ because } 5^3 = 125.$$

The 3 in $\sqrt[3]{125}$ is called the **index**. In $\sqrt{36}$, the index is understood to be 2. That is, we could write $\sqrt[2]{36}$ for the square root. Examples of other roots are,

$$\sqrt[4]{81} = 3 \text{ because } 3^4 = 81,$$

$$\sqrt[5]{-32} = -2 \text{ because } (-2)^5 = -32,$$

and

$$\sqrt[5]{\frac{1}{32}} = \frac{1}{2} \text{ because } \left(\frac{1}{2}\right)^5 = \frac{1}{32}.$$

The symbol $\sqrt[n]{b}$ is read "the nth root of b." **If $a = \sqrt[n]{b}$, then $a^n = b$.** We need to know the conditions on a and b that will guarantee that $\sqrt[n]{b}$ is a real number.

As we discussed in Section 9.1, $\sqrt{-4}$ is not classified as a real number because the square of a real number is positive or zero. Such numbers are called **complex numbers** and will be discussed in the next course in algebra. In the general case,

$$\sqrt[n]{b} \text{ is not a real number if } n \text{ is even and } b \text{ is negative.}$$

In other words, even roots of negative numbers are **not** real numbers.

> ### Example 1: Roots

a. $\sqrt[3]{-27} = -3$ because $(-3)^3 = -27$

b. $\sqrt[4]{0.0016} = 0.2$ because $(0.2)^4 = 0.0016$

c. $\sqrt[5]{0.00001} = 0.1$ because $(0.1)^5 = 0.00001$

d. $\sqrt[6]{-64}$ is not a real number.

Rational (or Fractional) Exponents

The following definitions show how radicals can be expressed using fractional exponents. We will find that algebraic operations with radicals are generally easier to perform when fractional exponents are used.

$b^{1/n}$

If n is a positive integer and $\sqrt[n]{b}$ is a real number, then

$$\sqrt[n]{b} = b^{1/n}.$$

Thus, by definition,

$$\sqrt{25} = (25)^{1/2} = 5$$

and

$$\sqrt[3]{8} = (8)^{1/3} = 2.$$

> ### Example 2: Simplify Fractional Exponents

Simplify the following expressions

a. $(49)^{1/2}$

 Solution: $(49)^{1/2} = \sqrt{49} = 7$

b. $(-216)^{1/3}$

 Solution: $(-216)^{1/3} = \sqrt[3]{-216} = -6$

c. $25^{\frac{1}{2}}$

 Solution: $25^{\frac{1}{2}} = \sqrt{25} = 5$

Continued on next page ...

d. $\left(\dfrac{16}{81} \right)^{\frac{1}{4}}$

$\qquad$ **Solution:** $\quad \left(\dfrac{16}{81} \right)^{\frac{1}{4}} = \sqrt[4]{\dfrac{16}{81}} = \dfrac{2}{3}$

$\bullet \bullet$

Now consider the problem of evaluating the expression $8^{\frac{2}{3}}$. We need the following definition for using rational exponents.

$b^{m \backslash n}$

If b is a nonnegative real number and m and n are integers with n > 0, then,

$$b^{m/n} = \left(b^{1/n} \right)^{m} = \left(\sqrt[n]{b} \right)^{m} \ \text{or} \ b^{m/n} = \left(b^{m} \right)^{1/n}.$$

Using this definition, we have,

$$8^{2/3} = \left(8^{1/3} \right)^{2} = \left(2 \right)^{2} = 4 \ \text{or} \ 8^{2/3} = \left(8^{2} \right)^{1/3} = \left(64 \right)^{1/3} = 4$$

and $\qquad 32^{3/5} = \left(32^{1/5} \right)^{3} = \left(2 \right)^{3} = 8 \ \text{or} \ 32^{3/5} = \left(32^{3} \right)^{1/5} = \left(32,768 \right)^{1/5} = 8.$

Now that we have $b^{m/n}$ defined for rational exponents, we can consider simplifying expressions such as

$$x^{2/3} \cdot x^{1/6} \ \text{and} \ 36^{-1/2} \ \text{and} \ 2^{3/4} \cdot 2^{1/4}.$$

All the previous properties of exponents apply to rational exponents (See Chapter 6).

Example 3: Simplify $\bullet \bullet \bullet \bullet \bullet \bullet \bullet \bullet \bullet \bullet \bullet \bullet \bullet \bullet \bullet \bullet \bullet \bullet$

Using the properties of exponents, simplify each expression. Assume that all variables are positive.

a. $x^{2/3} \cdot x^{1/6}$

$\qquad$ **Solution:** $\quad x^{2/3} \cdot x^{1/6} = x^{2/3 + 1/6} = x^{4/6 + 1/6} = x^{5/6}$ $\qquad$ Add the exponents.

b. $\dfrac{a^{3/4}}{a^{1/2}}$

$\qquad$ **Solution:** $\quad \dfrac{a^{3/4}}{a^{1/2}} = a^{3/4 - 1/2} = a^{3/4 - 2/4} = a^{1/4}$ $\qquad$ Subtract the exponents.

Continued on next page ...

609

c. $36^{-1/2}$

Solution: $36^{-1/2} = \dfrac{1}{36^{1/2}} = \dfrac{1}{6}$ Use the property of negative exponents: $a^{-n} = \dfrac{1}{a^n}$.

d. $\left(2y^{1/4}\right)^3 \cdot \left(3y^{1/8}\right)^2$

Solution: $\left(2y^{1/4}\right)^3 \cdot \left(3y^{1/8}\right)^2 = 2^3\,y^{3/4} \cdot 3^2\,y^{1/4}$ Multiply the exponents.

$= 8y^{3/4} \cdot 9y^{1/4}$ Simplify.

$= 72y^{3/4\,+\,1/4}$ $8 \cdot 9 = 72$.

$= 72y$ Add the exponents.

e. $2^{3/4} \cdot 2^{1/4}$

Solution: $2^{3/4} \cdot 2^{1/4} = 2^{3/4\,+\,1/4}$ Keep the base 2.

$= 2^1$ Add the exponents.

$= 2$

9.6 Exercises

Write each of the expressions in 1 – 10 as radical expressions, then simplify.

1. $16^{1/2}$ **2.** $49^{1/2}$ **3.** $-8^{1/3}$ **4.** $216^{1/3}$

5. $81^{1/4}$ **6.** $16^{1/4}$ **7.** $(-32)^{1/5}$ **8.** $243^{1/5}$

9. $0.0004^{1/2}$ **10.** $0.09^{1/2}$

Simplify the expressions in Exercises 11 – 30.

11. $\left(\dfrac{4}{9}\right)^{1/2}$ **12.** $\left(\dfrac{1}{27}\right)^{1/3}$ **13.** $\left(\dfrac{8}{125}\right)^{1/3}$ **14.** $\left(\dfrac{16}{9}\right)^{1/2}$

15. $36^{-1/2}$ **16.** $64^{-1/3}$ **17.** $8^{-1/3}$ **18.** $100^{-1/2}$

19. $81^{-1/4}$ **20.** $16^{-1/4}$ **21.** $4^{3/2}$ **22.** $64^{2/3}$

23. $(-27)^{2/3}$ **24.** $36^{3/2}$ **25.** $9^{5/2}$ **26.** $8^{5/3}$

27. $16^{-3/4}$ **28.** $25^{-3/2}$ **29.** $(-8)^{-2/3}$ **30.** $(-125)^{-2/3}$

Using the properties of exponents, simplify each expression in Exercises 31 – 60. Assume that all variables are positive.

31. $x^{1/4} \cdot x^{1/3}$ **32.** $x^{1/4} \cdot x^{1/4}$ **33.** $5^{1/2} \cdot 5^{-3/2}$ **34.** $2^{-2/3} \cdot 2^{1/3}$

35. $x^{1/5} \cdot x^{2/5}$ **36.** $x^{1/8} \cdot x^{5/8}$ **37.** $\dfrac{x^{3/4}}{x^{1/4}}$ **38.** $\dfrac{x^{5/7}}{x^{3/7}}$

39. $\dfrac{x^{4/5}}{x^{2/5}}$ **40.** $\dfrac{x^{2/3}}{x^{2/3}}$ **41.** $\dfrac{2^{2/3}}{2^{-1/3}}$ **42.** $\dfrac{a^{5/4}}{a^{-1/4}}$

43. $a^{2/3} \cdot a^{1/2}$ **44.** $y^{3/4} \cdot y^{1/2}$ **45.** $x^{3/4} \cdot x^{-1/8}$ **46.** $y^{1/2} \cdot y^{-5/6}$

47. $\dfrac{x^{2/5}}{x^{1/2}}$ **48.** $\dfrac{x^{1/2}}{x^{2/3}}$ **49.** $\dfrac{6^2}{6^{1/2}}$ **50.** $\dfrac{10^3}{10^{4/3}}$

51. $\dfrac{x^{3/4}}{x^{-1/2}}$ **52.** $\dfrac{a^{2/5}}{a^{-1/10}}$ **53.** $\dfrac{x^{5/3}}{x^{-1/3}}$ **54.** $\dfrac{y^{1/2}}{y^{-2/3}}$

55. $\left(7x^{1/3}\right)^2 \cdot \left(4x^{1/2}\right)$ **56.** $\left(2x^{1/2}\right)^3 \cdot \left(3x^{1/3}\right)^2$ **57.** $\left(9x^2\right)^{1/2} \cdot \left(8x^3\right)^{1/3}$

58. $\left(x^{2/3}\right)^3 \cdot \left(25x^4\right)^{1/2}$ **59.** $\left(16x^8\right)^{1/4} \cdot \left(9x^2\right)^{-1/2}$ **60.** $\left(2x^{3/4}\right)^2 \cdot \left(3x^{3/2}\right)^{-2}$

Hawkes Learning Systems: Introductory Algebra

Rational Exponents

9.7 Distance Between Two Points and Midpoint Formula

Objectives

After completing this section, you will be able to:

1. *Find the distance between two points.*
2. *Determine if triangles are right triangles given the coordinates of the vertices by using the distance formula and the Pythagorean Theorem.*
3. *Show that specified geometric relationships are true by using the distance formula.*
4. *Find the perimeters of triangles given the coordinates of the vertices by using the distance formula.*
5. *Find the coordinates of the midpoints of line segments.*

Distance Between Two Points on a Vertical or Horizontal Line

A formula particularly useful in fields of study where geometry is involved, such as surveying, machining, and engineering, is that for finding the distance between two points. The formula is related to the Cartesian coordinate system discussed in Chapter 4.

We first discuss the distance between two points that are on a horizontal line or on a vertical line as indicated by the following two questions.

1. What is the distance between the points $P_1(2, 3)$ and $P_2(6, 3)$? (The y-coordinates are equal.)
2. What is the distance between the points $P_3(-1, -4)$ and $P_4(-1, 1)$? (The x-coordinates are equal.) See Figure 9.3.

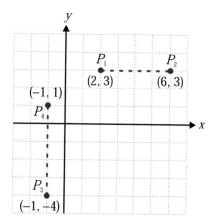

Figure 9.3

Since P_1 and P_2 lie on a horizontal line and have the same y-coordinate, the distance between the points is the difference between the x-coordinates:

$$\text{distance } (P_1 \text{ to } P_2) = 6 - 2 = 4.$$

Since P_3 and P_4 lie on a vertical line and have the same x-coordinate, the distance between the points is the difference between the y-coordinates:

$$\text{distance } (P_3 \text{ to } P_4) = 1 - (-4) = 5.$$

For the distance from P_1 to P_2, why not reverse the coordinates and take

$$\text{distance } (P_1 \text{ to } P_2) = 2 - 6 = -4?$$

The reason is that the term **distance** is taken to mean a nonnegative number (positive or 0). To guarantee a nonnegative number, take the absolute value of the difference. By taking the absolute value, the order of subtraction is no longer important.

The distance between two points, as indicated in the following formulas, is represented by d.

The Distance Between Two Points on a Horizontal or Vertical Line

For $P_1(x_1, y_1)$ and $P_2(x_2, y_1)$ on a horizontal line,

$$d = |x_2 - x_1| \ (or \ d = |x_1 - x_2|)$$

For $P_1(x_1, y_1)$ and $P_2(x_1, y_2)$ on a vertical line,

$$d = |y_2 - y_1| \ (or \ d = |y_1 - y_2|)$$

Example 1: Distance • • • • • • • • • • • • • • • • • • •

a. Find the distance, d, between the two points $(5, 7)$ and $(-3, 7)$.

Solution: Since the points are on a horizontal line (they have the same y-coordinate), we find the absolute value of the difference between the x-coordinates.

$$d = |-3 - 5| = |-8| = 8$$

or $\quad d = |5 - (-3)| = |8| = 8$

b. Find the distance, d, between the two points $(2, 8)$ and $\left(2, -\dfrac{1}{2}\right)$.

Solution: Since the points are on a vertical line (they have the same x-coordinate), we find the absolute value of the difference between the y-coordinates.

$$d = \left|8 - \left(-\frac{1}{2}\right)\right| = \left|8\frac{1}{2}\right| = 8\frac{1}{2}$$

or $\quad d = \left|-\dfrac{1}{2} - 8\right| = \left|-8\dfrac{1}{2}\right| = 8\dfrac{1}{2}$

• •

The Pythagorean Theorem

Among the most interesting and useful applications of squares and square roots are those dealing with **right triangles** and the **Pythagorean Theorem**. (Pythagoras was a famous Greek mathematician who lived in the 6th century BC.)

A **right triangle** is a triangle in which one angle is a right angle (measures 90°). Two of the sides are perpendicular and are called **legs**. The longest side is called the **hypotenuse** and is opposite the 90° angle.

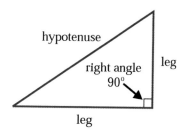

Figure 9.4

The Pythagorean Theorem

In a right triangle, the square of the hypotenuse is equal to the sum of the squares of the two legs.

$$c^2 = a^2 + b^2$$

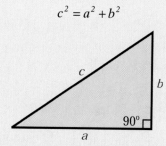

Example 2: The Pythagorean Theorem ● ● ● ● ● ● ● ● ● ● ●

What is the length of the hypotenuse of a right triangle if one leg is 8 cm long and the other leg is 6 cm long?

Solution: By the Pythagorean Theorem,

$$c^2 = 8^2 + 6^2$$

$$c^2 = 64 + 36$$

$$c^2 = 100$$

$$c = \sqrt{100} = 10 \text{ cm}$$

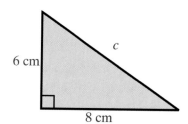

The hypotenuse is 10 cm long. (Note that only the positive square root of 100 is used because the negative square root does not make sense as a solution to the problem.)

● ●

Distance Between Two Points Not on a Vertical or Horizontal Line

If two points do not lie on a vertical line or a horizontal line, we apply the Pythagorean Theorem and calculate the length of the hypotenuse of the right triangle formed as shown in Figure 9.5 In Figure 9.5, the distance between the two points $P_1(-1, 2)$ and $P_2(5, 4)$ is the hypotenuse and the third point is at the vertex $(5, 2)$ where the right angle is located.

$$d^2 = a^2 + b^2$$

$$d^2 = 6^2 + 2^2 = 36 + 4 = 40$$

$$d = \sqrt{40} = 2\sqrt{10}$$

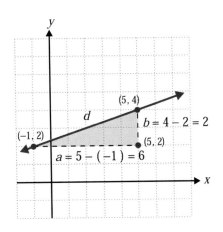

Figure 9.5

We could write directly $d = \sqrt{a^2 + b^2}$. Carrying this idea one step further, we can write a formula for d involving the coordinates of two general points $P_1(x_1, y_1)$ and $P_2(x_2, y_2)$. (See Figure 9.6.)

$$d = \sqrt{(x_2 - x_1)^2 + (y_2 - y_1)^2}$$

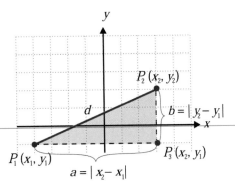

Figure 9.6

Formula for the Distance Between Two Points

The formula for the distance between two points $P_1(x_1, y_1)$ and $P_2(x_2, y_2)$ is

$$d = \sqrt{(x_2 - x_1)^2 + (y_2 - y_1)^2}$$

The absolute values, $\left| x_2 - x_1 \right|$ and $\left| y_2 - y_1 \right|$ are not needed in the formula because each of the expressions, $x_2 - x_1$ and $y_2 - y_1$, is squared. So, even though the differences might be negative, the squares, $(x_2 - x_1)^2$ and $(y_2 - y_1)^2$ will always be nonnegative. Also, the order of subtraction may be reversed because the squares will give the same value.

NOTES

In the actual calculation of *d*, be sure to add the squares before taking the square root. In general,

$$\sqrt{a^2 + b^2} \neq a + b$$

For example, suppose that $d = \sqrt{3^2 + 4^2} = \sqrt{9 + 16}$. Then, adding first gives

$$\sqrt{9 + 16} = \sqrt{25} = 5 \qquad \text{RIGHT}$$

Taking square roots first gives the wrong result:

$$\sqrt{9 + 16} = \sqrt{9} + \sqrt{16} = 3 + 4 = 7 \quad \text{WRONG}$$

Example 3: Distance Between Two Points • • • • • • • • • • •

a. Find the distance between the two points $(3, 4)$ and $(-2, 7)$.

Solution: $d = \sqrt{(3 - (-2))^2 + (4 - 7)^2}$

$= \sqrt{5^2 + (-3)^2}$

$= \sqrt{25 + 9}$

$= \sqrt{34}$

b. Find the distance between the two points $\left(\dfrac{1}{2}, \dfrac{2}{3} \right)$ and $\left(\dfrac{3}{4}, \dfrac{5}{3} \right)$.

Solution: $d = \sqrt{\left(\dfrac{3}{4} - \dfrac{1}{2} \right)^2 + \left(\dfrac{5}{3} - \dfrac{2}{3} \right)^2}$

$= \sqrt{\left(\dfrac{1}{4} \right)^2 + \left(\dfrac{3}{3} \right)^2}$

$= \sqrt{\dfrac{1}{16} + 1}$

$= \sqrt{\dfrac{1}{16} + \dfrac{16}{16}}$

$= \sqrt{\dfrac{17}{16}} = \dfrac{\sqrt{17}}{4}$

Continued on next page ...

c. Show that the triangle determined by the points $A(-2, 1)$, $B(3, 4)$, and $C(1, -4)$ is an isosceles triangle. (An isosceles triangle has two equal sides. So, the object is to determine whether or not two sides have the same length.)

Solution: The length of the line segment AB is the distance between the points A and B. We denote this distance by $|AB|$. Thus, to show that the triangle ABC is isosceles, we need to show that $|AB| = |AC|$ or that $|AB| = |BC|$ or $|AC| = |BC|$. If none of these relationships are true then the triangle does not have two equal sides and is not isosceles.

$$|AB| = \sqrt{(-2-3)^2 + (1-4)^2} = \sqrt{(-5)^2 + (-3)^2}$$
$$= \sqrt{25+9} = \sqrt{34}$$

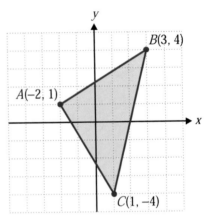

$$|AC| = \sqrt{(-2-1)^2 + (1-(-4))^2}$$
$$= \sqrt{(-3)^2 + (5)^2}$$
$$= \sqrt{9+25} = \sqrt{34}$$

Since $|AB| = |AC|$, the triangle is isosceles.

The Midpoint Formula

Another formula that involves the coordinates of points in the plane is that for finding the midpoint of the segment joining the two given points. This point is found by averaging the corresponding x- and y-coordinates.

Formula for the Midpoint Between Two Points

The formula for the midpoint between two points $P_1(x_1, y_1)$ and $P_2(x_2, y_2)$ is

$$\left(\frac{x_1 + x_2}{2}, \frac{y_1 + y_2}{2}\right).$$

Example 4: Midpoint ●

Find the coordinates of the midpoint of the line segment joining the two points $P_1(-1, 2)$ and $P_2(4, 6)$.

Solution: The midpoint is $\left(\dfrac{-1+4}{2}, \dfrac{2+6}{2} \right) = \left(\dfrac{3}{2}, 4 \right)$

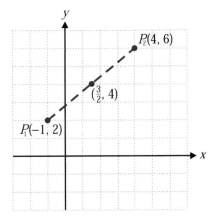

● ●

9.7 Exercises

For Exercises 1 – 24, (a) find the distance between the two given points and (b) find the coordinates of the midpoints of the line segment joining the given points.

1. $(-3,6),(-3,2)$ **2.** $(5,7),(7,-3)$ **3.** $(4,-3),(7,-3)$

4. $(-1,2),(5,2)$ **5.** $(3,1),\left(-\dfrac{1}{2},1\right)$ **6.** $\left(\dfrac{4}{3},1\right),\left(\dfrac{4}{3},-\dfrac{2}{3}\right)$

7. $(3,1),(2,0)$ **8.** $(4,6),(5,-2)$ **9.** $(1,5),(-1,2)$

10. $(0,0),(-3,4)$ **11.** $(2,-7),(-3,5)$ **12.** $(5,-3),(7,-3)$

13. $\left(\dfrac{3}{7},\dfrac{4}{7}\right),(0,0)$ **14.** $(-5,2),(1,1)$ **15.** $(4,1),(7,5)$

16. $(10,7),(1,7)$ **17.** $(-10,3),(2,-2)$ **18.** $\left(\dfrac{7}{3},2\right),\left(-\dfrac{2}{3},1\right)$

19. $(-3,2),(3,-6)$ **20.** $\left(\dfrac{3}{4},6\right),\left(\dfrac{3}{4},-2\right)$ **21.** $(4,0),(0,-3)$

22. $(0,-2),(4,-3)$ **23.** $\left(\dfrac{4}{5},\dfrac{2}{7}\right),\left(-\dfrac{6}{5},\dfrac{2}{7}\right)$ **24.** $(6,8),(2,5)$

25. Use the distance formula and the Pythagorean Theorem to decide if the triangle determined by the points $A(1,-2)$, $B(7,1)$, and $C(5,5)$ is a right triangle.

26. Use the distance formula and the Pythagorean Theorem to decide if the triangle determined by the points $A(-5,-1)$, $B(2,1)$, and $C(-1,6)$ is a right triangle.

In Exercises 27 and 28, show that the triangle determined by the given points is an isosceles triangle (has two equal sides).

27. $A(1,1),B(5,9),C(9,5)$ **28.** $A(1,-4),B(3,2),C(9,4)$

In Exercises 29 and 30, show that the triangle determined by the given points is an equilateral triangle (all sides equal).

29. $A(1,0),B(3,\sqrt{12}),C(5,0)$ **30.** $A(0,5),B(0,-3),C(\sqrt{48},1)$

In Exercises 31 and 32, show that the diagonals of the rectangle *ABCD* are equal.

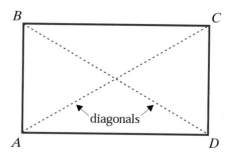

31. $A(2,-2), B(2,3), C(8,3), D(8,-2)$ **32.** $A(-1,1), B(-1,4), C(4,4), D(4,1)$

In Exercises 33 – 36, use the Pythagorean Theorem to determine whether or not each of the triangles is a right triangle.

33.

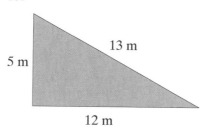

13 m

5 m

12 m

34.

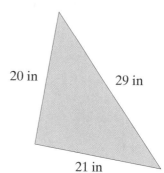

20 in

29 in

21 in

35.

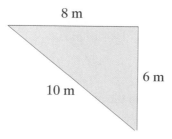

8 m

6 m

10 m

36.

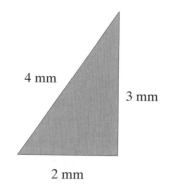

4 mm

3 mm

2 mm

In Exercises 37 – 42, use your calculator to find the answers.

37. A square is said to be inscribed in a circle if each corner of the square lies on the circle. (Use $\pi = 3.14$.)
   a. Find the circumference and area of a circle with diameter 30 feet.
   b. Find the perimeter and area of a square inscribed in the circle.

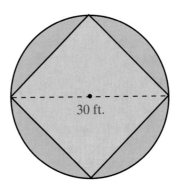

30 ft.

38. The shape of a baseball infield is a square with sides 90 feet long.
   a. Find the distance (to the nearest tenth of a foot) from home plate to second base.
   b. Find the distance (to the nearest tenth of a foot) from first base to third base.

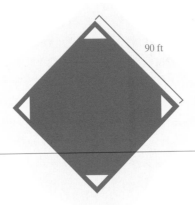

90 ft

39. The distance from home plate to the pitcher's mound is 60.5 feet.
   a. Is the pitcher's mound exactly half way between home plate and second base?
   b. If not, which base is it closer to, home plate or second base?
   c. Do the two diagonals of the square intersect at the pitcher's mound?

40. The GE Building in New York is 850 feet tall (70 stories). At a certain time of day, the building casts a shadow 100 feet long. Find the distance from the top of the building to the tip of the shadow (to the nearest tenth of a foot)

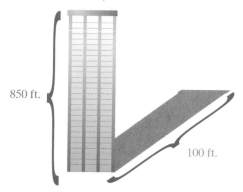

850 ft.

100 ft.

41. To create a square inside a square, a quilting pattern requires four triangular pieces like the one shaded in the figure shown here. If the square in the center measures 12 centimeters on a side, and the two legs of each traingle are of equal length, how long are the legs of each triangle, to the nearest tenth of a centimeter?

12 cm

42. If an airplane passes directly over your head at an altitude of 1 mile, how far (to the nearest hundredth of a mile) is the airplane from your position after it has flown 2 miles farther at the same altitude?

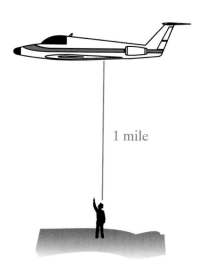

1 mile

In Exercises 43 – 45, find the perimeter of the triangle determined by the given points.

43. $A(-5, 0), B(3, 4), C(0, 0)$ **44.** $A(-6, -1), B(-3, 3), C(6, 4)$

45. $A(-2, 5), B(3, 1), C(2, -2)$

46. Suppose that $\left(\dfrac{3}{4}, -5\right)$ is the midpoint of the line segment AB where $A(2, y)$ and $B(x, -3)$. Find the values of x and y.

47. Suppose that $(2, 7)$ is the midpoint of the line segment AB where $A(-1, 1)$ and $B(x, y)$. Find the values of x and y.

Writing and Thinking About Mathematics

48. If three positive integers satisfy the Pythagorean Theorem, they are called a **Pythagorean Triple.** For example, 3, 4, and 5 are a Pythagorean Triple because $3^2 + 4^2 = 5^2$. There are an infinite number of such triples. To see how some triples can be found, fill out the following table and verify that the last three numbers are indeed Pythagorean Triples.

u	v	$2uv$	$u^2 - v^2$	$u^2 + v^2$
2	1	4	3	5
3	2	___	___	___
5	2	___	___	___
4	3	___	___	___
7	1	___	___	___
6	5	___	___	___

Hawkes Learning Systems: Introductory Algebra

Distance Between Two Points and the Midpoint Formula

Chapter 9 Index of Key Ideas and Terms

Rational Numbers page 567

A **rational number** is any number that can be written in the

form $\dfrac{a}{b}$ where a and b are integers and $b \neq 0$.

In decimal form, all rational numbers can be written as terminating decimals or infinite repeating decimals.

Perfect Squares page 568

The square of an integer is called a **perfect square**.

Square Root Terminology page 568

The symbol $\sqrt{}$ is called a **radical sign**.
The number under the radical sign is called the **radicand**.
The complete expression, such as $\sqrt{49}$, is called a **radical** or **radical expression**.

Square Root page 569

If b is a nonnegative real number and a is a real number such that $a^2 = b$, then a is called a **square root** of b.
If a is nonnegative, then we write $\sqrt{b} = a$ and $-\sqrt{b} = -a$.

Cube Root page 570

If a and b are real numbers such that $a^3 = b$, then a is called the **cube root** of b.
We write $\sqrt[3]{b} = a$.

Irrational Numbers page 571

Irrational numbers are real numbers that are not rational numbers. Written in decimal form, irrational numbers are infinite nonrepeating decimals.

Real Numbers page 572

Real numbers are numbers that are either rational or irrational.

Summary of Square Roots page 574

In general, if a is a positive real number, then

1. $\sqrt{a}$ is rational if a is a perfect square,

2. $\sqrt{a}$ is irrational if a is not a perfect square,

3. $\sqrt{-a}$ is nonreal.

Properties of Square Roots

If a and b are positive real numbers, then

1. $\sqrt{ab} = \sqrt{a}\sqrt{b}$

2. $\sqrt{\dfrac{a}{b}} = \dfrac{\sqrt{a}}{\sqrt{b}}$

Simplest Form

A square root is considered to be in simplest form when the radicand has no perfect square as a factor.
A cube root is considered to be in simplest form when the radicand has no perfect cube as a factor.

Square Root of x^2

If x is a real number, then $\sqrt{x^2} = |x|$.
Note: If $x \geq 0$ is given, then we can write $\sqrt{x^2} = x$.

Square Roots of Expressions with Even and Odd Exponents

For any real number x and positive integer m,

$$\sqrt{x^{2m}} = |x^m| \qquad \text{and} \qquad \sqrt{x^{2m+1}} = x^m\sqrt{x}$$

Note that the absolute value sign is necessary only if m is odd.
Also, note that for $\sqrt{x^{2m+1}}$ to be defined, x cannot be negative.

Adding Like Radicals

Multiplying Radicals

Rationalizing Denominators with One Term

Rationalizing Denominators with a Sum or Difference

Extraneous Solution

To Solve an Equation with a Square Root Radical Expression

1. Isolate the radical expression on one side of the equation.
2. Square both sides of the equation.
3. Solve the new equation.
4. Check each solution of the new equation in the original equation and eliminate any extraneous solutions.

n^{th} Roots   page 607

The symbol $\sqrt[n]{b}$ is read "the nth root of b."

$\sqrt[n]{b}$ is not a real number if n is even and b is negative.

Fractional Exponents   pages 608 - 609

If n is a positive integer and $\sqrt[n]{b}$ is a real number, then $\sqrt[n]{b} = b^{\frac{1}{n}}$.

If b is a nonnegative number and m and n are integers with $n > 0$, then

$$b^{\frac{m}{n}} = \left(b^{\frac{1}{n}} \right)^m = \left(\sqrt[n]{b} \right)^m$$

Distance Between Two Points on a Horizontal or Vertical Line   pages 612 - 614

For $P_1(x_1, y_1)$ and $P_2(x_2, y_1)$ on a horizontal line, $d = |x_2 - x_1|$.

For $P_1(x_1, y_1)$ and $P_2(x_1, y_2)$ on a vertical line, $d = |y_2 - y_1|$.

The Pythagorean Theorem   page 615

In a right triangle, the square of the hypotenuse is equal to
the sum of the squares of the two legs.

Formula for the Distance Between Two Points   page 616

The formula for the distance between two points

$P_1(x_1, y_1)$ and $P_2(x_2, y_2)$ is $d = \sqrt{(x_2 - x_1)^2 + (y_2 - y_1)^2}$.

The Midpoint Formula   page 618

The formula for the midpoint between two points

$P_1(x_1, y_1)$ and $P_2(x_2, y_2)$ is $\left(\dfrac{x_1 + x_2}{2}, \dfrac{y_1 + y_2}{2} \right)$.

Chapter 9 Review

For a review of the topics and problems from Chapter 9, look at the
following lessons from *Hawkes Learning Systems: Introductory Algebra*

Evaluating Radicals
Simplifying Radicals
Addition and Subtraction with Radicals
Multiplying Radicals
Rationalizing Denominators
Division of Radicals
Solving Radical Expressions
Rational Exponents
Distance Between Two Points and the Midpoint Formula

Chapter 9 Test

1. Write the rational number $\dfrac{3}{7}$ in the form of an infinite repeating decimal.

2. Write the irrational number $\sqrt{13}$ in decimal form accurate to 7 decimal places. (Use your calculator.)

Simplify the radical expressions in Exercises 3 – 14.

3. $\sqrt{144}$

4. $\sqrt{25x^2}$

5. $-\sqrt{81x^4y^3}$

6. $\sqrt[3]{27a^3}$

7. $\sqrt[3]{24x^6y^{12}}$

8. $-\sqrt{\dfrac{16x^2}{y^8}}$

9. $\sqrt{\dfrac{4a^5}{9b^{10}}}$

10. $5\sqrt{8}+3\sqrt{2}$

11. $2\sqrt{27x^2}+5\sqrt{48x^2}-4\sqrt{3x^2}$

12. $\dfrac{\sqrt{96}+12}{24}$

13. $\dfrac{21-\sqrt{98}}{7}$

14. $\sqrt[3]{-40}+\sqrt[3]{108}$

Multiply the expressions in Exercises 15 – 18 and simplify if possible.

15. $3\sqrt{5}\cdot 5\sqrt{5}$

16. $\sqrt{15}\cdot\left(\sqrt{3}+\sqrt{5}\right)$

17. $\left(5+\sqrt{11}\right)\left(5-\sqrt{11}\right)$

18. $\left(6+\sqrt{2}\right)\left(5-\sqrt{2}\right)$

Rationalize each denominator in Exercises 19 – 22 and simplify if possible.

19. $\dfrac{3}{\sqrt{18}}$

20. $\sqrt{\dfrac{4}{75}}$

21. $\dfrac{2}{\sqrt{3}-1}$

22. $\dfrac{\sqrt{2}}{\sqrt{3}+\sqrt{2}}$

Solve the equations in Exercises 23 – 25. Check all solutions in the original equation.

23. $\sqrt{5x-1}=13$

24. $\sqrt{4x+2}=\sqrt{12-x}$

25. $2x-5=\sqrt{13-6x}$

Simplify the expressions in Exercises 26 – 30. Assume that all variables are positive.

26. $16^{3/4}$

27. $\left(\dfrac{4}{9}\right)^{-3/2}$

28. $a^{3/4}\cdot a^{-1/2}$

29. $\dfrac{y^{3/4}}{y^{-1/2}}$

30. $\left(16x^2\right)^{1/2}\left(27x^3\right)^{2/3}$

For Exercises 31 and 32, (a) find the distance between the two points and (b) find the coordinates of the midpoint of the line segment joining the two points.

31. $(-5, 7)(3, -1)$ **32.** $(2, 5)(-4, -6)$

33. Use the Pythagorean Theorem to determine whether or not a triangle with sides 6 in., 8 in. and 10 in. is a right triangle. Sketch a picture of the triangle.

34. A ladder that is 25 feet long is leaning against a building. If the bottom of the ladder is 6 feet from the base of the building, how far up the building (to the nearest tenth of a foot) is the ladder reaching?

25 ft

6 ft

Cumulative Review: Chapters 1 – 9

Complete the square by adding the correct term in Exercises 1 and 2. Then factor as indicated.

1. $x^2 + 8x +$ _____ $= ($ $)^2$ **2.** $x^2 -$ _____ $+ 49 = ($ $)^2$

Perform the indicated operations in Exercises 3 and 4 and simplify.

3. $\dfrac{2x+10}{x^2-x-6} \cdot \dfrac{x^2+6x+8}{x+5}$ **4.** $\dfrac{x}{x+6} - \dfrac{2x-5}{x^2+4x-12}$

Graph the equations given in Exercises 5 and 6.

5. $y = \dfrac{3}{4}x - 2$ **6.** $2x + 3y = 12$

Simplify the radical expressions in Exercises 7 – 14.

7. $-\sqrt{63x^4}$ **8.** $\sqrt{\dfrac{36a^3}{25b^6}}$ **9.** $\sqrt{48x^{10}y^{20}}$ **10.** $\sqrt{54a^8b^{14}}$

11. $\sqrt[3]{40}$ **12.** $\sqrt[3]{-250}$ **13.** $\dfrac{8+\sqrt{12}}{2}$ **14.** $\dfrac{20-\sqrt{48}}{8}$

In Exercises 15 – 28, perform the indicated operations and simplify.

15. $2\sqrt{50} - \sqrt{98} + 3\sqrt{8}$ **16.** $11\sqrt{2} + \sqrt{50}$ **17.** $4\sqrt{20} - \sqrt{45}$

18. $\sqrt[3]{-54} + 5\sqrt[3]{2}$ **19.** $\sqrt{50x^4} - 2\sqrt{8x^4}$ **20.** $\sqrt{12} - 5\sqrt{3} + \sqrt{48}$

21. $\sqrt{24} \cdot \sqrt{6}$ **22.** $2\sqrt{30} \cdot \sqrt{20}$ **23.** $2\sqrt{48} \cdot \sqrt{2}$

24. $\sqrt{14} \cdot \sqrt{21}$ **25.** $\left(\sqrt{3}-5\right)\left(\sqrt{3}+5\right)$ **26.** $\left(\sqrt{2}+3\right)\left(\sqrt{2}+5\right)$

27. $\left(\sqrt{x}+\sqrt{2}\right)\left(\sqrt{x}-\sqrt{2}\right)$ **28.** $\left(2\sqrt{x}+3\right)\left(\sqrt{x}-8\right)$

Rationalize each denominator in Exercises 29 – 34 and simplify if possible.

29. $\dfrac{9}{\sqrt{3}}$ **30.** $\sqrt{\dfrac{3}{20}}$ **31.** $\sqrt{\dfrac{5}{27}}$

32. $-\sqrt{\dfrac{9}{2y}}$ **33.** $\dfrac{2}{\sqrt{3}-5}$ **34.** $\dfrac{1}{\sqrt{2}+3}$

Solve the equations in Exercises 35 – 42.

35. $x(x+4) - (x+4)(x-3) = 0$

36. $(2x+1)(x-4) = 5$

37. $\dfrac{x-1}{2x+1} + \dfrac{1}{x+2} = \dfrac{1}{2}$

38. $x+3 = \sqrt{x+15}$

39. $\sqrt{7x+2} = 3$

40. $\sqrt{4x-3} = 5$

41. $\sqrt{2x-9} = 7$

42. $\sqrt{5x-3} = \sqrt{2x+3}$

Solve the systems of equations in Exercises 43 and 44.

43. $\begin{cases} 2x - y = 7 \\ 3x + 2y = 0 \end{cases}$

44. $\begin{cases} 4x + 3y = 15 \\ 2x - 5y = 1 \end{cases}$

Set up equations and solve the problems in Exercises 45 – 50.

45. Two automobiles start from the same place at the same time and travel in opposite directions. One travels 12 mph faster than the other. At the end of $3\frac{1}{2}$ hours, they are 350 miles apart. Find the rate of each automobile.

46. The owner of a candy store sells candy for $1.50 per pound and $2.00 per pound. How many pounds of each must he use to obtain 20 pounds of a mixture of candy which sells for $1.80 per pound?

47. A man and his daughter can paint their cabin in 3 hours. Working alone, it would take the daughter 8 hours longer than it would the father. How long would it take them each working alone?

48. For lunch, Jason had one burrito and 2 tacos. Matt had 2 burritos and 3 tacos. If Jason spent $3.35 and Matt spent $5.90, find the price of one burrito and the price of one taco.

49. A boat is being pulled to the shore from a dock. When the rope to the boat is 40 meters long, the boat is 30 meters from the dock. What is the height of the dock (to the nearest tenth of a meter) above the deck of the boat?

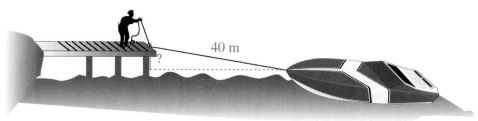

40 m

50. A ball is dropped from the top of a building that is 240 feet tall. Use the formula $h = -16t^2 + v_0t + h_0$ to answer the following questions.

a. When will the ball be 144 feet above the ground?

b. How far will the ball have dropped at that time?

c. When will the ball hit the ground?

Quadratic Equations

Did You Know?

The quadratic formula is a general method for solving second-degree equations of the form $ax^2 + bx + c = 0$, where a, b, and c can be any real numbers. The quadratic formula is a very old formula; it was known to Babylonian mathematicians around 2000 B.C. However, Babylonian and, later, Greek mathematicians always discarded negative solutions of quadratic equations because they felt that these solutions had no physical meaning. Greek mathematicians always tried to interpret their algebraic problems from a geometrical viewpoint and hence the development of the geometric method of "completing the square".

Consider the following equation: $x^2 + 6x = 7$. A geometric figure is constructed having areas x^2, $3x$ and $3x$.

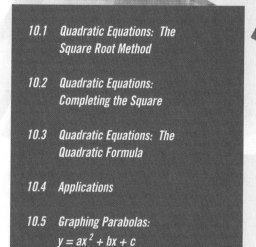

Note that to make the figure a square, one must add a 3 by 3 section (area = 9). Thus, 9 must be added to both sides of the equation to restore equality. Therefore,

$$x^2 + 6x = 7 \qquad \text{Original equation}$$
$$x^2 + 6x + 9 = 7 + 9 \qquad \text{Adding 9 to complete the square}$$
$$x^2 + 6x + 9 = 16$$

So, the square with side $x + 3$ now has an area of 16 square units. Therefore, the sides must be of length 4 units, which means $x + 3 = 4$. Hence, $x = 1$.

Note that actually the solution set of the original equation is $\{1, -7\}$, since $(-7)^2 + 6(-7) = 49 - 42 = 7$. Thus, the Greek mathematicians "lost" the negative solution because of their strictly geometric interpretation of quadratic equations. There were, therefore, many quadratic equations that the Greek mathematicians could not solve because both solutions were negative numbers or nonreal (complex) numbers. Negative solutions to equations were almost completely ignored until the early 1500's, during the Renaissance.

10

10.1 **Quadratic Equations: The Square Root Method**

10.2 **Quadratic Equations: Completing the Square**

10.3 **Quadratic Equations: The Quadratic Formula**

10.4 **Applications**

10.5 **Graphing Parabolas:** $y = ax^2 + bx + c$

"What is the difference between method and device? A method is a device you can use twice."

George Pòlya in *How to Solve It*

Quadratic equations were discussed in Chapter 7, and the solutions were found by factoring. However, in many "real-life" situations, factoring is simply not practical because the coefficients do not "cooperate" and the solutions to the equations involve square roots that yield irrational numbers or nonreal complex numbers.

Three techniques, other than factoring, for solving quadratic equations are presented in this chapter. The third technique, the **quadratic formula**, is essentially a combination of the first two in formula form. You will study and use this formula in almost every mathematics course you ever take after this one.

10.1   Quadratic Equations: The Square Root Method

Objectives

After completing this section, you will be able to:

1. Solve quadratic equations using the definition of square root.

2. Solve applications related to right triangles and the Pythagorean Theorem.

Review of Solving Quadratic Equations by Factoring

Factoring polynomials and solving quadratic equations by factoring were discussed in Chapter 7. While this chapter deals with three new techniques for solving quadratic equations, the method of solving by factoring is so basic and important that an example is given here for emphasis and review. (**Note**: Now would be a good time to review all of Chapter 7.)

Example 1: Factoring Method ● ● ● ● ● ● ● ● ● ● ● ● ● ● ● ● ● ●

Solve the following quadratic equations by factoring.

a. $x^2 + 7x = 18$

Solution:

$$x^2 + 7x = 18 \qquad \text{Original equation.}$$

$$x^2 + 7x - 18 = 0 \qquad \text{Standard form with 0 on one side.}$$

$$(x+9)(x-2) = 0 \qquad \text{Factor.}$$

$$x + 9 = 0 \quad \text{or} \quad x - 2 = 0 \qquad \text{Set each factor equal to 0.}$$

$$x = -9 \qquad\qquad x = 2 \qquad \text{Solve each linear equation.}$$

The solutions are −9 and 2.   *Continued on next page ...*

b. $2x^2 - x = 3$

Solution:		
	$2x^2 - x = 3$	Original equation.
	$2x^2 - x - 3 = 0$	Standard form with 0 on one side.
	$(2x - 3)(x + 1) = 0$	Factor.
	$2x - 3 = 0$ or $x + 1 = 0$	Set each factor equal to 0.
	$x = \dfrac{3}{2} \qquad x = -1$	Solve each linear equation.

The solutions are $\dfrac{3}{2}$ and -1.

● ●

Solving Quadratic Equations by Using the Square Root Method

Solving quadratic equations by factoring is a very good technique, and we want to use it much of the time. However, there are quadratic expressions that are not easily factored and some that are not factorable at all using real numbers. For example, consider solving the equation $x^2 - 5$ by factoring. As the following steps show, the factors involve square roots that are not integers.

$x^2 = 5$	
$x^2 - 5 = 0$	Get 0 on one side.
$x^2 - \left(\sqrt{5}\right)^2 = 0$	Difference of two squares with $5 = \left(\sqrt{5}\right)^2$.
$\left(x + \sqrt{5}\right)\left(x - \sqrt{5}\right) = 0$	Factor.
$x + \sqrt{5} = 0$ or $x - \sqrt{5} = 0$	Set each factor equal to 0.
$x = -\sqrt{5} \qquad x = \sqrt{5}$	There are two irrational solutions.

A simpler and more direct way to solve this equation is by **taking square roots of both sides**, as shown in the following statement.

Square Root Method

For a quadratic equation in the form $x^2 = c$ where c is nonnegative,

$$x = \sqrt{c} \ \text{ or } \ x = -\sqrt{c}.$$

This can be written as $x = \pm\sqrt{c}.$

Example 2: The Square Root Method • • • • • • • • • • • • • •

Solve the following quadratic equations by using the square root method. Write the radicals in simplest form.

a. $3x^2 = 51$

    Solution: $3x^2 = 51$ Divide both sides by 3 so that the coefficient of x^2 is 1.

                 $x^2 = 17$

                 $x = \pm\sqrt{17}$ Keep in mind that the expression $x = \pm\sqrt{17}$ represents the two equations $x = \sqrt{17}$ and $x = -\sqrt{17}$.

b. $(x+4)^2 = 21$

    Solution: $(x+4)^2 = 21$

                 $x+4 = \pm\sqrt{21}$

                 $x = -4 \pm\sqrt{21}$ There are two solutions: $-4+\sqrt{21}$ and $-4-\sqrt{21}$.

c. $(x-2)^2 = 50$

    Solution: $(x-2)^2 = 50$

                 $x-2 = \pm\sqrt{50}$ Simplify the radical $\left(\sqrt{50} = \sqrt{25}\sqrt{2} = 5\sqrt{2}\right)$.

                 $x = 2 \pm 5\sqrt{2}$ There are two solutions: $2+5\sqrt{2}$ and $2-5\sqrt{2}$.

d. $(2x+4)^2 = 72$

    Solution: $(2x+4)^2 = 72$

                 $2x+4 = \pm\sqrt{72}$

                 $2x = -4 \pm 6\sqrt{2}$ Simplify the radical $\left(\sqrt{72} = \sqrt{36}\sqrt{2} = 6\sqrt{2}\right)$.

                 $x = \dfrac{-4 \pm 6\sqrt{2}}{2}$

                 $x = \dfrac{\cancel{2}\left(-2 \pm 3\sqrt{2}\right)}{\cancel{2}}$ Factor and simplify.

                 $x = -2 \pm 3\sqrt{2}$ There are two solutions: $-2+3\sqrt{2}$ and $-2-3\sqrt{2}$.

e. $(x+7)^2 + 10 = 8$

    Solution: $(x+7)^2 + 10 = 8$

                 $(x+7)^2 = -2$

    There is no real solution. The square of a real number cannot be negative.

• •

The Pythagorean Theorem

The Pythagorean Theorem was discussed in Section 9.7 when the formula for the distance between two points was developed. In this chapter, the Pythagorean Theorem is used in more general situations and applications to find the lengths of various sides of right triangles. The theorem is stated again for easy reference and to emphasize its importance.

Remember, a **right triangle** is a triangle in which one angle is a right angle (measures 90°). Two of the sides are perpendicular and are called **legs**. The longest side is called the **hypotenuse** and is opposite the 90° angle.

The Pythagorean Theorem

In a right triangle, the square of the hypotenuse is equal to the sum of the squares of the two legs.

$$c^2 = a^2 + b^2$$

Example 3: The Pythagorean Theorem • • • • • • • • • • • •

a. If the hypotenuse of a right triangle is 15 cm long and one leg is 10 cm long, what is the length of the other leg?

Solution: Using the Pythagorean Theorem,

$$a^2 + 10^2 = 15^2$$
$$a^2 + 100 = 225$$
$$a^2 = 125$$
$$a = \sqrt{125} = \sqrt{25} \cdot \sqrt{5} = 5\sqrt{5} \ (\approx 11.18 \, \text{cm})$$

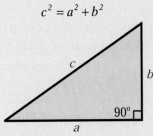

The other leg is $5\sqrt{5}$ cm long (or approximately 11.18 cm long).

(The negative solution to the quadratic equation is not considered because length is not negative.)

Continued on next page ...

b. The diagonal of a rectangle is twice the width. The length of the rectangle is 6 feet. Find the width of the rectangle.

Solution: Let $x =$ width of the rectangle
and $2x =$ length of the diagonal.

Then, by using the Pythagorean Theorem, we have

$$(2x)^2 = x^2 + 6^2$$
$$4x^2 = x^2 + 36$$
$$3x^2 = 36$$
$$x^2 = 12$$
$$x = \sqrt{12} = \sqrt{4} \cdot \sqrt{3} = 2\sqrt{3} \ (\approx 3.46)$$

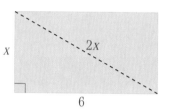

The width of the rectangle is $2\sqrt{3}$ feet or approximately 3.46 feet. (The negative solution to the quadratic equation is not considered because length is not negative.)

c. Use the Pythagorean Theorem to decide if a triangle with sides of length 8 in., 12 in., and 14 in. is a right triangle.

Solution: If the triangle is a right triangle, then the sum of the squares of the two shorter sides must equal the square of the longest side.

$$8^2 + 12^2 = 64 + 144 = 208$$
$$14^2 = 196$$

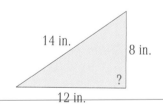

So, $8^2 + 12^2 \neq 14^2$, and the triangle is **not** a right triangle.

●●●●●●●●●●●●●●●●●●●●●●●●●●●●●●●●●●●

10.1 Exercises

Solve the quadratic equations in Exercises 1 – 10 by factoring.

1. $x^2 = 11x$ **2.** $x^2 - 10x + 16 = 0$ **3.** $x^2 = -15x - 36$

4. $2x^2 + 36x + 34 = 0$ **5.** $9x^2 + 6x - 15 = 0$ **6.** $5x^2 + 17x = -6$

7. $(x+3)(x-1) = 4x$ **8.** $(x-7)(x-2) = 6$ **9.** $(2x-3)(2x+1) = 3x - 6$

10. $(x-2)(5x+4) = 3x^2 - 15x - 12$

Solve the quadratic equations in Exercises 11 – 40 by using the square root method. Write the radicals in simplest form.

11. $x^2 = 121$ **12.** $x^2 = 81$ **13.** $3x^2 = 108$

14. $5x^2 = 245$ **15.** $x^2 = 35$ **16.** $x^2 = 42$

17. $x^2 - 62 = 0$ **18.** $x^2 - 75 = 0$ **19.** $x^2 - 45 = 0$

20. $x^2 - 98 = 0$ **21.** $3x^2 = 54$ **22.** $5x^2 = 60$

23. $9x^2 = 4$ **24.** $4x^2 = 25$ **25.** $(x-1)^2 = 4$

26. $(x+3)^2 = 9$ **27.** $(x+2)^2 = -25$ **28.** $(x-5)^2 = 36$

29. $(x+1)^2 = \dfrac{1}{4}$ **30.** $(x-9)^2 = -\dfrac{9}{25}$ **31.** $(x-3)^2 = \dfrac{4}{9}$

32. $(x-2)^2 = \dfrac{1}{16}$ **33.** $(x-6)^2 = 18$ **34.** $(x+8)^2 = 75$

35. $2(x-7)^2 = 24$ **36.** $3(x+11)^2 = 60$ **37.** $(3x+4)^2 = 27$

38. $(2x+1)^2 = 48$ **39.** $(5x-2)^2 = 63$ **40.** $(4x-3)^2 = 125$

Use the Pythagorean Theorem to decide if each of the triangles determined by the three sides given in Exercises 41 – 44 is a right triangle.

41. 8 cm, 15 cm, 17 cm   **42.** 4 cm, 5 cm, 6 cm

43. $\sqrt{2}$ in., $\sqrt{2}$ in., 2 in.   **44.** 3 ft., 6 ft., $3\sqrt{5}$ ft.

In each of the Exercises 45 – 48, right triangles are shown with two sides given. Determine the length of the missing side.

45. $a = 9, b = 12, c = ?$

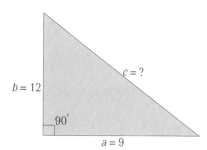

46. $a = 10, b = 24, c = ?$

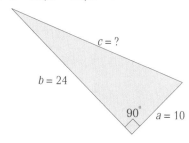

47. $a = 6, c = 12, b = ?$

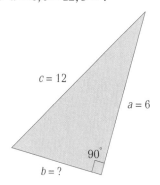

48. $b = 10, c = 30, a = ?$

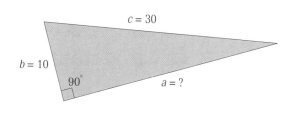

49. The hypotenuse of a right triangle is twice the length of one of the legs. The length of the other leg is $4\sqrt{3}$ feet. Find the length of the leg and the hypotenuse.

50. One leg of a right triangle is three times the length of the other. The length of the hypotenuse is 20 cm. Find the lengths of the legs.

51. The two legs of a right triangle are the same length. The hypotenuse is 6 centimeters long. Find the length of the legs.

52. The two legs of a right triangle are the same length. The hypotenuse is $4\sqrt{2}$ meters long. Find the length of the legs.

53. The top of a telephone pole is 40 feet above the ground and a guy wire is to be stretched from the top of the pole to a point on the ground 10 feet from the base of the pole. How long (to the nearest tenth of a foot) should the wire be if, for connecting purposes, 6 feet of wire is to be added after the distance is calculated?

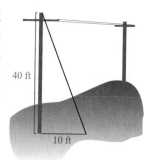

40 ft

10 ft

54. The library is located "around the corner" from the bank. If the library is 100 yards from the street corner and the bank is 75 yards around the corner on another street, what is the distance between the library and the bank "as the crow flies?"

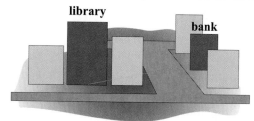

library

bank

55. A ball is dropped from the top of a building that is known to be 144 feet high. The formula for finding the height of the ball at any time is $h = 144 - 16t^2$ where t is measured in seconds. How many seconds will it take for the ball to hit the ground?

56. An oil spill from a ruptured pipe line is circular in shape and covers an area of about 8 square miles. About how many miles long (to the nearest tenth of a mile) is the diameter of the oil spill? (Use $\pi = 3.14$.)

Calculator Problems

Use a calculator to solve the equations in Exercises 57 – 64. Round off your answers to the nearest hundredth.

57. $x^2 = 647$

58. $x^2 = 378$

59. $19x^2 = 523$

60. $14x^2 = 795$

61. $6x^2 = 17.32$

62. $15x^2 = 229.63$

63. $2.1x^2 = 35.82$

64. $4.7x^2 = 118.34$

Hawkes Learning Systems: Introductory Algebra

SQE: The Square Root Method

10.2 Quadratic Equations: Completing the Square

Objectives

After completing this section, you will be able to:

1. Determine the constant terms that will make incomplete trinomials perfect square trinomials.

2. Solve quadratic equations by completing the square.

Completing the Square

In Section 7.4, we discussed factoring perfect square trinomials and how to add terms to binomials so that the resulting trinomial will be a perfect square. The procedure is called **completing the square**. The discussion here is how to solve quadratic equations by completing the square.

As review, we complete the square for $x^2 + 12x$ and $a^2 - 20a$. Both binomials have a leading coefficient of 1.

$$x^2 + 12x + \underline{\hspace{1cm}} = (\hspace{1cm})^2$$

The number 36 will complete the square since

$$\frac{1}{2}(12) = 6 \text{ and } 6^2 = 36.$$

So,

$$x^2 + 12x + \underline{36} = (x + 6)^2$$

Similarly, for

$$a^2 - 20a + \underline{\hspace{1cm}} = (\hspace{1cm})^2$$

we have

$$\frac{1}{2}(-20) = -10 \text{ and } (-10)^2 = 100$$

So,

$$a^2 - 20a + \underline{100} = (a - 10)^2$$

If the leading coefficient is **not 1**, we proceed to factor out the leading coefficient as follows:

$$2x^2 + 20x + \underline{\qquad} = 2(\qquad)^2$$

$$2\left(x^2 + 10x + \underline{\qquad}\right) = 2(\qquad)^2 \quad \text{Factor out 2.}$$

$$2\left(x^2 + 10x + \underline{25}\right) = 2(x+5)^2 \quad \text{Complete the square of the expression inside the parentheses.}$$

$$2x^2 + 20x + 50 = 2(x+5)^2 \quad \text{The number to be added to the original expression is } 2 \cdot 25 = 50.$$

Solving Quadratic Equations by Completing the Square

The following procedure is used in the examples to solve quadratic equations by completing the square.

To Solve Quadratic Equations by Completing the Square

1. Arrange terms with variables on one side and constants on the other.

2. Divide each term by the coefficient of x^2. (We want the leading coefficient to be 1.)

*3. Find the number that completes the square of the quadratic expression and add this number to **both** sides of the equation.*

4. Find the positive and negative square roots of both sides.

5. Solve for x. Remember, usually there will be two solutions.

Example 1: Completing the Square

Solve the following quadratic equations by completing the squares.

a. $x^2 - 6x + 4 = 0$

Solution:

$$x^2 - 6x + 4 = 0$$

$$x^2 - 6x = -4 \quad \text{Add } -4 \text{ to both sides of the equation.}$$

$$x^2 - 6x + 9 = -4 + 9 \quad \text{Add 9 to both sides of the equation. The left side is now a } \textbf{perfect square trinomial.}$$

$$\frac{1}{2}(-6) = -3 \text{ and } (-3)^2 = 9$$

$$(x-3)^2 = 5 \quad \text{Simplify.}$$

$$x - 3 = \pm\sqrt{5} \quad \text{Find square roots of both sides.}$$

$$x = 3 \pm \sqrt{5} \quad \text{Solve for } x.$$

Continued on next page ...

b. $x^2 + 5x = 7$

   Solution: $x^2 + 5x = 7$

$$x^2 + 5x + \frac{25}{4} = 7 + \frac{25}{4}$$ Complete the square on the left: $\frac{1}{2} \cdot 5 = \frac{5}{2}$

$$\text{and} \left(\frac{5}{2}\right)^2 = \frac{25}{4}.$$

$$\left(x + \frac{5}{2}\right)^2 = \frac{53}{4}$$ Simplify: $\left(7 + \frac{25}{4} = \frac{28}{4} + \frac{25}{4} = \frac{53}{4}\right).$

$$x + \frac{5}{2} = \pm\sqrt{\frac{53}{4}}$$ Find square roots.

$$x = -\frac{5}{2} \pm \sqrt{\frac{53}{4}}$$ Solve for x.

$$x = -\frac{5}{2} \pm \frac{\sqrt{53}}{2}$$ Special property of square roots: $\sqrt{\frac{a}{b}} = \frac{\sqrt{a}}{\sqrt{b}}$

for $a > 0$ and $b > 0$.

$$\text{or } x = \frac{-5 \pm \sqrt{53}}{2}$$ Combine the fractions.

c. $6x^2 + 12x - 9 = 0$

   Solution: $6x^2 + 12x - 9 = 0$

$$6x^2 + 12x = 9$$ Add 9 to both sides of the equation.

$$\frac{6x^2}{6} + \frac{12x}{6} = \frac{9}{6}$$ Divide each term by 6 **so that the leading coefficient will be 1.**

$$x^2 + 2x = \frac{3}{2}$$ The leading coefficient is 1.

$$x^2 + 2x + 1 = \frac{3}{2} + 1$$ Complete the square: $\frac{1}{2} \cdot 2 = 1$ and $1^2 = 1.$

$$(x + 1)^2 = \frac{5}{2}$$ Simplify.

$$x + 1 = \pm\sqrt{\frac{5}{2}}$$ Find square roots.

$$x = -1 \pm \sqrt{\frac{5}{2}}$$ Solve for x.

$$x = -1 \pm \frac{\sqrt{5}}{\sqrt{2}} \cdot \frac{\sqrt{2}}{\sqrt{2}}$$ Rationalize the denominator.

Continued on next page ...

$$x = -1 \pm \frac{\sqrt{10}}{2} \qquad \text{Simplify.}$$

$$\text{or } x = \frac{-2 \pm \sqrt{10}}{2} \qquad \text{Combine the fractions with a common}$$

denominator.

d. $2x^2 + 5x - 8 = 0$

Solution: $2x^2 + 5x - 8 = 0$

$$2x^2 + 5x = 8 \qquad \text{Add 8 to both sides of the equation.}$$

$$\frac{2x^2}{2} + \frac{5x}{2} = \frac{8}{2} \qquad \text{Divide each term by 2 \textbf{so that the leading}}$$
coefficient will be 1.

$$x^2 + \frac{5}{2}x = 4 \qquad \text{Simplify.}$$

$$x^2 + \frac{5}{2}x + \frac{25}{16} = 4 + \frac{25}{16} \qquad \text{Complete the square: } \frac{1}{2} \cdot \frac{5}{2} = \frac{5}{4} \text{ and } \left(\frac{5}{4}\right)^2 = \frac{25}{16}.$$

$$\left(x + \frac{5}{4}\right)^2 = \frac{89}{16} \qquad \text{Simplify } \left(4 + \frac{25}{16} = \frac{64}{16} + \frac{25}{16} = \frac{89}{16}\right).$$

$$x + \frac{5}{4} = \pm\sqrt{\frac{89}{16}} \qquad \text{Find the square roots.}$$

$$x = -\frac{5}{4} \pm \frac{\sqrt{89}}{4} \qquad \text{Solve for } x. \text{ Also, } \sqrt{\frac{89}{16}} = \frac{\sqrt{89}}{\sqrt{16}} = \frac{\sqrt{89}}{4}.$$

$$\text{or } x = \frac{-5 \pm \sqrt{89}}{4} \qquad \text{Combine the fractions.}$$

e. $x^2 + 8x = -4$

Solution: $x^2 + 8x = -4$

$$x^2 + 8x + 16 = -4 + 16 \qquad \text{Complete the square.}$$

$$(x + 4)^2 = 12 \qquad \text{Simplify.}$$

$$x + 4 = \pm\sqrt{12} \qquad \text{Find square roots.}$$

$$x = -4 \pm \sqrt{12} \qquad \text{Solve for } x.$$

$$x = -4 \pm 2\sqrt{3} \qquad \sqrt{12} = \sqrt{4} \cdot \sqrt{3} = 2\sqrt{3} \text{ by special property of}$$

radicals $\sqrt{ab} = \sqrt{a} \cdot \sqrt{b}$ for $a > 0$ and $b > 0$.

● ●

Solving quadratic equations by completing the square is a good technique and it provides practice with several algebraic operations. However, do not forget that, in general, factoring is the preferred technique whenever the factors are relatively easy to find. Some of the following exercises allow the options of solving by factoring or by completing the square.

10.2 Exercises

Add the correct constants in Exercises 1 – 10 and factor the resulting trinomial as indicated. (Constant factors of the x^2 term must be dealt with in Exercises 3, 4, 9, and 10.)

1. $x^2 + 12x + \underline{} = ()^2$

2. $x^2 - 14x + \underline{} = ()^2$

3. $2x^2 - 16x + \underline{} = 2()^2$

4. $5x^2 + 10x + \underline{} = 5()^2$

5. $x^2 - 3x + \underline{} = ()^2$

6. $x^2 + 5x + \underline{} = ()^2$

7. $x^2 + x + \underline{} = ()^2$

8. $x^2 - 7x + \underline{} = ()^2$

9. $2x^2 + 4x + \underline{} = 2()^2$

10. $3x^2 + 18x + \underline{} = 3()^2$

Solve each quadratic equation in Exercises 11 – 30 by completing the square.

11. $x^2 + 6x - 7 = 0$

12. $x^2 + 8x + 12 = 0$

13. $x^2 - 4x - 45 = 0$

14. $x^2 - 10x + 21 = 0$

15. $x^2 - 3x - 40 = 0$

16. $x^2 + x - 42 = 0$

17. $3x^2 + x - 4 = 0$

18. $2x^2 + x - 6 = 0$

19. $4x^2 - 4x - 3 = 0$

20. $3x^2 + 11x + 10 = 0$

21. $x^2 + 6x + 3 = 0$

22. $x^2 - 4x - 3 = 0$

23. $x^2 + 2x - 5 = 0$

24. $x^2 + 8x + 2 = 0$

25. $3x^2 - 2x - 1 = 0$

26. $2x^2 + 5x + 2 = 0$

27. $x^2 + x - 3 = 0$

28. $x^2 - 5x + 1 = 0$

29. $2x^2 + 3x - 1 = 0$

30. $3x^2 - 4x - 2 = 0$

Solve the quadratic equations in Exercises 31 – 50 by factoring or completing the square.

31. $x^2 - 9x + 2 = 0$

32. $x^2 - 8x - 20 = 0$

33. $x^2 + 7x - 14 = 0$

34. $x^2 + 5x + 3 = 0$

35. $x^2 - 11x - 26 = 0$

36. $x^2 + 6x - 4 = 0$

37. $2x^2 + x - 2 = 0$ **38.** $6x^2 - 2x - 2 = 0$ **39.** $3x^2 - 6x + 3 = 0$

40. $5x^2 + 15x - 5 = 0$ **41.** $4x^2 + 20x - 8 = 0$ **42.** $2x^2 - 7x + 4 = 0$

43. $6x^2 - 8x + 1 = 0$ **44.** $5x^2 - 10x + 3 = 0$ **45.** $2x^2 + 7x + 4 = 0$

46. $3x^2 - 5x - 3 = 0$ **47.** $4x^2 - 2x - 3 = 0$ **48.** $2x^2 - 9x + 7 = 0$

49. $3x^2 + 8x + 5 = 0$ **50.** $5x^2 + 11x - 1 = 0$

Hawkes Learning Systems: Introductory Algebra

 SQE: Completing the Square

10.3   Quadratic Equations: The Quadratic Formula

Objectives

After completing this section, you will be able to:

1. Write quadratic equations in standard form.

2. Identify the coefficients of quadratic equations in standard form.

3. Solve quadratic equations by using the quadratic formula.

The Quadratic Formula: $x = \dfrac{-b \pm \sqrt{b^2 - 4ac}}{2a}$

Now we are interested in developing a formula that will be useful in solving quadratic equations of any form. **This formula will always work**, but do not forget the factoring and completing the square techniques because in some cases they are easier to apply than the formula.

General Form of a Quadratic Equation

*The **general quadratic equation** is*

$$ax^2 + bx + c = 0$$

where a, b, and c are real constants and a ≠ 0.

We want to solve the general quadratic equation for x in terms of the constant coefficients $a, b,$ and c. The technique is to **complete the square** (Section 10.2).

Development of the Quadratic Formula

$ax^2 + bx + c = 0$	Begin with the general quadratic equation.
$ax^2 + bx = -c$	Add $-c$ to both sides.
$\dfrac{ax^2}{a} + \dfrac{bx}{a} = \dfrac{-c}{a}$	Divide each term by a so that the leading coefficient will be 1.
$x^2 + \dfrac{b}{a}x = \dfrac{-c}{a}$	Rewrite: $\dfrac{bx}{a} = \dfrac{b}{a}x.$
$x^2 + \dfrac{b}{a}x + \left(\dfrac{b}{2a}\right)^2 = \left(\dfrac{b}{2a}\right)^2 + \dfrac{-c}{a}$	Complete the square: $\dfrac{1}{2}\left(\dfrac{b}{a}\right) = \dfrac{b}{2a}.$

$$\left(x + \frac{b}{2a}\right)^2 = \frac{b^2}{4a^2} + \frac{-c}{a} \qquad \text{Simplify.}$$

$$\left(x + \frac{b}{2a}\right)^2 = \frac{b^2}{4a^2} + \frac{-c \cdot 4a}{a \cdot 4a} \qquad \text{Common denominator is } 4a^2.$$

$$\left(x + \frac{b}{2a}\right)^2 = \frac{b^2 - 4ac}{4a^2} \qquad \text{Simplify.}$$

$$x + \frac{b}{2a} = \pm\sqrt{\frac{b^2 - 4ac}{4a^2}} \qquad \text{Find the square roots of both sides.}$$

$$x + \frac{b}{2a} = \pm\frac{\sqrt{b^2 - 4ac}}{\sqrt{4a^2}} \qquad \text{Use the relationship } \sqrt{\frac{a}{b}} = \frac{\sqrt{a}}{\sqrt{b}} \text{ if } a, b > 0.$$

$$x + \frac{b}{2a} = \pm\frac{\sqrt{b^2 - 4ac}}{2a} \qquad \text{Simplify.}$$

$$x = \frac{-b}{2a} \pm \frac{\sqrt{b^2 - 4ac}}{2a} \qquad \text{Solve for } x.$$

$$x = \frac{-b \pm \sqrt{b^2 - 4ac}}{2a} \qquad \text{THE QUADRATIC FORMULA}$$

Or, we can write the two solutions separately as:

$$x = \frac{-b + \sqrt{b^2 - 4ac}}{2a} \quad \text{and} \quad x = \frac{-b - \sqrt{b^2 - 4ac}}{2a}$$

Quadratic Formula

The solutions of the general quadratic equation $ax^2 + bx + c = 0$, $a \neq 0$, *are*

$$x = \frac{-b \pm \sqrt{b^2 - 4ac}}{2a}$$

NOTES

Special Note:
The expression $b^2 - 4ac$ is called the **discriminant**.
If

$b^2 - 4ac = 0 \rightarrow$ there is only one real solution.
$b^2 - 4ac > 0 \rightarrow$ there are two real solutions.
$b^2 - 4ac < 0 \rightarrow$ there are no real solutions. (The square root of a negative number is not a real number.)

A discussion of negative discriminants is given in later courses in algebra.

Solving Quadratic Equations by Using the Quadratic Formula

Now the solutions to quadratic equations can be found by going directly to the formula.

Example 1: The Quadratic Formula ● ● ● ● ● ● ● ● ● ● ● ● ● ●

Solve the following quadratic equations using the quadratic formula:

$$x = \frac{-b \pm \sqrt{b^2 - 4ac}}{2a}$$

a. $2x^2 + x - 2 = 0$

Solution: $a = 2, b = 1, c = -2$

$$x = \frac{-1 \pm \sqrt{1^2 - 4(2)(-2)}}{2 \cdot 2} = \frac{-1 \pm \sqrt{1 + 16}}{4} = \frac{-1 \pm \sqrt{17}}{4}$$

In many practical applications of quadratic equations, we want to know a decimal approximation of the solutions. Using a calculator, we find the following approximate values to the solutions of the equation $2x^2 + x - 2 = 0$:

$$x = \frac{-1 + \sqrt{17}}{4} \approx 0.78078 \quad \text{and} \quad x = \frac{-1 - \sqrt{17}}{4} \approx -1.2808$$

b. $-3x^2 = -5x + 1$

Solution: $-3x^2 = -5x + 1$ First rewrite the equation so that one side is 0.

$-3x^2 + 5x - 1 = 0$

Now we see that $a = -3, b = +5$, and $c = -1$.

Substituting in the quadratic formula gives

$$x = \frac{-5 \pm \sqrt{5^2 - 4(-3)(-1)}}{2(-3)} = \frac{-5 \pm \sqrt{25 - 12}}{-6} = \frac{-5 \pm \sqrt{13}}{-6} = \frac{5 \pm \sqrt{13}}{6}$$

This shows that the quadratic formula works correctly even though the leading coefficient a is negative. We could also multiply all the terms on both sides of the equation by -1 and solve the new equation. The solutions will be the same.

$-3x^2 + 5x - 1 = 0$

$3x^2 - 5 + 1 = 0$ Multiply every term by -1.

Continued on next page ...

Now, $a = 3, b = -5$, and $c = 1$.

$$x = \frac{-(-5) \pm \sqrt{(-5)^2 - 4(3)(1)}}{2 \cdot 3} = \frac{5 \pm \sqrt{25 - 12}}{6} = \frac{5 \pm \sqrt{13}}{6}$$

Again, if needed, a calculator will give approximate values for the solutions:

$$x = \frac{5 + \sqrt{13}}{6} \approx 1.4343 \quad \text{and} \quad x = \frac{5 - \sqrt{13}}{6} \approx 0.2324$$

c. $\dfrac{1}{6}x^2 - x + \dfrac{1}{2} = 0$

Solution: $6 \cdot \dfrac{1}{6}x^2 - 6 \cdot x + 6 \cdot \dfrac{1}{2} = 6 \cdot 0$ Multiply each term by 6, the least common denominator.

$$x^2 - 6x + 3 = 0$$ Integer coefficients are much easier to use in the formula.

So $a = 1, b = -6$, and $c = 3$

$$x = \frac{-(-6) \pm \sqrt{(-6)^2 - 4(1)(3)}}{2 \cdot 1} = \frac{6 \pm \sqrt{36 - 12}}{2}$$

$$= \frac{6 \pm \sqrt{24}}{2} = \frac{6 \pm 2\sqrt{6}}{2} = \frac{2\left(3 \pm \sqrt{6}\right)}{2}$$

$$= 3 \pm \sqrt{6}$$

d. $2x^2 - 25 = 0$

Solution: The square root method could be applied by adding 25 to both sides, dividing by 2, and then taking square roots. The result is the same by using the quadratic formula with $b = 0$.

$$2x^2 - 25 = 0 \ \left(\text{or } 2x^2 + 0x - 25 = 0 \right)$$

So, $a = 2, b = 0$, and $c = -25$

$$x = \frac{-(0) \pm \sqrt{0^2 - 4(2)(-25)}}{2 \cdot 2} = \frac{\pm\sqrt{200}}{4} = \frac{\pm 10\sqrt{2}}{4} = \frac{\pm 5\sqrt{2}}{2}$$

e. $(3x - 1)(x + 2) = 4x$

Solution: $(3x - 1)(x + 2) = 4x$

$$3x^2 + 5x - 2 = 4x$$ Multiply on the left hand side.

$$3x^2 + x - 2 = 0$$ Subtract $4x$ from both sides so that one side is 0.

Continued on next page ...

Now we see that $a = 3, b = 1$, and $c = -2$.

Substituting in the quadratic formula gives

$$x = \frac{-1 \pm \sqrt{1^2 - 4(3)(-2)}}{2 \cdot 3} = \frac{-1 \pm \sqrt{25}}{6} = \frac{-1 \pm 5}{6}$$

$$x = \frac{-1+5}{6} = \frac{4}{6} = \frac{2}{3} \quad \text{and} \quad x = \frac{-1-5}{6} = \frac{-6}{6} = -1$$

Special Note: Whenever the solutions are rational numbers, the equation can be solved by factoring. In this example, we could have solved as follows:

$$3x^2 + x - 2 = 0$$

$$(3x - 2)(x + 1) = 0$$

$$3x - 2 = 0 \text{ and } x + 1 = 0$$

$$x = \frac{2}{3} \qquad x = -1$$

f. $4x^2 + 12x + 9 = 0$

Solution: $a = 4, b = 12, c = 9$

$$x = \frac{-12 \pm \sqrt{12^2 - 4(4)(9)}}{2 \cdot 4}$$

$$= \frac{-12 \pm \sqrt{144 - 144}}{8} = \frac{-12 \pm \sqrt{0}}{8} = -\frac{12}{8} = -\frac{3}{2}$$

Note that when the discriminant is 0, there is only one solution. This equation could also be solved by factoring in the following manner:

$4x^2 + 12x + 9 = 0$ The quadratic expression is a **perfect square trinomial**.

$(2x + 3)^2 = 0$

$2x + 3 = 0$ The factor $2x+3$ is repeated.

$2x = -3$

$x = -\frac{3}{2}$ $-\frac{3}{2}$ is called a **double solution** or a **double root**.

Quadratic equations should be solved by factoring whenever possible. Factoring is generally the easiest method. This fact is illustrated in Examples 1e and 1f. However, the quadratic formula will always give the solutions whether they are rational, irrational, or nonreal (as shown in Example 2a). Thus, the quadratic formula is a very useful tool. In future courses in mathematics, every text and instructor will assume that you know or can easily remember the quadratic formula.

Example 2: Nonreal Solutions ● ● ● ● ● ● ● ● ● ● ● ● ● ● ● ●

Solve the following equation: $x^2 + x + 1 = 0$

Solution: $a = 1, \; b = 1, \; c = 1$

$$x = \frac{-1 \pm \sqrt{1^2 - 4(1)(1)}}{2 \cdot 1} = \frac{-1 \pm \sqrt{1-4}}{2} = \frac{-1 \pm \sqrt{-3}}{2}$$

There is no real solution. This example illustrates the fact that not every equation has real solutions. None of the exercises in this text has this kind of answer. Such solutions are called **nonreal complex numbers** and will be discussed in detail in the next course in algebra.

● ●

Practice Problems

Solve the equations using the quadratic formula.

1. $x^2 + 5x + 1 = 0$

2. $4x^2 - x = 1$

3. $x^2 - 4x - 4 = 0$

4. $-2x^2 + 3 = 0$ (**Note:** Here $b = 0$.)

5. $5x^2 - x = 0$ (**Note:** Here $c = 0$.)

Answers to Practice Problems: 1. $x = \dfrac{-5 \pm \sqrt{21}}{2}$ **2.** $x = \dfrac{1 \pm \sqrt{17}}{8}$ **3.** $x = 2 \pm 2\sqrt{2}$ **4.** $x = \dfrac{0 \pm 2\sqrt{6}}{-4}$ or $x = \dfrac{\pm \sqrt{6}}{2}$

5. $x = 0, \dfrac{1}{5}$ (This problem could be solved by factoring.)

10.3 Exercises

Rewrite each of the quadratic equations in Exercises 1 – 10 in the form $ax^2 + bx + c$ with $a > 0$; then identify the constants a, b, and c.

1. $x^2 - 3x = 2$ **2.** $x^2 + 2 = 5x$ **3.** $x = 2x^2 + 6$ **4.** $5x^2 = 3x - 1$

5. $4x + 3 = 7x^2$ **6.** $x = 4 - 3x^2$ **7.** $4 = 3x^2 - 9x$ **8.** $6x + 4 = 3x^2$

9. $x^2 + 5x = 3 - x^2$ **10.** $x^2 + 4x - 1 = 2x + 3x^2$

Solve the quadratic equations in Exercises 11 – 35 by using the quadratic formula

11. $x^2 - 4x - 1 = 0$ **12.** $x^2 - 3x + 1 = 0$ **13.** $x^2 - 3x = 4$ **14.** $x^2 + 5x = 2$

15. $-2x^2 + x = -1$ **16.** $-3x^2 + x = -1$ **17.** $5x^2 + 3x - 2 = 0$ **18.** $-2x^2 + 5x - 1 = 0$

19. $9x^2 = 3x$ **20.** $4x^2 - 81 = 0$ **21.** $x^2 - 7 = 0$ **22.** $2x^2 + 5x - 3 = 0$

23. $x^2 + 4x = x - 2x^2$ **24.** $x^2 - 2x + 1 = 2 - 3x^2$ **25.** $3x^2 + 4x = 0$

26. $4x^2 - 10 = 0$ **27.** $\dfrac{2}{5}x^2 + x - 1 = 0$ **28.** $2x^2 + 3x - \dfrac{3}{4} = 0$

29. $-\dfrac{x^2}{2} - 2x + \dfrac{1}{3} = 0$ **30.** $\dfrac{x^2}{3} - x - \dfrac{1}{5} = 0$ **31.** $(2x + 1)(x + 3) = 2x + 6$

32. $(x + 5)(x - 1) = -3$ **33.** $\dfrac{5x + 2}{3x} = x - 1$ **34.** $-6x^2 = -3x - 1$

35. $4x^2 = 7x - 3$

Solve the quadratic equations in Exercises 36 – 54 by using any method (factoring, completing the square, or the quadratic formula).

36. $2x^2 + 7x + 3 = 0$ **37.** $5x^2 - x - 4 = 0$ **38.** $9x^2 - 6x - 1 = 0$

39. $3x^2 - 7x + 1 = 0$ **40.** $2x^2 - 2x - 1 = 0$ **41.** $10x^2 = x + 24$

42. $9x^2 + 12x = -2$ **43.** $3x^2 - 11x = 4$ **44.** $5x^2 = 7x + 5$

45. $-4x^2 + 11x - 5 = 0$ **46.** $-4x^2 + 12x - 9 = 0$ **47.** $10x^2 + 35x + 30 = 0$

48. $6x^2 + 2x = 20$ **49.** $25x^2 + 4 = 20$ **50.** $3x^2 - 4x + \dfrac{1}{3} = 0$

51. $\dfrac{3}{4}x^2 - 2x + \dfrac{1}{8} = 0$ **52.** $\dfrac{11}{2}x + 1 = 3x^2$ **53.** $\dfrac{3}{7}x^2 = \dfrac{1}{2}x + 1$

54. $\dfrac{35}{4x} = x - 1$

Solve the quadratic equations in Exercises 55 – 59 and write the answers in decimal form accurate to four decimal places.

55. $x^2 - x - 1 = 0$ **56.** $3x^2 - 8x - 13 = 0$ **57.** $(2x + 1)(2x - 1) = 3x$

58. $\dfrac{-x-1}{3x} = x + 1$ **59.** $5x^2 - 100 = 0$

Hawkes Learning Systems: Introductory Algebra

 SQE: The Quadratic Formula

10.4   Applications

Objectives

After completing this section, you will be able to:

1. Solve applied problems using quadratic equations.

Application problems are designed to teach you to read carefully, to think clearly, and to translate from English to algebraic expressions and equations. The problems do not tell you directly to add, subtract, multiply, divide, or square. You must decide on a method of attack based on the wording of the problem and your previous experience and knowledge. Study the following examples and read the explanations carefully. They are similar to some, but not all, of the problems in the exercises. Example 1 makes use of the Pythagorean Theorem that was discussed in Sections 9.7 and 10.1.

Example 1: The Pythagorean Theorem ● ● ● ● ● ● ● ● ● ● ● ● ● ●

The length of a rectangle is 7 feet longer than the width. If one diagonal measures 13 feet, what are the dimensions of the rectangle?

Solution: Draw a diagram for problems involving geometric figures whenever possible.

Let $x =$ width of the rectangle

and $x + 7 =$ length of the rectangle.

Then, by the Pythagorean Theorem,

$$\left(x+7\right)^2 + x^2 = 13^2$$

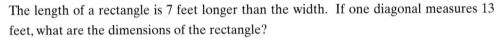

$$x^2 + 14x + 49 + x^2 = 169$$

$$2x^2 + 14x + 49 - 169 = 0$$

$$2x^2 + 14x - 120 = 0$$

$$2\left(x^2 + 7x - 60\right) = 0$$

$$2\left(x - 5\right)\left(x + 12\right) = 0$$

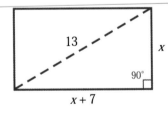

$$x - 5 = 0 \ \text{ or } \ x + 12 = 0$$

$$x = 5 \qquad \cancel{x = -12} \qquad \text{A negative number does not fit}$$

Thus, $x = 5$ and $x + 7 = 12$.   the conditions of the problem.

Continued on next page ...

Check: $5^2 + 12^2 \overset{?}{=} 13^2$

$25 + 144 \overset{?}{=} 169$

$169 = 169$

The width is 5 ft. and the length is 12 ft.

● ●

Example 2: Work ●

Working for a janitorial service, a woman and her daughter can clean a building in 5 hours. If the daughter were to do the job by herself, she would take 24 hours longer than her mother would take. How long would it take her mother to clean the building without the daughter's help?

(**Note:** For reference, this problem is similar to those discussed in Section 8.6 on work. The difference is that in this section the resulting equations are quadratic.)

Solution: Let x = time for mother alone
and $x + 24$ = time for daughter alone.

	Hours to Complete	Part Completed in 1 Hour
Mother	x	$\dfrac{1}{x}$
Daughter	$x + 24$	$\dfrac{1}{x+24}$
Together	5	$\dfrac{1}{5}$

part done by mother in 1 hour	+	part done by daughter in 1 hour	=	part done working together in 1 hour
$\dfrac{1}{x}$	+	$\dfrac{1}{x+24}$	=	$\dfrac{1}{5}$

$$\frac{1}{x}(5x)(x+24) + \frac{1}{x+24}(5x)(x+24) = \frac{1}{5}(5x)(x+24)$$

↑   ↑   ↑

Multiply each term by the LCM of the denominators.

Continued on next page ...

$$5(x+24)+5x = x(x+24)$$

$$5x+120+5x = x^2+24x$$

$$0 = x^2+24x-10x-120$$

$$0 = x^2+14x-120$$

$$0 = (x-6)(x+20)$$

$$x-6=0 \;\; \text{or} \;\; x+20=0$$

$$x=6 \qquad \cancel{x=-20}$$

The mother could do the job alone in 6 hours.

● ●

Example 3: Distance-Rate-Time

A small plane travels at a speed of 200 mph in still air. Flying with a tailwind, the plane is clocked over a distance of 960 miles. Flying against a headwind, it takes 2 hours more time to complete the return trip. What was the wind velocity?

Solution: The basic formula is $d = rt$ (distance = rate · time).

Also, $t = \dfrac{d}{r}$ and $r = \dfrac{d}{t}$.

Let x = wind velocity.
Then $200 + x$ = speed of airplane going with the wind (tailwind)
and $200 - x$ = speed of airplane returning against the wind
                        (headwind)
             960 = distance each way.

We know distance and can represent rate (or speed), so the formula

$t = \dfrac{d}{r}$ is used for representing time.

	Rate	Time	Distance
Going	$200 + x$	$\dfrac{960}{200+x}$	960
Returning	$200 - x$	$\dfrac{960}{200-x}$	960

Continued on next page ...

$$\underbrace{\frac{\text{time}}{\text{returning}}} \quad - \quad \underbrace{\frac{\text{time}}{\text{going}}} \quad = \quad \underbrace{\frac{\text{difference}}{\text{in time}}}$$

$$\frac{960}{200-x} \quad - \quad \frac{960}{200+x} \quad = \quad 2$$

$$\frac{960}{200-x}(200-x)(200+x)+\frac{960}{200+x}(200-x)(200+x)=2(200-x)(200+x)$$

$$960(200+x)-960(200-x)=2(40,000-x^2)$$

$$192,000+960x-192,000+960x=80,000-2x^2$$

$$2x^2+1920x-80,000=0$$

$$x^2+960x-40,000=0$$

$$(x-40)(x+1000)=0$$

$$x-40=0 \ \text{ or } \ x+1000=0$$

$$x=40 \qquad \quad \cancel{x=-1000}$$

The wind velocity was 40 mph.

Check: $\dfrac{960}{200-40}-\dfrac{960}{200+40}=2$

$$\frac{960}{160}-\frac{960}{240}=2$$

$$6-4=2$$

$$2=2$$

● ●

Example 4: Geometry ● ● ● ● ● ● ● ● ● ● ● ● ● ● ● ● ● ●

a. A square piece of cardboard has a small square, 2 in. by 2 in., cut from each corner. The edges are then folded up to form a box with a volume of 5000 cu. in. What are the dimensions of the box? (**Hint:** The volume is the product of the length, width, and height: $V=lwh$.)

Continued on next page ...

Solution: Draw a diagram illustrating the information.

Let x = one side of the square.

$$2(x-4)(x-4) = 5000$$

$$2(x^2 - 8x + 16) = 5000$$

$$x^2 - 8x + 16 = 2500$$

$$x^2 - 8x - 2484 = 0$$

$$(x - 54)(x + 46) = 0$$

$$x - 54 = 0 \quad \text{or} \quad x + 46 = 0$$

$$x = 54 \qquad \cancel{x = -46}$$

$$x - 4 = 50$$

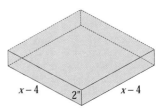

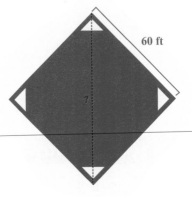

The dimensions of the box are 50 in. by 50 in. by 2 in.

b. A softball field is in the shape of a square with sides of 60 feet. What is the distance (to the nearest tenth of a foot) the catcher must throw from home plate to second base?

60 ft

Solution: Since the distance from home plate to second base is the hypotenuse of a right triangle with sides of length 60 feet, we can use the Pythagorean Theorem as follows:

Continued on next page ...

$$x^2 = 60^2 + 60^2$$

$$x^2 = 3600 + 3600$$

$$x^2 = 7200$$

$x = \sqrt{7200}$ We want only the positive square root.

$x = 84.9$ ft   Accurate to the nearest tenth of a foot.

The catcher must throw about 84.9 feet.

● ●

10.4 Exercises

Find the solutions to each of the following problems by setting up and solving a quadratic equation.

1. The sum of a positive number and its square is 132. Find the number.

2. The dimensions of a rectangle can be represented by two consecutive even integers. The area of the rectangle is 528 square centimeters. Find the width and the length of the rectangle. (**Hint**: area = length · width.)

3. A rectangle has a length 5 meters less than twice its width. If the area is 63 square meters, find the dimensions of the rectangle. (**Hint**: area = length · width.)

4. The sum of a positive number and its square is 992. Find the number.

5. The area of a rectangular field is 198 square meters. If it takes 58 meters of fencing to enclose the field, what are the dimensions of the field? (**Hint**: The length plus the width is 29 meters.)

6. The length of a rectangle exceeds the width by 5 cm. If both were increased by 3 cm, the area would be increased by 96 cm^2. Find the dimensions of the original rectangle.
[**Hint**: The original rectangle has area $w(w + 5)$.]

7. The Wilsons have a rectangular swimming pool that is 10 ft. longer than it is wide. The pool is completely surrounded by a concrete deck that is 6 ft. wide. The total area of the pool and the deck is 1344 ft^2. Find the dimensions of the pool.

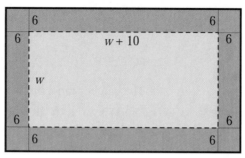

8. The product of two positive consecutive odd integers exceeds their sum by 287. Find the positive integers.

9. The sum of the squares of two positive consecutive integers is 221. Find the positive integers.

10. The difference between two positive numbers is 9. If the smaller number is added to the square of the larger number, the result is 147. Find the numbers.

11. The difference between a positive number and 3 is four times the reciprocal of the number. Find the number. (**Hint:** The reciprocal of x is $\dfrac{1}{x}$.)

12. The sum of a positive number and 5 is fourteen times the reciprocal of the number. Find the number. (**Hint:** The reciprocal of x is $\dfrac{1}{x}$.)

13. Each side of a square is increased by 10 cm. The area of the resulting square is 9 times the area of the original square. Find the length of the sides of the original square.

14. If 5 meters are added to each side of a square, the area of the resulting square is four times the area of the original square. Find the length of the sides of the original square.

15. The diagonal of a rectangle is 13 meters. The length is 2 meters more than twice the width. Find the dimensions of the rectangle.

16. The length of a rectangle is 4 meters more than its width. If the diagonal is 20 meters, what are the dimensions of the rectangle?

17. A right triangle has two equal sides. If the hypotenuse is 12 cm, what is the length of the equal sides?

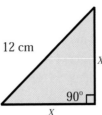

12 cm

X

90°

X

18. Mr. Prince owns a 15 unit apartment complex. If all units are rented, the rent for each apartment is $200 per month. Each time the rent is increased by $20, he will lose 1 tenant. What is the rental rate if he receives $3120 monthly in rent? (**Hint**: Let x = number of empty units.)

19. The Ski Club is planning to charter a bus to a ski resort. The cost will be $900 and each member will share the cost equally. If the club had 15 more members, the cost per person would be $10 less. How many are in the club now?

(**Hint**: If x = number in club now, $\dfrac{900}{x}$ = cost per person.)

20. A sporting goods store owner estimates that if he sells a certain model of basketball shoes for x dollars a pair, he will be able to sell $125 - x$ pairs. Find the price if his sales are $3750. Is there more than one possible answer?

21. Mr. Green traveled to a city 200 miles from his home to attend a meeting. Due to car trouble, his average speed returning was 10 mph less than his speed going. If the total driving time for the round trip was 9 hours, at what rate of speed did he travel to the city?

	Rate	Time	Distance
Going	x	?	200
Returning	$x - 10$	?	200

22. A motorboat takes a total of 2 hours to travel 8 miles downstream and 4 miles back on a river that is flowing at a rate of 2 mph. Find the rate of the boat in still water.

23. A small motorboat travels 12 mph in still water. It takes 2 hours longer to travel 45 miles going upstream than it does going downstream. Find the rate of the current. (**Hint**: $12 + c$ = rate going downstream and $12 - c$ = rate going upstream.)

24. Recently Mr. and Mrs. Roberts spent their vacation in San Francisco, which is 540 miles from their home. Being a little reluctant to return home, the Roberts took 2 hours longer on their return trip and their average speed was 9 mph slower than when they were going. What was their average rate of speed as they traveled from home to San Francisco?

25. The Blumin Garden Club planned to give their president a gift of appreciation costing $120 and to divide the cost evenly. In the meantime, 5 members dropped out of the club. If it now costs each of the remaining members $2 more than originally planned, how many members initially participated in the gift buying?

(**Hint**: If x = number in club initially, $\dfrac{120}{x}$ = cost per member.)

$120

26. A rectangular sheet of metal is 6 in. longer that it is wide. A box is to be made by cutting out 3 in. squares at each corner and folding up the sides. If the box has a volume of 336 in.^{3}, what were the original dimensions of the sheet metal? (See Example 4.)

27. A box is to be made out of a square piece of cardboard by cutting out 2 in. squares at each corner and folding up the sides. If the box has a volume of 162 in.^{3}, how big was the piece of cardboard? (See Example 4.)

28. A man and his son can paint their cabin in 3 hours. Working alone it would take the son 8 hours longer than it would the father. How long would it take the father to paint the cabin alone?

29. Two pipes can fill a tank in 8 minutes if both are turned on. If only one is used it would take 30 minutes longer for the smaller pipe to fill the tank than the larger pipe. How long will it take this small pipe to fill the tank?

30. A farmer and his son can plow a field with two tractors in 4 hours. If it would take the son 6 hours longer than the father to plow the field alone, how long would it take each if they worked alone?

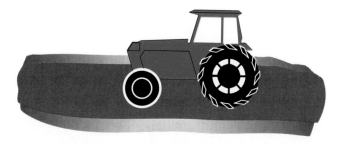

31. A ball is thrown upward with an initial velocity of 32 feet per second from the edge of a cliff near the beach. The cliff is 50 feet above the beach and the height of the ball can be found by using the equation $h = -16t^2 + 32t + 50$ where t is measured in seconds.

  a. When will the ball be 66 feet above the beach?
  b. When will the ball be 30 feet above the beach?
  c. In about how many seconds will the ball hit the beach?

32. A ladder is 30 feet long and you want to place the base of the ladder 10 feet from the base of a building. About how far up the building (to the nearest tenth of a foot) will the ladder reach?

30 ft

10 ft

Writing and Thinking About Mathematics

33. Suppose that you are to solve an applied problem and the solution leads to a quadratic equation. You decide to use the quadratic formula to solve the equation. Explain what restrictions you must be aware of when you use the formula.

Hawkes Learning Systems: Introductory Algebra

Applications (Quadratic Equations)

10.5 Graphing Parabolas: $y = ax^2 + bx + c$

In Chapters 4 and 5, we discussed linear equations and linear functions in great detail. For example,

$$y = 2x + 5 \text{ is a linear function}$$

and its graph is a straight line with slope 2 and y-intercept 5. The equation can also be written in the function form

$$f(x) = 2x + 5$$

In general, we can write linear functions in the form

$$y = mx + b \quad \text{or} \quad f(x) = mx + b$$

where the slope is m and the y-intercept is b.

Quadratic Functions

Now consider the function

$$y = x^2 - 2x - 3$$

This function is **not** linear, so its graph is **not** a straight line. The nature of the graph can be seen by plotting several points, as shown in Figure 10.1.

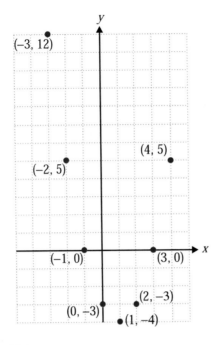

x	$x^2 - 2x - 3 = y$
-3	$(-3)^2 - 2(-3) - 3 = 12$
-2	$(-2)^2 - 2(-2) - 3 = 5$
-1	$(-1)^2 - 2(-1) - 3 = 0$
0	$(0)^2 - 2(0) - 3 = -3$
1	$(1)^2 - 2(1) - 3 = -4$
2	$(2)^2 - 2(2) - 3 = -3$
3	$(3)^2 - 2(3) - 3 = 0$
4	$(4)^2 - 2(4) - 3 = 5$

Figure 10.1

The points do not lie on a straight line. The graph can be seen by drawing a smooth curve called a **parabola** through the points, as shown in Figure 10.2. The point (1, –4) is the "turning point" of the curve and is called the **vertex** of the parabola. The line $x = 1$ is the **line of symmetry** or the **axis of symmetry** for the parabola. The curve is a "mirror image" of itself on either side of the line $x = 1$.

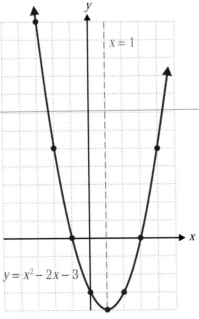

The graph of $y = x^2 - 2x - 3$ is a parabola.

$x = 1$ is the line of symmetry.

The point $(1, -4)$ is the vertex. This is the lowest point on the curve and -4 is the **minimum value** for y.

Figure 10.2

Parabolas (or parabolic arcs) occur frequently in describing information in real life. For example, the paths of projectiles (thrown balls, artillery shells, arrows) and a revenue function in business can be illustrated using parabolas.

Quadratic Function

A **quadratic function** is a function of the form

$$y = ax^2 + bx + c$$

where a, b, and c are real constants and $a \neq 0$.

The graph of every quadratic function is a parabola. The position of the parabola, its shape, and whether it "opens up" or "opens down" can be determined by investigating the function and its coefficients.

The following information about quadratic functions is given here without proof. A thorough development is part of the next course in algebra.

General Information on Quadratic Functions

For the **quadratic function** $y = ax^2 + bx + c$:

1. If $a > 0$, the parabola "opens upward."
2. If $a < 0$, the parabola "opens downward."

3. $x = -\dfrac{b}{2a}$ is the **line of symmetry**.

4. The **vertex** (turning point) occurs where $x = -\dfrac{b}{2a}$. Substitute this value for x in the function and find the y-value of the vertex. The vertex is the lowest point on the curve if the parabola opens upward or it is the highest point on the curve if the parabola opens downward.

Several graphs related to the basic form

$$y = ax^2 \quad \text{where } b = 0 \text{ and } c = 0$$

are shown in Figure 10.3.

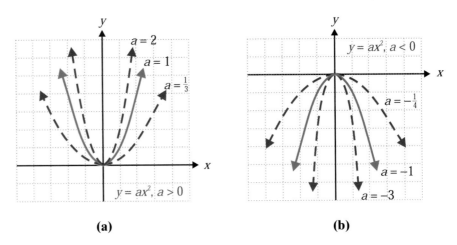

(a)   **(b)**

Figure 10.3

In the cases illustrated in Figure 10.3, where $y = ax^2$, we have $b = 0$, and

$$x = -\frac{b}{2a} = -\frac{0}{2a} = 0$$

So, the vertex occurs at the origin, $(0, 0)$. In addition, the axis of symmetry is the y-axis (the line $x = 0$).

Example 1: Graphing Quadratic Functions ● ● ● ● ● ● ● ● ● ● ●

For each quadratic function, find (**i**) its vertex (**ii**) its line of symmetry and (**iii**) x-intercepts. Then plot a few specific points and graph the parabola

a. $y = x^2 - 1$

Solution: $a = 1, b = 0$, and $c = -1$

i. The vertex is at $x = -\frac{0}{2 \cdot 1} = 0$. Substituting 0 for x gives $y = 0^2 - 1 = -1$.

The vertex is the point $(0, -1)$.

ii. The line of symmetry is $x = -\frac{0}{2(1)} = 0$.

Continued on next page ...

iii. The x-intercepts can be found by setting $y = 0$ and solving the resulting quadratic equation. Thus,

$$x^2 - 1 = 0$$
$$x^2 = 1$$
$$x = \pm\sqrt{1}$$
$$x = \pm 1$$

Thus, the graph crosses the x-axis at the points $(1, 0)$ and $(-1, 0)$. These points and others can be found as shown in the table.

x	$x^2 - 1 = y$
-2	$(-2)^2 - 1 = 3$
-1	$(-1)^2 - 1 = 0$
0	$(0)^2 - 1 = -1$
1	$(1)^2 - 1 = 0$
2	$(2)^2 - 1 = 3$

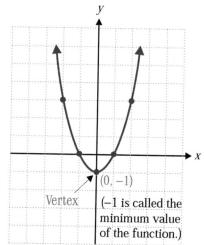

Vertex $(0, -1)$

(-1 is called the minimum value of the function.)

b. $f(x) = x^2 - 6x + 1$

Solution: $a = 1, b = -6, c = 1$

i. The vertex is at $x = -\dfrac{-6}{2 \cdot 1} = 3$. Substituting 3 for x gives

$y = 3^2 - 6 \cdot 3 + 1 = -8$. The vertex is the point $(3, -8)$.

ii. The line of symmetry is $x = -\dfrac{-6}{2(1)} = 3$.

Continued on next page ...

iii. The x-intercepts can be found by setting $y = 0$ and solving the resulting quadratic equation. Thus, using the quadratic formula, we have

$$x^2 - 6x + 1 = 0$$

$$x = \frac{-(-6) \pm \sqrt{(-6)^2 - 4(1)(1)}}{2 \cdot 1} = \frac{6 \pm \sqrt{32}}{2} = \frac{6 \pm 4\sqrt{2}}{2} = 3 \pm 2\sqrt{2}$$

The x-intercepts are at $\left(3 + 2\sqrt{2}, 0\right)$ and $\left(3 - 2\sqrt{2}, 0\right)$.

Estimating these values with a calculator gives the x-intercepts at about $(5.83, 0)$ and $(0.17, 0)$

Thus, the graph crosses the x-axis at about $(5.83, 0)$ and $(0.17, 0)$.

Other points can be found as shown in the table.

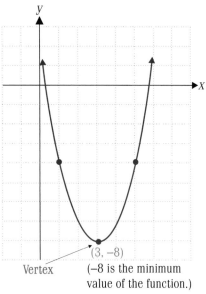

(3, −8)

Vertex   (−8 is the minimum
                value of the function.)

x	$x^2 - 6x + 1 = y$
1	$(1)^2 - 6(1) + 1 = -4$
3	$(3)^2 - 6(3) + 1 = -8$
5	$(5)^2 - 6(5) + 1 = -4$
$3 + 2\sqrt{2}$	$(3 + 2\sqrt{2})^2 - 6(3 + 2\sqrt{2}) + 1 = 0$
$3 - 2\sqrt{2}$	$(3 - 2\sqrt{2})^2 - 6(3 - 2\sqrt{2}) + 1 = 0$

c. $y = -2x^2 + 4x$

Solution: i. The vertex is at $x = -\dfrac{4}{2(-2)} = 1$. Substituting 1 for x gives

$$y = -2 \cdot 1^2 + 4 \cdot 1 = 2.$$

The vertex is the point $(1, 2)$.

ii. The line of symmetry is $x = -\dfrac{4}{2(-2)} = 1$.

iii. The x-intercepts can be found by setting $y = 0$ and solving the resulting quadratic equation. Thus, by factoring, we have

Continued on next page ...

$$-2x^2 + 4x = 0$$
$$-2x(x-2) = 0$$
$$x = 0 \text{ and } x = 2$$

The x-intercepts are at $(0, 0)$ and $(2, 0)$. Thus, the graph crosses the x-axis at the points $(0, 0)$ and $(2, 0)$.

These points and others can be found as shown in the table.

x	$-2x^2 + 4x = y$
0	$-2(0)^2 + 4(0) = 0$
1	$-2(1)^2 + 4(1) = 2$
2	$-2(2)^2 + 4(2) = 0$

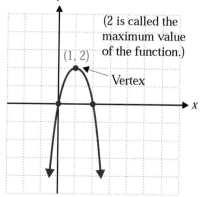

(2 is called the maximum value of the function.)

(1, 2)

Vertex

d. Find the dimensions of the rectangle with maximum area if the perimeter is 40 meters.

Solution: Let A = area

l = length

w = width.

$$A = lw \quad \text{and} \quad 2l + 2w = 40$$
$$2w = 40 - 2l$$
$$w = 20 - l$$

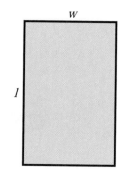

So, substituting for w,

$$A = l(20 - l)$$
$$A = 20l - l^2$$
$$A = -l^2 + 20l$$

This is a quadratic function with $a = -1$, $b = 20$, and $c = 0$. The maximum area occurs at the vertex of the corresponding parabola where

$$l = -\frac{b}{2a} = -\frac{20}{2(-1)} = 10$$

and $\quad w = 20 - l = 20 - 10 = 10$

This maximum occurs when $l = 10$ m and $w = 10$ m. (The rectangle is, in fact, a square and the area $= 100$ m^2.)

Using a Calculator to Graph Quadratic Functions

We illustrate how to graph a quadratic function on a TI-83 Plus graphing calculator by graphing the function $y = x^2 + 2x - 3$ in the following step-by-step manner.

Step 1: Press the key marked (Y=).

Step 2: Enter the function to be graphed: $\backslash Y_1 = X^2 + 2X - 3$.

(**Note:** The variable X is found by pressing (X,T,θ,n).
X^2 is then found by pressing (x^2).)

(Or, you can enter $\backslash Y_1 = X \wedge 2 + 2X - 3$.)

Step 3: Press (GRAPH) in the upper right hand corner of the keyboard.
The graph of the parabola should appear on the display as shown here.

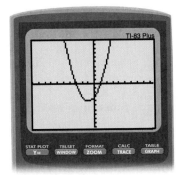

If the graph does not appear as shown, you may need to adjust your window as follows:

a. Press (WINDOW) and set the numbers as needed, or

b. Press (ZOOM) and then press (6) to get the standard window to get the window from −10 to 10 on the x-axis and from −10 to 10 on the y-axis.

To estimate the location of various points on the graph of the parabola, you can press (TRACE) and then use the left and right arrows to move along the parabola. The function $Y_1 = X^2 - 2X - 3$ will appear at the top of the display and the coordinates of the point indicated by the cursor will appear at the bottom of the display. An example is shown here.

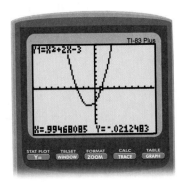

Example 2: Graphing with a Calculator • • • • • • • • • • •

Use your graphing calculator to graph the quadratic function $y = 2x^2 - x - 15$. Then use the trace key to estimate the location of three points on the graph. (As shown in the following diagrams, set the **WINDOW** so that **xmin** = −10, **xmax** = 10, **ymin** = −20 and **ymax** = 10.)

Solution: **Step 1:** Press the key marked (Y=).

Step 2: Enter the function to be graphed: $\backslash Y_1 = 2X^2 - X - 15$.

Step 3: Press (GRAPH) in the upper right hand corner of the keyboard.

The graph of the parabola should appear on the display as shown here.

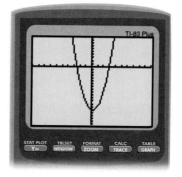

Now, press the (TRACE) key and move the cursor around to estimate the location of points. Three such locations are shown here.

• •

10.5 Exercises

For each quadratic function in Exercises 1 – 20, find (**a**) its vertex, (**b**) its line of symmetry, (**c**) locate (or estimate) its x-intercepts, and (**d**) graph the function. (**Note:** If solving the quadratic equation results in nonreal solutions, then the graph does not cross the x-axis.)

1. $y = x^2 + 4$
2. $y = x^2 - 6$
3. $y = 8 - x^2$
4. $y = -2 - x^2$
5. $y = x^2 - 2x - 3$
6. $y = x^2 - 4x + 5$
7. $y = x^2 + 6x$
8. $y = x^2 - 8x$
9. $y = -x^2 - 4x + 2$
10. $y = -x^2 - 5x + 4$
11. $y = 2x^2 - 10x + 3$
12. $y = 2x^2 - 12x - 5$
13. $y = x^2 + 7x - 4$
14. $y = x^2 + 5x + 4$
15. $y = -x^2 + x - 3$
16. $y = -x^2 - 7x + 3$
17. $y = 2x^2 + 7x - 4$
18. $y = 2x^2 - x - 3$
19. $y = 3x^2 + 5x + 2$
20. $y = 3x^2 - 9x + 5$

21. The perimeter of a rectangle is 60 yards. What are the dimensions of the rectangle with maximum area?

22. The perimeter of a rectangle is 56 feet. What are the dimensions of the rectangle with maximum area?

In Exercises 23 – 26 use the formula $h = -16t^2 + v_0 t + h_0$ where h is the height of the object after time, t; v_0 is the initial velocity; and h_0 is the initial height.

23. A ball is thrown vertically upward from the ground with an initial velocity of 112 ft. per sec.
   a. When will the ball reach its maximum height?
   b. What will be the maximum height?

24. A ball is thrown vertically upward from the ground with an initial velocity of 104 ft. per sec.
   a. When will the ball reach its maximum height?
   b. What will be the maximum height?

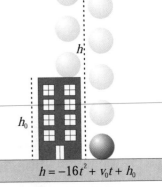

$$h = -16t^2 + v_0 t + h_0$$

25. A stone is projected vertically upward from a platform that is 32 ft. high at a rate of 128 ft. per sec.
   a. When will the stone reach its maximum height?
   b. What will be the maximum height?

26. A stone is projected vertically upward from a platform that is 20 ft. high at a rate of 160 ft. per sec.
    a. When will the stone reach its maximum height?
    b. What will be the maximum height?

In business, the term **revenue** represents income. The revenue (income) is found by multiplying the number of units sold times the price per unit. Revenue = (price) · (units sold).

27. A store owner estimates that by charging x dollars each for a certain lamp, he can sell $40 - x$ lamps each week. What price will yield maximum revenue?

28. When fishing reels are priced at p dollars each, local consumers will buy $36 - p$ fishing reels. What price will yield maximum revenue?

29. A manufacturer produces calculators. She estimates that by selling them for x dollars each, she will be able to sell $80 - 2x$ calculators each week.
    a. What price will yield a maximum revenue?
    b. What will be the maximum revenue?

30. A manufacturer produces radios. He estimates that by selling them for x dollars each, he will be able to sell $100 - x$ radios each month.
    a. What price will yield a maximum revenue?
    b. What will be the maximum revenue?

Writing and Thinking About Mathematics

31. Use your graphing calculator to graph each of the following quadratic functions.
    a. Graph each equation one at a time.
    b. Graph all three parabolas at the same time (all three curves should appear on the display).
        i. $y = x^2 + 1$ ii. $y = x^2 + 3$ iii. $y = x^2 - 4$

32. In your own words, state the effect on the graph of $y = x^2$ by changing the constant term k in the function $y = x^2 + k$.

33. Use your graphing calculator to graph each of the following quadratic functions

    a. Graph each equation one at a time.

    b. Graph all three parabolas at the same time (all three curves should appear on the display).

      i. $y = (x-3)^2$ **ii.** $y = (x-5)^2$ **iii.** $y = (x+2)^2$

34. In your own words, state the effect on the graph of $y = x^2$ by changing the constant term h in the function $y = (x-h)^2$.

35. Discuss the general relationship of the graph of a function of the form $y = (x-h)^2 + k$ to the graph of the function $y = x^2$.

Hawkes Learning Systems: Introductory Algebra

Graphing Parabolas

Chapter 10 Index of Key Ideas and Terms

For a quadratic equation in the form $x^2 = c$ where c is nonnegative, $x = \sqrt{c}$ or $x = -\sqrt{c}$. This can be written as $x = \pm\sqrt{c}$.

In a right triangle, the square of the hypotenuse is equal to the sum of the squares of the legs.

Symbolically, $c^2 = a^2 + b^2$, where c is the length of the hypotenuse and a and b are the lengths of the legs.

To solve quadratic equations by completing the square:

1. Arrange the terms with variables on one side and constants on the other.
2. Divide each term by the coefficient of x^2. (We want the leading coefficient to be 1.)
3. Find the number that completes the square of the quadratic expression and add this number to **both** sides of the equation.
4. Find the positive and negative square roots of both sides.
5. Solve for x. Remember, usually there will be two solutions.

The general quadratic equation is $ax^2 + bx + c = 0$, where a, b, and c are real constants and $a \neq 0$.

The solutions of the general quadratic equation $ax^2 + bx + c = 0$, $a \neq 0$

are $x = \dfrac{-b \pm \sqrt{b^2 - 4ac}}{2a}$.

Quadratic Functions

page 669

A quadratic function is a function of the form

$$y = ax^2 + bx + c$$

where $a, b,$ and c are real constants and $a \neq 0$.

General Information on Quadratic Functions

page 669

For the quadratic function $y = ax^2 + bx + c$

1. If $a > 0$, the parabola "opens upward."

2. If $a < 0$, the parabola "opens downward."

3. $x = -\dfrac{b}{2a}$ is the line of symmetry.

4. The vertex (turning point) occurs where $x = -\dfrac{b}{2a}$. Substitute this value for x in the function and find the y-value of the vertex. The vertex is the lowest point on the curve if the parabola opens upward or it is the highest point on the curve if the parabola opens downward.

Chapter 10 Review

For a review of the topics and problems from Chapter 10, look at the following lessons from *Hawkes Learning Systems: Introductory Algebra*

SQE: The Square Root Method
SQE: Completing the Square
SQE: The Quadratic Formula
Applications (Quadratic Equations)
Graphing Parabolas

Chapter 10 Test

Solve the equations in Exercises 1 and 2.

1. $(x+5)^2 = 49$ **2.** $8x^2 = 96$

3. Find the missing side in the right triangle shown here.

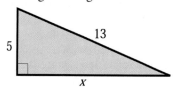

Add the correct constant in Exercises 4 – 6 in order to make the trinomial factorable as indicated.

4. $x^2 - 24x +$ _____ $= ($ $)^2$ **5.** $x^2 + 9x +$ _____ $= ($ $)^2$

6. $3x^2 + 9x +$ _____ $= 3($ $)^2$

Solve the quadratic equations in Exercises 7 – 9 by factoring.

7. $x^2 - 3x + 2 = 0$ **8.** $5x^2 + 12x = 0$ **9.** $3x^2 - x - 10 = 0$

Solve the quadratic equations in Exercises 10 – 12 by completing the square.

10. $x^2 + 6x + 8 = 0$ **11.** $x^2 - 2x - 5 = 0$ **12.** $2x^2 + 3x - 3 = 0$

Solve the quadratic equations in Exercises 13 – 15 using the quadratic formula.

13. $3x^2 + 8x + 2 = 0$ **14.** $2x^2 = 4x + 3$ **15.** $\frac{1}{2}x^2 - x = 2$

Solve the quadratic equations in Exercises 16 – 18 using any method.

16. $x^2 + 6x = 2$ **17.** $3x^2 - 7x + 2 = 0$ **18.** $2x^2 - 3x = 1$

For each quadratic function in Exercises 19 – 21, find (**a**) its vertex, (**b**) its line of symmetry, (**c**) locate (or estimate) its x-intercepts, and (**d**) graph the function.
(**Note**: If solving the quadratic equation results in nonreal solutions, then the graph does not cross the x-axis.)

19. $y = x^2 - 5$ **20.** $y = -x^2 + 4x$ **21.** $y = 2x^2 + 3x + 1$

First, determine an equation for each of the Exercises 22 – 25, and then solve the problem.

22. The diagonal of a square is 36 inches long. Find the length of each side.

23. A toy rocket is projected upward from the ground with an initial velocity of 144 feet per second. The height of the rocket can be found by using the equation $h = -16t^2 + 144t$ where t is measured in seconds.
  a. In about how many seconds will the rocket be 96 feet above the ground?
  b. What is the maximum height of the rocket?
  c. In about how many seconds will the rocket hit the ground?

24. A small boat travels 8 mph in still water. It takes 10 hours longer to travel 60 miles upstream than it takes to travel 60 miles downstream. Find the rate of the current.

25. A rectangular sheet of metal is 3 inches longer than it is wide. A box is to be made by cutting out 4 inch squares at each corner and folding up the sides. If the box has a volume of 720 cubic inches, what are the dimensions of the sheet of metal?
(**Hint:** Sketch a diagram of the rectangle and a diagram of the box.)

Cumulative Review: Chapters 1 – 10

Simplify each expression in Exercises 1 and 2.

1. $\left(-3x^2y\right)^2$

2. $\dfrac{\left(2x^2\right)\left(-6x^3\right)}{3x^{-1}}$

In Exercises 3 and 4, simplify by combining like terms.

3. $4x + 3 - 2(x + 4) + (2 - x)$

4. $4x - [3x - 2(x + 5) + 7 - 5x]$

Solve each of the equations in Exercises 5 – 16.

5. $5(x - 8) - 3(x - 6) = 0$　**6.** $\dfrac{5}{4}x + 4 = \dfrac{3}{4}x + 7$　**7.** $x^2 - 8x + 15 = 0$

8. $x^2 + 5x - 36 = 0$　**9.** $\dfrac{1}{x} + \dfrac{2}{3x} = 1$　**10.** $\dfrac{3}{x-1} + \dfrac{5}{x+1} = 2$

11. $(x - 4)^2 = 9$　**12.** $(x + 3)^2 = 6$　**13.** $x^2 + 4x - 2 = 0$

14. $2x^2 - 3x - 4 = 0$　**15.** $x - 2 = \sqrt{x + 18}$　**16.** $x + 3 = 2\sqrt{2x + 6}$

In Exercises 17 – 20, solve each inequality and graph the solution set on a real number line.

17. $15 - 3x + 1 \geq x + 4$

18. $4(6 - x) \geq -2(3x + 1)$

19. $2(x - 5) - 4 < x - 3(x - 1)$

20. $x - (2x + 5) \leq 9 - (4 - x)$

Solve the system of equations in Exercises 21 – 23.

21. $\begin{cases} x + 5y = 9 \\ 2x - 3y = 1 \end{cases}$　**22.** $\begin{cases} 6x - y = 11 \\ 2x - 3y = -11 \end{cases}$　**23.** $\begin{cases} 2x - 5y = 11 \\ 5x - 3y = -1 \end{cases}$

In Exercises 24 and 25, solve each formula for the indicated letter.

24. $P = a + 2b$, solve for b

25. $A = \dfrac{h}{2}(a + b)$, solve for a

Factor the expressions in Exercises 26 – 30.

26. $x^2 - 12x + 36$

27. $25x^2 - 49$

28. $3x^2 + 10x + 3$

29. $3x^2 + 6x - 72$

30. $x^3 - 3x^2 - 4x - 32$

In Exercises 31 – 36, perform the indicated operations and simplify.

31. $\left(5x^2+3x-7\right)+\left(2x^2-3x-7\right)$

32. $\left(2x^2-7x+3\right)+\left(5x^2-3x+4\right)$

33. $\dfrac{x^2}{x^2-4}\cdot\dfrac{x^2-3x+2}{x^2-x}$

34. $\dfrac{3x}{x^2-6x-7}\div\dfrac{2x^2}{x^2-8x+7}$

35. $\dfrac{2x}{x^2-25}-\dfrac{x+1}{x+5}$

36. $\dfrac{4x}{x^2-3x-4}+\dfrac{x-2}{x^2+3x+2}$

Simplify the radical expressions in Exercises 37 – 42 and then use a calculator to find the value of each expression accurate to four decimal places.

37. $\sqrt{63}$

38. $\sqrt{243}$

39. $\sqrt[3]{250}$

40. $\sqrt{\dfrac{25}{12}}$

41. $\sqrt{147}+\sqrt{48}$

42. $\left(2\sqrt{3}+1\right)\left(\sqrt{3}-4\right)$

Simplify the expression in Exercises 43 – 46.

43. $4x^{\frac{2}{3}}\cdot2x^{\frac{1}{3}}$

44. $\left(4x^3\right)^2\left(2x^2\right)^{-1}$

45. $\left(\dfrac{m^{1/2}n^{3/4}}{m^{-1/2}n^{1/4}}\right)^{1/3}$

46. $\left(\dfrac{25x^2y^{1/2}}{36x^4y^{1/3}}\right)^{1/2}$

In Exercises 47 – 49, graph the linear function and state its slope and its y-intercept.

47. $y=\dfrac{2}{3}x+4$

48. $4x-3y=6$

49. $2x+4y=-5$

For each quadratic function in Exercises 50 – 52, find (**a**) its vertex, (**b**) its line of symmetry, (**c**) locate (or estimate) its x-intercepts, and (**d**) graph the function.
(**Note:** If solving the quadratic equation results in nonreal solutions, then the graph does not cross the x-axis.)

50. $y=x^2-8x+7$

51. $y=2(x+3)^2-6$

52. $y=x^2+x+1$

53. Write the equation $5x+2y=6$ in slope-intercept form. Find the slope and the y-intercept.

54. Given the two points $(-5, 2)$ and $(3, 5)$, find:

    a. the slope of the line through the two points,

    b. the distance between the two points,

    c. the midpoint of the line segment joining the points, and

    d. an equation of the line through the two points.

55. Given the function $f(x) = 5x - 3$, find (a) $f(-1)$, (b) $f(6)$, and (c) $f(0)$.

56. Given the function $F(x) = 3x^2 - 2x + 7$, find (a) $F(0)$, (b) $F(-2)$, and (c) $F(5)$.

57. Anna has money in two separate savings accounts; one pays 5% and one pays 8%. The amount invested at 5% exceeds the amount invested at 8% by $800. The total annual interest is $209. How much is invested at each rate?

58. Ben can clean the weeds from a vacant lot in 10 hours. Beth can do the same job in 8 hours. How long would it take if they worked together?

59. The speed of a boat in still water is 8 mph. It travels 6 miles upstream and returns. The total trip takes 1 hr. 36 min. Find the speed of the current.

60. How many liters of a solution that is 40% insecticide must be mixed with 40 liters of a solution that is 25% insecticide to produce a solution that is 30% insecticide?

61. The length of a rectangle is one inch less than twice the width. The length of the diagonal is 17 inches. Find the width and length of the rectangle.

62. George bought 7 pairs of socks. Some cost $2.50 per pair and some cost $3.00 per pair. The total cost was $19.50. How many pairs of each type did he buy?

63. The length of a rectangle is five more than twice the width. The area is 52 cm^2. Find the width and the length.

64. At a small college, the ratio of women to men is 9:7. If there is a total enrollment of 6400 students, how many women and how many men are enrolled?

65. Louise has $2700 invested. Part of the money is invested at 5.5% and the remainder is invested at 7.2%. Her total annual interest is $168.90. How much is invested at each rate?

66. Glen traveled 6 miles downstream and returned. His speed downstream was 5 mph faster than his speed upstream. If the total trip took 1 hour, find his speed in each direction.

67. A ball is thrown upward from the edge of a building with an initial velocity of 32 feet per second. The building is 120 feet tall and the height of the ball above the ground can be found by using the equation $h = -16t^2 + 96t + 120$ where t is measured in seconds.
   a. When will the ball be 200 feet above the ground?
   b. When will the ball be 60 feet above the ground?
   c. In about how many seconds will the ball hit the ground?

68. A ladder is 40 feet long and the base of the ladder is to be set 15 feet from the base of a building under construction. About how far up the building (to the nearest tenth of a foot) will the ladder reach?

Appendix

GRAPHING LINEAR INEQUALITIES AND SYSTEMS OF LINEAR INEQUALITIES

Graphing Linear Inequalities

After completing this section, you will be able to:

1. Understand the concept of a half-plane.

2. Graph linear inequalities.

We discussed and graphed linear equations in Chapter 4 including slopes and intercepts and the basic forms of linear equations, namely

$$m = \frac{y_2 - y_1}{x_2 - x_1}$$
the slope of a line through two points

$$y = mx + b$$
the slope-intercept form, and

$$y - y_1 = m(x - x_1)$$
the point-slope form.

A review of these topics and the use of your calculator in graphing linear equations would be helpful at this time.

In this section, we will develop the concepts of analyzing and graphing **linear inequalities**. We begin with some terminology. A straight line separates a plane into two **half-planes**. The points on one side of the line are in one of the half-planes, and the points on the other side of the line are in the other half-plane. The line itself is called the **boundary line**. If the boundary line is included with a half-plane, then the half-plane is said to be **closed**. If the boundary line is not included, then the half-plane is said to be **open**. (Note the similarity between the terminology for open and closed intervals.) For example, all the points that satisfy the linear inequality

$$5x - 3y < 15$$

lie in an open half-plane on one side of the line

$$5x - 3y = 15$$

There are two basic methods of deciding which side of the line is the graph of the solution set of the inequality. In both methods, the boundary line must be graphed first.

Two Methods for Graphing Linear Inequalities

First, graph the boundary line (*dashed if the inequality is $<$ or $>$, solid if the inequality is $\leq$ or $\geq$*).

Method 1

    a. Test any one point obviously on one side of the line.

    b. If the test-point satisfies the inequality, shade the half-plane on that side of the line. Otherwise, shade the other half-plane.

Method 2

    a. Solve the inequality for y (assuming that the line is not vertical).

    b. If the solution shows $y <$ or $y \leq$, then shade the half-plane below the line.

    c. If the solution shows $y >$ or $y \geq$, then shade the half-plane above the line.

(**Note:** If the boundary line is vertical, then it is of the form $x = a$ and Method 1 should be used.)

Figure A.1 shows both (a) an open half-plane and (b) a closed half-plane with the line $5x - 3y = 15$ as the boundary line.

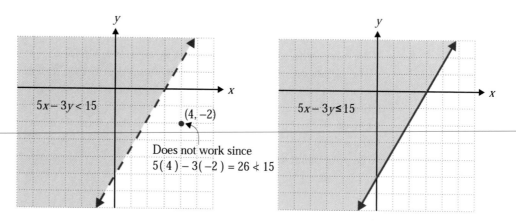

The points on the line $5x - 3y = 15$ are not included so the line is dashed. The half-plane is open.

(a)

The points on the line $5x - 3y = 15$ are included so the line is solid. The half-plane is closed.

(b)

Figure A.1

Example 1: Graphing Linear Inequalities • • • • • • • • • •

Graph the following inequalities.

a. Graph the half-plane that satisfies the inequality $2x + y \leq 6$.

Solution: Method 1 is used in this example.

Step 1: Graph the line $2x + y = 6$ as a solid line.

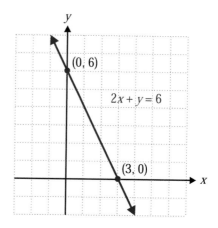

Step 2: Test any point on one side of the line.
In this example, we have chosen $(0, 0)$.

$$2 \cdot 0 + 0 \leq 6$$
$$0 \leq 6$$

This is a true statement.

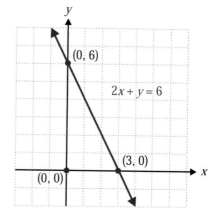

Continued on next page ...

Step 3: Shade the points on the same side as the point $(0, 0)$.

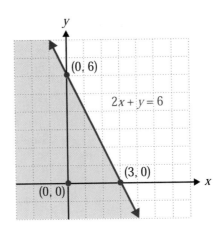

b. Graph the solution set to the inequality $y > 2x$.

Solution: Since the inequality is already solved for y, Method 2 is easy to apply.

Step 1: Graph the line $y = 2x$ as a dashed line.

Step 2: By Method 2, the graph consists of those points above the line. Shade the half-plane above the line.

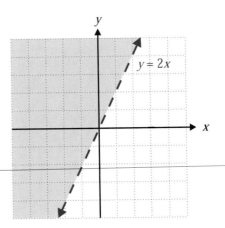

c. Graph the half-plane that satisfies the inequality $y > 1$.

Solution: Again, the inequality is already solved for y and Method 2 is used.

Step 1: Graph the horizontal line $y = 1$ as a dashed line.

Step 2: By Method 2, shade the half-plane above the line.

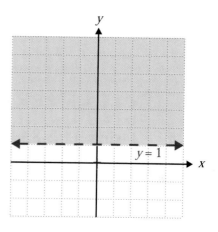

d. Graph the solution set to the inequality $x \leq 0$.

Solution: The boundary line is a vertical line and Method 1 is used.

Step 1: Graph the line $x = 0$ as a solid line. Note that this is the y-axis.

Step 2: Test the point $(-2, 1)$.

$-2 \leq 0$

This statement is true.

Step 3: Shade the half-plane on the same side of the line as $(-2, 1)$. This half-plane consists of the points with x-coordinate 0 or negative.

Continued on next page ...

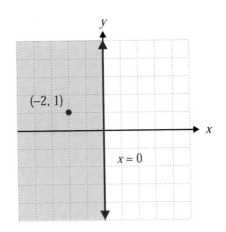

Using a TI-83 Plus Graphing Calculator to Graph Linear Inequalities

The first step in using the TI-83 Plus (or any other graphing calculator) to graph a linear inequality is to solve the inequality for y. This is necessary because this is the way that the boundary line equation can be graphed as a function. Thus, Method 2 for graphing the correct half-plane is appropriate.

Note that when you press the Y= key on the calculator, a slash (\ ) appears to the left of the Y expression as in \Y1=. This slash is actually a command to the calculator to graph the corresponding function as a solid line or curve. If you move the cursor to position over the slash and hit ENTER, you will find the following options appearing:

If the slash (which is actually four dots if you look closely) becomes a set of three dots, then the corresponding graph of the function will be dotted. By setting the shading above the dots, the corresponding graph on the display will show shading above the line or curve. By setting the shading below the dots, the corresponding graph on the display will show shading below the line or curve. The actual shading occurs only when the slash is four dots. So, the calculator is not good for determining whether the boundary curve is included or not. The following examples illustrate two situations.

Example 2: Graphing Using a TI-83 Plus

a. Graph the linear inequality $2x + y \geq 3$.

> **Solution:** **Step 1:** Solving the inequality for y gives: $y \geq -2x + 3$.
>
> **Step 2:** Press the key (**Y=**) and enter the function: $\backslash Y_1 = -2X + 3$.
>
> **Step 3:** Go to the \ and hit (**ENTER**) two times so that the display appears as follows:

> **Step 4:** Press (**GRAPH**) and (assuming that the window is okay) the following graph should appear on the display.

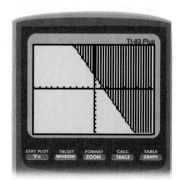

Continued on next page ...

b. Graph the linear inequality $-3x + 2y < -4$

Solution: **Step 1:** Solving the inequality for y gives: $y < \dfrac{3}{2}x - 2$.

Step 2: Press the key ⬤ Y= and enter the function: $\backslash Y_1 = (3/2)X{-}2$.

Step 3: Go to the $\backslash$ and hit ⬤ ENTER three times so that the display appears as follows:

Step 4: Press GRAPH and (assuming that the window is okay) the following graph should appear on the display.

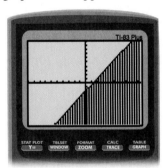

● ●

Practice Problems

1. Which of the following points satisfy the inequality $x + y < 3$?

    a. $(2, 1)$ **b.** $\left(\dfrac{1}{2}, 3\right)$ **c.** $(0, 5)$ **d.** $(-5, 2)$

2. Which of the following points satisfy the inequality $x - 2y \geq 0$?

    a. $(2, 1)$ **b.** $(1, 3)$ **c.** $(4, 2)$ **d.** $(3, 1)$

3. Which of the following points satisfy the inequality $x < 3$?

    a. $(1, 0)$ **b.** $(0, 1)$ **c.** $(4, -1)$ **d.** $(2, 3)$

Answers to Practice Problems: 1. (d) **2.** (a), (c), (d) **3.** (a), (b), (d)

Exercises A.1

Graph each of the linear inequalities in Exercises 1 – 30.

1. $y > 3x$ **2.** $x - y > 0$ **3.** $y \le 2x - 1$ **4.** $y < 4 - x$

5. $y > -7$ **6.** $2y - x \ge 2$ **7.** $2x - 3 \le 0$ **8.** $4x + y < 2$

9. $5x - 2y \ge 4$ **10.** $3y - 8 \le 2$ **11.** $y < -\dfrac{1}{4}x$ **12.** $2y - 3x \le 4$

13. $2x - y < 1$ **14.** $4x + 3y \le 6$ **15.** $2x - 3y \ge -3$ **16.** $2y + 5x \ge 0$

17. $2x + 5y \ge 10$ **18.** $4x + 2y \le 10$ **19.** $3x < 4 - y$ **20.** $x < 2y + 3$

21. $2x - 3y < 9$ **22.** $x \le 4 - 3y$ **23.** $5x - 2y \ge 4$ **24.** $3x + 6 \ge 2y$

25. $4y \ge 3x - 8$ **26.** $\dfrac{1}{2}x + y < 3$ **27.** $2x + \dfrac{1}{3}y < 2$ **28.** $2x - \dfrac{1}{2}y > -1$

29. $\dfrac{1}{2}x + \dfrac{2}{3}y \ge \dfrac{5}{6}$ **30.** $\dfrac{1}{4}x - \dfrac{1}{2}y \le \dfrac{3}{4}$

Use your graphing calculator to graph each of the linear inequalities in Exercises 31 – 40.

31. $y > 2x$ **32.** $x - y \le 4$ **33.** $x + 2y > 7$ **34.** $3x - 2y \ge 10$

35. $2x - y \le 6$ **36.** $y \ge -2$ **37.** $x - 3y \ge 5$ **38.** $2x + 8y \ge 0$

39. $2x + 4y < 9$ **40.** $4x - 3y > 15$

Hawkes Learning Systems: Introductory Algebra

Graphing Linear Inequalities

A.2

Graphing Systems of Linear Inequalities

After completing this section, you will be able to:

1. Solve systems of linear inequalities graphically.

In some branches of mathematics, in particular a topic called (interestingly enough) game theory, the solution to a very sophisticated problem can involve the set of points that satisfy a system of several **linear inequalities**. In business these ideas relate to problems such as minimizing the cost of shipping goods from several warehouses to distribution outlets. In this section we will consider graphing the solution sets to only two inequalities. We will leave the problem solving techniques to another course.

First, we review the ideas related to systems of equations discussed in Chapter 5. Systems of two linear equations were solved by using three methods: graphing, substitution, and addition. We found that such systems can be

(**a**) **consistent** (one point satisfies both equations, the lines intersect in one point),
(**b**) **inconsistent** (no point satisfies both equations, the lines are parallel), or
(**c**) **dependent** (an infinite number of points satisfy both equations, the lines are the same).

Figure A.2 shows an example of each case.

Systems (Figure A.2)

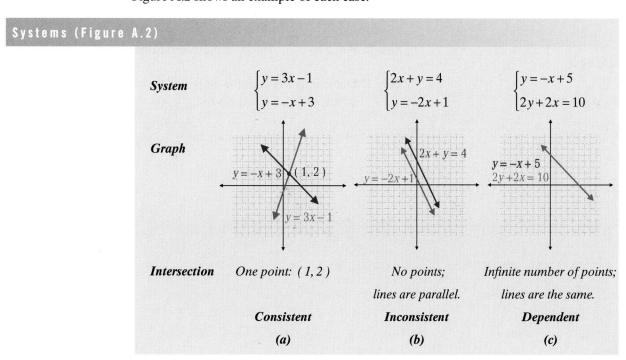

System	$\begin{cases} y = 3x - 1 \\ y = -x + 3 \end{cases}$	$\begin{cases} 2x + y = 4 \\ y = -2x + 1 \end{cases}$	$\begin{cases} y = -x + 5 \\ 2y + 2x = 10 \end{cases}$
Intersection	*One point: (1, 2)*	*No points;* *lines are parallel.*	*Infinite number of points;* *lines are the same.*
	Consistent	**Inconsistent**	**Dependent**
	(a)	*(b)*	*(c)*

In this section, we will develop techniques for graphing (and therefore solving) **systems of two linear inequalities**. The solution set (if there are any solutions) to a system of two linear inequalities consists of the points in the intersection of two half-planes and portions of boundary lines indicated by the inequalities. From Section A.1, we know that a straight line separates a plane into two **half-planes**. The line itself is called the **boundary line**, and the boundary line may be included (the half-plane is **closed**) or the boundary line may not be included (the half-plane is **open**). The following procedure may be used to solve a system of linear inequalities.

To Solve a System of Two Linear Inequalities

1. *Graph both half-planes.*

2. *Shade the region that is common to both of these half-planes.* (*This region is called the **intersection** of the two half-planes.*)

3. *To check, pick one test-point in the intersection and verify that it satisfies both inequalities.*

(***Note**: If there is no intersection, then the system is inconsistent and has no solution.*)

Example 1: Graphing Systems of Linear Inequalities

a. Graph the points that satisfy the system of inequalities $\begin{cases} x \leq 2 \\ y \geq -x + 1 \end{cases}$.

Solution: **Step 1:** For $x \leq 2$, the points are to the left of and on the line $x = 2$.

Step 2: For $y \geq -x + 1$, the points are above and on the line $y = -x + 1$.

Step 3: Shade only the region with points that satisfy both inequalities. In this case, we test the point $(0, 3)$:

$0 \leq 2$ A true statement

$3 \geq -0 + 1$ A true statement

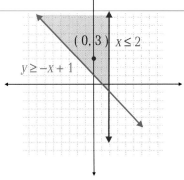

b. Solve the system of linear inequalities graphically: $\begin{cases} 2x + y \le 6 \\ x + y < 4 \end{cases}$.

Solution: **Step 1:** Solve each inequality for y : $\begin{cases} y \le -2x + 6 \\ y < -x + 4 \end{cases}$.

Step 2: For $y \le -2x + 6$, the points are below and on the line $y = -2x + 6$.

Step 3: For $y < -x + 4$, the points are below but not on the line $y = -x + 4$.

Step 4: Shade only the region with points that satisfy both inequalities. Note that the line $y = -x + 4$ is dashed. In this case, we test the point $(0, 0)$.

$2 \cdot 0 + 0 \le 6$ A true statement

$0 + 0 < 4$ A true statement

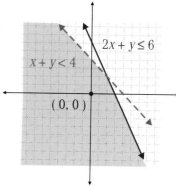

Using a TI-83 Plus Graphing Calculator to Graph Systems of Linear Inequalities

To graph a system of linear inequalities with a TI-83 Plus graphing calculator, first solve each inequality for y and then enter both of the corresponding functions after pressing the ⬭ Y= ⬭ key. By setting the graphing symbol to the left of Y1 and Y2 to the desired form then pressing GRAPH, the desired region will be graphed as a cross-hatched area on the display (assuming that the window is set correctly). The following example shows how this can be done.

Example 2: Graphing Systems of Linear Inequalities ● ● ● ● ●

Use a TI-83 Plus graphing calculator to graph the following system of linear inequalities:

$$\begin{cases} 2x + y < 4 \\ 2x - y \le 0 \end{cases}.$$

Solution: **Step 1:** Solve each inequality for y: $\begin{cases} y < -2x + 4 \\ y \ge 2x \end{cases}.$

(**Note**: Solving $2x - y \le 0$ for y can be written as $2x \le y$ and then as $y \ge 2x$.)

Step 2: Press the ⬭ Y= ⬭ key and enter both functions and the corresponding symbols as they appear here:

Step 3: Press **GRAPH**. The display should appear as follows. The solution is the cross-hatched region.

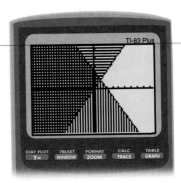

● ●

Exercises A.2

In Exercises 1 – 20, solve the systems of two linear inequalities graphically.

1. $\begin{cases} y > 2 \\ x \geq -3 \end{cases}$ **2.** $\begin{cases} 2x + 5 < 0 \\ y \geq 2 \end{cases}$ **3.** $\begin{cases} x < 3 \\ y > -x + 2 \end{cases}$ **4.** $\begin{cases} y \leq -5 \\ y \geq x - 5 \end{cases}$

5. $\begin{cases} x \leq 3 \\ 2x + y > 7 \end{cases}$ **6.** $\begin{cases} 2x - y > 4 \\ y < -1 \end{cases}$ **7.** $\begin{cases} x - 3y \leq 3 \\ x < 5 \end{cases}$ **8.** $\begin{cases} 3x - 2y \geq 8 \\ y \geq 0 \end{cases}$

9. $\begin{cases} x - y \geq 0 \\ 3x - 2y \geq 4 \end{cases}$ **10.** $\begin{cases} y \geq x - 2 \\ x + y \geq -2 \end{cases}$ **11.** $\begin{cases} 3x + y \leq 10 \\ 5x - y \geq 6 \end{cases}$ **12.** $\begin{cases} y \geq 2x - 5 \\ 3x + 2y > -3 \end{cases}$

13. $\begin{cases} 3x + 4y \geq -7 \\ y < 2x + 1 \end{cases}$ **14.** $\begin{cases} 2x - 3y \geq 0 \\ 8x - 3y < 36 \end{cases}$ **15.** $\begin{cases} x + y < 4 \\ 2x - 3y < 3 \end{cases}$ **16.** $\begin{cases} 2x + 3y < 12 \\ 3x + 2y > 13 \end{cases}$

17. $\begin{cases} x + y \geq 0 \\ x - 2y \geq 6 \end{cases}$ **18.** $\begin{cases} y \geq 2x + 3 \\ y \leq x - 2 \end{cases}$ **19.** $\begin{cases} x + 3y \leq 9 \\ x - y \geq 5 \end{cases}$ **20.** $\begin{cases} x - y \geq -2 \\ x + 2y < -1 \end{cases}$

Use a graphing calculator to solve the systems of linear inequalities in Exercises 21 – 30.

21. $\begin{cases} y \geq 0 \\ 3x - 5y \leq 10 \end{cases}$ **22.** $\begin{cases} 3x + 2y \leq 15 \\ 2x + 5y \geq 10 \end{cases}$ **23.** $\begin{cases} 4x - 3y \geq 6 \\ 3x - y \leq 3 \end{cases}$ **24.** $\begin{cases} y \leq 0 \\ 3x + y \leq 11 \end{cases}$

25. $\begin{cases} 3x - 4y \geq -6 \\ 3x + 2y \leq 12 \end{cases}$ **26.** $\begin{cases} 3y \leq 2x + 2 \\ x + 2y \leq 11 \end{cases}$ **27.** $\begin{cases} x + y \leq 8 \\ 3x - 2y \geq -6 \end{cases}$ **28.** $\begin{cases} x + y \leq 7 \\ 2x - y \leq 8 \end{cases}$

29. $\begin{cases} y \leq x \\ y < 2x + 1 \end{cases}$ **30.** $\begin{cases} x - y \geq -2 \\ 4x - y < 16 \end{cases}$

Writing and Thinking About Mathematics

31. The material in the text and the exercises do not discuss cases in solving a system of two linear inequalities in which the boundary lines are parallel. Describe, in your own words, what you think the solutions in such cases might be. After you have done this, solve the following systems graphically and discuss how these cases relate to your previous analysis.

a. $\begin{cases} y \leq 2x - 5 \\ y \geq 2x + 3 \end{cases}$ **b.** $\begin{cases} y \leq -x + 2 \\ y \geq -x - 1 \end{cases}$ **c.** $\begin{cases} y \leq \dfrac{1}{2}x + 3 \\ y \geq \dfrac{1}{2}x - 3 \end{cases}$

Hawkes Learning Systems: Introductory Algebra

 Systems of Linear Inequalities

A.3

Pi

As discussed in the text on page 5, π is an irrational number, and so the decimal form of π is an infinite nonrepeating decimal. Mathematicians even in ancient times realized that π is a constant value obtained from the ratio of a circle's circumference to its diameter, but they had no sense that it might be an irrational number. As early as about 1800 B.C. the Babylonians gave π a value of 3, and around 1600 B.C. the ancient Egyptians were using the approximation of 256/81, what would be decimal value of about 3.1605. In the third century B.C. the Greek mathematician Archimedes used polygons approximating a circle to determine that the value of π must lie between 223/71(=3.1408) and 22/7(=3.1429). He was thus accurate to two decimal places. About seven hundred years later, in the fourth century A.D. Chinese mathematician Tsu Chung-Chi refined Archimedes' method and expressed the constant as 355/113, which was correct to six decimal places. By 1610, Ludolph van Ceulen of Germany had also used a polygon method to find π accurate to 35 decimal places.

Knowing that the decimal expression of π would not terminate, mathematicians still sought a repeating pattern in its digits. Such a pattern would mean that π was a rational number and that there would be some ratio of two whole numbers that would produce the correct decimal representations. Finally, in 1767, Johann Heinrich Lambert provided a proof to show that π is indeed irrational and thus is nonrepeating as well as nonterminating.

Since Lambert's proof, mathematicians have still made an exercise of calculating π to more and more decimal places. The advent of the computer age in this century has made that work immeasurably easier, and on occasion you will still see newspaper articles pronouncing that mathematics researchers have reached a new high in the number of decimal places in their approximations. In 1988 that number was 201,326,000 decimal places. Within 1 year that record was more than doubled, and most recent approximations of π now reach beyond two hundred billion decimal places! For your understanding, appreciation and interest, the value of π is given in the table on the next page to a mere 3742 decimal places as calculated by a computer program. To show π calculated to one billion decimal places would take every page of nearly 400 copies of this text!

The Value of π

π =

3.141592653589793238462643383279502884197169399375105820974944592307816406286
20899862803482534211706798214808651328230664709384460955058223172535940812848
1117450284102701938521105559644622948954930381964428810975665933446128475648 2
33786783165271201909145648566923460348610454326648213393607260249141273724587
00660631558817488152092096282925409171536436789259036001133053054882046652138
41469519415116094330572703657595919530921861173819326117931051185480744623799
62749567351885752724891227938183011949129833673362440656643086021394946395224
73719070217986094370277053921717629317675238467481846766940513200056812714526
35608277857713427577896091736371787214684409012249534301465495853710507922796
89258923542019956112129021960864034418159813629774771309960518707211349999998
37297804995105973173281609631859502445945534690830264252230825334468503526193
11881710100031378387528865875332083814206171776691473035982534904287554687311
59562863882353787593751957781857780532171226806613001927876611195909216420198
93809525720106548586327886593615338182796823030195203530185296899577362259941
38912497217752834791315155748572424541506959508295331168617278558890750983817
54637464939319255060400927701671139009848824012858361603563707660104710181942
95559619894676783744944825537977472684710404753464620804668425906949129331367
70289891521047521620569660240580381501935112533824300355876402474964732639141
99272604269922796782354781636009341721641219924586315030286182974555706749838
50549458858692699569092721079750930295532116534498720275596023648066549911988
18347977535663698074265425278625518184175746728909777727938000816470600161452
49192173217214772350141441973568548161361157352552133475741849468438523323907
39410335454776241686251898356948556209921922218427255025425688767179049460165
34668049886272327917860857843838279679766814541009538837863609506800642251252
05117392984896084128488626945604241965285022210661186306744278622039194945047
12371378696095563643719172874677646575739624138908658326459958133904780275900 9
94657640789512694683983525957098258226205224894077267194782684826014769909026
40136394437455305068203496252451749399651431429809190659250937221696461515709
85838741059788595977297549893016175392846813826868386894277415599185592524595
39594310499725246808459872736446958486538367362226260991246080512438843904512
44136549762780797715691435997700129616089441694868555848406353422072225828488
64815845602850601684273945226746767889525213852254995466672782398645659611635
48862305774564980355936345681743241125150760694794510965960940252288797108931
45669136867228748940560101503308617928680920874760917824938589009714909675985
26136554978189312978482168299894872265880485756401427047755513237964145152374
62343645428584447952658678210511413547357395231134271661021359695362314429524
84937187110145765403590279934037420073105785390621983874478084784896833214145
71386875194350643021845319104848100537061468067491927819119793995206141966342
87544406437451237181921799983910159195618146751426912397489409071864942319615
67945208095146550225231603881930142093762137855956638937787083039069792077346
72218256259966150142150306803844773454920260541466592520149744285073251866600
21324340881907104863317346496514539057962685610055081066587969981635747363840
52571459102897064140110971206280439039759515677157700420337869936007230558763
17635942187312514712053292819182618612586732157919841484882916447060957527069
57220917567116722910981690915280173506712748583222871835209353965725121083579
15136988209144421006751033467110314126711136990865851639831501970165151168517
14376576183515565088490998985998238734552833163550764791853589322618548963213
29330898570642046752590709154814165498594616371802709819943099244889575712828
905923233260972997120844335732654893823911932597...

Chapter 1

Exercises 1.1, Pages 12 – 14

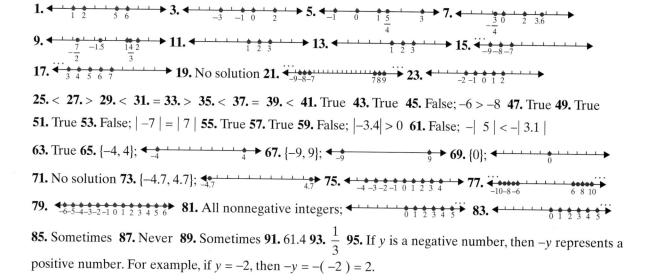

25. < **27.** > **29.** < **31.** = **33.** > **35.** < **37.** = **39.** < **41.** True **43.** True **45.** False; $-6 > -8$ **47.** True **49.** True
51. True **53.** False; $|-7| = |7|$ **55.** True **57.** True **59.** False; $|-3.4| > 0$ **61.** False; $-|5| < -|3.1|$
63. True **65.** $\{-4, 4\}$; **67.** $\{-9, 9\}$; **69.** $\{0\}$;
71. No solution **73.** $\{-4.7, 4.7\}$; **75.** **77.**
79. **81.** All nonnegative integers; **83.**
85. Sometimes **87.** Never **89.** Sometimes **91.** 61.4 **93.** $\dfrac{1}{3}$ **95.** If y is a negative number, then $-y$ represents a positive number. For example, if $y = -2$, then $-y = -(-2) = 2$.

Exercises 1.2, Pages 19 – 21

1. 13 **3.** -4 **5.** 0 **7.** 5 **9.** -13 **11.** -8 **13.** -10 **15.** 0 **17.** 17 **19.** -29 **21.** -9 **23.** -3 **25.** 22 **27.** -54 **29.** -7 **31.** -16
33. -26 **35.** 0 **37.** -32 **39.** 5 **41.** -83 **43.** 12 **45.** -32 **47.** -2 is a solution **49.** -4 is a solution **51.** -6 is a solution
53. 18 is a solution **55.** -10 is a solution **57.** -2 is not a solution **59.** -72 is not a solution **61.** Sometimes
63. Never **65.** Never **67.** Sometimes **69.** Always **71.** 84 **73.** $-97,714$ **75.** -6143

Exercises 1.3, Pages 26 – 28

1. -11 **3.** 6 **5.** -47 **7.** 0 **9.** 52 **11.** 5 **13.** -10 **15.** 12 **17.** 3 **19.** -16 **21.** 24 **23.** -15 **25.** -16 **27.** -57 **29.** -54 **31.** 1
33. 8 **35.** -26 **37.** 1 **39.** -8 **41.** -10 **43.** -6 **45.** -139 **47.** $-1 > -7$ **49.** $10 > -10$ **51.** $-6 < 6$ **53.** $-4 > -5$
55. $-37 < -34$ **57.** -3 is a solution **59.** 3 is a solution **61.** 4 is not a solution **63.** -10 is a solution **65.** 12 is a
solution **67.** -9 is a solution **69.** 16 is not a solution **71.** $1044 > -39$ **73.** $-15,254 > -35,090$ **75.** 2316 points
77. $6°$ below 0 (or $-6°$)

Exercises 1.4, Pages 33 – 35

1. –12 **3.** 56 **5.** 57 **7.** 56 **9.** –30 **11.** 26 **13.** –60 **15.** –24 **17.** –288 **19.** 0 **21.** 4 **23.** –6 **25.** –3 **27.** 13 **29.** 0
31. undefined **33.** –11 **35.** –4 **37.** negative **39.** negative **41.** negative **43.** 0 **45.** undefined **47.** True **49.** True
51. True **53.** False; $17 + (-3) > (-14) + (-4)$ **55.** True **57.** –12 is a solution **59.** –72 is a solution **61.** –8 is not a
solution **63.** 5 is a solution **65.** –4 is a solution **67.** –34,459,110 **69.** –2671 **71.** Dividing 0 by any number other
than 0 gives a quotient of 0.

Exercises 1.5, Pages 43 – 44

1. 5^2 **3.** 7^2 **5.** 11^2 **7.** 8^2 or 4^3 or 2^6 **9.** 2^3 **11.** 5^3 **13.** 3^5 **15.** 10^2 **17.** 64 **19.** 100 **21.** 343 **23.** 10,000 **25.** 1 **27.** 225
29. 27,000 **31.** 10,000 **33.** 36 **35.** –64 **37.** 15,625 **39.** 262,144 **41.** –225 **43.** –8 **45.** 196 **47.** $2 \cdot 5 \cdot 7$
49. 43 is prime **51.** $3^2 \cdot 5^2$ **53.** $3 \cdot 11^2$ **55.** $2 \cdot 3^2 \cdot 5$ **57.** $2^2 \cdot 5 \cdot 7$ **59.** $2^3 \cdot 5^3$ **61. a.** 36 **b.** 16 **63.** –25 **65.** –10 **67.** –45
69. –137 **71.** 152 **73.** –189 **75.** 35 **77.** –36 **79.** $(3^2 - 9) = 0$ and division by 0 is undefined.

Exercises 1.6, Pages 49 – 50

1. a. $-1, 0, 4$ **b.** $-3.56, -\dfrac{5}{8}, -1, 0, 0.7, 4, \dfrac{13}{2}$ **c.** $-\sqrt{8}, \pi, \sqrt[3]{25}$ **d.** All are real numbers **3.** $3 + 7$ **5.** $4 \cdot 19$
7. $30 + 48$ **9.** $(2 \cdot 3) \cdot x$ **11.** $(3 + x) + 7$ **13.** $0 \cdot 6 = 0$ **15.** $x + 7$ **17.** $2x - 24$ **19.** Commutative property of
multiplication **21.** Additive identity **23.** Commutative property of addition **25.** Commutative property of
multiplication **27.** Commutative property of multiplication **29.** Commutative property of addition;
$19 + 3 = 3 + 19 = 22$ **31.** Associative property of multiplication; $(2 \cdot 7) \cdot 4 = 2 \cdot (7 \cdot 4) = 56$ **33.** Associative
property of addition; $(2(3) + 14) + 3 = 2(3) + (14 + 3) = 23$ **35.** Commutative property of multiplication;
$11 \cdot 4 = 4 \cdot 11 = 44$ **37.** Distributive property; $3(-2 + 15) = 3(-2) + 3(15) = 39$ **39.** Commutative property of
multiplication; $(-2 + 2)(-2 - 4) = (-2 - 4)(-2 + 2) = 0$ **41.** Commutative property of addition;
$3 + (4 + 4) = (4 + 4) + 3 = 11$ **43.** $6(11) = 66$ and $6 \cdot 3 + 6 \cdot 8 = 18 + 48 = 66$ **45.** $10(-7) = -70$ and
$10 \cdot 2 - 10 \cdot 9 = -70$ **47.** $5(-2) = -10$ and $5(14) - 5(16) = 70 - 80 = -10$

Chapter 1 Test, Pages 55 – 56

1. a. The integers are –10 and 0 **b.** The rational numbers are $-10, \dfrac{-3}{4}, 0, \dfrac{7}{6}, 3\dfrac{4}{9}$ and 7.121212... **c.** The
irrational numbers are $-\pi$ and $-\sqrt{5}$ **d.** All are real numbers **2. a.** True. Any integer n can be represented as
$\dfrac{n}{1}$ which is a rational number, since the numerator and the denominator are integers. **b.** False, because
rational numbers also include non-integers, such as $\dfrac{2}{5}$. **3.** Because in the second expression the negative sign
is inside the parenthesis, hence –10 is squared to get a positive value while the first expression is negative:
$-10^2 = -100$ and $(-10)^2 = 100$

4. a. < **b.** > **c.** = **5.**

6.

7. $y = 7$ or $y = -7$ **8.** 2, 3, 5, 7, 11, 13, 17, 19, 23, 29, 31, 37, 41, 43 and 47 **9.** $\{-2, -1, 0, 1, 2\}$;

10. $\{..., -10, -9, 9, 10, ...\}$;

11. 13 **12.** −90 **13.** −18 **14.** 19 **15.** 1 **16.** 162 **17.** 7 **18.** 0 **19.** 8^2 or 4^3 or 2^6 **20. a.** −25 **b.** −265 **c.** −44 **21. a.** $2^4 \cdot 5$ **b.** $5^2 \cdot 7$ **22. a.** Multiplicative Identity **b.** Commutative Property of Addition **c.** Associative Property of Addition **d.** Zero Factor Law **e.** Distributive Property **f.** Commutative Property of Multiplication **23.** −60 **24.** $|x| = x$ if x is positive or 0. $|x| = -x$ if x is negative. **25. a.** −19,487,171 **b.** 0 **26.** Gain of 152 points and average at the end of the week is 11,822 points.

Chapter 2

Exercises 2.1, Pages 65 – 67

1. $\dfrac{5}{6} = \dfrac{5}{6} \cdot \dfrac{8}{8} = \dfrac{40}{48}$ **3.** $\dfrac{0}{9} = \dfrac{0}{9} \cdot \dfrac{7b}{7b} = \dfrac{0}{63b}$ **5.** $\dfrac{2}{5}$ **7.** $\dfrac{3x}{7}$ **9.** $\dfrac{3ab}{25}$ **11.** $\dfrac{-2}{3y}$ **13.** $\dfrac{-1}{2x^2}$ **15.** $\dfrac{-12y}{35}$ **17.** $\dfrac{3}{8}$ **19.** 35 **21.** $\dfrac{-1}{6}$

23. $\dfrac{81}{100}$ **25.** $\dfrac{-7}{4}$ **27.** $\dfrac{50}{9}$ **29.** $\dfrac{4}{3y}$ **31.** $\dfrac{-21}{8}$ **33.** undefined **35.** $\dfrac{22}{a^2}$ **37.** $\dfrac{25}{98p^3q^2}$ **39.** $\dfrac{16}{5}$ **41.** $\dfrac{25}{32}$ **43.** 2 inches

45. a. more than 60 **b.** less than 60 **c.** 72 **47.** $\dfrac{7}{20}$ and $\dfrac{13}{20}$ **49.** $\dfrac{135}{32}$ ft. or $4\dfrac{7}{32}$ ft. **51.** $\dfrac{-25}{24}$ **53.** Division by zero is undefined.

Exercises 2.2, Pages 76 – 79

1. a. 1, 2, 3, 6 **b.** 6, 12, 18, 24, 30, 36 **3. a.** 1, 3, 5, 15 **b.** 15, 30, 45, 60, 75, 90 **5.** 120 **7.** $40xy$ **9.** $210x^2y^2$ **11.** $\dfrac{7}{9}$

13. $\dfrac{5}{23}$ **15.** $-\dfrac{1}{6}$ **17.** $\dfrac{4}{3}$ **19.** $-\dfrac{28}{25x}$ **21.** $\dfrac{83}{60}$ **23.** $\dfrac{5}{42}$ **25.** $\dfrac{33}{70}$ **27.** $\dfrac{3}{4x}$ **29.** $\dfrac{5}{24x}$ **31.** 0 **33.** $\dfrac{12+y}{2y}$ **35.** $\dfrac{x-2}{14}$

37. $\dfrac{6a-25}{30}$ **39.** $\dfrac{5b+22}{20}$ **41.** $\dfrac{11}{30}$ **43.** $\dfrac{31}{24}$ **45.** $\dfrac{7}{2}$ **47.** $\dfrac{-23}{21y}$ **49.** $\dfrac{7}{5}$ **51.** $\dfrac{1}{18,009,460}$ **53.** $\dfrac{92}{19}$ **55.** 344 yd.

57. a. $\dfrac{13}{30}$ **b.** $1170

Exercises 2.3, Pages 89 – 90

1. 0.86; eighty-six hundredths **3.** 5.1; five and one tenth **5.** −18.06; negative eighteen and six hundredths **7.** 0.087 **9.** −5.14 **11.** 7.0021 **13.** 91.1 **15.** 0.50 **17.** 67.057 **19.** 317.23 **21.** 263.51 **23.** −55.58 **25.** −152.83 **27.** 108.72 **29.** −7.626 **31.** 15.1 **33.** 0.375 **35.** 0.05 **37.** 0.4375 **39.** $-0.\overline{54}$ **41.** $1.\overline{185}$ **43.** $\dfrac{13}{10}$ **45.** $\dfrac{1323}{250}$

47. $\dfrac{69}{40}$ **49. a.** For all positive numbers, the product will be less than the other number.

b. The product will equal the other number when the other number is 0. **c.** The product will be more than the other number when the other number is negative.

Answers

Exercises 2.4, Pages 94 – 99

1. $-23 + 13 - 6$; -16 lbs **3.** $4 + 3 - 9$; $-2°$ **5.** $47 - 22 + 8 - 45$; $-\$12$ **7.** $14 - 6 + 11 - 15$; $4°$
9. $187 - 241 + 82 + 26$; $\$54$ **11.** 4 yd. gain **13.** 8th floor **15.** $\$150$ **17.** -13 **19.** -11 **21.** -33 **23.** -15
25. -26 **27.** $\$220$ **29.** $-18°F$, $-3°F$ **31.** $-\$8$ **33.** $-14{,}777$ ft. **35.** 85 years old **37.** -48 **39.** 78.57 **41.** 26.6 calls
43. 76.0 **45.** 2.2 books **47.** 42 calls

Exercise 2.5, Pages 106 – 107

1. -5, $\dfrac{1}{6}$, and 8 are like terms; $7x$ and $9x$ are like terms **3.** $-x^2$ and $2x^2$ are like terms; $5xy$ and $-6xy$ are like
terms; $3x^2y$ and $5x^2y$ are like terms. **5.** 24, 8.3 and -6 are like terms; $1.5xyz$, $-1.4xyz$ and xyz are like terms.
7. 64 **9.** -121 **11.** $15x$ **13.** $3x$ **15.** $-2n$ **17.** $5y^2$ **19.** $12x^2$ **21.** $7x + 2$ **23.** $x - 3y$ **25.** $8x^2 + 3y$ **27.** $2n + 3$
29. $7a - 8b$ **31.** $8x + y$ **33.** $2x^2 - x$ **35.** $-2n^2 + 2n$ **37.** $3x^2 - xy + y^2$ **39.** $2x$ **41.** $-y$ **43. a.** $3x + 4$ **b.** 16
45. a. $3.6x^2$ **b.** 57.6 **47. a.** $\dfrac{17x}{8}$ **b.** 8.5 **49. a.** 0 **b.** 0 **51. a.** $-2x - 8$ **b.** -16 **53. a.** $9y + 2$ **b.** 29 **55. a.** $10a + 13$ **b.** -7
57. a. $3ab + b^2 + b^3$ **b.** 6 **59. a.** $8a$ **b.** -16 **61. a.** $6a$ **b.** $9a^2$ **c.** $\dfrac{2}{3a}$ **d.** $\dfrac{1}{9a^2}$ **e.** 1 **63. a.** $-14ab$ **b.** $49a^2b^2$ **c.** $\dfrac{2}{7ab}$
d. $\dfrac{1}{49a^2b^2}$ **e.** 1 **65. a.** $3a$ **b.** $2.25a^2$ **c.** $\dfrac{4}{3a}$ **d.** $\dfrac{4}{9a^2}$ **e.** -1

Exercise 2.6, Pages 112 – 113

1. 4 times a number **3.** 1 more than twice a number **5.** 5.3 less than 7 times a number **7.** -2 times the
difference between a number and 8 **9.** 5 times the sum of twice a number and 3 **11.** 6 times the difference
between a number and 1 **13.** 3 times a number plus 7; 3 times the sum of a number and 7 **15.** The product of 7
and a number minus 3; 7 times the difference between a number and 3 **17.** $x + 6$ **19.** $x - 4$ **21.** $3x - 5$ **23.** $\dfrac{x - 3}{7}$
25. $3(x - 8)$ **27.** $3y - 5$ **29.** $8 - 2x$ **31.** $8(x - 6) + 4$ **33.** $3x - 5x$ **35.** $2(x - 7) - 6$ **37.** $2(17 + x) + 9$ **39. a.** $x - 6$
b. $6 - x$ **41.** $\$4.95x$ **43.** $7t + 3$ **45.** $6t + 3$ **47.** $3x + 8$ **49.** $c + 0.2c = 1.2c$

Chapter 2 Test, Pages 117 – 118

1. $\dfrac{5}{9} = \dfrac{5}{9} \cdot \dfrac{4}{4} = \dfrac{20}{36}$ **2.** $\dfrac{2}{3a} = \dfrac{2}{3a} \cdot \dfrac{4a}{4a} = \dfrac{8a}{12a^2}$ **3.** $\dfrac{-4y}{3}$ **4.** $\dfrac{7b}{11a^2}$ **5.** 96 **6.** $90x^2y^2$ **7.** $\dfrac{-9}{2}$ **8.** $\dfrac{3}{5}$ **9.** $\dfrac{9}{10}$ **10.** $\dfrac{-2}{y}$
11. $\dfrac{29}{40}$ **12.** $\dfrac{15 + n}{5n}$ **13.** Seventeen and three thousandths **14.** 83.15 **15.** 36.53 **16.** -16.5 **17.** 17.952 **18.** 1.35
19. $7 - 5x$ **20.** $4.4y$ **21.** $7a^2 - 5a + 10$ **22. a.** $\dfrac{19y}{24}$ **b.** $\dfrac{19}{8}$ **23. a.** $2x - 16$ **b.** -20
24. $6x - 3$ **25.** $2(x + 5)$ **26.** $2(x + 5) - 4$ **27.** -9 **28.** 79.0 **29.** $\dfrac{-33}{16}$ **30. a.** -0.7 **b.** $\$15.80$ **31. a.** more **b.** less
c. $\$80$ **32. a.** 15 gallons **b.** No, $1 short.

Chapter 2 Cumulative Review, Pages 119 – 121

1. 20 **2.** –14 **3.** 9 **4.** –15 **5.** –126 **6.** 459 **7.** 39 **8.** –124 **9. a.** 17 **b.** –26 **10. a.** 81 **b.** –512
11. a. $2^3 \cdot a^2 \cdot b$ **b.** $3^2 \cdot 5^2 \cdot a \cdot b^3$ **12. a.** $2^2 \cdot 3 \cdot 5$ **b.** $3^2 \cdot 17$ **13.** 108 **14.** $72xy^2$ **15.** Associative property of multiplication **16.** Commutative property of multiplication **17.** Distribution property **18.** Commutative property of addition **19.** Associative property of addition **20.** Multiplicative identity **21.** –2.3, 2.3

22. **23.** 19 **24.** 18 **25.** 90 **26.** 12 **27.** $\dfrac{43}{40}$ **28.** $-\dfrac{1}{36}$ **29.** $\dfrac{22}{15y}$ **30.** $\dfrac{1}{4a}$ **31.** $\dfrac{18+x}{6x}$

32. $\dfrac{5}{6}$ **33.** $\dfrac{7}{5}$ **34.** $\dfrac{2}{9}$ **35.** $\dfrac{1}{6}$ **36.** $\dfrac{-7}{45}$ **37.** $\dfrac{263}{576}$ **38.** $\dfrac{3}{5}$ **39.** 0 **40.** $\dfrac{-7}{16}$ **41.** 17.27 **42.** –15.31 **43.** 14.6

44. 58.19 **45.** 5.292 **46. a.** $|x+y|+|x-y|$ **b.** 10 **47. a.** $3|x|+5|y|-2(x-y)$ **b.** 11 **48. a.** $2x+2$ **b.** 12
49. a. y^3+2y^2-3y-3 **b.** 3 **50. a.** $2x-3$ **b.** $\dfrac{x}{6}+3x$ **c.** $2(x+10)-13$ **51. a.** The sum of 6 and 4 times a

number **b.** 18 times the difference of a number and 5 **52.** $-\dfrac{5}{51}$ **53.** –5 **54.** $\dfrac{10}{3}$ cups (or $3\dfrac{1}{3}$ cups)

55. 21.7 **56. a.** 1000 **b.** 2500 **57.** $5.9x$ **58.** $\dfrac{a}{6}$ **59.** $\dfrac{14x}{15}$ **60. a.** $10x$ **b.** $25x^2$ **c.** $\dfrac{2}{5x}$ **d.** $\dfrac{1}{25x^2}$ **e.** 1 **61. a.** 0 **b.** $-36y^2$

c. $\dfrac{1}{3y}$ **d.** $\dfrac{-1}{36y^2}$ **e.** –1

Chapter 3

Exercises 3.1, Pages 135 – 139

1. $y=6$ **3.** $n=-10$ **5.** $y=5.2$ **7.** $x=3$ **9.** $x=25$ **11.** $a=0$ **13.** $x=\dfrac{1}{2}$ **15.** $x=20$ **17.** $x=-4$ **19.** $y=-2$;
21. $a=1$ **23.** $x=-1$ **25.** $a=6.2$ **27.** 25 m, 34 m, 12 m **29.** 16.125 m **31.** $x=2$ in. **33.** 36 in.
35. Yes; **37.** 0.25 years (or 3 months) **39.** 24% annual interest rate, \$4,400 **41.** \$855 **43.** \$1690

45. 3.5 hrs. **47.** $x=-17.214$ **49.** $x=246$

Exercises 3.2, Pages 144 – 148

1. Use the distributive property; Add 12 to both sides; Simplify; Add x to both sides; Simplify; Divide both sides by 4; Simplify **3.** Multiply both sides by 30; Use the distributive property; Simplify; Add –5 to both sides; Simplify; Add $-12a$ to both sides; Simplify; Divide both sides by –2; Simplify. **5.** $x=-3$ **7.** $x=-0.12$ **9.** $x=-5$
11. $y=-4$ **13.** $n=0$ **15.** $x=0$ **17.** $z=-1$ **19.** $y=\dfrac{1}{5}$ **21.** $x=-4$ **23.** $x=-3$ **25.** $x=-21$ **27.** $y=0$

29. $x = -5$ **31.** $n = \dfrac{1}{6}$ **33.** $n = 2$ **35.** $x = \dfrac{7}{60}$ **37.** $x = \dfrac{8}{5}$ **39.** $x = \dfrac{2}{3}$ **41.** $l = 64$ ft **43.** 10 cm **45.** 12 ft, 38 ft **47.** \$490 **49.** $x = -9.293$ **51.** $x = 4.58$

Exercises 3.3, Pages 154 – 158

1. $x - 5 = 13 - x$; 9 **3.** $36 = 2x + 4$; 16 **5.** $7x = 2x + 35$; 7 **7.** $3x + 14 = 6 - x$; -2 **9.** $\dfrac{2x}{5} = x + 6$; -10

11. $4(x - 5) = x + 4$; 8 **13.** $\dfrac{2x + 5}{11} = 4 - x$; 3 **15.** $x - 21 = 8x$; -3 **17.** $x + x + 4 = 24$; 10 cm, 10 cm

19. $x + x + 2x - 3 = 45$; 12 cm, 12 cm, 21 cm **21.** $3x + 1500 = 12{,}000$; \$3500 **23.** $n + n + 2 = 60$; 29, 31

25. $2n + 3(n + 1) = 83$; 16, 17 **27.** $n + 2 + n + 4 - n = 66$; 60, 62, 64 **29.** $n + n + 2 + n + 4 = n + 2 + 168$; 82, 84, 86

31. $2l + 2(75) = 410$; 130 yards **33.** $\dfrac{2l}{3} = 18$; 27 ft. **35.** $x + x + 2 + x + 6 = 29$; 7 ft, 9 ft, 13 ft

37. $c + c + 49.50 = 125.74$; calculator: \$38.12, textbook: \$87.62 **39.** $2n + 3(n + 2) = 2(n + 4) + 7$; 3, 5, 7

41. Find two consecutive intergers whose sum is 33; 16, 17 **43.** Find 3 consecutive even integers such that the

sum of the first and the third is 3 times the second; $-2, 0, 2$ **45.** The quotient of a number and 2 increased by $\dfrac{1}{3}$

is equal to 3 times the number divided by 4; $\dfrac{4}{3}$

Exercises 3.4, Pages 167 – 171

1. 91% **3.** 137% **5.** 37.5% **7.** 150% **9.** 0.69 **11.** 0.113 **13.** 0.005 **15.** 0.82 **17.** $\dfrac{7}{20}$ **19.** $\dfrac{13}{10}$ **21. a.** 32% **b.** 28%
c. 8% **23. a.** 59.7% **b.** 16.7% **25.** 61.56 **27.** 40 **29.** 2180 **31.** 125% **33.** 80 **35. a.** 8% **b.** 10% **c.** b **37. a.** \$990
b. 22% **39.** \$1952.90 **41.** 1.5% **43.** 40 **45. a.** \$700 **b.** 33.33% **47. a.** \$7134.29 **b.** \$6609.24 **49. a.** 10% **b.** 11.11%
c. The first percentage is a percentage of his original weight while the second percentage is a percentage of
his weight after he lost weight. **51. a.** Because the 6% is against the selling price, not the \$141,000 the couple
wanted. **b.** The selling price is 100%. Therefore, \$141,000 is 94% of the selling price (selling price(100%) –
realtor fee (6%) = amount for couple(94%)). The selling price should be \$150,000.

Exercises 3.5, Pages 179 – 183

1. \$120 **3.** 72 days **5.** \$10,000 **7.** 176 ft/sec **9.** 4 milliliters **11.** \$1120 **13.** 14 **15.** 336 in. or 28 ft. **17.** \$337.50 **19.** \$2400
21. $b = P - a - c$ **23.** $m = \dfrac{F}{a}$ **25.** $w = \dfrac{A}{l}$ **27.** $n = \dfrac{R}{p}$ **29.** $P = A - I$ **31.** $m = 2A - n$ **33.** $t = \dfrac{I}{Pr}$ **35.** $b = \dfrac{P - a}{2}$

37. $\beta = 180 - \alpha - \gamma$ **39.** $h = \dfrac{V}{lw}$ **41.** $b = \dfrac{2A}{h}$ **43.** $\pi = \dfrac{A}{r^2}$ **45.** $g = \dfrac{mv^2}{2K}$ **47.** $y = \dfrac{6 - 2x}{3}$ **49.** $x = \dfrac{11 - 2y}{5}$

51. $b = \dfrac{2A - hc}{h}$ or $b = \dfrac{2A}{h} - c$ **53.** $x = \dfrac{8R + 36}{3}$ **55.** $y = -x - 12$ **57.** $C = 0.12x + 25$ **59.** $C = 358n - 5400$ **61. a.** 1
b. -1 **c.** 1.5 **d.** -2

Exercises 3.6, Pages 191 – 194

1. e **3.** k **5.** c **7.** d **9.** g **11.** l **13.** m **15.** h **17.** 70 cm; 300 cm^2 **19.** 54 mm; 126 mm^2 **21.** 31.4 in; 78.5 in^2
23. 1436.03 in^3 **25.** 729 m^3 **27.** 56 in **29.** 63 cm^2 **31.** 6154.4 ft^3. **33.** 125 ft^3 **35.** 272 cm^2 **37.** 529.875 in^3
39. 28 cm; 48 cm^2 **41.** 212.04 m^2 **43.** 130.08 in^2 **45.** 6,334,233.21 cm^2

Exercises 3.7, Pages 204 – 206

1. False; $3 > -3$ **3.** True **5.** True **7.** True **9.** True **11.** half-open interval; **13.** half-open

interval; **15.** half-open interval; **17.** half-open interval;

19. open interval; **21.** $x \geq -\dfrac{11}{4}$;

23. $x > -3$; **25.** $y \leq 5.15$; **27.** $x > 0.9$;

29. $y < -\dfrac{8}{3}$; **31.** $x > -\dfrac{3}{2}$; **33.** $x \leq 8$;

35. $x > \dfrac{9}{2}$; **37.** $x \geq -3$; **39.** $x \leq -9$;

41. $-2 < x < 6$; **43.** $1 < x < 2$; **45.** $-\dfrac{8}{5} \leq x < \dfrac{6}{5}$;

47. $-\dfrac{21}{2} \leq x \leq 3$; **49.** $\dfrac{20}{3} \leq x < \dfrac{33}{4}$;

51. $44 \leq x \leq 100$ **53.** $59°$ to $149°$ F **55.** The second side is between 11 cm and 35 cm and the third side is

between 6 cm and 18 cm. **57.** $86 \leq x \leq 100$ **59.** 171 adult tickets **61.** less than 100 miles

Chapter 3 Test, Pages 213 – 215

1. $x = -3$ **2.** $x = 5$ **3.** $x = 0$ **4.** $x = -\dfrac{9}{8}$; **5.** $x = 20$; **6.** $x = -2$ **7.** 111.6 **8.** 32% **9.** $337.50 **10.** The $6000 investment
is the better investment since it has a higher percent of profit, 6%, than the $10,000 investment (5%).

11. 3 m, 4 m, 5 m **12.** $m = \dfrac{N - p}{rt}$ **13.** $y = \dfrac{7 - 5x}{3}$ **14.** $-20°$C **15.** $x = \dfrac{26}{3}$ **16. a.** perimeter = 27.42 ft

b. area = 50.13 ft^2 **17.** 4186.67 cm^2 **18.** $x \leq -\dfrac{10}{3}$; **19.** $x < -20$;

20. $x > \dfrac{-13}{3}$; **21.** $\dfrac{-9}{4} \leq x \leq -1$; **22.** $(2y + 5) + y = -22$; $-9, -13$
23. $2n + 3(n + 1) = 83$; 16, 17 **24.** $74 + 2w = 122$ **a.** 24 in. **b.** 888 in.^{2} **25.** $0.75x = 547.50$ **a.** $730, **b.** $605.35
26. $3(n + 2) = n + (n + 1) + 27$; $n = 25, 27, 29$ **27. a.** 200 ft^3 **b.** 7.4 yd^3 **28.** 100.48 in.^{3}

Chapter 3 Cumulative Review, Pages 216 – 219

1. True **2.** True **3.** False; $\frac{3}{4} \le |-1|$ **4.** 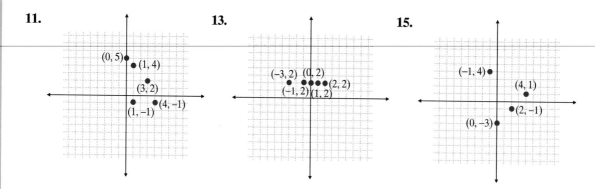 **5.** **6.** See page 31 **7.** 2, 3, 5, 7, 11,

13, 17, 19, 23, 29, 31, 37, 41, 43, 47 **8. a.** $5^2 \cdot 2 \cdot 3$ **b.** 5^3 **c.** $2^4 \cdot 3 \cdot 5$ **9.** 164 **10.** 13 **11.** 37 **12.** $\frac{49}{40}$ **13.** $-\frac{31}{5}$ **14.** 3150

15. $-2y - 12; -18$ **16.** $4x + 7; -1$ **17.** $x^2 + 13x; -22$ **18.** $x - 7; -9$ **19.** $9 - 2x$ **20.** $3(x + 10)$ **21.** $24x + 5$ **22.** $6T + E$

$+ 3F$ **23.** $x = -1$ **24.** $x = 3$ **25.** $x = -11$ **26.** $x = -4$ **27.** $x = \frac{31}{8}$ **28.** $y = \frac{24}{5}$ **29.** $y = \frac{15}{2}$ **30.** $y = 0$ **31.** $v = \frac{h + 16t^2}{t}$

32. $r = \frac{A - P}{Pt}$ **33.** $y = \frac{5x - 14}{3}$ **34.** $133\frac{1}{3}$ **35.** 22% **36.** 16.38 **37.** $x \le 2$;

38. $x \ge \frac{13}{3}$; **39.** $x > \frac{9}{28}$; **40.** $-1.1 \le x \le 7.4$;

41. a. 45.7 m **b.** 139.25 m^2 **42.** 1177.5 in^3 **43.** $80 **44.** $506.94 **45.** $x = 11$ **46.** 18, 20, 22 **47.** $-2, -1, 0, 1$ **48.** 33

miles **49.** The value of the furniture sold was $12,500, Jay's and Kay's salaries were $1,250. **50.** 10.71 in. **51.** 86

$\le x \le 100$ **52.** $10,000 **53.** These are equally good investments since the percent of profit (8%) is the same for

both **54. a.** 5 in., 12 in., 13 in. **b.** 30 in^2

Chapter 4

Exercises 4.1, Pages 230 – 236

1. $\{(-5, 1), (-3, 3), (-1, 1), (1, 2), (2, -2)\}$ **3.** $\{(-3, -2), (-1, -3), (-1, 3), (0, 0), (2, 1)\}$ **5.** $\{(-4, 4),\ (-3, -4),$

$(0, -4), (0, 3), (4, 1)\}$ **7.** $\{(-6, 2), (-1, 6), (0, 0), (1, -6), (6, 3)\}$ **9.** $\{(-5, 0), (-2, 2), (-1, -4), (0, 6), (2, 0)\}$

11. **13.** **15.**

17.

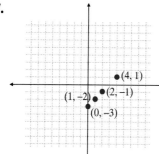

19.

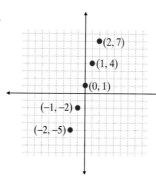

21.

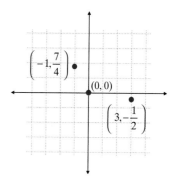

23.

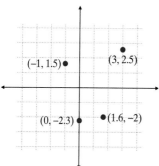

25. b, c, d **27.** a, c **29.** a, c, d **31.** $(0, -4), (2, -2), (4, 0), (1, -3)$

33. $(0, 3), (2, 2), (6, 0), (-2, 4)$ **35.** $(0, -8), (1, -4), (2, 0), (1, -4)$ **37.** $(0, 2), (-1, \frac{8}{3}), (3, 0), (6, -2)$

39. $\left(0, -\frac{7}{4}\right), (1, -1), \left(\frac{7}{3}, 0\right), \left(3, \frac{1}{2}\right)$ **41.** $0, -1, -6, 2$

43. $-3, 1, -7, \frac{7}{4}$

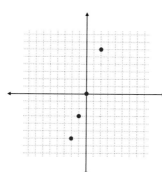

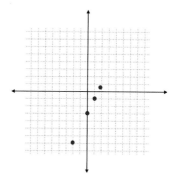

45. $7, \frac{7}{3}, 10, 6$

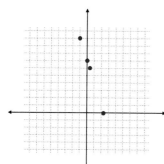

47. $2, 4, -1, -1$

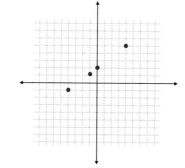

49. $-\frac{9}{5}, 3, -3, \frac{4}{3}$

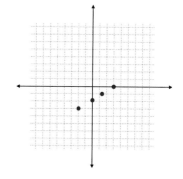

51. $-5, 2, -10, \dfrac{8}{5}$

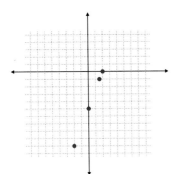

53. $-1.5, 0, 3.8, -1$

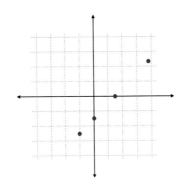

55. $-0.8, -2.4, -1, -7.2$

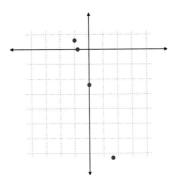

57. a. $-4, 14, 23, 32, 41, 50, 59$ **b.**

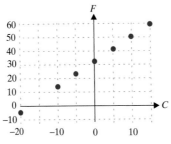

59. Answers will vary.

Exercises 4.2, Pages 246 – 248

1. a **3.** d **5.** f **7.**

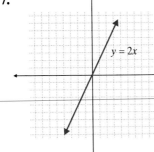

$y = 2x$

9.

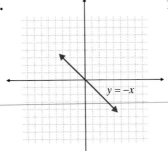

$y = -x$

11.

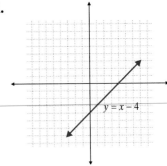

$y = x - 4$

13.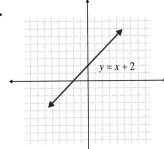

15.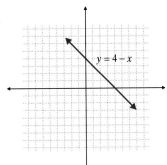

17.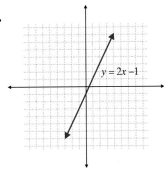

19.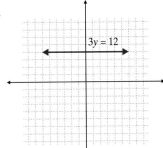

21.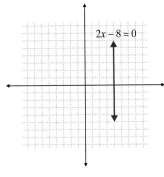

23.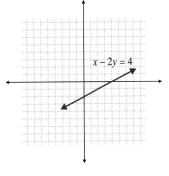

25.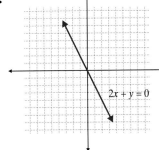

27.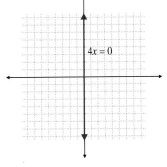

29.

31.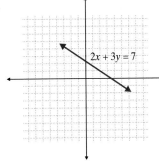

33.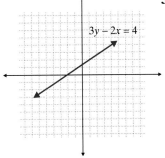

35.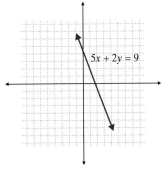

Answers

37.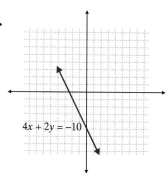

$4x + 2y = -10$

39.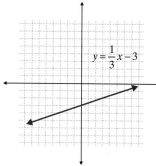

$y = \dfrac{1}{3}x - 3$

41.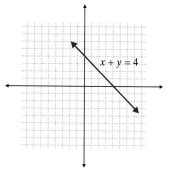

$x + y = 4$

43.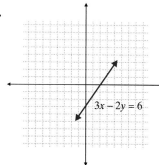

$3x - 2y = 6$

45.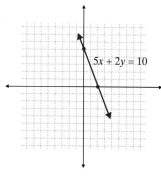

$5x + 2y = 10$

47.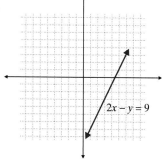

$2x - y = 9$

49.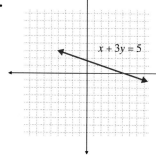

$x + 3y = 5$

51.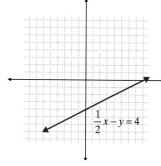

$\dfrac{1}{2}x - y = 4$

53.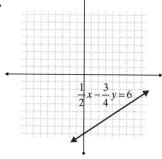

$\dfrac{1}{2}x - \dfrac{3}{4}y = 6$

55.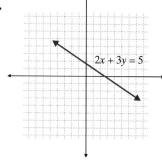

$2x + 3y = 5$

57.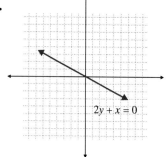

$2y + x = 0$

59.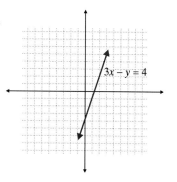

$3x - y = 4$

61.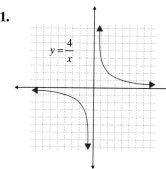

63.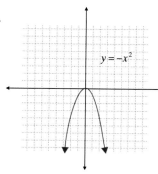

65.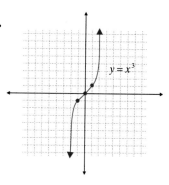

Exercises 4.3, Pages 263 – 264

1. -5 **3.** $\dfrac{8}{7}$ **5.** $\dfrac{1}{8}$ **7.** $\dfrac{3}{10}$ **9.** $\dfrac{5}{4}$ **11.** $y = \dfrac{2}{3}x$ **13.** $y = -\dfrac{3}{4}x - 3$

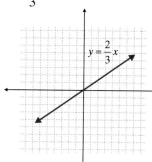

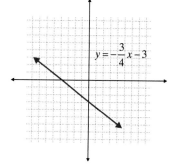

15. $y = -\dfrac{5}{3}x + 3$ **17.** $y = 2x - 1$ **19.** $y = -\dfrac{1}{4}x - 5$

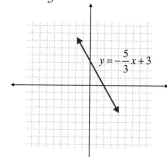

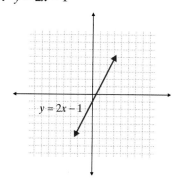

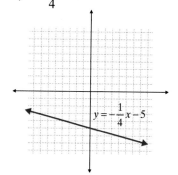

21. $y = \dfrac{3}{2}x - 4$ **23.** $y = x + 5$ **25.** $y = \dfrac{2}{5}x - 6$

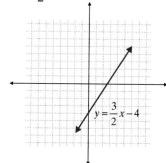

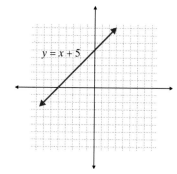

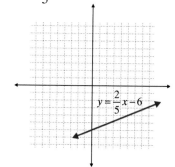

Answers

27. $m = 3, b = 4$

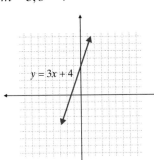

$y = 3x + 4$

29. $m = -\dfrac{2}{3}, b = 1$

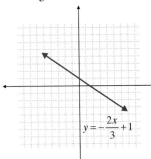

$y = -\dfrac{2x}{3} + 1$

31. $m = 2, b = -3$

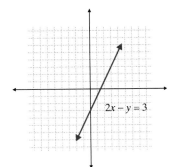

$2x - y = 3$

33. $m = -4, b = 0$

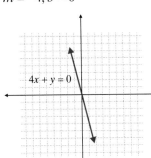

$4x + y = 0$

35. $m = -\dfrac{1}{4}, b = -2$

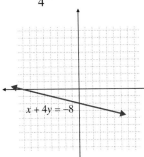

$x + 4y = -8$

37. $m = \dfrac{2}{5}, b = -2$

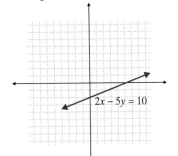

$2x - 5y = 10$

39. $m = -\dfrac{7}{2}, b = 2$

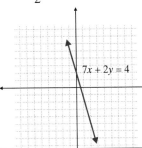

$7x + 2y = 4$

41. $m = -\dfrac{3}{8}, b = -2$

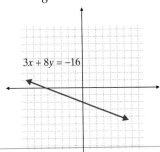

$3x + 8y = -16$

43. $m = -\dfrac{2}{3}, b = \dfrac{8}{3}$

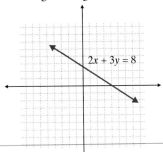

$2x + 3y = 8$

45. $m = \dfrac{5}{2}, b = -\dfrac{3}{2}$

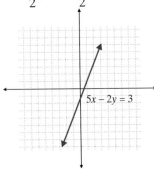

$5x - 2y = 3$

47. $m = 0$

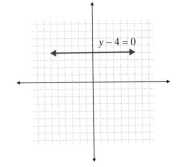

$y - 4 = 0$

49. m is undefined

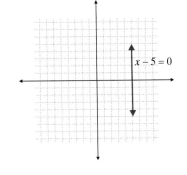

$x - 5 = 0$

51. m is undefined

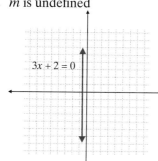

$3x + 2 = 0$

53. $m = 0$

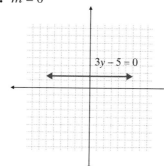

$3y - 5 = 0$

55. m is undefined

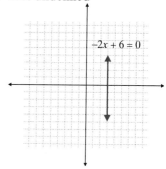

$-2x + 6 = 0$

Exercises 4.4, Pages 272 – 274

1. $y = 2x + 3$ **3.** $y = -\dfrac{2}{5}x + \dfrac{13}{5}$ **5.** $y = \dfrac{3}{4}x + \dfrac{17}{4}$ **7.** $y = -1$ **9.** $x = 2$ **11.** $y = \dfrac{3}{2}x - \dfrac{5}{6}$ **13.** $y = -\dfrac{1}{3}x + \dfrac{7}{3}$

15. $y = -\dfrac{2}{3}x + 6$ **17.** $y = 2$ **19.** $y = \dfrac{1}{3}x + \dfrac{11}{3}$ **21.** $x = -2$ **23.** $y = -\dfrac{3}{10}x + \dfrac{4}{5}$ **25.** $y = 5$ **27.** $x = -1$

29. $y = 3x$ **31.** $y = 1$ **33.** $y = 2x - 2$ **35.** $c = 0.15x + 30$ **37. a.** $I = .09x + 200$

b.

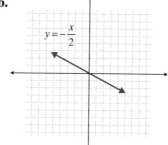

c. The slope is percentage of sales on which the clerk earns an income. She receives $9 for every $100 in sales.

39. a. $A = -200p + 2400$ **b.** $m = -200$ **c.** Attendence decreases by 200 people for every dollar increase in price. **d.** Since p represents admission charged to enter the park, p only makes sense for positive values of p.

41. a. $y = -\dfrac{1}{4}x + \dfrac{7}{4}$

b.

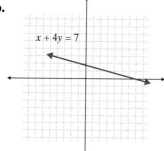

$x + 4y = 7$

43. a. $y = -\dfrac{1}{2}x$

b.

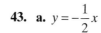

$y = -\dfrac{x}{2}$

45. a. $y = \dfrac{3}{2}x + \dfrac{1}{2}$

b.

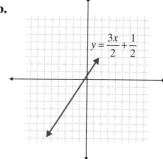

$y = \dfrac{3x}{2} + \dfrac{1}{2}$

Exercises 4.5, Pages 282 – 286

1. a. $\{\,(\,-4,0\,),(\,-2,4\,),(\,1,2\,),(\,2,5\,),(\,6,-3\,)\,\}$ **b.** $D = \{\,-4,-2,1,2,6\,\}$, $R = \{\,-3,0,2,4,5\,\}$ **c.** It is a function.
3. a. $\{\,(\,-5,-4\,),(\,-4,-2\,),(\,-2,-2\,),(\,1,-2\,),(\,2,1\,)\,\}$ **b.** $D = \{\,-5,-4,-2,1,2\,\}$, $R = \{\,-4,-2,1\,\}$
c. It is a function. **5. a.** $\{\,(\,-4,-2\,),(\,-4,1\,),(\,-1,-1\,),(\,-1,3\,),(\,3,-4\,)\,\}$ **b.** $D = \{\,-4,-1,3\,\}$, $R = \{\,-4,-2,-1,1,3\,\}$
c. It is not a function. **7. a.** $\{\,(\,-5,3\,),(\,-5,-5\,),(\,1,-2\,),(\,1,2\,),(\,0,5\,)\,\}$ **b.** $D = \{\,-5,1,0\},\, R = \{\,-5,-2,2,3,5\,\}$
c. It is not a function. **9. a.** $\{\,(\,-5,5\,),(\,-3,1\,),(\,0,-3\,),(\,3,-1\,),(\,4,2\,)\,\}$ **b.** $D = \{\,-5,-3,0,3,4\},\, R = \{\,-3,-1,1,2,5\,\}$
c. It is a function.

11. Function

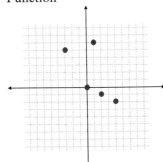

13. Function

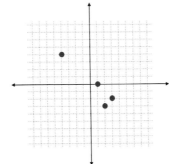

15. Function

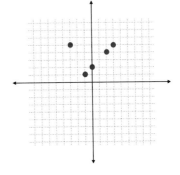

17. Function

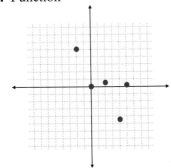

19. Function

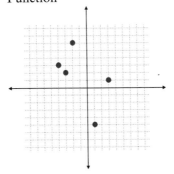

21. $D: -6 \le x \le 6$; $R: 0 \le y \le 6$ **23.** $D: -6 < x \le 6$; $R: \{-6,-4,-2,0,2,4\}$ **25.** Not a function **27.** Not a function
29. $D:$ All real numbers; $R: -1.5 \le y \le 1.5$ **31.** Not a function.
33. $f(\,-2\,) = -3, f(\,-1\,) = -1, f(\,0\,) = 1, f(\,1\,) = 3, f(\,5\,) = 11$
35. $g(\,-2\,) = 10, g(\,-1\,) = 6, g(\,0\,) = 2, g(\,1\,) = -2, g(\,5\,) = -18$
37. $h(\,-2\,) = 4, h(\,-1\,) = 1, h(\,0\,) = 0, h(\,1\,) = 1, h(\,5\,) = 25$
39. $F(\,-2\,) = 16, F(\,-1\,) = 9, F(\,0\,) = 4, F(\,1\,) = 1, F(\,5\,) = 9$
41. $H(\,-2\,) = 8, H(\,-1\,) = 7, H(\,0\,) = 0, H(\,1\,) = -7, H(\,5\,) = 85$
43. $P(\,-2\,) = 9.5, P(\,-1\,) = 6, P(\,0\,) = 3.5, P(\,1\,) = 2, P(\,5\,) = 6$
45. $f(\,-2\,) = -44, f(\,-1\,) = -15, f(\,0\,) = -2, f(\,1\,) = 1, f(\,5\,) = 33$

Chapter 4 Test, Pages 290 – 292

1. b and c **2.** $(-3, 2), (-2, -1), (0, -3), (1, 2), (3, 4), (5, 0)$

3.

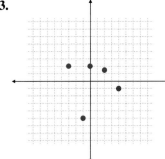

4.

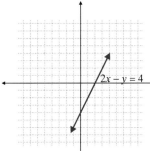

5. y-intercept $= \dfrac{9}{4}$; x-intercept $= 3$

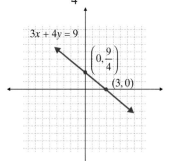

6.

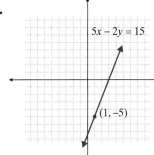

7. $m = -\dfrac{3}{5}, b = -3$

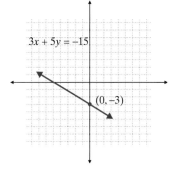

8. $m = -\dfrac{7}{4}$ **9.** $y - 2 = \dfrac{1}{4}(x + 3)$ or $y = \dfrac{1}{4}x + \dfrac{11}{4}$ **10.** $y = -x + 4; m = -1$ **11.** $y = -2, m = 0$ **12. a.** $C = 4t + 9$

b. Because t represents the number of hours worked and one cannot work a negative number of hours. **c.** For every hour a person rents the carpet cleaning machine he will be charged \$4. **13.** These lines are parallel since they have the same slope, but different y-intercepts. **14.** $y = 2x - 8$ **15.** $D = \{-1, 0, 2, 5\}; R = \{-3, 4, 6\}$ **16.** It is not function because the domain element, 2, has more than one range element, –5 and 0. **17. a.** –7 **b.** 14 **18. a.** 22 **b.** 7 **19.** It is a function. $D = \{-6, -4, -3, 3, 7\}; R = \{-4, -3, 0, 1, 3\}$ **20.** It is a function. $D: -2 \le x \le 2$; $R: 0 \le y \le 4$ **21.** It is not a function **22.** It is a function. $D:$ All non-negative numbers $R: y \ge 0$

Chapter 4 Cumulative Review, Pages 293 – 296

1. $x + 15$ **2.** $3x + 8$ **3.** 540 **4.** $120a^2b^3$ **5.** 2, 16 **6.** –4,–9 **7.** –4, 14 **8.** 4, –12 **9.** False, $-15 < 5$ **10.** True

11. False, $\dfrac{7}{8} \ge \dfrac{7}{10}$ **12.** True **13.** $x = \dfrac{8}{5}$ **14.** $x = -\dfrac{1}{2}$ **15.** $x = \dfrac{3}{8}$ **16.** $x = -3$ **17.** $d = \dfrac{C}{\pi}$ **18.** $y = \dfrac{10 - 3x}{5}$

19. $w = \dfrac{P - 2l}{2}$ **20.** $h = \dfrac{3v}{\pi r^2}$ **21.** $x \ge -6$ ⟵┼─┼─●─┼─┼─┼─┼→ -6 **22.** $-7 < x < 4$ ⟵○─┼─┼─┼─┼─○→ -7 4

23. $x \ge 72$ ⟵─────●──────→ 72 **24.** $-8 < x \le \dfrac{112}{9}$ ⟵○─────●→ -8 $\dfrac{112}{9}$ **25.** ⟵┼●●●●●●●●●●→ -4–3–2–1 0 1 2 3 4

26. ⟵●●┼┼┼┼┼┼┼●●→ -8–7 7 8 **27.** $D = \{-3, -2, -1, 1, 3\}; R = \{0, 1, 3\}$; It is a function

721

Answers

28. $D = \{-3, -1, 1, 4\}$; $R = \{-2, -1, 1, 3, 4\}$; It is not a function. **29.** $D = \{-2, 0, 2, 3, 4\}$;
$R = \{-1, 0, 1, 2, 3, 4\}$; It is not a function **30.** $D: -5 < x \le -1$ and $0 < x \le 6$ $R = \{-4, -2, 2, 4, 5\}$; It is a function.

31. a.

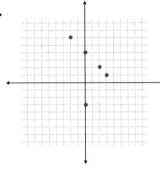

b. $D = \{-2, 0, 2, 3\}$;
$\quad R = \{-3, 1, 2, 4, 6\}$

c. It is not a function

32. a.

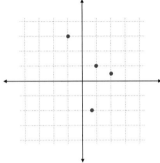

b. $D = \{-1, 1, \dfrac{2}{3}, 2\}$;

$\quad R = \{-2, 1, \dfrac{1}{2}, 3\}$

c. It is a function

33. a, b, c **34.** a, b, c, d

35.

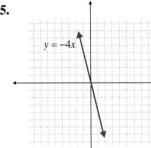

$y = -4x$

36.

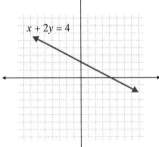

$x + 2y = 4$

37.

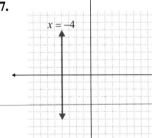

$x = -4$

38.

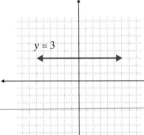

$y = 3$

39.

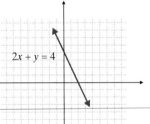

$2x + y = 4$

40.

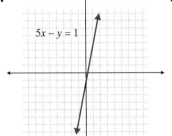

$5x - y = 1$

41. $m = 2, b = 3$

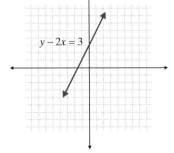

$y - 2x = 3$

42. $m = -\dfrac{2}{5}, b = 2$

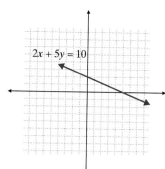
$2x + 5y = 10$

43. $2x + 3y = 16$ **44.** $4x + 3y = 18$

45. $y = -1$

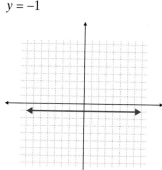

46. $y = -\dfrac{1}{4}x + \dfrac{13}{4}$

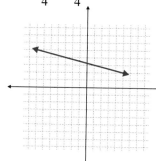

47. $x = -3$

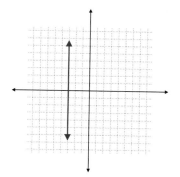

48. $y = \dfrac{3}{2}x - 11$

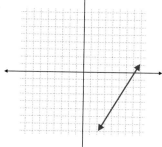

49. $2x + y = 0$ **50.** $x - 2y = -6$ **51. a.** -11 **b.** 22

52. a. -1 **b.** 76 **53.** $20, 22, 24$ **54.** length $= 25$ cm, width $= 16$ cm **55.** -19 **56.** 54 to 100 **57.** less than 140 miles **58.** $15, 17, 19$

Chapter 5

Exercises 5.1, Pages 306 – 309

1. c **3.** a, c **5.** $(0, 2)$ **7.** $(4, 2)$ **9.** $m_1 = -2, b_1 = 3, m_2 = -2, b_2 = \dfrac{5}{2}$ **11.** $m_1 = 3, b_1 = -8, m_2 = 3, b_2 = -6$

Answers

13. Consistent, $(2, 0)$

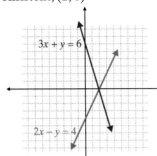

$3x + y = 6$
$2x - y = 4$

15. Inconsistent

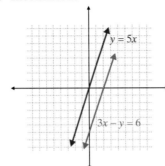

$y = 5x$
$3x - y = 6$

17. Consistent; $(2, 3)$

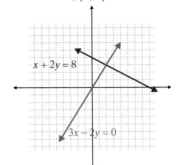

$x + 2y = 8$
$3x - 2y = 0$

19. Dependent

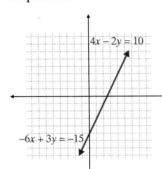

$4x - 2y = 10$
$-6x + 3y = -15$

21. Inconsistent

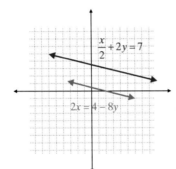

$\frac{x}{2} + 2y = 7$
$2x = 4 - 8y$

23. Consistent; $\left(\frac{1}{2}, 1\right)$

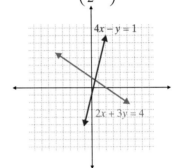

$4x - y = 1$
$2x + 3y = 4$

25. Inconsistent

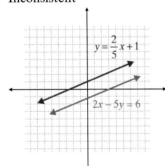

$y = \frac{2}{5}x + 1$
$2x - 5y = 6$

27. Dependent

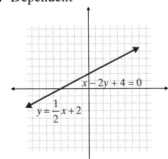

$x - 2y + 4 = 0$
$y = \frac{1}{2}x + 2$

29. Consistent $(3, 0)$

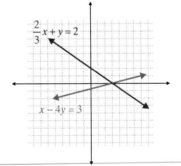

$\frac{2}{3}x + y = 2$
$x - 4y = 3$

31. Consistent; $(1, 2)$

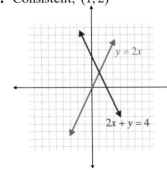

$y = 2x$
$2x + y = 4$

33. Consistent; $(-3, -2)$

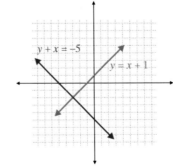

$y + x = -5$
$y = x + 1$

35. Consistent; $(-1, 2)$

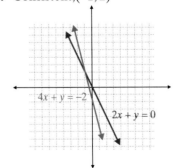

$4x + y = -2$
$2x + y = 0$

37. Consistent; $(1, -1)$

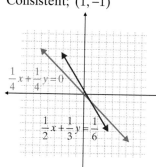

$$\frac{1}{4}x + \frac{1}{4}y = 0$$

$$\frac{1}{2}x + \frac{1}{3}y = \frac{1}{6}$$

39. $x = 20, y = 5$

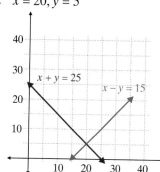

$x + y = 25$

$x - y = 15$

41. $x = 2, y = 8$

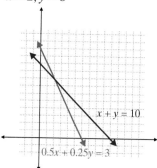

$x + y = 10$

$0.5x + 0.25y = 3$

43. $(1, 4)$

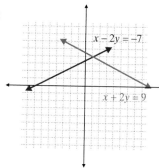

$x - 2y = -7$

$x + 2y = 9$

45. $\left(\frac{3}{2}, 2\right)$

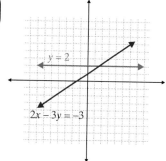

$y = 2$

$2x - 3y = -3$

Exercises 5.2, Pages 314 – 316

1. $(2, 4)$ **3.** $(1, -2)$ **5.** $(-6, -2)$ **7.** $(4, 1)$ **9.** Inconsistent **11.** $(3, 2)$ **13.** Dependent **15.** $(2, 2)$ **17.** $\left(\frac{1}{2}, -4\right)$
19. $\left(2, \frac{5}{2}\right)$ **21.** $\left(-2, \frac{1}{2}\right)$ **23.** $(2, -10)$ **25.** $\left(\frac{5}{2}, \frac{1}{8}\right)$ **27.** $(4, -1)$ **29.** $\left(\frac{13}{5}, \frac{-39}{5}\right)$ **31.** $(10, 4)$ **33.** $(9, -3)$
35. $(6, -4)$ **37.** $(-7.5, 0)$ **39.** $(20, 5)$ **41.** $(2, 8)$

Exercises 5.3, Pages 323 – 326

1. $x = 3, y = -1$ **3.** $x = 1, y = -\frac{3}{2}$ **5.** Inconsistent **7.** $x = 1, y = -5$ **9.** Dependent **11.** $x = 4, y = 3$
13. $x = -3, y = -1$ **15.** $x = 2, y = -1$ **17.** $x = -2, y = -3$ **19.** $x = 5, y = -6$ **21.** $x = 10, y = 2$ **23.** $x = 7, y = 0$
25. $x = 5, y = 6$ **27.** $x = -3, y = 7$ **29.** $x = \frac{44}{17}, y = -\frac{2}{17}$ **31.** $x = 6, y = 4$ **33.** $x = 4, y = 7$ **35.** $x = \frac{1}{2}, y = \frac{2}{3}$
37. $x = -\frac{45}{7}, y = \frac{92}{7}$ **39.** $y = 5x - 7, m = 5, b = -7$ **41.** $y = \frac{1}{9}x + \frac{13}{9}; \quad m = \frac{1}{9}, b = \frac{13}{9}$
43. $y = \frac{11}{4}x - \frac{49}{4}; \quad m = \frac{11}{4}, b = -\frac{49}{4}$

45. $x = 3, y = -1$

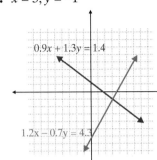

47. $x = 1.2, y = 1.6$

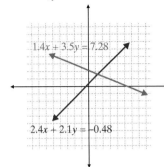

49. $x = 6.726, y = 0.134$

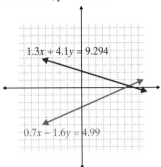

51. $x = \$4000$ at 10%; $y = \$6000$ at 6% **53.** 40 liters of 30% solution (x); 60 liters of 40% solution (y)

Exercises 5.4, Pages 332 – 337

1. $\begin{cases} x + y = 56 \\ x - y = 10 \end{cases}$; $x = 33, y = 23$ **3.** $\begin{cases} x + y = 36 \\ 2x + 3y = 87 \end{cases}$; $x = 21, y = 15$ **5.** $\begin{cases} \dfrac{1}{3}(x + y) = 4 \\ \dfrac{1}{2}(x - y) = 4 \end{cases}$; Rate of boat $(x) = 10$ mph,

Rate of current $(y) = 2$ mph **7.** $\begin{cases} x + y = 3\dfrac{1}{2} \\ 52x + 56y = 190 \end{cases}$ He traveled $1\dfrac{1}{2}$ hrs at the first rate (x) and 2 hours at the

second rate (y). **9.** $\begin{cases} x - y = 10 \\ \dfrac{200}{x} + \dfrac{200}{y} = 9 \end{cases}$ where x is his going rate and y is his return rate. His rate of speed to the

city was 50 mph. **11.** $\begin{cases} x = 4y \\ 3(x + y) = 105 \end{cases}$ Steve (x) traveled at 28 mph and Fred (y) traveled at 7 mph

13. $\begin{cases} x = y + 8 \\ 3(x + y) = 324 \end{cases}$ Mary's speed (x) was 58 mph and Linda's speed (y) was 50 mph. **15.** $\begin{cases} x + y = \dfrac{8}{5} \\ 10x = 6y \end{cases}$;

$x = 6$ miles. He jogged about 12 miles. **17.** $\begin{cases} x = \dfrac{7}{2}y \\ 2(x - y) = 580; \end{cases}$ Speed of airliner $(x) = 406$ mph and speed of

private plane $(y) = 116$ mph. **19.** $\begin{cases} q + d = 27 \\ 0.25q + 0.1d = 5.4; \end{cases}$ 18 quarters and 9 dimes. **21.** $\begin{cases} A - 2 = \dfrac{1}{2}(B - 2) \\ A + 8 = \dfrac{2}{3}(B + 8); \end{cases}$ Anna

is 12 years old and Beth is 22 years old. **23.** $\begin{cases} l = 2w - 1 \\ 2(l + 4) + 2(w + 4) = 116; \end{cases}$ Length is 33 m and width is 17 m.

25. $\begin{cases} -1 = m(-2) + b \\ -7 = m(6) + b; \end{cases}$ $y = -\dfrac{3}{4}x - \dfrac{5}{2}$; **27.** $\begin{cases} p + n = 182 \\ 0.01p + 0.05n = 3.90; \end{cases}$ 52 nickels and 130 pennies

29. $\begin{cases} a + s = 3500 \\ 3.5a + 2.5s = 9550 \end{cases}$ 800 adults and 2700 students attended. **31.** $\begin{cases} l = 3w \\ 2(l + w) = 260; \end{cases}$ Length is 97.5 m and

width is 32.5 m **33.** $\begin{cases} g + r = 12,500 \\ 2g + 3.5r = 36,250; \end{cases}$ 5000 general admission and 7500 reserved tickets were sold.

35. $\begin{cases} 70c + 160a = 620 \\ 2c + a = 7; \end{cases}$ 1 adult ticket is \$3 and 1 child's ticket is \$2. **37.** $\begin{cases} 2x + y = 55 \\ x + 2y = 68; \end{cases}$ $x = \$14$ for shirts $y =$

\$27 for pair of slacks **39.** $\begin{cases} 4x + 3y = 58 \\ 8x + 7y = 126; \end{cases}$ 7 for Model X, 10 for Model Y. **41.** $\begin{cases} a + b = 13 \\ 10b + a = 10a + b + 45; \end{cases}$ $a = 4; b =$

9; The number is 49.

Exercises 5.5, Pages 343 – 345

1. \$5500 at 6%, \$3500 at 10% **3.** \$7400 at 5.5%, \$2600 at 6% **5.** \$450 at 8%, \$650 at 10% **7.** \$3500 in each or \$7000 total. **9.** \$20,000 at 24%, \$11000 at 18% **11.** \$800 at 5%, \$2100 at 7% **13.** \$8500 at 9%, \$3500 at 11% **15.** 20 pounds of 20%, 30 pounds of 70% **17.** 20 ounces of 30%, 30 ounces of 20% **19.** 450 pounds of 35%, 1350 pounds of 15% **21.** 20 pounds of 40%, 30 pounds of 15%

Chapter 5 Test, Pages 348 – 349

1. c **2.** d **3.** Consistent, $\left(\dfrac{4}{3}, \dfrac{-14}{3} \right)$ **4.** Consistent, $(4, 1)$ **5.** Consistent; $x = -8, y = -20$

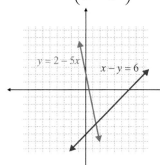

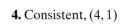

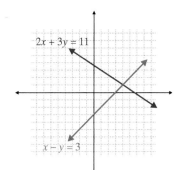

6. Consistent; $x = -2, y = 6$ **7.** Consistent; $x = -\dfrac{3}{14}, y = \dfrac{13}{7}$ **8.** Consistent; $x = \dfrac{43}{12}, y = -\dfrac{11}{24}$ **9.** Consistent; $x = -3, y = 5$ **10.** Inconsistent **11.** Consistent; $x = 3, y = -6$ **12.** $a = 8, b = -1$ **13.** Speed of the boat = 14 mph; Speed of the current = 2 mph **14.** Pen price = \$0.79; Pencil price = \$0.08 **15.** \$1600 at 8% and \$960 at 6% **16.** 1600 lbs. of 83%, 400 lbs. of 68% **17.** 13 in. by 17 in. **18.** nickels = 45, quarters = 60

Chapter 5 Cumulative Review, Pages 350 – 353

1. -165 **2.** 28 **3.** 55 **4.** 8750 **5.** $x = \dfrac{1}{5}$ **6.** $x = \dfrac{11}{4}$ **7.** $y = -7$ **8.** $a = \dfrac{25}{11}$ **9.** $x \geq 13$;

10. $-5 \leq x \leq -4.2$; **11.** $t = \dfrac{v - k}{g}$ **12.** $y = \dfrac{3x - 6}{4}$

Answers

13. $m = -\dfrac{1}{2}, b = \dfrac{11}{4}$;

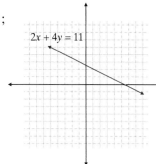

2x + 4y = 11

14. $x - 3y = -5$ **15.** b, c, d **16.** $x = \dfrac{13}{7}, y = \dfrac{44}{7}$

17. Consistent; $(2, 0)$ **18.** Consistent; $(-1, 4)$ **19.** Consistent; $\left(\dfrac{13}{10}, -\dfrac{9}{10} \right)$ **20.** Inconsistent

21. Consistent; $(-6, 2)$ **22.** Inconsistent **23.** Dependent **24.** Consistent; $(2, -4)$ **25.** Consistent; $(2, 3)$

26. Consistent; $(-1, -6)$ **27.** Inconsistent **28.** Dependent **29.** Consistent; $(8, 5)$ **30.** Consistent; $\left(\dfrac{22}{21}, \dfrac{13}{21} \right)$

31. Consistent; $(1, 2)$ **32.** Consistent; $\left(\dfrac{54}{11}, \dfrac{-5}{11} \right)$ **33.** $y = -7x + 20$ **34.** $y = -\dfrac{4}{3}x + \dfrac{7}{3}$ **35. a.** the measures of the angles are 20° and 70° **b.** the measures of the angles are 40° and 140°. **36.** 10 of the 25 cent stamps and 5 of the 34 cent stamps **37.** speed of the boat = 10 mph and speed of the current = 2 mph **38.** length = 14 yards and width = 11 yards **39.** They produced 4 of Model A and 18 of Model B. **40.** Eastbound train traveled at 40 mph and the westbound train traveled at 45 mph **41.** $N = 950t - 850$ **42.** { $(-5, -2), (-3, -2), (-2, 4), (1, -1), (2, 4)$ }; $D = \{ -5, -3, -2, 1, 2 \}; R = \{ -2, -1, 4 \}$; It is a function. **43.** Not a function **44.** Function **45. a.** −22 **b.** 23 **c.** 3.5 **46. a.** −10 **b.** −1 **c.** 80

Chapter 6

Exercises 6.1, Pages 365 – 366

1. 27; product rule **3.** 512; product rule or 0 exponent rule **5.** $\dfrac{1}{16}$; negative exponent rule **7.** $\dfrac{1}{216}$; negative exponent rule **9.** 24 **11.** 54 **13.** −54 **15.** $\dfrac{4}{9}$; negative exponent rule **17.** $\dfrac{-5}{4}$; negative exponent rule **19.** x^4; product rule **21.** y^{11}; product rule **23.** $\dfrac{1}{y^2}$; negative exponent rule **25.** $\dfrac{5}{y^4}$; negative exponent rule **27.** $\dfrac{-10}{x^3}$; negative exponent rule **29.** $\dfrac{1}{y^2}$; negative exponent rule and 0 exponent rule **31.** 2; 0 exponent rule **33.** 729; quotient rule **35.** $\dfrac{1}{10000}$; quotient rule and negative exponent rule **37.** x^2; quotient rule **39.** x^2; quotient rule **41.** x^4; quotient rule **43.** $\dfrac{1}{x^4}$; quotient rule and negative exponent rule **45.** x^6; quotient rule **47.** x^2; quotient rule **49.** y^2; quotient rule **51.** $3x^3$ product rule or 0 exponent rule **53.** $10x^4$; product rule **55.** $36x^3$; product rule or 0 exponent rule **57.** $-14x^5$; product rule **59.** $-12x^6$; product rule **61.** $4y$; quotient rule **63.** $3y^2$; quotient rule **65.** $-2y^2$; quotient rule **67.** $-7x^2$; quotient rule **69.** 10^3 product and quotient rules

71. $1; 0$ exponent rule

73. $\dfrac{15x^3}{y^2}$; product and negative exponent rules **75.** $\dfrac{-2y^6}{x^4}$; quotient and negative exponent rules **77.** $12a^3b^9c$; product rule **79.** $-4a^{10}b^3c$; quotient rule **81.** 1 **83.** $.390625$ **85.** $99,376.2576$

Exercises 6.2, Pages 377 – 379

1. -81 **3.** 81 **5.** $1,000,000$ **7.** $36x^6$ **9.** $-108x^6$ **11.** -3 **13.** $\dfrac{-2y^6}{27x^{15}}$ **15.** $\dfrac{16x^2}{y^4}$ **17.** $\dfrac{9x^4}{y^6}$ **19.** $\dfrac{y^2}{x^2}$ **21.** $\dfrac{y^{10}}{4x^2}$

23. $\dfrac{1}{64a^6b^9}$ **25.** $25x^2y^4$ **27.** $16a^4b^4$ **29.** $\dfrac{1}{8a^3b^6}$ **31.** $\dfrac{y^8}{16x^8}$ **33.** $\dfrac{8a^6}{b^9}$ **35.** $\dfrac{49y^4}{x^6}$ **37.** $\dfrac{24y^5}{x^7}$ **39.** $\dfrac{4x^3}{3}$ **41.** 8.6×10^4

43. 3.62×10^{-2} **45.** 1.83×10^7 **47.** 0.042 **49.** $7,560,000$ **51.** $851,500,000$ **53.** $3 \times 10^2 \cdot 1.5 \times 10^{-4}; 4.5 \times 10^{-2}$

55. $3 \times 10^{-4} \cdot 2.5 \times 10^{-6}; 7.5 \times 10^{-10}$ **57.** $\dfrac{3.9 \times 10^3}{3 \times 10^{-3}}; 1.3 \times 10^6$ **59.** $\dfrac{1.25 \times 10^2}{5 \times 10^4}; 2.5 \times 10^{-3}$

61. $\dfrac{2 \times 10^{-2} \cdot 3.9 \times 10^3}{1.3 \times 10^{-2}}; 6 \times 10^3$ **63.** $\dfrac{5 \times 10^{-3} \cdot 6.5 \times 10^2 \cdot 3.3 \times 10^0}{1.1 \times 10^{-3} \cdot 2.5 \times 10^3}; 3.9 \times 10^0$

65. $\dfrac{(1.4 \times 10^{-2})(9.22 \times 10^2)}{(3.5 \times 10^3)(2.0 \times 10^6)}; 1.844 \times 10^{-9}$ **67.** 1.67×10^{-24} grams **69.** 5.866×10^{12} miles; 1.34918×10^{13} miles

71. 5.98×10^{27} grams **73.** 6×10^8 ounces **75.** $10,000,000,000; 19,025.87519$ years

Exercises 6.3, Pages 384 – 385

1. Monomial **3.** Not a polynomial **5.** Trinomial **7.** Not a polynomial **9.** Binomial **11.** $4y$; first degree monomial. $a_1 = 4, a_0 = 0$ **13.** $x^3 + 3x^2 - 2x$; third degree trinomial. $a_3 = 1, a_2 = 3, a_1 = -2, a_0 = 0$
15. $-2x^2$; second degree monomial. $a_2 = -2, a_1 = 0, a_0 = 0$ **17.** 0; monomial of no degree. $a_0 = 0$
19. $6a^5 - 7a^3 - a^2$; fifth degree trinomial. $a_5 = 6, a_4 = 0, a_3 = -7, a_2 = -1, a_1 = 0, a_0 = 0$ **21.** $2y^3 + 4y$; third degree binomial. $a_3 = 2, a_2 = 0, a_1 = 4, a_0 = 0$ **23.** 4; monomial of degree 0. $a_0 = 4$ **25.** $2x^3 + 3x^2 - x + 1$; third degree polynomial. $a_3 = 2, a_2 = 3, a_1 = -1, a_0 = 1$ **27.** $4x^4 - x^2 + 4x - 10$; fourth degree polynomial. $a_4 = 4, a_3 = 0, a_2 = -1, a_1 = 4, a_0 = -10$ **29.** $9x^3 + 4x^2 + 8x$; third degree trinomial. $a_3 = 9, a_2 = 4, a_1 = 8, a_0 = 0$
31. $p(-1) = -16$ **33.** $p(3) = -41$ **35.** $p(-2) = 81$ **37.** $p(-1) = 13$ **39.** $p(2) = 24$

Exercises 6.4, Pages 390 – 391

1. $3x^2 + 7x + 2$ **3.** $2x^2 + 11x - 7$ **5.** $3x^2$ **7.** $x^2 - 5x + 17$ **9.** $-x^2 + 2x - 7$ **11.** $4x^2 + 2x - 7$ **13.** $-x^2 + 7x - 3$
15. $4x^3 + 7$ **17.** $2x^3 + 11x^2 - 4x - 3$ **19.** $5x^2 - 5x - 1$ **21.** $x^2 + x + 6$ **23.** $-x^4 - 2x^3 - 16$ **25.** $-3x^2 - 6x - 2$
27. $9x^2 + 3x$ **29.** $3x^4 - 7x^3 + 3x^2 - 5x + 9$ **31.** $-4x^2 + 6x - 11$ **33.** $-2x^3 + 3x^2 - 2x - 8$ **35.** $2x^3 + 4x^2 + 3x - 10$
37. $8x^4 + 6x^2 + 15$ **39.** $2x^3 - 5x^2 + 11x - 11$ **41.** $4x - 13$ **43.** $-7x + 9$ **45.** $2x + 17$ **47.** $3x^3 + 13x^2 + 9$
49. $8x^2 - x - 2$ **51.** $3x - 19$ **53.** $7x - 1$ **55.** $4x^2 - x$ **57.** $x^2 + 11x - 8$ **59.** $7x^3 - x^2 - 7$

Answers

1. $-6x^5 - 15x^3$ **3.** $-20x^4 + 30x^2$ **5.** $-y^5 + 8y - 2$ **7.** $-4x^8 + 8x^7 - 12x^4$ **9.** $-35t^5 + 14t^4 + 7t^3$ **11.** $-x^4 - 5x^2 + 4x$
13. $6x^2 - x - 2$ **15.** $9a^2 - 25$ **17.** $-10x^2 + 39x - 14$ **19.** $x^3 + 5x^2 + 8x + 4$ **21.** $x^2 + x - 12$ **23.** $a^2 - 2a - 48$
25. $x^2 - 3x + 2$ **27.** $3t^2 - 3t - 60$ **29.** $x^3 + 11x^2 + 24x$ **31.** $2x^2 - 7x - 4$ **33.** $6x^2 + 17x - 3$ **35.** $4x^2 - 9$
37. $16x^2 + 8x + 1$ **39.** $y^3 + 2y^2 + y + 12$ **41.** $3x^2 - 8x - 35$ **43.** $-16x^3 + 50x^2 + 25x - 14$ **45.** $12x^4 + 28x^3 + 47x^2 +$
$34x + 48$ **47. a.** $4t^2 + 17t - 42$ **b.** 8 **49. a.** $5a^2 + 17a - 12$ **b.** 42 **51. a.** $3x^2 + 2x - 16$ **b.** 0 **53. a.** $6x^2 - x - 35$
b. -13 **55. a.** $25y^2 + 20y + 4$ **b.** 144 **57. a.** $y^3 - 4y^2 + 2y - 8$ **b.** -12 **59. a.** $9x^2 - 16$ **b.** 20 **61. a.** $x^3 - 8$ **b.** 0
63. a. $25a^2 - 60a + 36$ **b.** 16 **65. a.** $3x^3 - 2x^2 + 26x + 9$ **b.** 77 **67. a.** $t^3 - 6t^2 + 11t - 6$ **b.** 0 **69. a.** $y^4 + 2y^3 + y^2 - 4$
b. 32 **71.** $-3x - 21$ **73.** $3a^2 - 17a + 11$ **75.** $2y^2 - 61$

1. $x^2 - 9$, difference of two squares **3.** $x^2 - 10x + 25$, perfect square trinomial **5.** $x^2 - 36$, difference of two
squares **7.** $x^2 + 16x + 64$, perfect square trinomial **9.** $2x^2 + x - 3$, **11.** $9x^2 - 24x + 16$, perfect square trinomial
13. $4x^2 - 1$, difference of two squares **15.** $9x^2 - 12x + 4$, perfect square trinomial **17.** $x^2 + 6x + 9$, perfect square
trinomial **19.** $x^2 - 10x + 25$, perfect square trinomial **21.** $25x^2 - 81$, difference of two squares **23.** $4x^2 + 28x + 49$,
perfect square trinomial **25.** $81x^2 - 4$, difference of two squares **27.** $10x^4 - 11x^2 - 6$ **29.** $49x^2 + 14x + 1$, perfect
square trinomial **31.** $5x^2 + 11x + 2$ **33.** $4x^2 + 13x - 12$ **35.** $3x^2 - 25x + 42$ **37.** $x^2 + 10x + 25$ **39.** $x^4 - 1$
41. $x^4 + 6x^2 + 9$ **43.** $x^6 - 4x^3 + 4$ **45.** $x^4 + 3x^2 - 54$ **47.** $x^2 - \dfrac{4}{9}$ **49.** $x^2 - \dfrac{9}{16}$ **51.** $x^2 + \dfrac{6}{5}x + \dfrac{9}{25}$

53. $x^2 - \dfrac{5}{3}x + \dfrac{25}{36}$ **55.** $x^2 - \dfrac{1}{4}x - \dfrac{1}{8}$ **57.** $x^2 + \dfrac{5}{6}x + \dfrac{1}{6}$ **59.** $x^2 - 1.96$ **61.** $x^2 - 5x + 6.25$ **63.** $x^2 - 4.6225$

65. $x^2 + 2.48x + 1.5376$ **67.** $2.0164x^2 + 27.264x + 92.16$ **69.** $129.96x^2 - 12.25$ **71.** $93.24x^2 + 142.46x - 104.04$

73. a. $f(x) = 15x^4 - 30x^5 + 15x^6$ **b.** 0.234375 **75. a.** $A(x) = 4x^2 + 140x + 1000$ **b.** $A(x) = 4x^2 + 140x$

1. $x^2 + 2x + \dfrac{3}{4}$ **3.** $2x^2 - 3x - \dfrac{3}{5}$ **5.** $2x + 5$ **7.** $x + 6 - \dfrac{3}{x}$ **9.** $2x + 3 - \dfrac{3}{2x}$ **11.** $2x - 3 - \dfrac{1}{x}$ **13.** $x + 3y - \dfrac{11y}{7x}$

15. $\dfrac{3x^2}{4} - 2y - \dfrac{y^2}{x}$ **17.** $\dfrac{5x}{8} - 1 + \dfrac{2}{y}$ **19.** $\dfrac{8x^2}{9} - xy + \dfrac{5}{9y}$ **21.** $12 + \dfrac{2}{23}$ **23.** $7 + \dfrac{24}{59}$ **25.** $x - 2$ **27.** $a - \dfrac{15}{a - 2}$

29. $y - 5 - \dfrac{22}{y + 4}$ **31.** $5y - 11 + \dfrac{48}{y + 5}$ **33.** $4c - 5 + \dfrac{1}{2c + 3}$ **35.** $2m + 1 - \dfrac{3}{5m - 3}$ **37.** $x - 2 + \dfrac{10}{x + 5}$ **39.** $y^2 - 7y + 12$

41. $3a^2 + 1 + \dfrac{2}{4a - 1}$ **43.** $x^2 + 7x + 35 + \dfrac{170}{x - 5}$ **45.** $x^2 - 3x + 9$ **47.** $2x + \dfrac{3x + 3}{x^2 - 2}$ **49.** $4x^2 + 6x + 9$

51. $4x + 3 + \dfrac{8}{x - 1}$ **53.** $3x^2 + 2x + 5$ **55.** $2a + 3 - \dfrac{4a}{a^2 + 2}$

Chapter 6 Test, Pages 419 – 420

1. $-10a^5$ **2.** 1 **3.** $\dfrac{4x^3}{y^7}$ **4.** $\dfrac{x}{3y^2}$ **5.** $\dfrac{x^2}{4y^2}$ **6.** $4x^2y^4$ **7. a.** 135000 **b.** 0.0000027 **8. a.** 1.25×10^8 **b.** 5.2×10^{-4}

9. $8x^2 + 3x$; second degree binomial **10.** $-x^3 + 3x^2 + 3x - 1$; third degree polynomial **11.** $5x^5 + 2x^4 - 11x + 3$; fifth degree polynomial **12. a.** 20 **b.** -110 **13.** $-3x + 2$ **14.** $20x - 8$ **15.** $5x^3 - x^2 + 6x + 5$ **16.** $-2x^2 + x - 6$ **17.** $7x^4 + 14x^2 + 4$ **18.** $7x^3 - 2x^2 - 7x + 1$ **19.** $15x^7 - 20x^6 + 15x^5 - 40x^4 - 10x^2$ **20.** $49x^2 - 9$; difference of two squares **21.** $16x^2 + 8x + 1$; perfect square trinomial **22.** $36x^2 - 60x + 25$; perfect square trinomial

23. $12x^2 + 24x - 15$ **24.** $6x^3 - 69x^2 + 189x$ **25.** $7x^2 + 7x + 14$ **26.** $10x^4 + 4x^3 - 15x^2 - 41x - 14$ **27.** $2x + \dfrac{3}{2} - \dfrac{3}{x}$ **28.** $\dfrac{5}{3} + 2b + \dfrac{b^2}{a}$ **29.** $x - 6 - \dfrac{2}{2x + 3}$ **30.** $x - 9 + \dfrac{15x - 12}{x^2 + x - 3}$

Chapter 6 Cumulative Review, Pages 421 – 424

1. $(x + 15)$ **2.** $(3x + 8)$ **3.** 540 **4.** $120a^2b^3$ **5.** 16, 2 **6.** $-9, -4$ **7.** 14, -4 **8.** 4, -12 **9.** $5x^2 + 16x$; second degree binomial; $a_2 = 5, a_1 = 16, a_0 = 0$ **10.** $5x - 18$; first degree binomial; $a_1 = 5, a_0 = -18$ **11.** $x - 9$; first degree binomial; $a_1 = 1, a_0 = -9$ **12.** $x^4 - 2x^3 + 4x^2 - 10x + 40$; fourth degree polynomial; $a_4 = 1, a_3 = -2, a_2 = 4$ $a_1 = -10, a_0 = 40$ **13.** $x = \dfrac{8}{5}$ **14.** $x = -\dfrac{1}{2}$ **15.** $x = \dfrac{3}{8}$ **16.** $x = -3$ **17.** $d = \dfrac{C}{\pi}$ **18.** $y = \dfrac{10 - 3x}{5}$ **19.** $x \geq -6$;

20. $-7 < x < 4$;

21. $y = \dfrac{2}{3}x + \dfrac{8}{3}$

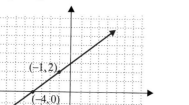
(−1, 2) (−4, 0)

22. $5x + 8y = -17$

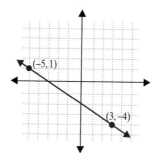
(−5, 1) (3, −4)

23. $2x + 5y = 35$

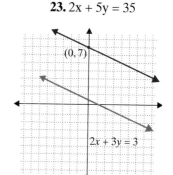
(0, 7) $2x + 3y = 3$

24. a. 22 **b.** 57 **25.** $x = 0, y = 6$ **26.** $x = -\dfrac{12}{7}, y = \dfrac{31}{7}$ **27.** $x = 1, y = 7$ **28.** $x = -\dfrac{94}{25}, y = -\dfrac{57}{25}$ **29.** $2x$

30. $64x^6y^3$ **31.** $\dfrac{49x^{10}}{y^4}$ **32.** $\dfrac{36x^4}{y^{10}}$ **33.** $\dfrac{a^6}{b^4}$ **34.** $\dfrac{x}{3y^4}$ **35.** 1 **36.** $\dfrac{x^8}{9y^6}$ **37. a.** 0.00000028 **b.** 35,100

38. $1.5 \times 10^{-3} \cdot 4.2 \times 10^3$; 6.3×10^0 **39.** $\dfrac{8.4 \times 10^2}{2.1 \times 10^{-4}}$; 4×10^6 **40.** $\dfrac{5 \times 10^{-3} \times 7.7 \times 10}{1.1 \times 10^{-2} \times 3.5 \times 10^3}$; 1×10^{-2} **41. a.** $x^2 + 4x$; **b.** second degree binomial **c.** 21 **d.** 5 **42. a.** $3x^2 + 8x$ **b.** second degree binomial **c.** 51 **d.** 35

43. a. $-x^4 + 3x^3 + x^2 - 2x$ **b.** fourth degree polynomial **c.** 3 **d.** -965 **44. a.** $-x^3 - 6x^2 + 2x - 4$ **b.** third degree

polynomial **c.** -79 **d.** -39 **45.** $x^3 - 2x^2 - 6x - 7$ **46.** $-x^2 + x - 5$ **47.** $3x^3 + 4x^2 - 7x - 3$ **48.** $-x^3 + 7x^2 - 6$
49. $6x^2 - 7x + 1$ **50.** $-x - 17$ **51.** $-3x^3 + 12x^2 - 3x$ **52.** $5x^5 + 10x^4$ **53.** $x^2 - 36$ **54.** $x^2 + x - 12$
55. $9x^2 + 42x + 49$ **56.** $4x^2 - 1$ **57.** $x^4 - 25$ **58.** $x^4 - 4x^2 + 4$ **59.** $-x^2 + 5x - 2$ **60.** $x^2 - 8x + 16$

61. $2x^2 - x - 36$ **62.** $5x^2 - 27x - 18$ **63.** $6x^2 + x - 12$ **64.** $9x^2 - 64$ **65.** $4x^3 - 3x^2 - x$ **66.** $4x - 7 + \dfrac{3}{x}$

67. $\dfrac{13x}{5} + 2y + \dfrac{y^2}{x}$ **68.** $\dfrac{1}{7} - 3y + \dfrac{1}{xy} - 4y^2$ **69.** $x - 2$ **70.** $x + 2 + \dfrac{4}{4x - 3}$ **71.** $2x^2 - x + 3 - \dfrac{2}{x + 3}$ **72.** $20, 22, 24$

73. 16 cm by 25 cm **74.** -19 **75.** $38, 38, 42$ **76.** $a = 3, b = -4$ **77.** 35,000 at 8% & 65,000 at 6% **78.** 6 ounces of
10% mixture & 2 ounces of 30% mixture

Chapter 7

Exercises 7.1, Pages 434 – 435

1. 5 **3.** 8 **5.** 1 **7.** $10x^3$ **9.** $13ab$ **11.** $-14cd^2$ **13.** $12xy$ **15.** x^4 **17.** $-4y$ **19.** $3x^3$ **21.** $2x^2y$ **23.** $m + 9$
25. $x - 6$ **27.** $b + 1$ **29.** $3y + 4x + 1$ **31.** $11(x - 11)$ **33.** $4y(4y^2 + 3)$ **35.** $-8(a + 2b)$ **37.** $-3a(2x - 3y)$
39. $2x^2y(8x^2 - 7)$ **41.** $-14x^2y(y^2 + 1)$ **43.** $5(x^2 - 3x - 1)$ **45.** $4m^2(2x^3 - 3y + z)$ **47.** $51x^3y^5(x^2y - 1)$
49. $x^4y^2(15 + 24x^2y^4 - 32x^3y)$ **51.** $(y + 3)(7y^2 + 2)$ **53.** $(x - 4)(3x + 1)$ **55.** $(x - 2)(4x^3 - 1)$
57. $(2y + 3)(10y - 7)$ **59.** $(x - 2)(a - b)$ **61.** $(b + c)(x + 1)$ **63.** $(x^2 + 6)(x + 3)$ **65.** $(x + 6y)(x - 4)$
67. $(y - 4)(5x + z)$ **69.** $(2x - 3y)(z - 8)$ **71.** $(a + 3)(5y + x)$ **73.** $(4y + 3)(x - 1)$ **75.** $(x - 1)(y + 1)$
77. Not factorable **79.** Not factorable

Exercises 7.2, Pages 440 – 442

1. $\{1, 15\}, \{-1, -15\}, \{3, 5\}, \{-3, -5\}$ **3.** $\{1, 20\}, \{-1, -20\}, \{4, 5\}, \{-4, -5\}, \{2, 10\}, \{-2, -10\}$ **5.** $\{1, -6\}, \{6, -1\}, \{2, -3\}, \{3,$
$-2\}$ **7.** $\{1, 16\}, \{-1, -16\}, \{4, 4\}, \{-4, -4\}, \{8, 2\}, \{-8, -2\}$ **9.** $\{1, -10\}, \{10, -1\}, \{5, -2\}, \{2, -5\}$ **11.** $4, 3$ **13.** $-7, 2$ **15.** $8, -1$
17. $-6, -6$ **19.** $-5, -4$ **21.** $x + 1$ **23.** $p - 10$ **25.** $a + 6$ **27.** $(x - 4)(x + 3)$ **29.** $(y + 6)(y - 5)$ **31.** Not Factorable
33. Not Factorable **35.** $(x + 6)(x - 3)$ **37.** $(y - 12)(y - 2)$ **39.** $(x - 9)(x + 3)$ **41.** $x(x + 7)(x + 3)$
43. $5(x - 4)(x + 3)$ **45.** $10(y - 3)(y + 2)$ **47.** $4p^2(p + 1)(p + 8)$ **49.** $2x^2(x - 9)(x + 2)$ **51.** $2(x^2 - x - 36)$
53. $(a - 36)(a + 6)$ **55.** $3y^3(y - 8)(y + 1)$ **57.** $(x - 3y)(x + y)$ **59.** $20(a + b)(a + b)$ **61.** $x + 5$ inches
63. $x(10 - 2x)(40 - 2x)$; The height is x, the width is $10 - 2x$, and the length is $40 - 2x$. **65.** This is not an error,
but the trinomial is not completely factored. The completely factored form of this trinomial is
$2(x + 2)(x + 3)$.

Exercises 7.3, Pages 452 – 453

1. $(x + 2)(x + 3)$ **3.** $(2x - 5)(x + 1)$ **5.** $(6x + 5)(x + 1)$ **7.** $-(x - 2)(x - 1)$ **9.** $(x - 5)(x + 2)$

11. $-(x-14)(x+1)$ **13.** Not Factorable **15.** $-x(2x+1)(x-1)$ **17.** $(t-1)(4t+1)$ **19.** $(5a-6)(a+1)$
21. $(7x-2)(x+1)$ **23.** $(4x+3)(x+5)$ **25.** $(x-2)(x+8)$ **27.** $2(2x-5)(3x-2)$ **29.** $(3x-1)(x-2)$
31. $(3x-1)(3x-1)$ **33.** Not Factorable **35.** $(3y-4)(4y+3)$ **37.** $5(x^2+9)$ **39.** $2b(4a-3)(a-2)$
41. Not Factorable **43.** $(4x-1)(4x-1)$ **45.** $(8x-3)(8x-3)$ **47.** $2(3x-5)(x+2)$ **49.** $5(2x+3)(x+2)$
51. $-2(9x^2-36x+4)$ **53.** $x^2(7x^2-5x+3)$ **55.** $-2(m-2)(6m+1)$ **57.** $3x(2x-1)(x+2)$ **59.** $9xy^3(x^2+x+1)$
61. $3x(2x-9)(2x-9)$ **63.** $4xy(3y-4)(4y-3)$ **65.** $7y^2(y-4)(3y-2)$ **67.** $(a+3)(x+y)$
69. $(5+b)(x^2+y^2)$ **71. a.** 640 ft, 384 ft **b.** 144 ft, 400 ft **c.** 7 seconds; $0=-16(t+7)(t-7)$

Exercises 7.4, Page 459

1. $(x-1)(x+1)$ **3.** $(x-7)(x+7)$ **5.** $(x+2)^2$ **7.** $(x-6)^2$ **9.** $(4x-3)(4x+3)$ **11.** $(3x-1)(3x+1)$
13. $(5-2x)(5+2x)$ **15.** $(x-7)^2$ **17.** $(x+3)^2$ **19.** $3(x-3y)(x+3y)$ **21.** $x(x-y)(x+y)$
23. $\left(x-\dfrac{1}{2}\right)\left(x+\dfrac{1}{2}\right)$ **25.** $\left(x-\dfrac{3}{4}\right)\left(x+\dfrac{3}{4}\right)$ **27.** $(x-1)(x+1)(x^2+1)$ **29.** $2(x-8)^2$ **31.** $a(y+1)^2$
33. $4y^2(x-3)^2$ **35.** Not Factorable **37.** Not Factorable **39.** Not Factorable **41.** $x^2-6x+9=(x-3)^2$

43. $x^2-4x+4=(x-2)^2$ **45.** $x^2+8x+16=(x+4)^2$ **47.** $x^2-18x+81=(x-9)^2$ **49.** $x^2+x+\dfrac{1}{4}=\left(x+\dfrac{1}{2}\right)^2$

51. $x^2-9x+\dfrac{81}{4}=\left(x-\dfrac{9}{2}\right)^2$ **53.** $x^2+5x+\dfrac{25}{4}=\left(x+\dfrac{5}{2}\right)^2$ **55.** $x^2-3x+\dfrac{9}{4}=\left(x-\dfrac{3}{2}\right)^2$

Exercises 7.5, Page 467

1. $x=2,3$ **3.** $x=-2,\dfrac{9}{2}$ **5.** $x=-3$ **7.** $x=-5$ **9.** $x=0,2$ **11.** $x=-1,4$ **13.** $x=0,-3$ **15.** $x=4,-3$

17. $x=4,2$ **19.** $x=-4,3$ **21.** $x=-\dfrac{1}{2},3$ **23.** $x=-\dfrac{2}{3},2$ **25.** $x=4,-\dfrac{1}{2}$ **27.** $x=-2,\dfrac{4}{3}$ **29.** $x=\dfrac{3}{2}$

31. $x=3,-1$ **33.** $x=-2,-8$ **35.** $x=-5,2$ **37.** $x=-5,7$ **39.** $x=-6,2$ **41.** $x=-1,\dfrac{2}{3}$ **43.** $x=4,-\dfrac{3}{2}$

45. $x=0,\dfrac{8}{5}$ **47.** $x=\pm2$ **49.** $x=-1$ **51.** $x=2$ **53.** $x=3$ **55.** $x=-5,10$ **57.** $x=-2,-6$ **59.** $x=\dfrac{1}{2}$

61. $x=0,2,4$ **63.** $x=0,-\dfrac{1}{2},-\dfrac{2}{3}$ **65.** $x=\pm10$ **67.** $x=\pm5$ **69.** $x=-4$ **71.** $x=3$

Exercises 7.6, Pages 472 – 476

1. $x^2=7x; x=0,7$ **3.** $x^2=x+12; x=4$ **5.** $x(x+7)=78; x=-13,6.$ So the numbers are -13 and -6 or 13 and 6. **7.** $x^2+3x=54; x=6,$ So the number is 6. **9.** $x+(x+8)^2=124; x=3,$ So the numbers are 3 and 11. **11.** $x^2+(x-5)^2=97; x=-4,9,$ so the numbers are -9 and -4 or 9 and 4. **13.** $x(x+1)=72; x=8,$ so the numbers are 8 and 9. **15.** $x^2+(x+1)^2=85; x=6,$ so the numbers are 6 and 7. **17.** $(x+1)^2-x^2=17; x=8,$

so the numbers are 8 and 9. **19.** $x(x+2)=120$; $x=-12, 10$, so the numbers are -12 and -10 or 12 and 10.
21. $w(2w)=72$; $w=6$ in, width = 6 in and length = 12 in. **23.** $w(w+12)=85$; $w=5$, width = 5 m and
length = 17 m **25.** $l(l-4)=117$; $l=13$, width = 9 ft and length = 13 ft. **27.** $\frac{1}{2}(h+5)h=42$; $h=7$, height is 7
m and base is 12 m. **29.** $\frac{1}{2}(h-6)h=56$; $h=14$, height is 14 ft. **31.** $2(l+w)=20$; $lw=24$; width = 4 cm,
length = 6 cm. **33.** $x(x+13)=140$; $x=7$, so there are 7 trees in each row. **35.** $x(x+7)=144$; $x=9$, so there
are 9 rows. **37.** $(l+6)(l+1)=300$; $l=14$ m, so the rectangle is 14 m by 9 m. **39.** $w(52-2w)=320$; $w=10$
yd, 16 yd, so the corral is 10 yd by 32 yd or 16 yd by 20 yd. **41.** $w(3w)=48$; $w=4$; width = 4 ft and length = 12
ft. **43.** $x(x+8)=-16$; $x=-4$, so the numbers are -4 and 4.

Exercises 7.7, Pages 477 – 481

1. 2260.8 sq. in **3.** 5 in **5.** 800 lb **7.** 6 in **9.** 10 in **11.** 200 ft **13.** 4 seconds and 6 seconds **15.** 12 amps or 20 amps
17. $16 or $20 **19.** $4

Exercises 7.8, Pages 483

1. $(m+6)(m+1)$ **3.** $(x+9)(x+2)$ **5.** $(x-10)(x+10)$ **7.** $(m-3)(m+2)$ **9.** Not Factorable
11. $(8x-1)(8x+1)$ **13.** $(x+5)(x+5)$ **15.** $(x+12)(x-3)$ **17.** $(x+4)(x+9)$ **19.** $-5(x-6)(x-8)$
21. $-4(x-10)(x+5)$ **23.** $3(x-7)(x+7)$ **25.** $3x(x+3)(x+2)$ **27.** $4x(2x-5)(2x+5)$
29. $-(x-5)(3x-2)$ **31.** $2(2x-1)(x-3)$ **33.** $(2x-5)(6x-1)$ **35.** $(3x-7)(2x+5)$
37. $(4x-7)(2x+5)$ **39.** $(5x+6)(4x-9)$ **41.** $3(4x^2-20x-25)$ **43.** $-1(x-3)(3x+8)$
45. $-(2x-3)(4x-5)$ **47.** $(4x+5)(5x-4)$ **49.** $(6x-1)(3x-2)$ **51.** $-7x(5x-6)(5x+6)$
53. $x(21x^2-13x-2)$ **55.** $3x(3x-2)(4x+5)$ **57.** $2x(2x-1)(4x-11)$ **59.** $5(24x^2+2x+15)$
61. $(y-4)(x+3)$ **63.** $(x+2y)(x-6)$ **65.** $-(x^2-5)(x-8)$

Chapter 7 Test, Pages 488 – 489

1. 15 **2.** $8x^2y$ **3.** $2x^3$ **4.** $-7y^2$ **5.** $10x^3(2y+1)(y+1)$ **6.** $(x-5)(x-4)$ **7.** $-(x+7)(x+7)$
8. $6(x+1)(x-1)$ **9.** $2(6x-5)(x+1)$ **10.** $(x+3)(3x-8)$ **11.** $(4x-5y)(4x+5y)$ **12.** $x(x+1)(2x-3)$
13. $(2x-3)(3x-2)$ **14.** $(y+7)(2x-3)$ **15.** Not factorable **16.** $-3x(x^2-2x+2)$ **17.** 25; $(x-5)^2$
18. 64; $(x+8)^2$ **19.** $x=-2, \frac{5}{3}$ **20.** $x=8, -1$ **21.** $x=0, -6$ **22.** $x=5, -3$ **23.** $x=5, -\frac{3}{4}$

24. $x=-\frac{5}{4}, \frac{3}{2}$ **25.** $x=4, \frac{3}{2}$ **26.** $x=6, -4$ **27.** Length = 15 cm, Width = 11 cm **28.** $n=18, 19$ **29.** 12, 3
30. a. $A(x)=240-(x^2-3x)=-(x^2+3x-240)$ **b.** $P(x)=64$

Chapter 7 Cumulative Review, Pages 490 – 493

1. 120 **2.** $168x^2y$ **3.** $\dfrac{55}{48}$ **4.** $\dfrac{19}{60a}$ **5.** $\dfrac{3}{10}$ **6.** $\dfrac{75x}{23}$ **7.** $-8x^2y$ **8.** $4x^2y$ **9.** $-\dfrac{xy^4}{3}$ **10.** $7x^2y$

11. a. $D = \{2, 3, 5, 7.1\}$ **b.** $R = \{-3, -2, 0, 3.2\}$ **c.** It is a function because each first coordinate (domain) has only one corresponding second coordinate (range). **12. a.** 58 **b.** 4 **c.** $\dfrac{11}{4}$

13. $x + y + 2 = 0$

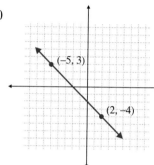

(−5, 3)

(2, −4)

14. $2x - y + 8 = 0$

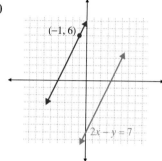

(−1, 6)

$2x - y = 7$

15. Function **16.** Not a function **17.** $x = \dfrac{3}{38},\ y = -\dfrac{153}{190}$ **18.** $x = -\dfrac{6}{7},\ y = \dfrac{17}{7}$ **19.** Consistent, $x = 3, y = -3$
20. Consistent, $x = 0, y = -1$ **21.** Inconsistent **22.** $(-2.076, -.692)$ and $(2.409, .803)$ **23.** $13x + 1$
24. $x^2 + 12x + 3$ **25.** $3x^2 - 5x + 3$ **26.** $4x^2 + 5x - 8$ **27.** $x^2 - 3x - 5$ **28.** $-4x^2 - 2x + 3$

29. $2x^2 + x - 28$ **30.** $-3x^2 - 17x + 6$ **31.** $x^2 + 12x + 36$ **32.** $4x^2 - 28x + 49$ **33.** $3x + 1 + \dfrac{22}{x - 2}$

34. $x^2 - 8x + 38 - \dfrac{142}{x + 4}$ **35.** $4(2x - 5)$ **36.** $6(x - 16)$ **37.** $(x - 6)(2x - 3)$ **38.** $(2x - 3)(3x + 4)$

39. $5(x - 2)$ **40.** $-4x(3x + 4)$ **41.** $8xy(2x - 3)$ **42.** $5x^2(2x^2 - 5x + 1)$ **43.** $(2x + 1)(2x - 1)$
44. $(y - 10)(y - 10)$ **45.** $3(x + 4y)(x - 4y)$ **46.** $(x - 9)(x + 2)$ **47.** $5(x + 4)(x + 4)$ **48.** Not Factorable
49. $(5x + 2)(5x + 2)$ **50.** $(x + 1)(3x + 2)$ **51.** $2x(x - 5)(x - 5)$ **52.** $4x(x^2 + 25)$ **53.** $(x + 2)(y + 3)$

54. $(x - 2)(a + b)$ **55.** $4; (x - 2)^2$ **56.** $81; (x + 9)^2$ **57.** $16; (x - 4)^2$ **58.** $10x; (x - 5)^2$ **59.** $5x; \left(x - \dfrac{5}{2}\right)^2$

60. $x = -1, 7$ **61.** $x = 0, -\dfrac{5}{3}$ **62.** $x = -3, \dfrac{3}{4}$ **63.** $x = 0, 7$ **64.** $x = -2, -6$ **65.** $x = -4, 7$ **66.** $x = 0, 1, -6$

67. $x = -10, 6$ **68.** $x = -5, 1$ **69.** $x = 0, -7$ **70.** $x = 0, -2$ **71.** $x = 0, 4$ **72.** $x = 5, -\dfrac{5}{2}$ **73.** $x = 1, -\dfrac{5}{4}$

74. $x = 0, -5, 2$ **75.** The speed of the boat is 5 mph. The speed of the current is 2 mph **76.** $7500 at 6%, $2500 at 8% **77.** Width = 13 in, Length = 17 in **78.** 12 and 3 **79.** 8 and 9 or −8 and −9 **80.** $t = 2.5, 3$ seconds

Chapter 8

Exercises 8.1, Pages 503 – 504

Answers

1. $x \neq 0$ **3.** $y \neq \dfrac{4}{3}$ **5.** $x \neq 0, 4$ **7.** $y \neq -2, 5$ **9.** No restriction **11.** $\dfrac{2}{75}$ **13.** 9 **15.** 0 **17.** $\dfrac{-3}{32}$ **19.** 1

21. $\dfrac{x-3}{x}, x \neq 0$ **23.** $\dfrac{x+2}{2}, x \neq 0$ **25.** $\dfrac{1}{2}, y \neq \dfrac{-3}{2}$ **27.** $\dfrac{x-2}{2x+1}, x \neq -\dfrac{1}{2}$ **29.** $\dfrac{1}{x-4}, x \neq 0, 4$ **31.** $-7, x \neq 2$

33. $1, x \neq \dfrac{-5}{3}$ **35.** $-1, x \neq -1, 1$ **37.** $\dfrac{x}{x+2}, x \neq -2$ **39.** $\dfrac{x+2}{x-5}, x \neq -5, 5$ **41.** $\dfrac{x-6}{x+3}, x \neq -3$

43. $\dfrac{-(x-3)}{2x^2+4x}, x \neq 0, 2, -2$ **45.** $\dfrac{4x+5}{x+2}, x \neq -\dfrac{5}{4}, -2$ **47.** Since $a = b, a - b = 0$. So the equation cannot be divided by $a - b$, since division by 0 is undefined.

Exercises 8.2, Pages 509 – 510

1. $\dfrac{2x}{x+2}$ **3.** $\dfrac{2x^2-6x}{x-1}$ **5.** $3x-6$ **7.** $\dfrac{2x-2}{x+1}$ **9.** $\dfrac{5x}{6}$ **11.** $\dfrac{x^2-2x}{(x-1)(x+1)}$ **13.** $\dfrac{x+3}{x+4}$ **15.** $\dfrac{3x+1}{x+1}$ **17.** $\dfrac{5-x}{(x+7)(x-2)}$

19. -1 **21.** $\dfrac{2x^2-x}{x+4}$ **23.** $\dfrac{4x}{x-3}$ **25.** $\dfrac{x-3}{2x-3}$ **27.** $\dfrac{3x-2}{3x+2}$ **29.** $\dfrac{x^2-x-12}{(3x+2)(2x+1)}$ **31.** $\dfrac{2x}{(x-4)(x-1)}$ **33.** $\dfrac{x^2-3x}{(x+1)^2}$

35. $\dfrac{x^2+5x}{2x+1}$ **37.** $\dfrac{(x+3)^2(x+5)^2}{(x-8)^2(x+1)(x-1)}$ **39.** $\dfrac{(2x-1)}{(6x+1)(2x+1)}$ **41.** $2x-5$ feet

Exercises 8.3, Pages 516 – 518

1. 4 **3.** 2 **5.** $\dfrac{x-3}{x+1}$ **7.** $\dfrac{1}{x-1}$ **9.** $\dfrac{x-2}{x+2}$ **11.** $\dfrac{x}{2x-1}$ **13.** $\dfrac{x-5}{x-4}$ **15.** $\dfrac{2}{x-3}$ **17.** $\dfrac{-2}{5(x-2)}$ **19.** $\dfrac{3x+8}{x(x+4)}$

21. $\dfrac{x^2-2x+8}{(x+4)(x-4)}$ **23.** $\dfrac{3}{x-2}$ **25.** $\dfrac{x+3}{x-5}$ **27.** $\dfrac{x^2+x+1}{(x+2)(x-1)}$ **29.** $\dfrac{x^2+11x+4}{(4-x)(4+x)}$ **31.** $\dfrac{4x+24}{(x+2)(x-2)}$

33. $\dfrac{9x+4}{12(2-x)}$ **35.** $\dfrac{-2x-17}{(x+5)(x-1)}$ **37.** $\dfrac{3x^2-20x}{(x+6)(x-6)}$ **39.** $\dfrac{2x^2+2x-1}{(x-7)(x-1)}$ **41.** $\dfrac{x^2+4x+2}{2(x+2)(x+3)}$ **43.** $\dfrac{x^2+3x}{-2(x+4)(x-4)}$

45. $\dfrac{-1}{(x+2)(x+1)(x-1)}$ **47.** $\dfrac{5x^2+18x}{(x+3)(x-3)(x+4)}$ **49.** $\dfrac{-2x^2-8x-3}{(x+2)^2(x+1)}$ **51.** $\dfrac{-2x^2+13x}{(x+4)(x-4)(x+1)}$

53. $\dfrac{3x^2-5x-7}{(2x-1)(x+3)(x-4)}$ **55.** $\dfrac{4x^2+17x}{(x+2)(x-2)(x+5)}$ **57.** $\dfrac{10x^2-17x-3}{(x+4)(x-4)}$ **59.** $\dfrac{x+1}{(x-2)(x-1)}$

Exercises 8.4, Pages 524 – 525

1. $\dfrac{8}{7}$ **3.** $\dfrac{10}{7}$ **5.** $\dfrac{1}{2}$ **7.** 3 **9.** $\dfrac{2x}{3y}$ **11.** $16xy$ **13.** $\dfrac{8}{7x^2y^3}$ **15.** $\dfrac{3x}{x-2}$ **17.** $\dfrac{2x^2+6x}{2x-1}$ **19.** $\dfrac{x+3}{x}$ **21.** $\dfrac{x+2}{4x}$

23. $\dfrac{2x}{3(x+6)}$ **25.** $\dfrac{x}{x-1}$ **27.** $\dfrac{x}{y+x}$ **29.** $\dfrac{2x+2}{x+2}$ **31.** $\dfrac{x+2}{x+3}$ **33.** $\dfrac{-5}{x+1}$ **35.** $\dfrac{29}{4(4x+5)}$ **37.** $\dfrac{x^2-3x-6}{x(x-1)}$

39. $\dfrac{x^2-4x-2}{(x+4)(x-4)}$ **41. a.** $\dfrac{8}{5}$ **b.** 1 **c.** $\dfrac{x^4+x^3+3x^2+2x+1}{x^3+x^2+2x+1}$

Exercises 8.5, Pages 533 – 536

1. $x = \frac{3}{2}$ **3.** $x = 7$ **5.** $x \neq 0; x = 10$ **7.** $x \neq -4; x = 20$ **9.** $x \neq 0; x = \frac{38}{3}$ **11.** $x \neq 0; x = \frac{3}{2}$ **13.** $x \neq 1; x = -3$

15. $x \neq 0, 2; x = 4$ **17.** $x \neq 4, -7; x = -62$ **19.** $x \neq 6; x = 5$ **21.** $x \neq 3; x = \frac{13}{5}$ **23.** $x \neq -4, 0; \ x = -6, 2$

25. $x \neq 0, -1; x = \frac{-3}{2}, 2$ **27.** $x \neq -4, 3; x = -11, 2$ **29.** $x \neq 3; x = 3;$ Therefore there is no solution.

31. $x \neq -3, -2; x = \frac{-3}{2}$ **33.** $x \neq 4, \frac{1}{2}; x = -3$ **35.** $x \neq \frac{1}{4}, -1; x = \frac{2}{3}$ **37.** $x \neq -5, 4; x = -2$ **39.** $x \neq -1, 1; x = -9, 2$

41. $x \neq 3, -1, x = -2$ **43.** $x \neq -5, -4; x = 3$ **45.** \$8.50 **47.** 144 bulbs **49.** 504 students **51.** 600 at bats **53.** 8.8 quarts

55. 40 ft. **57. a.** $\frac{2x^2 + 7x - 5}{x^2 - x}$ **b.** $x = \frac{5}{2}$ and $x \neq 1, 0$

Exercises 8.6, Pages 544 – 547

1. 21 and 27 **3.** 20 men **5.** $\frac{7}{9}$ **7.** 300 copies and 375 copies **9.** 45 mph, 60 mph **11.** speed of wind = 48 mph

13. 500 mph and 520 mph **15.** 480 mph **17.** 9 mph **19.** 14 mph **21.** 2 hrs 24 minutes **23.** 22.5 minutes
25. 45 minutes **27.** 1.5 hrs and 3 hrs **29.** 10 days, 15 days

Exercises 8.7, Pages 550 – 554

1. $\frac{7}{5}$ **3.** 36 **5.** 10 **7.** \$37.50 **9.** 190.4 lb **11.** $225 \, cm^3$ **13.** 6.28 ft **15.** 4 inches **17.** 139.219 ft^2 **19.** 192 lb per ft^2

21. 700 lb per ft^2 **23.** $5\frac{1}{3}$ ohms **25.** 384 rpm **27.** 36 teeth **29.** 4 ft

Chapter 8 Test, Pages 558 – 559

1. $\frac{x+3}{x}, x \neq 0, \frac{1}{2}$ **2.** $\frac{2x+1}{x+1}, x \neq -1, \frac{3}{4}$ **3.** $-\frac{x^2}{2}, x \neq 0, \frac{6}{5}$ **4.** -1 **5.** $\frac{27}{8}$ **6.** $\frac{7(4x+3)}{x(x+4)}, \ x \neq -\frac{2}{3}, -\frac{3}{4}, 0, -4$

7. $\frac{1}{2(2x+1)}, x \neq 1, -\frac{2}{3}, \frac{1}{2}, -\frac{1}{2}$ **8.** $\frac{(10x-13)}{(x-1)(x+1)(x-2)}; x \neq -1, 1, -2$ **9.** $\frac{8x^2 - 19x + 15}{(x+5)(x-2)(x-5)}, x \neq -5, 2, 5$

10. $\frac{(3x^2 - 8x - 27)}{(x-1)(x-4)(x+3)}, x \neq 1, 4, -3$ **11.** $\frac{x(x-3)}{2(x+1)}, x \neq 0, -1, -3, -5$ **12.** $\frac{1}{x+1}$ **13.** $\frac{6(x+3)}{x+18}$ **14.** $x = \frac{15}{26}$

15. $x = -\frac{5}{11}$ **16.** $x = 2$ **17. a.** $\frac{7x+11}{2x(x+1)}$ **b.** $x = \frac{1}{3}$ **18.** 65; 91 **19.** Lisa's speed = 57 mph; Kim's speed = 42 mph
20. 4 hrs. **21.** 1562.5 lbs **22.** 25,000 lbs **23.** 3 ohms **24.** 6 hrs.

Chapter 8 Cumulative Review, Pages 560 – 563

1. a. $\dfrac{19}{12}$ **b.** $-\dfrac{3}{14}$ **c.** $\dfrac{1}{4}$ **d.** $\dfrac{1}{12}$ **2.** 5 **3.** $3x^2 - 5x - 28$ **4.** $4x^2 - 20x + 25$ **5.** $6x - 23$ **6.** $10x + 4$ **7.** $t = \dfrac{A - P}{Pr}$

8. $r = \dfrac{C}{2\pi}$ **9.** $(2x + 5)(2x + 3)$ **10.** $2(x + 5)(x - 2)$ **11. a.** $D = \{-1, 0, 5, 6\}$ **b.** $R = \{0, 5, 2\}$ **c.** It is not a function

12. a. 8 **b.** -1 **c.** $-\dfrac{59}{27}$ **13.** $y = \dfrac{3}{4}x + \dfrac{11}{2}$; **14.** $y + 2x = -8$

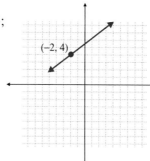

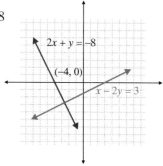

15. Function **16.** Function **17.** $x = \dfrac{37}{30}, y = \dfrac{119}{30}$ **18.** $x = -\dfrac{5}{2}, y = \dfrac{7}{4}$ **19.** Consistent $(5, 0)$ **20.** Dependent

21. Consistent $\left(\dfrac{9}{5}, \dfrac{39}{10}\right)$ **22.** $(-1.68, .326)$ and $(1.484, 1.594)$ **23.** $8x^2$ **24.** $5x^3y^2$

25. $\dfrac{3}{5}x + 2y + \dfrac{y^2}{x}$ **26.** $1 - 3y + \dfrac{8}{7}y^2$ **27.** $x - 15 - \dfrac{1}{x + 1}$ **28.** $2x^2 - x + 3 - \dfrac{2}{x + 3}$ **29.** $\dfrac{4x^2}{x - 3}, x \neq 3, -\dfrac{1}{2}$

30. Does not reduce, $x \neq 2$ **31.** $x - 3, x \neq 3$ **32.** $\dfrac{1}{x + 1}, x \neq 0, -1$ **33.** $-\dfrac{1}{3}, x \neq 4$ **34.** $\dfrac{2}{3}, x \neq -3$

35. $\dfrac{x}{x + 4}, x \neq -4, -3$ **36.** $\dfrac{x + 5}{2(x - 3)}, x \neq 3$ **37.** $x - y$ **38.** $\dfrac{4(4x - 7)}{x(x - 4)}$ **39.** $\dfrac{4}{(y + 2)(y + 3)}$ **40.** $\dfrac{x}{3x + 3}$

41. $\dfrac{x + 4}{3x}$ **42.** $\dfrac{3x(x + 2)^2}{x + 3}$ **43.** $\dfrac{2x + 1}{x - 1}$ **44.** $\dfrac{2x^2 + 14x - 8}{(x + 3)(x - 1)(x - 2)}$ **45.** $\dfrac{2(x + 2)}{(x - 1)(x + 4)}$ **46.** $\dfrac{x - 4}{(x + 2)(x - 2)}$

47. $\dfrac{19}{14}$ **48.** $\dfrac{x - 1}{x + 1}$ **49.** $\dfrac{3x}{2 - x}$ **50.** $-\dfrac{3}{5}$ **51.** $x = 6, -5$ **52.** $x = 2, 5$ **53.** $x = 2, -10$ **54.** $x = -45$

55. a. $\dfrac{9 - 2x}{x(x + 3)}$ **b.** $x = \dfrac{9}{2}$ **56. a.** $\dfrac{6x^2 - 16x - 20}{(x - 4)(x + 2)}$ **b.** $x = 7$ **57.** train's speed = 60 mph; speed of airplane = 230 mph

58. The father takes $2\dfrac{2}{3}$ hrs. and the daughter takes 8 hrs. **59.** 2.5 fc **60.** 60, 24 **61.** 3 mph **62.** $7\dfrac{9}{13}$ in.

Chapter 9

Exercises 9.1, Pages 577 – 579

1. 0.375 **3.** $0.\overline{27}$ **5.** $-0.\overline{857142}$ **7. a.** 5.6568 **b.** -4.2426 **c.** -1.7724 **d.** 2.7144 **e.** 2.7831
9. Rational **11.** Rational **13.** Irrational **15.** Nonreal **17.** Rational **19.** Rational **21.** Irrational
23. Rational **25.** Irrational **27.** 7.8990 **29.** -2.7082 **31.** 2.0811 **33.** -2.4934 **35.** 2.4145 **37.** 25 **39.** 6 **41.** 3
43. -2 **45.** 2 **47.** -6 **49.** 1.5 **51.** 8.0623 **53.** 6.1563 **55.** 3.4028 **57.** 2.9428 **59.** 6.3096 **61.** Because the
square of a real number is never negative. The same is also true for the fourth root of a negative number.

Exercises 9.2, Pages 586 – 587

1. $6|x|$ **3.** $\dfrac{1}{2}$ **5.** -1 **7.** 1 **9.** $2|x|\sqrt{2x}$ **11.** $-2\sqrt{14}$ **13.** $\dfrac{\sqrt{5}x^2}{3}$ **15.** $2\sqrt{3}$ **17.** $12\sqrt{2}$ **19.** $-6\sqrt{2}$ **21.** $-5\sqrt{5}$

23. $-4\sqrt[3]{2}$ **25.** $6x^2y\sqrt[3]{y^2}$ **27.** $-2\sqrt[3]{9}$ **29.** $-\dfrac{\sqrt{11}}{8}$ **31.** $\dfrac{2\sqrt{7}}{5}$ **33.** $\dfrac{4a^2\sqrt{2a}}{9b^8}$ **35.** $\dfrac{\sqrt[3]{3}}{2}$ **37.** $\dfrac{5\sqrt[3]{3}}{2}$ **39.** $\dfrac{1-\sqrt{3}}{2}$

41. $\dfrac{1+\sqrt{6}}{2}$ **43.** $\dfrac{4+\sqrt{5}}{5}$ **45.** $\dfrac{5-3\sqrt{3}}{2}$ **47.** $\dfrac{8-\sqrt{15}}{6}$ **49.** $\dfrac{3-2\sqrt{3}}{7}$ **51.** $5x\sqrt[3]{x}$ **53.** $\dfrac{5y^4}{3x^2}$ **55.** $\dfrac{4|x^7y^5|\sqrt{2x}}{13}$

Exercises 9.3, Pages 592 – 594

1. $8\sqrt{2}$ **3.** $7\sqrt{5}$ **5.** $-3\sqrt{10}$ **7.** $13\sqrt[3]{3}$ **9.** $-\sqrt{11}$ **11.** $3\sqrt{a}$ **13.** $7\sqrt{x}$ **15.** $4\sqrt{2}+3\sqrt{3}$ **17.** $8\sqrt{b}-4\sqrt{a}$
19. $13\left(\sqrt[3]{x}\right)-2\left(\sqrt[3]{y}\right)$ **21.** $5\sqrt{3}$ **23.** 0 **25.** $17\sqrt[3]{2}$ **27.** $2\sqrt{2}-6\sqrt{3}$ **29.** $6+\sqrt{5}$ **31.** $\sqrt{3}-4\sqrt{2}$
33. $5\sqrt[3]{2}-8\sqrt[3]{3}$ **35.** $4\sqrt{2x}$ **37.** $2y\sqrt{2y}$ **39.** $-4x\sqrt{3xy}$ **41.** $15x\sqrt{x}$ **43.** $-xy^2\sqrt{x}$ **45.** $4x^5y^{10}\sqrt{3}$
47. $-8x^8y^2$ **49.** $xy^2\sqrt[3]{2}\left(-2x^2y^2-2x^3y+3\right)$ **51.** $3\sqrt{2}-8$ **53.** 18 **55.** $-8\sqrt{3}$ **57.** $12+\sqrt{6}$
59. $2y+\sqrt{xy}$ **61.** $13-2\sqrt{2}$ **63.** $8-7\sqrt{6}$ **65.** $x-9$ **67.** -5 **69.** $x+2\sqrt{x}-15$

Exercises 9.4, Pages 599 – 600

1. $\dfrac{5\sqrt{2}}{2}$ **3.** $\dfrac{-3\sqrt{7}}{7}$ **5.** $2\sqrt{3}$ **7.** 3 **9.** 3 **11.** $\dfrac{1}{3}$ **13.** $\dfrac{2\sqrt{3}}{3}$ **15.** $\dfrac{3\sqrt{2}}{2}$ **17.** $\dfrac{\sqrt{x}}{x}$ **19.** $\dfrac{\sqrt{2xy}}{y}$ **21.** $\dfrac{\sqrt{2y}}{y}$

23. $\dfrac{3\sqrt{7}}{5}$ **25.** $-\dfrac{\sqrt{2y}}{5}$ **27.** $2+\sqrt{6}$ **29.** $-\dfrac{\left(3+\sqrt{5}\right)}{4}$ **31.** $\dfrac{-6\left(5+3\sqrt{2}\right)}{7}$ **33.** $\dfrac{\sqrt{3}}{23}\left(\sqrt{2}-5\right)$ **35.** $-\dfrac{7\left(1+3\sqrt{5}\right)}{44}$

37. $\dfrac{-\left(\sqrt{3}+\sqrt{5}\right)}{2}$ **39.** $5\left(\sqrt{2}-\sqrt{3}\right)$ **41.** $\dfrac{4\left(\sqrt{x}-1\right)}{\left(x-1\right)}$ **43.** $\dfrac{5\left(6-\sqrt{y}\right)}{36-y}$ **45.** $\dfrac{8\left(2\sqrt{x}-3\right)}{4x-9}$ **47.** $\dfrac{2\sqrt{y}\left(\sqrt{5y}+3\right)}{\left(5y-3\right)}$

49. $\dfrac{3\left(\sqrt{x}+\sqrt{y}\right)}{x-y}$ **51.** $\dfrac{x\left(\sqrt{x}-2\sqrt{y}\right)}{(x-4y)}$ **53.** $-\left(\sqrt{3}+1\right)\left(\sqrt{3}+2\right)$ **55.** $\dfrac{-\left(\sqrt{5}-2\right)\left(\sqrt{5}-3\right)}{4}$ **57.** $\dfrac{\left(\sqrt{x}+1\right)^2}{(x-1)}$

59. $\dfrac{\left(\sqrt{x}+2\right)\left(\sqrt{3x}-y\right)}{\left(3x-y^2\right)}$

Exercises 9.5, Page 605

1. $x = 33$ **3.** $x = 5$ **5.** No solution **7.** $x = 9$ **9.** $x = -3$ **11.** No solution **13.** $x = 6$ **15.** $x = -1$ **17.** $x = -22$
19. $x = 40$ **21.** $x = 2$ **23.** $x = 7$ **25.** $x = 4$ **27.** $x = -2$ **29.** $x = 2, x = 3$ **31.** $x = 2$ **33.** $x = 12$ **35.** $x = -2$ **37.** $x = 6$
39. $x = -\dfrac{1}{4}$ **41.** $x = 2, x = 5$ **43.** $x = -2, x = -3$ **45.** $x = -4, x = -6$

Exercises 9.6, Pages 610 – 611

1. $\sqrt{16}, 4$ **3.** $-\sqrt[3]{8}, -2$ **5.** $\sqrt[4]{81}, 3$ **7.** $\sqrt[5]{-32}, -2$ **9.** $\sqrt{0.0004}, 0.02$ **11.** $\dfrac{2}{3}$ **13.** $\dfrac{2}{5}$ **15.** $\dfrac{1}{6}$ **17.** $\dfrac{1}{2}$

19. $\dfrac{1}{3}$ **21.** 8 **23.** 9 **25.** 243 **27.** $\dfrac{1}{8}$ **29.** $\dfrac{1}{4}$ **31.** $x^{\frac{7}{12}}$ **33.** $\dfrac{1}{5}$ **35.** $x^{\frac{3}{5}}$ **37.** $x^{\frac{1}{2}}$ **39.** $x^{\frac{2}{5}}$ **41.** 2 **43.** $a^{\frac{7}{6}}$ **45.** $x^{\frac{5}{8}}$ **47.** $x^{-\frac{1}{10}}$

49. $6^{\frac{3}{2}}$ **51.** $x^{\frac{5}{4}}$ **53.** x^2 **55.** $196 \cdot x^{\frac{7}{6}}$ **57.** $6x^2$ **59.** $\dfrac{2x}{3}$

Exercises 9.7, Pages 620 – 624

1. a. 4 **b.** $(-3, 4)$ **3. a.** 3 **b.** $\left(\dfrac{11}{2}, -3\right)$ **5. a.** $\dfrac{7}{2}$ **b.** $\left(\dfrac{5}{4}, 1\right)$ **7. a.** $\sqrt{2}$ **b.** $\left(\dfrac{5}{2}, \dfrac{1}{2}\right)$ **9. a.** $\sqrt{13}$ **b.** $\left(0, \dfrac{7}{2}\right)$

11. a. 13 **b.** $\left(\dfrac{-1}{2}, -1\right)$ **13. a.** $\dfrac{5}{7}$ **b.** $\left(\dfrac{3}{14}, \dfrac{2}{7}\right)$ **15. a.** 5 **b.** $\left(\dfrac{11}{2}, 3\right)$ **17. a.** 13 **b.** $\left(-4, \dfrac{1}{2}\right)$ **19. a.** 10

b. $(0, -2)$ **21. a.** 5 **b.** $\left(2, \dfrac{-3}{2}\right)$ **23. a.** 2 **b.** $\left(\dfrac{-1}{5}, \dfrac{2}{7}\right)$ **25.** yes **27.** $|AB| = |CA| = 4\sqrt{5}$ **29.** Each side is 4 units.

31. Both diagonals are $\sqrt{61}$ units long **33.** Yes, $5^2 + 12^2 = 13^2$ **35.** Yes, $8^2 + 6^2 = 10^2$ **37. a.** $C = 2\pi r = 94.2$ ft,
$A = \pi r^2 = 706.5$ ft.^{2}, **b.** $P = 84.85$ ft, $A = 450$ ft.^{2} **39. a.** No **b.** Home plate **c.** No **41.** 8.5 cm **43.** $10 + 4\sqrt{5}$

45. $\sqrt{10} + \sqrt{41} + \sqrt{65}$ **47.** $(5, 13)$

Chapter 9 Test, Pages 628 – 629

1. $0.\overline{428571}$ **2.** 3.6055512 **3.** 12 **4.** $5|x|$ **5.** $-9x^2|y|\sqrt{y}$ **6.** $3a$ **7.** $2x^2y^4\sqrt[3]{3}$ **8.** $-\dfrac{4|x|}{y^4}$ **9.** $\dfrac{2a^2\sqrt{a}}{3|b^5|}$

10. $13\sqrt{2}$ **11.** $22|x|\sqrt{3}$ **12.** $\dfrac{\sqrt{6}+3}{6}$ **13.** $3 - \sqrt{2}$ **14.** $-2\sqrt[3]{5} + 3\sqrt[3]{4}$ **15.** 75 **16.** $3\sqrt{5} + 5\sqrt{3}$ **17.** 14

18. $28-\sqrt{2}$ **19.** $\dfrac{\sqrt{2}}{2}$ **20.** $\dfrac{2\sqrt{3}}{15}$ **21.** $1+\sqrt{3}$ **22.** $-2+\sqrt{6}$ **23.** $x=34$ **24.** $x=2$ **25.** No solution

26. 8 **27.** $\dfrac{27}{8}$ **28.** $a^{\frac{1}{4}}$ **29.** $y^{\frac{5}{4}}$ **30.** $36x^3$ **31. a.** $8\sqrt{2}$ **b.** $(-1,\,3)$ **32. a.** $\sqrt{157}$ **b.** $\left(-1,\,-\dfrac{1}{2}\right)$ **33.** Yes, this is a

right triangle. **34.** 24.3 ft

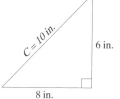

Chapter 9 Cumulative Review, Pages 630 – 632

1. $x^2+8x+16=(x+4)^2$ **2.** $x^2-14x+49=(x-7)^2$ **3.** $\dfrac{2(x+4)}{x-3}$ **4.** $\dfrac{x^2-4x+5}{(x+6)(x-2)}$

5.

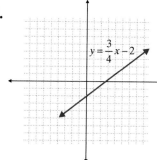

6.

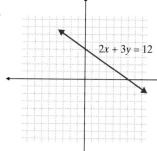

7. $-3\sqrt{7}x^2$ **8.** $\dfrac{6|a|\sqrt{a}}{5b^3}$ **9.** $4|x^5|y^{10}\sqrt{3}$

10. $3a^4|b^7|\sqrt{6}$ **11.** $2\sqrt[3]{5}$ **12.** $-5\sqrt[3]{2}$ **13.** $4+\sqrt{3}$ **14.** $\dfrac{5-\sqrt{3}}{2}$ **15.** $9\sqrt{2}$ **16.** $16\sqrt{2}$ **17.** $5\sqrt{5}$ **18.** $2\sqrt[3]{2}$

19. $x^2\sqrt{2}$ **20.** $\sqrt{3}$ **21.** 12 **22.** $20\sqrt{6}$ **23.** $8\sqrt{6}$ **24.** $7\sqrt{6}$ **25.** -22 **26.** $17+8\sqrt{2}$ **27.** $x-2$

28. $2x-13\sqrt{x}-24$ **29.** $3\sqrt{3}$ **30.** $\dfrac{\sqrt{15}}{10}$ **31.** $\dfrac{\sqrt{15}}{9}$ **32.** $\dfrac{-3\sqrt{2y}}{2y}$ **33.** $\dfrac{-\left(\sqrt{3}+5\right)}{11}$ **34.** $\dfrac{-\left(\sqrt{2}-3\right)}{7}$

35. $x=-4$ **36.** $x=\dfrac{9}{2},-1$ **37.** $x=4$ **38.** $x=1$ **39.** $x=1$ **40.** $x=7$ **41.** $x=29$ **42.** $x=2$ **43.** $x=2,y=-3$

44. $x=3,y=1$ **45.** 44 mph and 56 mph **46.** 8 pounds for \$1.50 and 12 pounds for \$2 **47.** 4 hrs and 12 hrs

48. Burrito price is \$1.75 and taco price is \$0.80 **49.** 26.5 m **50. a.** $\sqrt{6}$ or 2.45 sec **b.** 96 feet **c.** $\sqrt{15}$ or 3.87 seconds

Chapter 10

Exercises 10.1, Pages 639 – 641

1. $x=0,11$ **3.** $x=-12,-3$ **5.** $x=1,-\dfrac{5}{3}$ **7.** $x=-1,3$ **9.** $x=\dfrac{3}{4},1$ **11.** $x=\pm11$ **13.** $x=\pm6$ **15.** $x=\pm\sqrt{35}$

17. $x = \pm\sqrt{62}$ **19.** $x = \pm 3\sqrt{5}$ **21.** $x = \pm 3\sqrt{2}$ **23.** $x = \pm\dfrac{2}{3}$ **25.** $x = -1, 3$ **27.** No solution **29.** $x = -\dfrac{3}{2}, -\dfrac{1}{2}$

31. $x = \dfrac{7}{3}, \dfrac{11}{3}$ **33.** $x = 6 \pm 3\sqrt{2}$ **35.** $x = 7 \pm 2\sqrt{3}$ **37.** $x = \dfrac{-4 \pm 3\sqrt{3}}{3}$ **39.** $x = \dfrac{2 \pm 3\sqrt{7}}{5}$ **41.** Yes **43.** Yes

45. $c = 15$ **47.** $b = 6\sqrt{3}$ **49.** The length of the leg is 4 ft and the hypotenuse is 8 ft. **51.** $3\sqrt{2}$ cm

53. 47.2 ft **55.** 3 seconds **57.** $x = \pm 25.44$ **59.** $x = \pm 5.25$ **61.** $x = \pm 1.7$ **63.** $x = \pm 4.13$

Exercises 10.2, Pages 646 – 647

1. $x^2 + 12x + 36 = (x+6)^2$ **3.** $2x^2 - 16x + 32 = 2(x-4)^2$ **5.** $x^2 - 3x + \dfrac{9}{4} = \left(x - \dfrac{3}{2}\right)^2$ **7.** $x^2 + x + \dfrac{1}{4} = \left(x + \dfrac{1}{2}\right)^2$

9. $2x^2 + 4x + 2 = 2(x+1)^2$ **11.** $x = -7, 1$ **13.** $x = -5, 9$ **15.** $x = -5, 8$ **17.** $x = -\dfrac{4}{3}, 1$ **19.** $x = -\dfrac{1}{2}, \dfrac{3}{2},$

21. $x = -3 \pm \sqrt{6}$ **23.** $x = -1 \pm \sqrt{6}$ **25.** $x = -\dfrac{1}{3}, 1$ **27.** $x = \dfrac{-1 \pm \sqrt{13}}{2}$ **29.** $x = \dfrac{-3 \pm \sqrt{17}}{4}$

31. $x = \dfrac{9 \pm \sqrt{73}}{2}$ **33.** $x = \dfrac{-7 \pm \sqrt{105}}{2}$ **35.** $x = -2, 13$ **37.** $x = \dfrac{-1 \pm \sqrt{17}}{4}$ **39.** $x = 1$ **41.** $x = \dfrac{-5 \pm \sqrt{33}}{2}$

43. $x = \dfrac{4 \pm \sqrt{10}}{6}$ **45.** $x = \dfrac{-7 \pm \sqrt{17}}{4}$ **47.** $x = \dfrac{1 \pm \sqrt{13}}{4}$ **49.** $x = \dfrac{-5}{3}, -1$

Exercises 10.3, Pages 654 – 655

1. $x^2 - 3x - 2 = 0; a = 1, b = -3, c = -2$ **3.** $2x^2 - x + 6 = 0; a = 2, b = -1, c = 6$ **5.** $7x^2 - 4x - 3 = 0; a = 7, b = -4,$
$c = -3$ **7.** $3x^2 - 9x - 4 = 0; a = 3, b = -9, c = -4$ **9.** $2x^2 + 5x - 3 = 0; a = 2, b = 5, c = -3$ **11.** $x = 2 \pm \sqrt{5}$

13. $x = -1, 4$ **15.** $x = \dfrac{-1}{2}, 1$ **17.** $x = -1, \dfrac{2}{5}$ **19.** $x = 0, \dfrac{1}{3}$ **21.** $x = \pm\sqrt{7}$ **23.** $x = -1, 0$ **25.** $x = \dfrac{-4}{3}, 0$

27. $x = \dfrac{-5 \pm \sqrt{65}}{4}$ **29.** $x = \dfrac{-6 \pm \sqrt{42}}{3}$ **31.** $x = \dfrac{1}{2}, -3$ **33.** $x = \dfrac{4 \pm \sqrt{22}}{3}$ **35.** $x = 1, \dfrac{3}{4}$ **37.** $x = 1, \dfrac{-4}{5}$

39. $x = \dfrac{7 \pm \sqrt{37}}{6}$ **41.** $x = -\dfrac{3}{2}, \dfrac{8}{5}$ **43.** $x = 4, -\dfrac{1}{3}$ **45.** $x = \dfrac{11 \pm \sqrt{41}}{8}$ **47.** $x = \dfrac{-3}{2}, -2$ **49.** $x = \pm\dfrac{4}{5}$

51. $x = \dfrac{8 \pm \sqrt{58}}{6}$ **53.** $x = \dfrac{7 \pm \sqrt{385}}{12}$ **55.** $x = 1.6180, -0.6180$ **57.** $x = 1, -0.25$ **59.** $x = \pm 4.4721$

Exercises 10.4, Pages 661 – 666

1. $x^2 + x = 132; x = 11$ **3.** $w(2w - 5) = 63; w = 7$, the rectangle is 7 m by 9 m **5.** $x(29 - x) = 198; x = 11$, the field
is 11 m by 18 m **7.** $(w + 12)(w + 22) = 1344; w = 20$, the pool is 20 ft by 30 ft. **9.** $x^2 + (x+1)^2 = 221; x = 10$, the
numbers are 10 and 11. **11.** $x - 3 = \dfrac{4}{x}; x = 4$ **13.** $(x + 10)^2 = 9x^2; x = 5$, the side of the original square is 5 cm.

15. $w^2 + (2w+2)^2 = 169$; $w = 5$, the rectangle is 5 m by 12 m. **17.** $2x^2 = 144$; the legs of the triangle are $6\sqrt{2}$ cm each. **19.** $\dfrac{900}{x} - 10 = \dfrac{900}{x+15}$; There are 30 people in the club. **21.** $\dfrac{200}{x} + \dfrac{200}{x-10} = 9$; he traveled to the city at 50 mph. **23.** $\dfrac{45}{12-c} - \dfrac{45}{12+c} = 2$; the rate of the current is 3 mph **25.** $\dfrac{120}{x} + 2 = \dfrac{120}{x-5}$; there were initially 20 members. **27.** $2(x-4)^2 = 162$; $x = 13$, the cardboard was 13 in. by 13 in. **29.** $\dfrac{1}{x} + \dfrac{1}{x+30} = \dfrac{1}{8}$; $x = 10$; the small pipe would take 40 min. **31. a.** After 1 sec. **b.** 2.5 sec. **c.** 3 sec. **33.** a cannot be equal to zero and b^2 must be greater than or equal to $4ac$ to produce a real solution. Also, all the solutions found may not apply to the problem at hand. You must check that each answer makes sense in the context of the problem.

Exercises 10.5, Pages 676 – 678

1. a. $(0,4)$ **b.** $x = 0$ **c.** None

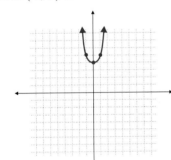

3. a. $(0,8)$ **b.** $x = 0$ **c.** $\pm 2\sqrt{2}$

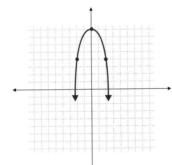

5. a. $(1,-4)$ **b.** $x = 1$ **c.** -1 and 3

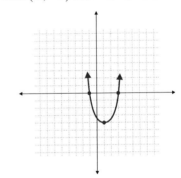

7. a. $(-3,-9)$ **b.** $x = -3$ **c.** $-6, 0$

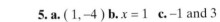
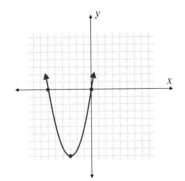

9. a. $(-2,6)$ **b.** $x = -2$ **c.** $-2 \pm \sqrt{6}$

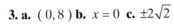
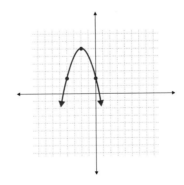

11. a. $\left(\dfrac{5}{2}, -\dfrac{19}{2}\right)$ **b.** $x = \dfrac{5}{2}$ **c.** $\dfrac{5 \pm \sqrt{19}}{2}$

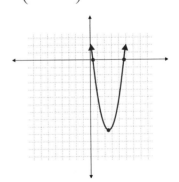

13. a. $\left(-\dfrac{7}{2}, -\dfrac{65}{4}\right)$ **b.** $x = -\dfrac{7}{2}$ **c.** $\dfrac{-7 \pm \sqrt{65}}{2}$ **15. a.** $\left(\dfrac{1}{2}, -\dfrac{11}{4}\right)$ **b.** $x = \dfrac{1}{2}$ **c.** None **17. a.** $\left(-\dfrac{7}{4}, -\dfrac{81}{8}\right)$ **b.** $x = -\dfrac{7}{4}$

c. $-4, \dfrac{1}{2}$

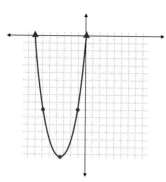
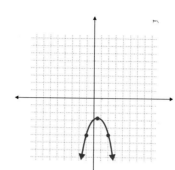
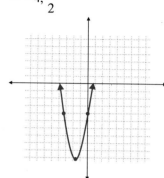

19. a. $\left(-\dfrac{5}{6}, -\dfrac{1}{12}\right)$ **b.** $x = -\dfrac{5}{6}$ **c.** $\dfrac{-2}{3}, -1$

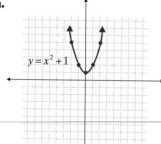
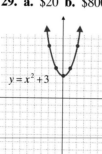

21. $A = x(30 - x); x = 15$, so the dimensions of the rectangle are 15 yd. by 15 yd. **23. a.** 3.5 seconds **b.** 196 feet
25. a. 4 seconds **b.** 288 feet **27.** $20 **29. a.** $20 **b.** $800
31. a.

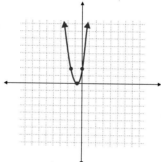
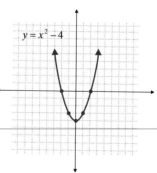

b.

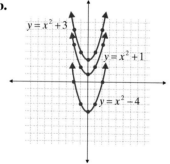

33. a.

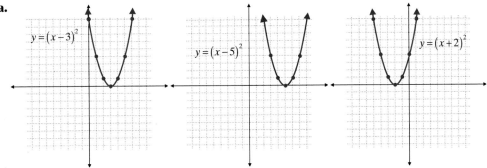

b.

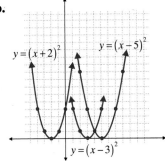

35. The graph $y = (x - h)^2 + k$ is the graph of $y = x^2$ shifted k units vertically and h units horizontally.

Chapter 10 Test, Pages 681 – 682

1. $x = -12, 2$ **2.** $\pm 2\sqrt{3}$ **3.** $x = 12$ **4.** $x^2 - 24x + 144 = (x - 12)^2$ **5.** $x^2 + 9x + \dfrac{81}{4} = \left(x + \dfrac{9}{2}\right)^2$

6. $3x^2 + 9x + \dfrac{27}{4} = 3\left(x + \dfrac{3}{2}\right)^2$ **7.** $x = 1, 2$ **8.** $x = \dfrac{-12}{5}, 0$ **9.** $x = -\dfrac{5}{3}, 2$ **10.** $x = -4, -2$ **11.** $x = 1 \pm \sqrt{6}$

12. $x = \dfrac{-3 \pm \sqrt{33}}{4}$ **13.** $x = \dfrac{-4 \pm \sqrt{10}}{3}$ **14.** $x = \dfrac{2 \pm \sqrt{10}}{2}$ **15.** $x = 1 \pm \sqrt{5}$ **16.** $x = -3 \pm \sqrt{11}$ **17.** $x = \dfrac{1}{3}, 2$

18. $x = \dfrac{3 \pm \sqrt{17}}{4}$

19. a. $(0, -5)$ **b.** $x = 0$ **c.** $\pm\sqrt{5}$ **20. a.** $(2, 4)$ **b.** $x = 2$ **c.** $0, 4$ **21. a.** $\left(\dfrac{-3}{4}, \dfrac{-1}{8}\right)$ **b.** $x = \dfrac{-3}{4}$ **c.** $-1, \dfrac{-1}{2}$

d.

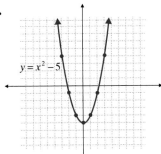

d.

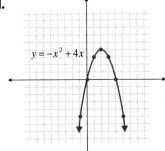

d.

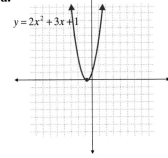

22. $2x^2 = 36^2$; $x = 18\sqrt{2}$ in. **23. a.** 0.725 sec and 8.275 sec **b.** 324 ft. **c.** 9 seconds **24.** $\dfrac{60}{8-x} - \dfrac{60}{8+x} = 10$; $x = 4$; the rate of current is 4 mph **25.** $4(w-5)(w-8) = 720$; $w = 20$, the metal is 20 in. by 23 in.

Chapter 10 Cumulative Review, Pages 683 – 686

1. $9x^4 y^2$ **2.** $-4x^6$ **3.** $x - 3$ **4.** $8x + 3$ **5.** $x = 11$ **6.** $x = 6$ **7.** $x = 3, 5$ **8.** $x = -9, 4$ **9.** $\dfrac{5}{3}$ **10.** $x = 0, 4$

11. $x = 1, 7$ **12.** $x = -3 \pm \sqrt{6}$ **13.** $x = -2 \pm \sqrt{6}$ **14.** $x = \dfrac{3 \pm \sqrt{41}}{4}$ **15.** $x = 7$ **16.** $x = -3, 5$

17. $x \le 3$ 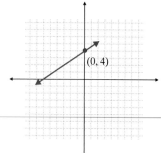 **18.** $x \ge -13$ **19.** $x < \dfrac{17}{4}$

20. $x \ge -5$ **21.** $x = \dfrac{32}{13}$, $y = \dfrac{17}{13}$ **22.** $x = \dfrac{11}{4}$, $y = \dfrac{11}{2}$ **23.** $x = -2, y = -3$

24. $b = \dfrac{P - a}{2}$ **25.** $a = \dfrac{2A}{h} - b$ **26.** $(x - 6)^2$ **27.** $(5x - 7)(5x + 7)$ **28.** $(x + 3)(3x + 1)$ **29.** $3(x - 4)(x + 6)$

30. $x^3 - 3x^2 - 4x - 32$ **31.** $7x^2 - 14$ **32.** $7x^2 - 10x + 7$ **33.** $\dfrac{x}{x + 2}$ **34.** $\dfrac{3(x - 1)}{2x(x + 1)}$ **35.** $\dfrac{-x^2 + 6x + 5}{x^2 - 25}$

36. $\dfrac{5x^2 + 2x + 8}{(x - 4)(x + 1)(x + 2)}$ **37.** $3\sqrt{7} = 7.9373$ **38.** $9\sqrt{3} = 15.5885$ **39.** $5\sqrt[3]{2} = 6.2996$ **40.** $\dfrac{5\sqrt{3}}{6} = 1.4434$

41. $11\sqrt{3} = 19.0526$ **42.** $2 - 7\sqrt{3} = -10.1244$ **43.** $8x$ **44.** $8x^4$ **45.** $m^{\frac{1}{3}} n^{\frac{1}{6}}$ **46.** $\dfrac{5y^{\frac{1}{12}}}{6x}$

47. slope $= \dfrac{2}{3}$; y-intercept $= (0, 4)$ **48.** slope $= \dfrac{4}{3}$ y-intercept $= (0, -2)$ **49.** slope $= -\dfrac{1}{2}$, y-intercept $= \left(0, -\dfrac{5}{4}\right)$

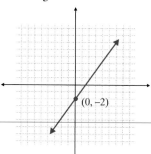

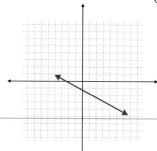

50. a. $(4, -9)$ **b.** $x = 4$ **c.** $(1, 7)$ **d.**

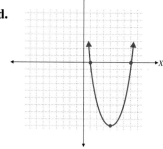

51. a. $(-3, -6)$ **b.** $x = -3$ **c.** $-3 \pm \sqrt{3}$

d.

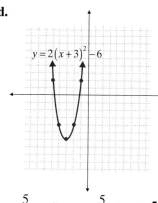

$y = 2(x+3)^2 - 6$

52. a. $\left(-\dfrac{1}{2}, \dfrac{3}{4}\right)$ **b.** $-\dfrac{1}{2}$ **c.** None

d.

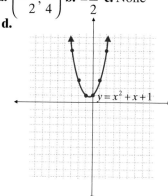

$y = x^2 + x + 1$

53. $y = -\dfrac{5}{2}x + 3$; $m = -\dfrac{5}{2}$, $b = 3$ **54. a.** $m = \dfrac{3}{8}$ **b.** $\sqrt{73}$ **c.** $\left(-1, \dfrac{7}{2}\right)$ **d.** $y - 2 = \dfrac{3}{8}(x+5)$ **55 a.** -8 **b.** 27 **c.** -3

56. a. 7 **b.** 23 **c.** 72 **57.** \$2100 at 5%, \$1300 at 8% **58.** $\dfrac{40}{9}$ hours **59.** 2 mph **60.** 20 liters **61.** 8 in. by 15 in.

62. 3 at \$2.50 and 4 at \$3.00 **63.** Width is 4 cm, length is 13 cm **64.** 3600 women, 2800 men

65. \$1500 at 5.5%, \$1200 at 7.2% **66.** 10 mph upstream, 15 mph downstream **67. a.** 1 second and 5 seconds **b.** 6.571 seconds **c.** 7.06 seconds **68.** 37.08 ft.

Appendix A.1, Page 695

1.

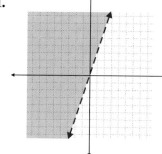

3.

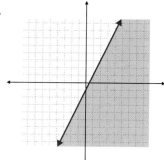

5.

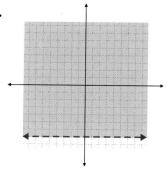

7.

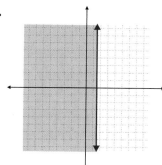

9.

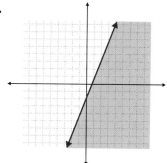

11.

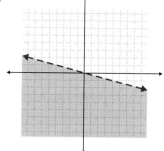

Answers

13.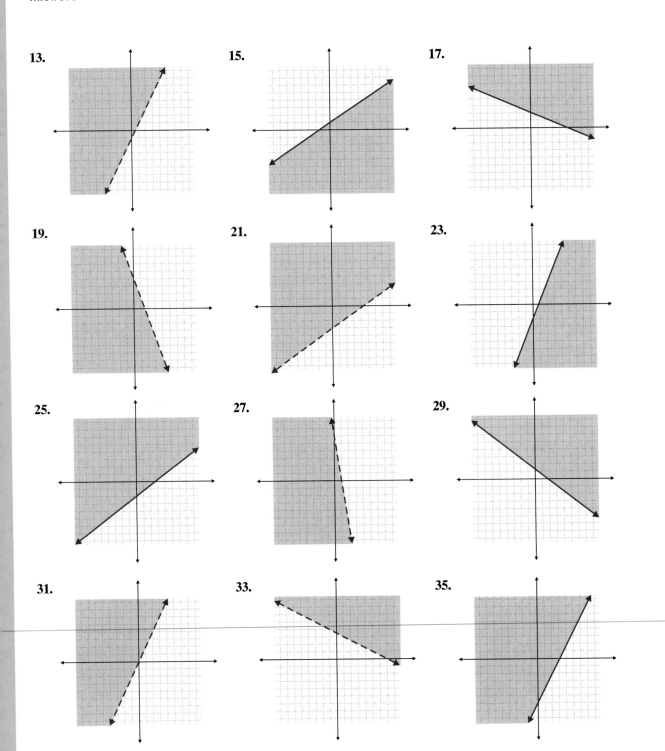

15.

17.

19.

21.

23.

25.

27.

29.

31.

33.

35.

37.

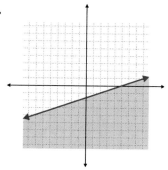

39.

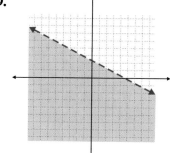

Appendix A.2, Page 701

1.

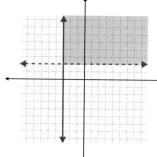

3.

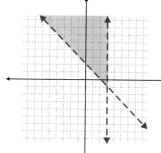

5.

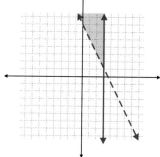

7.

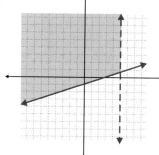

9.

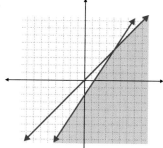

11.

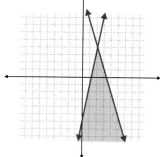

13.

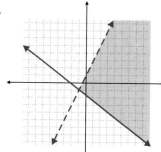

15.

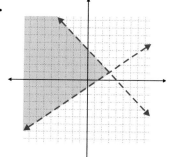

17.

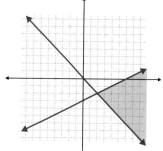

19.

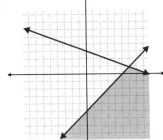

21.

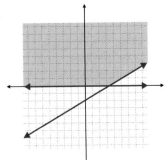

23.

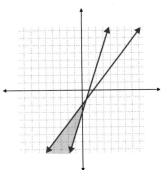

25.

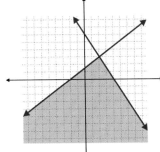

27.

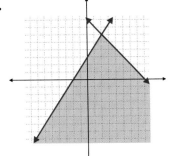

29.

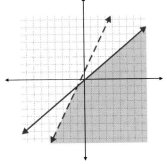

31. a.

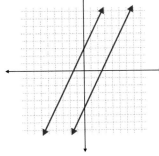

b.

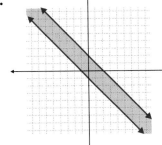

c.

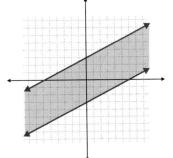

Index